AF775540

Compact Heat Exchangers – Analysis, Design and Optimization using FEM and CFD Approach

WILEY-ASME PRESS SERIES LIST

Primer on Engineering Standards: Expanded Textbook Edition	Greulich	April 2018
Robust Adaptive Control for Fractional-Order Systems with Disturbance and Saturation	Chen	December 2017
Stress in ASME Pressure Vessels, Boilers, and Nuclear Components	Jawad	October 2017
Robot Manipulator Redundancy Resolution	Zhang	October 2017
Combined Cooling, Heating, and Power Systems: Modeling, Optimization, and Operation	Shi	August 2017
Applications of Mathematical Heat Transfer and Fluid Flow Models in Engineering and Medicine	Dorfman	February 2017
Bioprocessing Piping and Equipment Design: A Companion Guide for the ASME BPE Standard	Huitt	December 2016
Geothermal Heat Pump and Heat Engine Systems: Theory and Practice	Chiasson	September 2016
Nonlinear Regression Modeling for Engineering Applications	Rhinehart	September 2016
Fundamentals of Mechanical Vibrations	Cai	May 2016
Introduction to Dynamics and Control of Mechanical Engineering Systems	To	March 2016

Compact Heat Exchangers – Analysis, Design and Optimization using FEM and CFD Approach

C. Ranganayakulu
Aeronautical Development Agency
Bangalore, India

K.N. Seetharamu
PES University
Bangalore, India

This Work is a co-publication between ASME Press and John Wiley & Sons Ltd.

WILEY

ASME PRESS

This edition first published 2018
© 2018 John Wiley & Sons Ltd

All rights reserved. No part of this publication may be reproduced, stored in a retrieval system, or transmitted, in any form or by any means, electronic, mechanical, photocopying, recording or otherwise, except as permitted by law. Advice on how to obtain permission to reuse material from this title is available at http://www.wiley.com/go/permissions.

The right of C. Ranganayakulu and K.N. Seetharamu to be identified as the authors of this work has been asserted in accordance with law.

Registered Offices
John Wiley & Sons, Inc., 111 River Street, Hoboken, NJ 07030, USA
John Wiley & Sons Ltd, The Atrium, Southern Gate, Chichester, West Sussex, PO19 8SQ, UK

Editorial Office
The Atrium, Southern Gate, Chichester, West Sussex, PO19 8SQ, UK

For details of our global editorial offices, customser services, and more information about Wiley products visit us at www.wiley.com.

Wiley also publishes its books in a variety of electronic formats and by print-on-demand. Some content that appears in standard print versions of this book may not be available in other formats.

Limit of Liability/Disclaimer of Warranty
While the publisher and authors have used their best efforts in preparing this work, they make no representations or warranties with respect to the accuracy or completeness of the contents of this work and specifically disclaim all warranties, including without limitation any implied warranties of merchantability or fitness for a particular purpose. No warranty may be created or extended by sales representatives, written sales materials or promotional statements for this work. The fact that an organization, website, or product is referred to in this work as a citation and/or potential source of further information does not mean that the publisher and authors endorse the information or services the organization, website, or product may provide or recommendations it may make. This work is sold with the understanding that the publisher is not engaged in rendering professional services. The advice and strategies contained herein may not be suitable for your situation. You should consult with a specialist where appropriate. Further, readers should be aware that websites listed in this work may have changed or disappeared between when this work was written and when it is read. Neither the publisher nor authors shall be liable for any loss of profit or any other commercial damages, including but not limited to special, incidental, consequential, or other damages.

Library of Congress Cataloging-in-Publication Data

Names: Ranganayakulu, C., author. | Seetharamu, K.N., author.
Title: Compact heat exchangers – analysis, design and optimization using FEM and CFD approach / By C. Ranganayakulu, K.N. Seetharamu.
Description: Hoboken, NJ : John Wiley & Sons, 2018. | Includes index. | Identifiers: LCCN 2017042879 (print) | LCCN 2017049499 (ebook) | ISBN 9781119424352 (pdf) | ISBN 9781119424376 (epub) | ISBN 9781119424185 (cloth)
Subjects: LCSH: Heat–Transmission. | Finite element method. | Computational fluid dynamics.
Classification: LCC QC320.22.F56 (ebook) | LCC QC320.22.F56 R36 2018 (print) | DDC 621.402/2–dc23
LC record available at https://lccn.loc.gov/2017042879

Cover Design: Wiley
Cover Image: © Devrimb/Gettyimages

Set in 10/12pt Warnock by SPi Global, Pondicherry, India

Printed and bound by CPI Group (UK) Ltd, Croydon, CR0 4YY

10 9 8 7 6 5 4 3 2 1

Contents

Preface

The importance of compact heat exchangers (CHEs) has been recognized in aerospace, automobile, process plants and other industries for 60 years or more. This importance is further demanded in the aerospace sector, with its requirements such as weight optimization, high compactness and high performance leading to the demand for cost-effective design and manufacturing techniques. While several books dealing mainly with heat exchangers have been published worldwide in English, no complete source of design data can be found on the many important aspects of CHE design, which an engineer can use as a comprehensive source of generalized design data for the most widely used fin surfaces. One of the first comprehensive books on the design of CHEs using primarily air or gases as working fluid was first published by Kays and London in 1967 with the SI unit edition appearing in 1984. This book is still widely used worldwide, with most design data referenced from 1967, for fin data experimentally generated. Because manufacturing technology has progressed significantly since the beginning of 21st century, many new and sophisticated forms of heat transfer surfaces have been used in CHEs. The design data for these surfaces is limited in the open literature and most of the aerospace industry CHE manufacturers keep their design data proprietary.

This book is an attempt to bring new concepts of design data generation numerically, which is cost-effective and more generic design data, which can be used more effectively by design and practising engineers. It is hoped that researchers and designers will find it of value, as well as academics and graduate students. The specialty of this book is numerical design data based on FEM and CFD. Numerical methods and techniques are introduced for estimation of performance deterioration such as flow non-uniformity, temperature non-uniformity and longitudinal heat conduction effects using FEM for CHE unit level and Colburn j factor and Fanning friction f factor data generation methods for various types of CHE fins using CFD at fin level. In addition, worked examples for single- and two-phase flow CHEs are provided, and the complete qualification tests are given for CHEs used in aerospace applications, typically, which are unavailable in open literature, as these are provided only in some standards. In order to keep the book to a reasonable size, some topics of relevance to CHE applications, which are available in other accessible books, such as *Compact Heat Exchangers* by Kays and London, 3rd edition, McGraw–Hill Book Company, New York [1984], *Compact Heat Exchangers* by J.E Hesselgreaves, Pergamon [2001] and *Plate-fin Heat Exchangers Guide to their Specification and Use* by M.A. Taylor, HTFS, UK, [1987] have been omitted, in particular those of transients (for regenerators), the effect of temperature dependent fluid properties and analytical solutions for flow in tubes.

The first three chapters of the book deal with the fundamentals of heat transfer, heat exchangers and numerical techniques of FEM and CFD. The application of FEM to compact heat exchangers is presented in Chapter 4, where heat conduction effects using FEM on evaporator tubes in pool boiling is also provided. Chapter 5 gives the complete information about CFD analysis of various fins of CHEs for generation of Colburn j and Fanning friction f factor correlations. In addition, the endurance life estimation of a typical CHE is also provided, based on FEM and CFD techniques. The basic concepts of sizing and design methodologies are provided for CHEs in Chapter 6 along with several worked examples for direct transfer type heat exchangers, boiling and condenser heat exchangers. Finally, in Chapter 7, the construction details and several qualifications tests are presented for qualification of CHEs especially for aerospace applications.

Many people have helped either directly or indirectly throughout the preparation of this book. Primarily, the authors are greatly indebted to their family members Lakshmi and Puji Krimmel (Dr C. Ranganayakulu) and Uma, Anil, Vani and Samartha Shastry (Prof. K.N. Seetharamu) for their unfailing support and patience during the development of this book. In particular, the authors wish to thank Aeronautical Development Agency and PES University, Bangalore for allowing us to work on this book. Dr C Ranganayakulu acknowledges his colleagues Mr R. Swaminathan, Mr A. Panigrahi and Mr Shahbaz Ulum for their support; in particular, Prof. Stephan Kabelac, Prof. Dieter Gorenflo, Prof. P. Nithiarasu and Prof. V. Vasudeva Rao for their guidance and encouragement; as well as Dr Saik L. Ismail, Dr A.C. Bhaskar, Dr R. Balasundar Rao, Dr M. Amaranatha Raju and Mr K.V. Ramana Murthy for their PhD works and Mr M.H. Prasad, Mr B. Mahesh and Mr C. Deepak Varma for book documentary works. Prof. Seetharamu also acknowledges Dr H.W. Lee, Dr Anvar Mydin, Dr G.A. Quadir, Dr Z.A. Zainal and in particular Dr T.R. Seetharam, Dr V. Krishna and Dr K.N.B. Murthy for their encouragement to write this book. Finally, the authors would like to thank the staff of John Wiley & Sons Ltd, UK, in particular Ms Anita Yadav and Mr Paul Beverley, for their constant support and encouragement during the preparation of the book.

C. Ranganayakulu
K.N. Seetharamu

Series Preface

The Wiley-ASME Press Series in Mechanical Engineering brings together two established leaders in mechanical engineering publishing to deliver high-quality, peer-reviewed books covering topics of current interest to engineers and researchers worldwide.

The series publishes across the breadth of mechanical engineering, comprising research, design and development, and manufacturing. It includes monographs, references and course texts.

Prospective topics include emerging and advanced technologies in Engineering Design; Computer-Aided Design; Energy Conversion & Resources; Heat Transfer; Manufacturing & Processing; Systems & Devices; Renewable Energy; Robotics; and Biotechnology.

1

Basic Heat Transfer

1.1 Importance of Heat Transfer

The subject of heat transfer is of fundamental importance in many branches of engineering. A *mechanical engineer* may be interested in knowing the mechanisms of heat transfer involved in the operation of equipment, such as boilers, condensers, air preheaters and economizers, and in thermal power plants, in order to improve performance. Refrigeration and air-conditioning systems also involve heat-exchanging devices, which need careful design. *Electrical engineers* are keen to avoid material damage due to hot spots, developed by improper heat transfer design in electric motors, generators and transformers. An *electronic engineer* is interested in knowing the most efficient methods of heat dissipation from chips and other semiconductor devices so that they can operate within safe operating temperatures. A *computer hardware engineer* wants to know the cooling requirements of circuit boards, as the miniaturization of computing devices is advancing rapidly. *Chemical engineers* are interested in heat transfer processes in various chemical reactions. A *metallurgical engineer* may need to know the rate of heat transfer required for a particular heat treatment process, such as the rate of cooling in a casting process, as this has a profound influence on the quality of the final product. *Aeronautical engineers* are interested in knowing the heat transfer rate in electronic equipment that uses compact heat exchangers for minimizing weight, in rocket nozzles and in heat shields used in re-entry vehicles. An *agricultural engineer* would be interested in the drying of food grains, food processing and preservation. *Civil engineers* need to be aware of the thermal stresses developed in quick-setting concrete, and the effect of heat and mass transfer on buildings and building materials. Finally, an *environmental engineer* is concerned with the effect of heat on the dispersion of pollutants in air, diffusion of pollutants in soils, thermal pollution in lakes and seas and their impact on life (Incropera et al. [1]).

The study of heat transfer can offer economical and efficient solutions to critical problems encountered in many branches of engineering. For example, we could consider the development of heat pipes that can transport heat at a much greater rate than copper or even silver rods of the same dimensions, even at almost isothermal conditions. The development of modern gas turbine blades, in which the gas temperature exceeds the melting point of the material of the blade, is possible by providing efficient cooling systems, and this is another example of the success of heat transfer design methods. The design of computer chips, which encounter heat fluxes of the same order those occurring in re-entry vehicles, especially when the surface temperature of the chips is limited to less than 100 °C, is another success story for heat transfer analysis.

Compact Heat Exchangers – Analysis, Design and Optimization using FEM and CFD Approach,
First Edition. C. Ranganayakulu and K.N. Seetharamu.
© 2018 John Wiley & Sons Ltd. Published 2018 by John Wiley & Sons Ltd.

Although there are many successful heat transfer designs, further developments are still necessary in order to increase the lifespan and efficiency of the many devices discussed above, which can lead to many more inventions. Also, if we are to protect our environment, it is essential to understand the many heat transfer processes involved and to take appropriate action, where necessary.

1.2 Heat Transfer Modes

Heat transfer is the exchange of thermal energy between physical systems. The rate of heat transfer is dependent on the temperatures of the systems and the properties of the intervening medium through which the heat is transferred. The three fundamental modes of heat transfer are conduction, convection and radiation. Heat transfer, the flow of energy in the form of heat, is a process by which a system changes its internal energy, hence it is of vital use in applications of the *first law of thermodynamics*. Conduction is also known as diffusion, not to be confused with diffusion related to the mixing of constituents of a fluid.

The direction of heat transfer is from a region of high temperature to another region of lower temperature, and is governed by the *second law of thermodynamics*. Heat transfer changes the internal energy of the systems from which and to which the energy is transferred. Heat transfer will occur in a direction that increases the entropy of the collection of systems. Heat transfer is that section of engineering science that studies the energy transport between material bodies due to a temperature difference (Bejan [2], Holman [3], Incropera and Dewitt [4], Sukhatme [5]). The three modes of heat transfer are

- Conduction
- Convection
- Radiation.

The conduction mode of heat transport occurs either because of an exchange of energy from one molecule to another, without the actual motion of the molecules, or because of the motion of any free electrons that are present. Therefore, this form of heat transport depends heavily on the properties of the medium and takes place in solids, liquids and gases if a difference in temperature exists.

Molecules present in liquids and gases have freedom of motion, and by moving from a hot to a cold region, they carry energy with them. The transfer of heat from one region to another, due to such macroscopic motion in a liquid or gas, added to the energy transfer by conduction within the fluid, is called *heat transfer by convection*. Convection may be free, forced or mixed. When fluid motion occurs because of a density variation caused by temperature differences, the situation is said to be a free or natural convection. When the fluid motion is caused by an external force, such as pumping or blowing, the state is defined as being one of forced convection. A mixed convection state is one in which both natural and forced convections are present. Convection heat transfer also occurs in boiling and condensation processes.

All bodies emit thermal radiation at all temperatures. This is the only mode that does not require a material medium for heat transfer to occur. The nature of thermal radiation is such that a propagation of energy, carried by *electromagnetic waves*, is emitted from the surface of the body. When these electromagnetic waves strike other body surfaces, a part is reflected, a part is transmitted and the remaining part is absorbed. All modes of heat transfer are generally present in varying degrees in a real physical problem. The important aspects in solving heat transfer problems are identifying the significant modes and deciding whether the heat transferred by other modes can be neglected.

1.3 Laws of Heat Transfer

It is important to quantify the amount of energy being transferred per unit time, and this requires the use of rate equations. For heat conduction, the rate equation is known as *Fourier's law*, which is expressed for one dimension as

$$q_x = -k\frac{dT}{dx} \tag{1.1}$$

where q_x is the heat flux in the x direction (W/m^2); k is the thermal conductivity (W/m · K), a property of material, and dT/dx is the temperature gradient (K/m).

For convective heat transfer, the rate equation is given by *Newton's law of cooling* as

$$q = h\left(T_w - T_a\right) \tag{1.2}$$

where q is the convective heat flux; (W/m^2); $(T_w - T_a)$ is the temperature difference between the wall and the fluid and h is the convection heat transfer coefficient, (W/m^{2}K).

The convection heat transfer coefficient frequently appears as a boundary condition in the solution of heat conduction through solids. We assume h to be known in many such problems. In the analysis of thermal systems, we can again assume an appropriate h if not available (e.g. heat exchangers, combustion chambers). However, if required, h can be determined via suitable experiments, although this is a difficult option.

The maximum flux that can be emitted by radiation from a black surface is given by the *Stefan–Boltzmann law*:

$$q = \sigma T_w^4 \tag{1.3}$$

where q is the radiative heat flux, (W/m^2); σ is the Stefan–Boltzmann constant (5.669×10^{-8}) in W/m^{2}K^4; and T_w is the surface temperature (K).

The heat flux emitted by a real surface is less than that of a black surface and is given by

$$q = \varepsilon \sigma T_w^4 \tag{1.4}$$

where ε is the radiative property of the surface and is referred to as the *emissivity*. The net radiant energy exchange between any two surfaces, 1 and 2, is given by

$$Q = F_\epsilon F_G \sigma A_1 \left(T_1^4 - T_2^4\right) \tag{1.5}$$

where F_ϵ is a factor that takes into account the nature of the two radiating surfaces; F_G is a factor that takes into account the geometric orientation of the two radiating surfaces and A_1 is the area of surface 1. When a heat transfer surface, at temperature T_1, is completely enclosed by a much larger surface at temperature T_2, the net radiant exchange can be calculated by

$$Q = qA_1 = \in_1 \sigma A_1 \left(T_1^4 - T_2^4\right) \tag{1.6}$$

With respect to the laws of thermodynamics, only the first law is of interest in heat transfer problems. The increase of energy in a system is equal to the difference between the energy transfer by heat to the system and the energy transfer by work done on the surroundings by the system, that is,

$$dE = dQ - dW \tag{1.7}$$

where Q is the total heat entering the system and $\boldsymbol{W}$ is the work done on the surroundings. Since we are interested in the rate of energy transfer in heat transfer processes, we can restate the first law of thermodynamics as follows:

The rate of increase of the energy of the system is equal to the difference between the rate at which energy enters the system and the rate at which the system does work on the surroundings, that is,

$$\frac{dE}{dt} = \frac{dQ}{dt} - \frac{dW}{dt} \tag{1.8}$$

where t is the time.

The important fluid properties associated with conduction phenomena are presented in Appendixes A2–A8.

1.4 Steady-State Heat Conduction

Heat conduction is the transfer of heat (internal energy) by microscopic collisions of particles and movement of electrons within a body. The microscopically colliding objects, which include molecules, atoms and electrons, transfer disorganized microscopic kinetic and potential energy, jointly known as internal energy. Conduction takes place in all phases of matter: solids, liquids, gases and plasmas. The rate at which energy is conducted as heat between two bodies is a function of the temperature difference (temperature gradient) between the two bodies and the properties of the conductive medium through which the heat is transferred. Thermal conduction is originally called diffusion.

Steady-state conduction is the form of conduction that happens when the temperature difference(s) driving the conduction are constant, so that (after an equilibration time), the spatial distribution of temperatures (temperature field) in the conducting object does not change any further. In steady-state conduction, the amount of heat entering any region of an object is equal to the amount of heat coming out (if this are not so, the temperature would be rising or falling, as thermal energy is tapped from or trapped in a region).

For example, a bar may be cold at one end and hot at the other, but after a state of steady-state conduction is reached, the spatial gradient of temperatures along the bar does not change any further, as time proceeds. Instead, the temperature at any given section of the rod remains constant, and this temperature varies linearly in space, along the direction of heat transfer.

1.4.1 One-Dimensional Heat Conduction

A one-dimensional approximation of the heat conduction equation is feasible for many physical problems, such as plane walls and fins (Bejan [2], Holman [3], Incropera and Dewitt [4], Ozisik [6]). In these problems, any major temperature variation is in one direction only and the variation in all other directions can be ignored. Other examples of one-dimensional heat transfer occur in cylindrical and spherical solids in which the temperature variation occurs only in the radial direction. In this section, such one-dimensional problems are considered for steady-state conditions, in which the temperature does not depend on time. Time-dependent and multidimensional problems will be discussed in later sections.

The steady-state heat conduction equation for a plane wall, shown in Figure 1.1, is

$$kA\frac{d^2T}{dx^2} = 0 \tag{1.9}$$

where k is the thermal conductivity and A is the cross-sectional area perpendicular to the direction of heat flow. The problem is complete with the following description of the boundary conditions.

At $x = 0$, $T = T_1$; and at $x = L$, $T = T_2$

The exact solution to Equation 1.9 is

$$kAT = C_1x + C_2 \tag{1.10}$$

On applying the appropriate boundary conditions to Equation 1.10, we obtain

$$C_2 = kAT_1 \tag{1.11}$$

And

$$C_1 = -\frac{kA(T_1 - T_2)}{L} \tag{1.12}$$

Therefore, substituting constants C_1 and C_2 into Equation 1.10 results in

$$T = -\frac{(T_1 - T_2)x}{L} + T_1 \tag{1.13}$$

Figure 1.1 Heat conduction through a homogeneous wall.

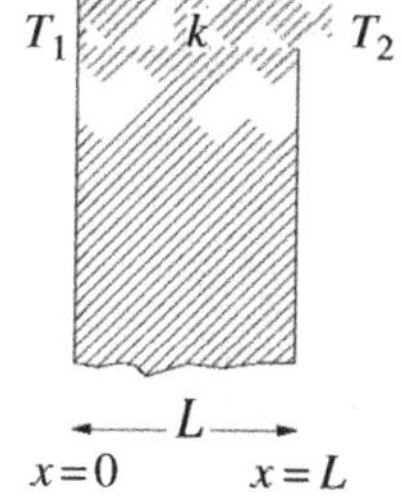

The above equation indicates that the temperature distribution within the wall is linear. The heat flow, Q, can be written as

$$Q = -kA\frac{dT}{dx} = \frac{kA(T_1 - T_2)}{L} \tag{1.14}$$

Example 1.1 The wall of an industrial furnace is constructed from 0.3 m thick fireclay brick having a thermal conductivity 1.7 W/m · K. Measurements made during steady-state operation reveal temperatures of 1400 and 1150 K at the inner and outer surfaces, respectively, as shown in Figure 1.2. What is the rate of heat loss through a wall that is 0.5 m by 1.2 m on a side?

Solution

Known: Steady-state conditions with prescribed wall thickness, area, thermal conductivity and surface temperatures.

Find: Wall heat loss.

Schematic: Figure 1.2

Assumptions:

Steady-state conditions
One-dimensional conduction through the wall
Constant thermal conductivity

Analysis: Since heat transfer through the wall is by conduction, the heat flux be determined from Fourier's law. Using Equation 1.1 gives

$$q_x = k\frac{\Delta T}{L} = 1.7\frac{\text{W}}{\text{m}}\cdot\text{K} \times \frac{250\,\text{K}}{0.15\,\text{m}} = 2833\,\text{W/m}^2$$

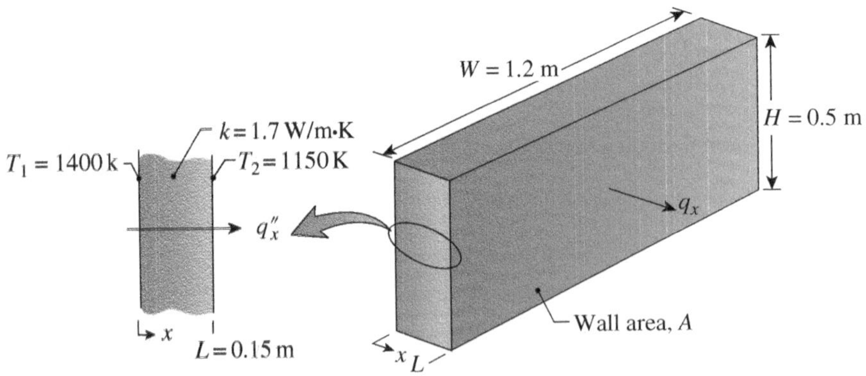

Figure 1.2 One-dimensional heat conduction slab.

The heat flux represents the rate of heat transfer through a section of unit area, and it is uniform (invariant) across the surface of the wall. The heat loss through the wall of area, $A = H \times W$ is then

$$(HW)\, q_x = (0.5\,\text{m} \times 1.2\,\text{m})2833\,\text{W/m}^2 = 1700\,\text{W}$$

Comments: Note the direction of heat flow and the distinction between heat flux and heat rate.

1.4.2 Three-Dimensional Heat Conduction Equation

The determination of temperature distribution in a medium (solid, liquid, gas or a combination of phases) is the main objective of a conduction analysis, that is, to know the temperature in the medium as a function of space at steady state and as a function of time during the transient state. Once this temperature distribution is known, the heat flux at any point within the medium, or on its surface, may be computed from Fourier's law, Equation 1.1. The knowledge of the temperature distribution within a solid can be used to determine the structural integrity via a determination of the thermal stresses and distortion. The optimization of the thickness of an insulating material and the compatibility of any special coatings or adhesives used on the material can be studied by knowing the temperature distribution and the appropriate heat transfer characteristics.

We shall now derive the conduction equation in Cartesian coordinates, as per Carslaw and Jaeger [7], by applying the energy conservation law to a differential control volume, as shown in Figure 1.3. The solution of the resulting differential equation, with prescribed boundary conditions, gives the temperature distribution in the medium.

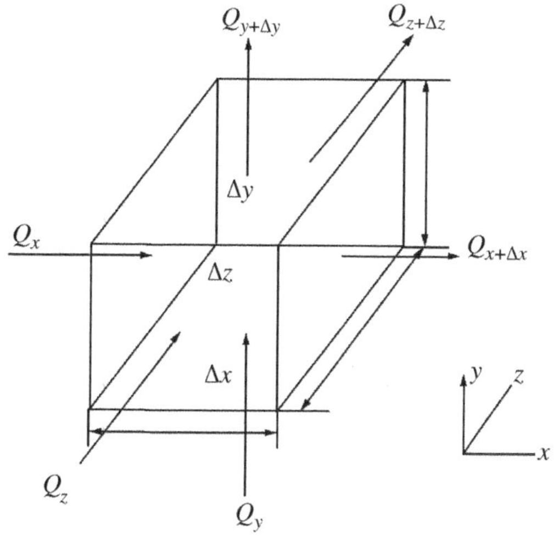

Figure 1.3 A differential control volume for heat conduction analysis.

A Taylor series expansion results in

$$
\begin{aligned}
Q_{x+dx} &= Q_x + \frac{\partial Q_x}{\partial x}\Delta x \\
Q_{y+dy} &= Q_y + \frac{\partial Q_y}{\partial y}\Delta y \\
Q_{z+dz} &= Q_z + \frac{\partial Q_z}{\partial z}\Delta z
\end{aligned}
\tag{1.15}
$$

Note that the second- and higher-order terms are neglected in the above equation. The heat generated in the control volume is $G\,\Delta x\,\Delta y\,\Delta z$ and the rate of change in energy storage is given as

$$
\rho\,\Delta x\,\Delta y\,\Delta z\,c_p\frac{\partial T}{\partial t} \tag{1.16}
$$

Now, with reference to Figure 1.3, we can write the energy balance as inlet energy + energy generated = energy stored + exit energy that is,

$$
G\,\Delta x\,\Delta y\,\Delta z + Q_x + Q_y + Q_z = \rho\Delta x\Delta y\Delta z\frac{\partial T}{\partial x} + Q_{x+dx} + Q_{y+dy} + Q_{z+dz} \tag{1.17}
$$

Substituting Equation 1.15 into the above equation, and rearranging, results in

$$
-\frac{\partial Q_x}{\partial x}\Delta x - \frac{\partial Q_y}{\partial y}\Delta y - \frac{\partial Q_z}{\partial z}\Delta z + G\,\Delta x\,\Delta y\,\Delta z = \rho\,\Delta x\,\Delta y\,\Delta z\,c_p\frac{\partial T}{\partial t} \tag{1.18}
$$

The total heat transfer Q in each direction can be expressed as

$$
\begin{aligned}
Q_x &= \Delta y\,\Delta z\,q_x = -k_x\Delta y\,\Delta z\frac{\partial T}{\partial x} \\
Q_y &= \Delta x\,\Delta z\,q_y = -k_y\Delta x\,\Delta z\frac{\partial T}{\partial x} \\
Q_z &= \Delta x\,\Delta y\,q_z = -k_z\Delta x\,\Delta y\frac{\partial T}{\partial z}
\end{aligned}
\tag{1.19}
$$

Substituting Equation 1.19 into Equation 1.18 and dividing by the volume, $\Delta x\ \Delta y\ \Delta z$, we get

$$
\frac{\partial}{\partial x}\left[k_x\frac{\partial T}{\partial x}\right] + \frac{\partial}{\partial y}\left[k_y\frac{\partial T}{\partial y}\right] + \frac{\partial}{\partial z}\left[k_z\frac{\partial T}{\partial z}\right] + G = \rho c_p\frac{\partial T}{\partial t} \tag{1.20}
$$

Equation 1.20 is the transient heat conduction equation for a stationary system expressed in Cartesian coordinates. The thermal conductivity, k, in the above equation is a vector. In its most general form, the thermal conductivity can be expressed as a tensor, that is,

$$
k = \begin{bmatrix} k_{xx} & k_{xy} & k_{xz} \\ k_{yx} & k_{yy} & k_{yz} \\ k_{zx} & k_{zy} & k_{zz} \end{bmatrix} \tag{1.21}
$$

Equations 1.20 and 1.21 are valid for solving heat conduction problems in anisotropic materials with a directional variation in the thermal conductivities. In many situations, however, the thermal conductivity can be taken as a non-directional property, that is, it is isotropic. In such materials, the heat conduction equation is written as (constant thermal conductivity)

$$\frac{\partial^2 T}{\partial x^2} + \frac{\partial^2 T}{\partial y^2} + \frac{\partial^2 T}{\partial z^2} + \frac{G}{k} = \frac{1}{\alpha}\frac{\partial T}{\partial t} \tag{1.22}$$

where $\alpha = k/\rho c_p$ is the *thermal diffusivity*, which is an important parameter in transient heat conduction analysis.

If the analysis is restricted only to steady-state heat conduction with no heat generation, the equation is reduced to

$$\frac{\partial^2 T}{\partial x^2} + \frac{\partial^2 T}{\partial y^2} + \frac{\partial^2 T}{\partial z^2} = 0 \tag{1.23}$$

For a one-dimensional case, the steady-state heat conduction equation is further reduced to

$$\frac{d}{dx}\left(k\frac{dT}{dx}\right) = 0 \tag{1.24}$$

The heat conduction equation for a cylindrical coordinate system is given by

$$\frac{1}{r}\frac{\partial}{\partial r}\left[k_r r\frac{\partial T}{\partial r}\right] + \frac{1}{r^2}\frac{\partial}{\partial \phi}\left[k_\phi\frac{\partial T}{\partial \phi}\right] + \frac{\partial}{\partial z}\left[k_z\frac{\partial T}{\partial z}\right] + G = \rho c_p\frac{\partial T}{\partial t} \tag{1.25}$$

where the heat fluxes can be expressed as

$$\begin{aligned} q_r &= -k_r\frac{\partial T}{\partial r} \\ q_\phi &= -\frac{k_\phi}{r}\frac{\partial T}{\partial \phi} \\ q_z &= -k_z\frac{\partial T}{\partial z} \end{aligned} \tag{1.26}$$

The heat conduction equation for a spherical coordinate system is given by

$$\frac{1}{r^2}\frac{\partial}{\partial r}\left[k_r r^2\frac{\partial T}{\partial r}\right] + \frac{1}{r^2 sin^2\theta}\frac{\partial}{\partial \phi}\left[k_\phi\frac{\partial T}{\partial \phi}\right] + \frac{1}{r^2\sin\theta}\frac{\partial}{\partial \theta}\left[k_\theta\sin\theta\frac{\partial T}{\partial \theta}\right] + G = \rho c_p\frac{\partial T}{\partial t} \tag{1.27}$$

where the heat fluxes can be expressed as

$$\begin{aligned} q_r &= -k_r\frac{\partial T}{\partial r} \\ q_\phi &= -\frac{k_\phi}{r\sin\theta}\frac{\partial T}{\partial \phi} \\ q_\theta &= -\frac{k_\theta}{r}\frac{\partial T}{\partial \theta} \end{aligned} \tag{1.28}$$

It should be noted that for both cylindrical and spherical coordinate systems, Equations 1.25 and 1.27 can be derived in a similar fashion as for Cartesian coordinates by considering the appropriate differential control volumes.

1.4.3 Boundary and Initial Conditions

The heat conduction equations discussed above will be complete for any problem only if the appropriate boundary and initial conditions are stated. With the necessary boundary and initial conditions, a solution to the heat conduction equations is possible. The boundary conditions for the conduction equation can be of two types or a combination of these two – the *Dirichlet* condition, in which the temperature on the boundaries is known and/or the *Neumann* condition, in which the heat flux is imposed (see Figure 1.4) as per Lewis et al. [8]:

Dirichlet condition

$$T = T_0 \text{ on } \Gamma_T \tag{1.29}$$

Neumann condition

$$q = -k\frac{\partial T}{\partial n} = C \text{ on } \Gamma_{qf} \tag{1.30}$$

In Equations 1.29 and 1.30, T_0 is the prescribed temperature; Γ is the boundary surface; n is the outward direction normal to the surface; and C is the constant flux given. The insulated, or adiabatic, condition can be obtained by substituting $C = 0$. The convective heat transfer boundary condition also falls into the *Neumann* category, and can be expressed as

$$-k\frac{\partial T}{\partial n} = h(T_w - T_a) \text{ on } \Gamma_{qc} \tag{1.31}$$

It should be observed that the heat conduction equation has second-order terms and hence requires two boundary conditions. Since time appears as a first-order term, only one initial value (i.e. at some instant of time all temperatures must be known) needs to be specified for the entire body, that is,

$$T = T_0 \text{ all over the domain } \Omega \text{ at } t = t_0 \tag{1.32}$$

where t_0 is a reference time.

The constant or variable temperature, conditions are generally easy to implement as temperature is a scalar. However, the implementation of surface fluxes is not as straightforward.

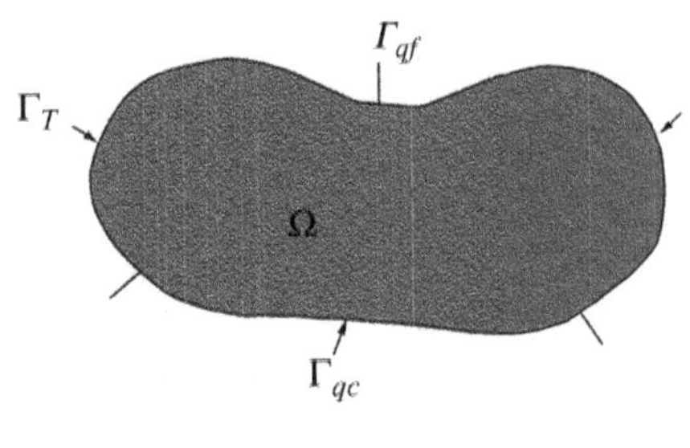

Figure 1.4 Boundary condition.

Equation 1.30 can be rewritten with the direction cosines of the outward normal as

$$k_x \frac{\partial T}{\partial x}\tilde{\imath} + k_y \frac{\partial T}{\partial y}\tilde{m} + k_z \frac{\partial T}{\partial z}\tilde{n} = C \text{ on } \Gamma_{qf} \tag{1.33}$$

Similarly, Equation 1.31 can be rewritten as

$$k_x \frac{\partial T}{\partial x}\tilde{\imath} + k_y \frac{\partial T}{\partial y}\tilde{m} + k_z \frac{\partial T}{\partial z}\tilde{n} = h(T - T_a) \text{ on } \Gamma_{qc} \tag{1.34}$$

where, $\tilde{\imath}$, $\tilde{m}$ and $\tilde{n}$ are the direction cosines of the appropriate outward surface normals.

The general energy equation for heat conduction, taking into account the spatial motion of the body is given by

$$\frac{\partial}{\partial x}\left[k_x \frac{\partial T}{\partial x}\right] + \frac{\partial}{\partial y}\left[k_y \frac{\partial T}{\partial y}\right] + \frac{\partial}{\partial z}\left[k_z \frac{\partial T}{\partial z}\right] + G = \rho c_p \left(\frac{\partial T}{\partial t} + u\frac{\partial T}{\partial x} + v\frac{\partial T}{\partial y} + w\frac{\partial T}{\partial z}\right) \tag{1.35}$$

where u, v and w are the components of the velocity in the three directions, x, y and z respectively.

1.5 Transient Heat Conduction Analysis

In the above, we have discussed steady-state heat conduction, in which the temperature in a solid body is assumed to be invariant with respect to time. However, many practical heat transfer applications are unsteady (transient) in nature and in such problems the temperature varies with respect to time. For instance, in many industrial plant components, such as boilers or refrigeration and air-conditioning equipment, the heat transfer process is transient during the initial stages of operation, so the analysis of transient heat conduction is very important.

1.5.1 Lumped Heat Capacity System

In this section, we consider the transient analysis of a body in which the temperature is assumed to be constant at any point within and on the surface of the body at any given instant of time. It is also assumed that the temperature of the whole body changes uniformly with time. Such an analysis is called a *lumped heat capacity* method and is a simple and approximate procedure in which no spatial variation in temperature is allowed. The change in temperature in such systems varies only with respect to time. It is therefore obvious that the lumped heat capacity analysis is limited to small-sized bodies and/or high thermal conductivity materials. Consider a body at an initial temperature T_0, immersed in a liquid maintained at a constant temperature T_a, as shown in Figure 1.5. At any instant in time, the convection heat loss from the surface of the body is at the expense of the internal energy of the body. Therefore, the internal energy of the body at any time will be equal to the heat convected to the surrounding medium, that is,

$$-\rho c_p V \frac{dT}{dt} = hA(T(t) - T_a) \tag{1.36}$$

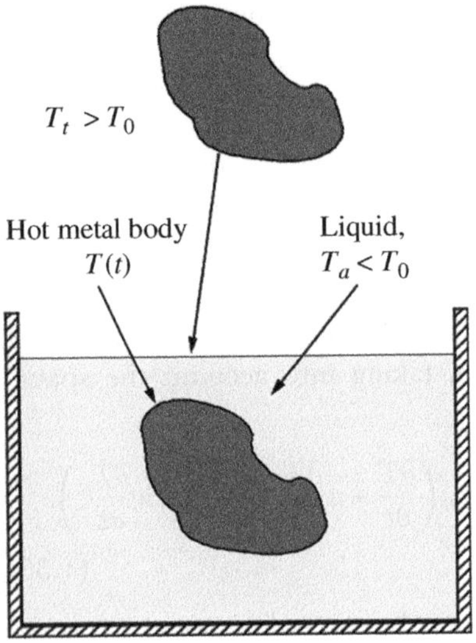

Figure 1.5 Lumped heat capacity system: A hot metal body is immersed in a liquid maintained at a constant temperature.

where ρ is the density, c_p is the specific heat and V is the volume of the hot metal body; A is the surface area of the body; h is the heat transfer coefficient between the body surface and the surrounding medium; t is the time; and $T(t)$ is the instantaneous temperature of the body.

Equation 1.36 is a first-order differential equation in time, which requires an initial condition to obtain a solution. As mentioned previously, the initial temperature of the body at time $t = 0$, is T_0. Applying the variable separation concept to Equation 1.36, we get

$$\frac{dT}{T(t) - T_a} = -\frac{hA}{\rho c_p V} dt \tag{1.37}$$

Integrating between temperatures T_0 and $T(t)$, we obtain

$$\int_{T_0}^{T(t)} \frac{dT}{T(t) - T_a} = -\int_0^t \frac{hA}{\rho c_p V} dt \tag{1.38}$$

Note that the temperature changes from T_0 to $T(t)$ as the time changes from 0 to t.

Integration of the above equation results in a transient temperature distribution as follows:

$$\ln\left(\frac{T - T_a}{T_0 - T_a}\right) = \frac{-hAt}{\rho c_p V} \tag{1.39}$$

or

$$\frac{T - T_a}{T_0 - T_a} = e^{\left[-\frac{hA}{\rho c_p V}\right] t} \tag{1.40}$$

The quantity $\rho c_p V/hA$ is referred to as the time constant of the system because it has the dimensions of time. When $t = \rho c_p V/hA$, it can be observed that the temperature difference $(T(t) - T_a)$ has a value of 36.78% of the initial temperature difference $(T_0 - T_a)$.

The lumped heat capacity analysis gives results within an accuracy of 5% when

$$\frac{h(V/A)}{k_s} < 0.1 \tag{1.41}$$

where k_s is the thermal conductivity of the solid. It should be observed that *(V/A)* represents a characteristic dimension of the body. The above non-dimensional parameter can thus be rewritten as hL/k_s, which is known as the *Biot number*. The Biot number represents a ratio between conduction resistances within the body to convection resistance at the surface of the hot body.

1.6 Heat Convection

Having discussed the conduction mode of heat transfer, we now look at the heat transfer mechanism due to a fluid motion. This method of heat transfer, which is caused by fluid motion, is referred to as *heat convection*. The study of fluid motion (fluid dynamics) is an important subject that has wide application in many engineering disciplines. Several industries use computer-based fluid dynamics analysis (computational fluid dynamics or CFD) tools for both design and analysis. For instance, aerospace applications, turbo-machines, weather forecasting, electronic cooling arrangements and flow in heat exchangers are just a few examples. There has been a vast increase in the use of CFD tools in engineering industries in the past three decades, mainly because of the ever-increasing computing power.

It is obvious that flow with a higher incoming velocity will transport heat at a higher rate. The flow rate is often characterized by a quantity called the *Reynolds number*, which is defined as

$$Re = \frac{\rho_a u_a L}{\mu_a} \tag{1.42}$$

where u_a is the average inlet velocity, L is a characteristic dimension, for example, the width or height of the channel, ρ_a is a reference (inlet) density and μ_a is a reference (inlet) dynamic viscosity of the fluid. If the Reynolds number is small and below a certain critical value, the flow is laminar, and if it is above this critical number, then the flow becomes turbulent. The critical Reynolds number for pipe and channel flows, based on the diameter or height, is approximately 2000.

If the flow is forced into the channel by means of an external device, for example, a pump, then the convection process is referred to as *forced convection*, and the Reynolds number is normally high (Jaluria [9]; Massarotti et al. [10]; Minkowycz et al. [11]). In such situations, the fluid motion created by the density (or temperature) difference (buoyancy-driven motion) is negligibly small compared to the forced motion of the fluid. However, at low and moderate Reynolds numbers, the motion created by the local density (or temperature) differences in the fluid is comparable to that of the forced flow. A situation in which the forced and density-difference driven motions are equally important is called *mixed convection* transport (Aung and Worku [12]). If the forced flow is suddenly stopped and the fluid is stagnant inside the channel, then the fluid motion will be entirely influenced by the local density (or temperature) differences until an equilibrium state is reached, that is, no local differences in density or temperature are present. Such a flow is often referred to as *natural, free* or *buoyancy-driven convection* (Nithiarasu *et al.* [13]). Also, the flow of fluid may be forced by external processes, or sometimes (in gravitational fields) by buoyancy forces caused when thermal energy expands the fluid (for example in a fire plume), thus influencing its own transfer. The latter process is often called "natural convection." All convective processes also move heat partly by diffusion. Another form of convection is forced convection. In this case, the fluid is forced to flow by the use of a pump, fan or other mechanical means.

The convection heat transfer mode is composed of two mechanisms. In addition to energy transfer due to *random molecular motion (diffusion)*, energy is also transferred by the *bulk* or *macroscopic motion* of the fluid. This fluid motion is associated with the fact that, at any instant, large numbers of molecules are moving collectively or as aggregates. Such motion, in the presence of a temperature gradient, contributes to the heat transfer. Because the molecules in the aggregate retain random motion, the total heat transfer is then due to a superposition of energy transport by the random motion of the molecules and by the bulk motion of the fluid. It is customary to use the term *convection* when referring to this cumulative transport and the term *advection* when referring to transport due to bulk fluid motion.

1.6.1 Flat Plate in Parallel Flow

Despite its simplicity, parallel flow over a flat plate (Figure 1.6) occurs in numerous engineering applications. As discussed earlier, laminar boundary layer development begins at the leading edge ($x = 0$) and transition to turbulence may occur at a downstream location (x_c) for which a critical Reynolds number $Re_{x,c}$ is achieved. We begin by considering conditions in the laminar boundary layer.

1.6.1.1 Laminar Flow Over an Isothermal Plate

The major convection parameters may be obtained by solving the appropriate form of the boundary layer equations. Assuming *steady, incompressible, laminar* flow with *constant fluid properties* and *negligible viscous dissipation* and recognizing that $dp/dx = 0$, the boundary layer equations reduce to

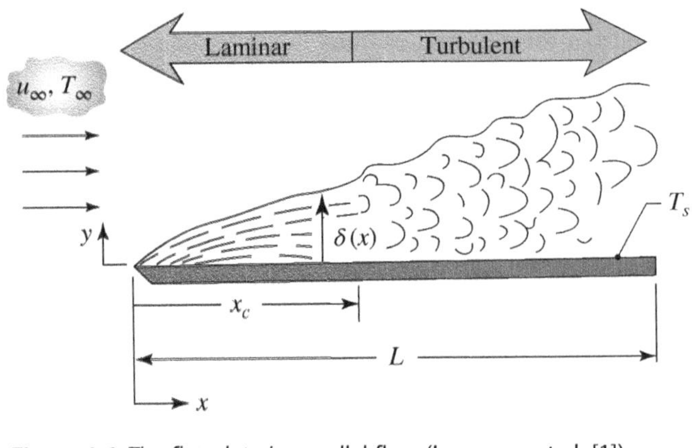

Figure 1.6 The flat plate in parallel flow (Incropera et al. [1]).

Continuity:

$$\frac{\partial u}{\partial x} + \frac{\partial u}{\partial y} = 0 \tag{1.43}$$

Momentum:

$$u\frac{\partial u}{\partial x} + v\frac{\partial u}{\partial y} = v\frac{\partial^2 u}{\partial y^2} \tag{1.44}$$

Energy:

$$u\frac{\partial T}{\partial x} + v\frac{\partial T}{\partial y} = \alpha\frac{\partial^2 T}{\partial y^2} \tag{1.45}$$

Species:

$$u\frac{\partial \rho_A}{\partial x} + v\frac{\partial \rho_A}{\partial y} = D_{AB}\frac{\partial^2 \rho_A}{\partial y^2} \tag{1.46}$$

Solution of these equations is simplified by the fact that for constant properties, conditions in the velocity (hydrodynamic) boundary layer are independent of temperature and species concentration. For the solution procedure to solve the hydrodynamic problem, see *Fundamentals of Heat and Mass Transfer* by Incropera et al [4] and Blasius [14]. The local Nusselt number is of the form

$$Nu_x \equiv \frac{h_x x}{k} = 0.332\, Re_x^{1/2}\, Pr^{1/3} \qquad Pr \geq 0.6 \tag{1.47}$$

Integrating and substituting from Equation 1.47, it follows that $\bar{h}_x = 2h_x$, Hence

$$\overline{Nu}_x \equiv \frac{\bar{h}_x x}{k} = 0.664\, Re_x^{1/2}\, Pr^{1/3} \qquad Pr \geq 0.6 \tag{1.48}$$

Employing the heat and mass transfer analogy, it follows that

$$\overline{Sh}_x \equiv \frac{\bar{h}_{m,x}\, x}{D} = 0.664\, Re_x^{1/2}\, Sc^{1/3} \qquad Sc \geq 0.6 \tag{1.49}$$

If the flow is laminar over the entire surface, the subscript x may be replaced by L.

We see that, for laminar flow over a flat plate, the average friction and convection coefficients from the leading edge to a point x on the surface are twice the local coefficients at that point. We also note that, in using these expressions, the effect of variable properties can be treated by evaluating all properties at the film temperature, $T_f = (T_s + T_\infty)/2$. For fluids of small Prandtl number, namely, liquid metals, Equation 1.47 does not apply. However, for this case, the thermal boundary layer development is much more rapid than that of the velocity boundary layer ($\delta_t \gg \delta$), and it is reasonable to assume uniform velocity ($u = u_\infty$) throughout the thermal boundary layer. From a solution to the thermal boundary layer equation based on this assumption [5], it can then be shown that

$$Nu_x = 0.565\, Pe_x^{1/2} \qquad Pr \leq 0.05, Pe_x \geq 100 \tag{1.50}$$

where $Pe_x \equiv Re_x Pr$ is the Peclet number. Despite the corrosive and reactive nature of liquid metals, their unique properties (low melting point and low vapour pressure, as well as high thermal capacity and high conductivity) render them attractive as coolants in applications requiring high heat transfer rates.

A single correlating equation, which applies for all Prandtl numbers, has been recommended by Ozisik [6]. For laminar flow over an isothermal plate, the local convection coefficient may be obtained from

$$Nu_x = \frac{0.3387 Re_x^{1/2} Pr^{1/3}}{\left[1 + (0.0468/Pr)^{2/3}\right]^{1/4}} \qquad Pe_x \geq 100 \tag{1.51}$$

with $\overline{Nu}_x = 2Nu_x$.

1.6.1.2 Turbulent Flow over an Isothermal Plate

From experiment [2], it is known that, for turbulent flows with Reynolds numbers up to approximately 10^8, the local friction coefficient is correlated to within 15% accuracy by an expression of the form

$$C_{f,x} = 0.0592 Re_x^{-1/5} \qquad Re_{x,c} \leq Re_x \leq 10^8 \tag{1.52}$$

Moreover, it is known that, to a reasonable approximation, the velocity boundary layer thickness may be expressed as

$$\delta = 0.37 x Re_x^{-1/5} \tag{1.53}$$

Comparing these results with those for the laminar boundary layer, Equations 1.48 and 1.51, we see that turbulent boundary layer growth is much more rapid (δ varies as $x^{4/5}$, in contrast to $x^{1/2}$ for laminar flow) and that the decay in the friction coefficient is more gradual ($x^{-1/5}$ vs $x^{-1/2}$). For turbulent flow, boundary layer development is influenced strongly by random fluctuations in the fluid and not by molecular diffusion. Hence relative boundary layer growth does not depend on the value of Pr or Sc, and Equation 1.53 may be used to obtain the thermal and concentration, as well as the velocity, boundary layer thicknesses. That is, for turbulent flow, $\delta \approx \delta_t \approx \delta_c$.

Using Equation 1.52 with the modified Reynolds (or Chilton and Colburn [15]) analogy, gives equations

$$\frac{C_f}{2} = St Pr^{2/3} = j_H \quad 0.6 < Pr < 60 \tag{1.54}$$

$$\frac{C_f}{2} = St_m Sc^{2/3} = j_m \quad 0.6 < Sc < 3000 \tag{1.55}$$

where j_H and j_m are the Colburn j factors for heat and mass transfer respectively. Then the local Nusselt number for turbulent flow is

$$Nu_x = St\, Re_x\, \mathrm{Pr} = 0.0296\, Re_x^{4/5}\, Pr^{1/3} \qquad 0.6 \leq Pr \leq 60 \tag{1.56}$$

and the local Sherwood number is

$$Sh_x = St_m\, Re_x Sc = 0.0296\, Re_x^{4/5}\, Sc^{1/3} \quad 0.6 \le Sc \le 3000 \tag{1.57}$$

Enhanced mixing causes the turbulent boundary layer to grow more rapidly than the laminar boundary layer and to have larger friction and convection coefficients.

Expressions for the average coefficients may now be determined. However, since the turbulent boundary layer is generally preceded by a laminar boundary layer, we first consider mixed boundary layer conditions.

1.6.1.3 Boundary Layer Development Over Heated Plate

We are especially interested in convection heat transfer, which occurs between a fluid in motion and a bounding surface when the two are at different temperatures. Consider fluid flow over the heated surface of Figure 1.7. A consequence of the fluid–surface interaction is the development of a region in the fluid through which the velocity varies from zero at the surface to a finite value u_∞ associated with the flow. This region of the fluid is known as the *hydrodynamic*, or *velocity, boundary layer*. Moreover, if the surface and flow temperatures differ, there will be a region of the fluid through which the temperature varies from T_s at $Y = 0$ to T_∞ in the outer flow. This region, called the thermal boundary layer, may be smaller, larger or the same size as that through which the velocity varies. In any case, if $T_s > T_\infty$, convection heat transfer will occur from the surface to the outer flow.

The convection heat transfer mode is sustained both by random molecular motion and by the bulk motion of the fluid within the boundary layer. The contribution due to random molecular motion (diffusion) dominates near the surface where the fluid velocity is low. In fact, at the interface between the surface and the fluid ($y = 0$), the fluid velocity is zero and heat is transferred by this mechanism only. The contribution due to bulk fluid motion originates from the fact that the boundary layer grows as the flow progresses in the x direction. In effect, the heat that is conducted into this layer is swept downstream and is eventually transferred to the fluid outside the boundary layer. Appreciation of boundary layer phenomena is essential to understanding convection heat transfer. It is for this reason that the discipline of fluid

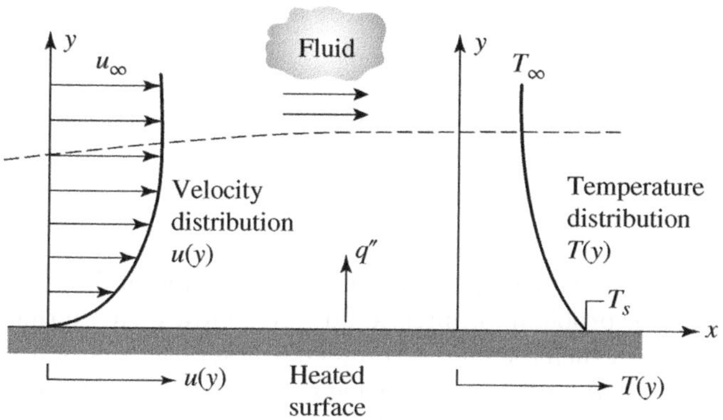

Figure 1.7 Boundary layer development in convection heat transfer.

Table 1.1 Typical values of the convection heat transfer coefficient.

Process	**$h(W/m^2 \cdot K)$**
Free convection	
Gases	2–25
Liquids	50–1000
Forced convection	
Gases	25–250
Liquids	100–20,000
Convection with phase change	
Boiling or condensation	2500–45,000

mechanics will play a vital role in our later analysis of convection. Regardless of the particular nature of the convection heat transfer process, the appropriate rate equation is of the form

$$q'' = h(T_s - T_\infty) \tag{1.58}$$

where q'', the convective heat flux (W/m^2), is proportional to the difference between the surface and fluid temperatures, T_s and T_∞, respectively. This expression is known as Newton's law of cooling, and the parameter h $(W/m^2 \cdot K)$ is termed the *convection heat transfer coefficient*. It depends on conditions in the boundary layer, which are influenced by surface geometry, the nature of the fluid motion and an assortment of fluid thermodynamic and transport properties. Any study of convection ultimately reduces to a study of the means by which h may be determined. In the solution of such problems we presume h to be known, using typical values given in Table 1.1. When Equation 1.58 is used, the convection heat flux is presumed to be positive if heat transferred from the surface $(T_s > T_\infty)$ and negative if heat is transferred to the surface $(T_\infty > T_s)$. However, if $T_\infty > T_s$, there is nothing to preclude us from expressing Newton's law of cooling as

$$q'' = h(T_\infty - T_s) \tag{1.59}$$

in which case heat transfer is positive if it is to the surface.

1.6.2 Internal Flow

Having acquired the means to compute convection transfer rates for external flow, we now consider the convection transfer problem for internal flow. In contrast, an internal flow, such as flow in a pipe, is one for which the fluid is confined by a surface. Hence the boundary layer is unable to develop without eventually being constrained. The internal flow configuration represents a convenient geometry for heating and cooling fluids used in chemical processing, environmental control and energy conversion technologies.

Our objectives are to develop an appreciation for the physical phenomena associated with internal flow and to obtain convection coefficients for flow conditions of practical importance. We begin by considering velocity (hydrodynamic) effects pertinent to internal flows, focusing on certain unique features of boundary layer development. Thermal boundary layer effects are considered next, and an overall energy balance is applied to determine fluid temperature variations in the flow direction. Finally, correlations for estimating the convection heat transfer coefficient are presented for a variety of internal flow conditions including the flow of supercritical fluids through tubes.

1.6.2.1 Hydrodynamic Considerations

When considering external flow, it is necessary to ask only whether the flow is laminar or turbulent. However, for an internal flow we must also be concerned with the existence of entrance and fully developed regions.

1.6.2.2 Flow Conditions

Consider laminar flow in a circular tube of radius r_o (Figure 1.8), where fluid enters the tube with a uniform velocity. We know that when the fluid makes contact with the surface, viscous effects become important, and a boundary layer develops with increasing x. This development occurs at the expense of a shrinking inviscid flow region and concludes with boundary layer merger at the centreline. Following this merger, viscous effects extend over the entire cross-section, and the velocity profile no longer changes with increasing x. The flow is then said to be fully developed, and the distance from the entrance at which this condition is achieved is termed the hydrodynamics entry length, $x_{fd,h}$. As shown in Figure 1.8, the fully developed velocity profile is parabolic for laminar flow in a circular tube. For turbulent flow, the profile is flatter, due to turbulent mixing in the radial direction.

When dealing with internal flows, it is important to be cognizant of the extent of the entry region, which depends on whether the flow is laminar or turbulent. The Reynolds number for flow in a circular tube is defined as

$$Re_D = \frac{\rho u_m D}{\mu} = \frac{u_m D}{\nu} \tag{1.60}$$

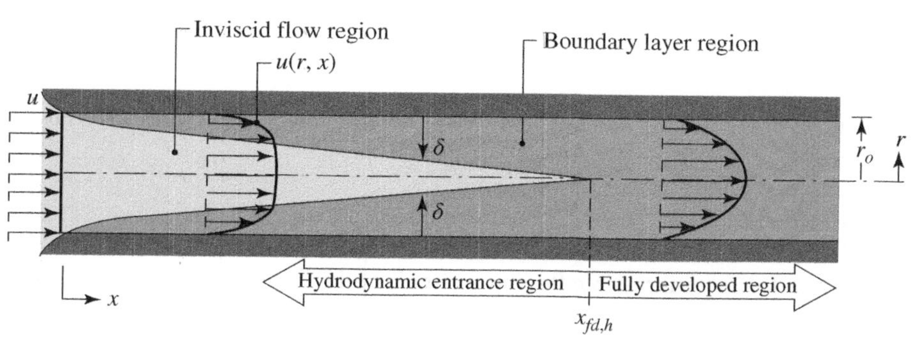

Figure 1.8 Laminar, hydrodynamic boundary layer development in a circular tube.

where u_m is the mean fluid velocity over the tube cross-section and D is the tube diameter. In a fully developed flow, the critical Reynolds number corresponding to the onset of turbulence is

$$Re_{D,C} \approx 2300 \quad (1.61)$$

although much larger Reynolds numbers ($Re_D \approx 10{,}000$) are needed to achieve fully turbulent conditions. The transition to turbulence is likely to begin in the developing boundary layer of the entrance region.

For laminar flow ($Re_D \approx 2300$), the hydrodynamic entry length may be obtained from an expression of the form (Incropera et al.[1])

$$\left(\frac{x_{fd,h}}{D}\right)_{lam} \approx 0.05\,Re_D \quad (1.62)$$

This expression is based on the presumption that fluid enters the tube from a rounded converging nozzle and is hence characterized by a nearly uniform velocity profile at the entrance (Figure 1.8). Although there is no satisfactory general expression for the entry length in turbulent flow, we know that it is approximately independent of Reynolds number and that, as a first approximation by Bejan [2],

$$10 \le \left(\frac{x_{fd,h}}{D}\right)_{turb} \le 60 \quad (1.63)$$

For the purposes of this text, we shall assume fully developed turbulent flow for $(x/D) > 10$.

1.6.2.3 Mean Velocity

The velocity varies over the cross-section and there is no well-defined free stream, it is necessary to work with a mean velocity u_m when dealing with internal flows. This velocity is defined such that, when multiplied by the fluid density ρ and the cross-sectional area of the tube A_c, it provides the rate of mass flow through the tube. Hence

$$\dot{m} = \rho u_m A_c \quad (1.64)$$

For steady, incompressible flow in a tube of uniform cross-sectional area, $\dot{m}$ and u_m are constants independent of x. From Equations 1.60 and 1.64 it is evident that, for flow in a circular tube ($A_c = \pi D^2/4$), the Reynolds number reduces to

$$Re_D = \frac{4\dot{m}}{\pi D \mu} \quad (1.65)$$

Since the mass flow rate may also be expressed as the integral of the mass flux (ρu) over the cross-section

$$\dot{m} = \int \rho u(r,x)\, dA_c \quad (1.66)$$

It follows that, for incompressible flow in a circular tube,

$$u_m = \frac{\int \rho u(r,x)\, dA_c}{\rho A_c} = \frac{2\pi\rho}{\rho\pi r_0^2}\int_0^{r_o} u(r,x)\, r\, dr = \frac{2}{r_0^2}\int_0^{r_o} u(r,x)\, r\, dr \quad (1.67)$$

The above expression may be used to determine u_m at any axial location x, from knowledge of the velocity profile $u(r)$ at that location.

1.6.2.4 Velocity Profile in the Fully Developed Region

The form of the velocity profile may readily be determined for the laminar flow of an incompressible, constant property fluid in the fully developed region of a circular tube. An important feature of hydrodynamic conditions in the fully developed region is that both the radial velocity component v and the gradient of the axial velocity component $(\partial u/\partial x)$ are everywhere zero.

$$v = 0 \text{ and } \left(\frac{\partial u}{\partial x}\right) = 0 \tag{1.68}$$

Hence the axial velocity component depends only on r, $u(x,r) = u(r)$.

The radial dependence of the axial velocity may be obtained by solving the appropriate form of the x-momentum equation. This form is determined by first recognizing that, for the conditions of Equation 1.68, the net momentum flux is everywhere zero in the fully developed region. Hence the momentum conservation requirement reduces to a simple balance between shear and pressure forces in the flow. For the annular differential elements of Figure 1.9, this force balance may be expressed as

$$\begin{aligned}\tau_r\,(2\pi r\,dx) - \left\{\tau_r(2\pi r\,dx) + \frac{d}{dr}\left[\tau_r(2\pi r\,dx)\right]dx\right\} \\ + p(2\pi r\,dr) - \left\{p(2\pi r\,dr) + \frac{d}{dx}\left[p(2\pi r\,dr)dx\right]\right\} = 0\end{aligned} \tag{1.69}$$

which reduces to

$$-\frac{d}{dr}(r\tau_r) = r\frac{dp}{dx} \tag{1.70}$$

with $y = r_o - r$, Newton's law of velocity, the equation

$$\tau_{xy} = \tau_{yx} = \mu\left[\frac{\partial u}{\partial y} + \frac{\partial r}{\partial x}\right]$$

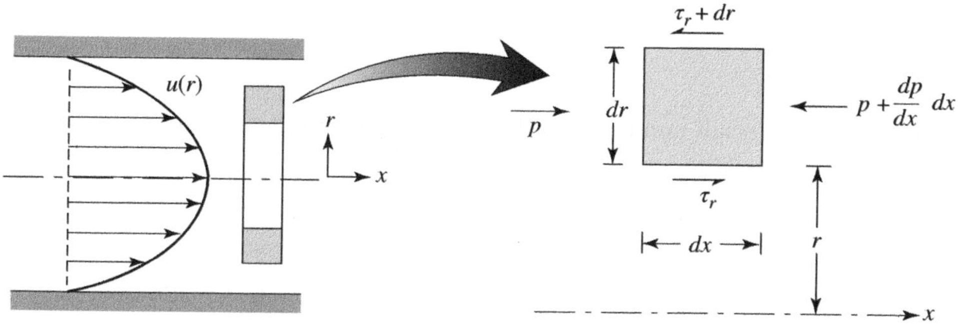

Figure 1.9 Force balance on a differential element for laminar, fully developed flow in a circular tube.

assumes the form

$$\tau_r = -\mu \frac{du}{dr} \tag{1.71}$$

and Equation 1.70 becomes

$$\frac{\mu}{r}\frac{d}{dr}\left(r\frac{du}{dr}\right) = \frac{dp}{dx} \tag{1.72}$$

Since the axial pressure gradient is independent of r, Equation 1.72 may be solved by integrating twice to obtain

$$r\frac{du}{dr} = \frac{1}{\mu}\left(\frac{dp}{dx}\right)\frac{r^2}{2} + C_1 \tag{1.73}$$

and

$$u(r) = \frac{1}{\mu}\left(\frac{dp}{dx}\right)\frac{r^2}{4} + C_1 \ln r + C_2 \tag{1.74}$$

The integration constant may be determined by invoking the boundary conditions

$$u(r_o) = 0 \ \text{ and } \ \frac{\partial u}{\partial r}\Big|_{r=0} = 0 \tag{1.75}$$

which respectively, impose the requirements of zero slip at the tube surface and radial symmetry about the centreline. It is a simple matter to evaluate the constants, and it follows that

$$u(r) = -\frac{1}{4\mu}\left(\frac{dp}{dx}\right) r_o^2 \left[1 - \left(\frac{r}{r_o}\right)^2\right] \tag{1.76}$$

Hence the fully developed velocity profile is parabolic. Note that the pressure gradient must always be negative.

The foregoing result may be used to determine the mean velocity of the flow. Substituting Equation 1.76 into Equation 1.68 and integrating, we obtain

$$u_m = -\frac{r_o^2}{8\mu}\frac{dp}{dx} \tag{1.77}$$

Substituting this result into Equation 1.69, the velocity profile is then

$$\frac{u(r)}{u_m} = 2\left[1 - \left(\frac{r}{r_o}\right)^2\right] \tag{1.78}$$

Since u_m can be computed from knowledge of the mass flow rate, Equation 1.77 can be used to determine the pressure gradient.

1.6.3 Forced Convection Relationships

The summary of forced convection relationships (from *Handbook of Heat Transfer* by Rohsenow et al. [16]) are presented here for easy reference.

- Laminar flow entrance length:

$$\frac{X/D}{Re_D} \approx 10^{-2} \tag{1.79}$$

- Skin friction coefficient definition:

$$C_{f,x} = \frac{\tau_W}{\frac{1}{2}\rho U^2} \tag{1.80}$$

- Laminar fully developed (Hagen–Poiseuille) flow between parallel plates with spacing *D*:

$$u(y) = \frac{3}{2}U\left[1 - \left(\frac{y}{D/2}\right)^2\right] \tag{1.81}$$

with

$$U = \frac{D^2}{12\mu}\left(-\frac{dP}{dx}\right) \tag{1.82}$$

- Laminar fully developed (Hagen–Poiseuille) flow in a tube with diameter *D*:

$$u = 2U\left[1 - \left(\frac{r}{r_0}\right)^2\right] \tag{1.83}$$

with

$$U = \frac{r_0^2}{8\mu}\left(-\frac{dP}{dx}\right) \tag{1.84}$$

- Hydraulic radius and diameter:

$$r_h = \frac{A}{p} \quad \text{hydraulic radius} \tag{1.85}$$

$$D_h = \frac{4A}{p} \quad \text{hydraulic diameter} \tag{1.86}$$

- Friction factor:

$$f = \begin{cases} \dfrac{\tau_w}{\frac{1}{2}\rho U^2} & (1.87) \\ \dfrac{24}{Re_{Dh}} \quad D_h = 2D \text{ parallel plates } (D = \text{spacing}) & (1.88) \\ \dfrac{16}{Re_{Dh}} \quad D_h = D \text{ round tube } (D = \text{diameter}) & (1.89) \end{cases}$$

$$\Delta P = f\frac{4L}{D_h}\left(\frac{1}{2}\rho U^2\right) \tag{1.90}$$

- Nusselt number:

$$N_u = \frac{hD}{k} = D\frac{\partial T/\partial r\backslash_{r=r_0}}{T_0 - T_m} \tag{1.91}$$

- Laminar thermal entrance length:

$$X_T \approx 10^{-2} Pr.D_h.Re_{Dh} \tag{1.92}$$

- Thermally developing Hagen–Poiseuille flow ($Pr = \infty$):
  - Round tube, isothermal wall:

$$Nu_x = \begin{cases} 1.077\,x_*^{-1/3} - 0.70 & x_* \leq 0.01 \\ 3.657 + 6.874\,(10^3 x_*)^{-0.488} e^{-57.2x_*} & x_* > 0.01 \end{cases} \tag{1.93}$$

$$Nu_{0-x} = \begin{cases} 1.615x_*^{-1/3} - 0.70 & x_* \leq 0.05 \\ 1.615x_*^{-1/3} - 0.20 & 0.005 < x_* \leq 0.03 \\ 3.657 + 0.0499/x_* & x_* > 0.03 \end{cases} \tag{1.94}$$

  - Round tube, uniform heat flux:

$$Nu_x = \begin{cases} 3.302\,x_*^{-1/3} - 1.0 & x_* \leq 0.00005 \\ 1.302x_*^{-1/3} - 0.5 & 0.0005 < x_* \leq 0.0015 \\ 4.364 + 8.68(10^3 x_*)^{-0.506} e^{-41x_*} & x_* > 0.001 \end{cases} \tag{1.95}$$

$$Nu_{0-x} = \begin{cases} 1.953\,x_*^{-1/3} & x_* \leq 0.03 \\ 4.364 + 0.0722/x_* & x_* > 0.03 \end{cases} \tag{1.96}$$

  - Parallel plates, isothermal surfaces:

$$Nu_{0-x} = \begin{cases} 1.233\,x_*^{-1/3} + 0.40 & x_* \leq 0.001 \\ 7.541 + 6.874\,(10^3 x_*)^{-0.488} e^{-245x_*} & x_* > 0.01 \end{cases} \tag{1.97}$$

$$Nu_{0-x} = \begin{cases} 1.849\,x_*^{-1/3} & x_* \leq 0.00005 \\ 1.849x_*^{-1/3} & 0.0005 < x_* \leq 0.006 \\ 7.541 + 0.0235/x_* & x_* > 0.006 \end{cases} \tag{1.98}$$

◦ Parallel plates, uniform heat flux:

$$Nu_x = \begin{cases} 1.490\, x_*^{-1/3} & x_* \le 0.0002 \\ 1.490\, x_*^{-1/3} - 0.40 & 0.0002 < x_* \le 0.0001 \\ 8.235 + 8.68(10^3 x_*)^{-0.506} e^{-164 x_*} & x_* > 0.001 \end{cases} \tag{1.99}$$

$$Nu_{0-x} = \begin{cases} 2.236 x_*^{-1/3} & x_* \le 0.001 \\ 2.236\, x_*^{-1/3} + 0.90 & 0.001 < x_* \le 0.01 \\ 8.235 + 0.0364/x_* & x_* > 0.01 \end{cases} \tag{1.100}$$

- Thermally and hydraulically developing flow:
  - Round tube, isothermal wall:

$$Nu_x = 7.55 + \frac{0.024\, x_*^{-1.14}\left(0.0179\, \mathrm{Pr}^{0.17}\, x_*^{-0.64} - 0.14\right)}{\left(1 + 0.0358\, \mathrm{Pr}^{0.17}\, x_*^{-0.64}\right)^2} \tag{1.101}$$

$$Nu_{0-x} = 7.55 + \frac{0.024\, x_*^{-1.14}}{1 + 0.0358\, \mathrm{Pr}^{0.17}\, x_*^{-0.64}} \tag{1.102}$$

$$\frac{\Delta P}{\frac{1}{2}\rho U^2} = 13.74(x_+)^{1/2} + \frac{1.25 + 64x_+ - 13.74(x_+)^{1/2}}{1 + 0.00021(x_+)^{-2}} \tag{1.103}$$

$$x_+ = \frac{x/D}{\mathrm{Re}_D} \tag{1.104}$$

  - Round tube, uniform heat flux:

$$\frac{Nu_x}{4.364\left[1 + (Gz/29.6)^2\right]^{1/6}} = \left\{1 + \left[\frac{Gz/19.04}{\left[1 + (Pr/0.0207)^{1/2}\right]\left[1 + (Gz/29.6)^2\right]^{1/3}}\right]^{3/2}\right\}^{1/3} \tag{1.105}$$

- Optimal channel sizes:
  - Laminar flow, parallel plates:

$$\frac{D_{opt}}{L} \approx 2.7 Be^{-1/4} \qquad Be = \frac{\Delta P L^2}{\mu \alpha} \tag{1.106}$$

$$\frac{q_{\max}}{HLW} \approx 0.60 \frac{k}{L^2}(T_{\max} - T_0) Be^{1/2} \tag{1.107}$$

- Staggered plates:

$$\frac{D_{opt}}{L} \approx 5.4Pr^{-1/4}\left(Re_L \frac{L}{b}\right)^{-1/2} \tag{1.108}$$

for the range

$$P_r = 0.72 \quad 10^2 \le \text{Re}_L \le 10^4 \quad 0.5 \le \frac{Nb}{L} \le 1.3$$

- Bundle of cylinders in cross flow:

$$\frac{S_{opt}}{D} \simeq 1.59 \frac{(H/D)^{0.52}}{\widetilde{P}^{0.13}.Pr^{0.24}} \quad \widetilde{P} = \frac{\Delta PD^2}{\mu \upsilon} \tag{1.109}$$

for the range

$$0.72 \le Pr \le 50 \quad 10^4 \le \widetilde{P} \le 10^8 \quad 25 \le \frac{H}{D} \le 200$$

$$\frac{S_{opt}}{D} \simeq 1.70 \frac{(H/D)^{0.52}}{R\,e_D^{0.26}.Pr^{0.24}} \tag{1.110}$$

$$\frac{T_D - T_\infty}{q\,D/kLW} \simeq \frac{4.5}{R\,e_D^{0.90}.Pr^{0.64}} \tag{1.111}$$

with

$$Re_D = U_\infty \frac{D}{\nu} \qquad 140 \le Re_D \le 14{,}000$$

- Array of pin fins with impinging flow:

$$\frac{S_{opt}}{L} \simeq 0.81\,Pr^{-0.25}.Re_L^{-0.32} \tag{1.112}$$

for the range

$$0.06 \le \frac{D}{L} \le 0.14 \quad 0.28 \le \frac{H}{L} \le 0.56 \quad 0.72 \le \text{Pr} \le 7$$

$$10 \le Re_D \le 700 \quad 90 \le Re_L \le 6000$$

- Turbulent duct flow:

$$\frac{D_{opt}/L}{\left(1 + t/D_{opt}\right)^{4/11}} = 0.071Pr^{-5/11}.Be^{-1/11} \tag{1.113}$$

$$\frac{q_{\max}}{HLW} \le 0.57\frac{k}{L^2}(T_{\max} - T_0)Pr^{4/99}\left(1 + \frac{t}{D_{opt}}\right)^{-67/99}.Be^{47/99} \tag{1.114}$$

with

$$Be = \frac{\Delta PL^2}{\mu\alpha}$$

for the range $10^4 \leq Re_{Dh} \leq 10^6$ $10^6 \leq Re_L \leq 10^8$ $10^{11} \leq Be \leq 10^{16}$

- Turbulent flow and entrance lengths:

$$\frac{X}{D} \simeq 10 \simeq \frac{X_T}{D} \tag{1.115}$$

- Turbulent flow friction factor:

$$f \simeq 0.046 Re_D^{-1/5} \quad 2 \times 10^4 \leq Re_D \leq 10^5 \tag{1.116}$$

- Turbulent flow heat transfer:

$$St.Pr^{2/3} \simeq \frac{f}{2} \tag{1.117}$$

for $Pr \geq 0.5$

$$Nu_D = \frac{hD}{k} = 0.023 Re_D^{4/5}.Pr^{1/3} \tag{1.118}$$

for $Pr \geq 0.50$

$$2 \times 10^4 \leq Re_D \leq 10^6$$

$$Nu_D = 0.023 \mathrm{Re}_D^{4/5}.\, \mathrm{Pr}^n \tag{1.119}$$

where n = 0.4 for heating the fluid and n = 0.3 for cooling the fluid in the range

$$\frac{L}{D} > 60 \quad 0.7 \leq Pr \leq 120 \quad 2500 \leq Re_D \leq 1.24 \times 10^5$$

$$Nu_D = 0.027 Re_D^{4/5}.Pr^{1/3}\left(\frac{\mu}{\mu_0}\right)^{0.14} \tag{1.120}$$

in the range

$$0.7 \leq Pr \geq 16{,}700 \quad Re_D \geq 10^4$$

Here

$\mu_0 = \mu(T_0)$ (T_0 is the wall temperature)
$\mu = \mu(T_m)$ (T_m is the bulk temperature)

$$Nu_D = \frac{(f/2)Re_D.Pr}{1.07 + 900/Re_D - 0.63/(1 + 10Pr) + 12.7(f/2)^{1/2}(Pr^{2/3} - 1)} \tag{1.121}$$

$$Nu_D = \frac{(f/2)Re_D.Pr}{1.07 + 12.7(f/2)^{1/2}(Pr^{2/3} - 1)} \tag{1.122}$$

where

$$0.5 \leq Pr \leq 10^6 \quad 4000 \leq \mathrm{Re}_D \leq 5 \times 10^6$$

and f from Figure 5.13 from reference [16].

$$Nu_D = \frac{(f/2)(\mathrm{Re}_D - 10^3)\,\mathrm{Pr}}{1 + 12.7(f/2)^{1/2}\left(\mathrm{Pr}^{2/3} - 1\right)} \tag{1.123}$$

where

$$0.5 \le \mathrm{Pr} \le 10^6 \quad 2300 \le Re_D \le 5 \times 10^6$$

and f from Figure 5.13 from reference [16].

$$Nu_D = 0.0214\left(Re_D^{0.8} - 100\right)Pr^{0.4} \tag{1.124}$$

where

$$0.5 \le \mathrm{Pr} \le 1.5 \quad 10^4 \le \mathrm{Re}_D \le 5 \times 10^6$$

$$Nu_D = 0.012\left(Re_D^{0.87} - 280\right)Pr^{0.4} \tag{1.125}$$

where

$$1.5 \le \mathrm{Pr} \le 500 \quad 3 \times 10^3 \le \mathrm{Re}_D \le 10^6$$

$$Nu_D = \begin{cases} 6.3 + 0.0167\, Re_D^{0.85} \,.\, Pr^{0.93} & q''_0 = constant \quad (1.126) \\ 4.8 + 0.0156\, Re_D^{0.85} \,.\, Pr^{0.93} & \mathrm{T}_0 = constant \quad (1.127) \end{cases}$$

where, for Equations (1.154) and (1.155),

$$0.004 \le Pr \le 0.1 \quad 10^4 \le Re_D \le 10^6$$

- Total heat transfer rate:

$$q = hA_w \Delta T_{lm} \tag{1.128}$$

- Isothermal wall:

$$\Delta T_{lm} = \frac{\Delta T_{in} - \Delta T_{out}}{In(\Delta T_{in}/\Delta T_{out})} \tag{1.129}$$

$$q = \dot{m}c_p \Delta T_{in}\left(1 - e^{-hA_w/\dot{m}\,c_p}\right) \tag{1.130}$$

- Uniform heat flux:

$$\Delta T_{lm} = \Delta T_{in} = \Delta T_{out} \tag{1.131}$$

1.7 Radiation

Thermal radiation is energy *emitted* by matter that is at a non-zero temperature. Although we will focus on radiation from solid surfaces, emission may also occur from liquids and gases. Regardless of the form of matter, the emission may be attributed to changes in the electron configurations of the constituent atoms or molecules. The energy of the radiation field is transported by electromagnetic waves (or alternatively, photons).

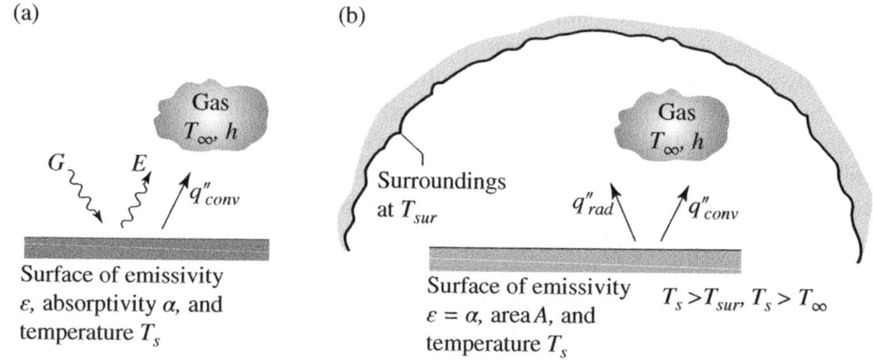

Figure 1.10 Radiation exchange: (a) at a surface and (b) between a surface and large surroundings.

While the transfer of energy by conduction or convection requires the presence of a material medium, radiation does not. In fact, radiation transfer occurs most efficiently in a vacuum.

Consider radiation transfer processes for the surface of Figure 1.10. Radiation that is emitted by the surface originates from the thermal energy of matter bounded by the surface, and the rate at which energy is released per unit area (W/m^2) is termed the surface emissive power E. There is an upper limit to the emissive power, which is prescribed by the *Stefan–Boltzmann law*

$$E_b = \sigma T_s^4 \tag{1.132}$$

where T_s is the absolute temperature (K) of the surface and σ is the *Stefan–Boltzmann constant* ($\sigma = 5.67 \times 10^{-8} \frac{W}{m^2} \cdot K^4$). Such a surface is called an ideal radiator or *blackbody*.

The heat flux emitted by a real surface is less than that of a blackbody at the same temperature and is given by

$$E = \varepsilon \sigma T_s^4 \tag{1.133}$$

where ε is a radiative property of the surface, termed the *emissivity*. With values in the range $0 \le \varepsilon \le 1$, this property provides a measure of how efficiently a surface emits energy relative to a blackbody. It depends strongly on the surface material and finish, and representative values are provided in Appendix A7.

Radiation may also be *incident* on a surface from its surroundings. The radiation may originate from a special source, such as the sun, or from other surfaces to which the surface of interest is exposed. Irrespective of the source(s), we designate the rate at which all such radiation is incident on a unit area of the surface as the *irradiation* G (Figure 1.10).

A portion, or all, of the irradiation may be absorbed by the surface, thereby increasing the thermal energy of the material. The rate at which radiant energy is absorbed per unit surface area may be evaluated from knowledge of a surface radiative property termed the absorptivity α. That is,

$$G_{abs} = \alpha G \tag{1.134}$$

where $0 \leq \alpha \leq 1$. If $\alpha < 1$ and the surface is *opaque*, portions of the irradiation are *reflected*. If the surface is semi-transparent, portions of the irradiation may also be *transmitted*. However, while absorbed and emitted radiation increase and reduce, respectively, the thermal energy of matter, reflected and transmitted radiation have no effect on this energy. Note that the value of α depends on the nature of the irradiation, as well as on the surface itself. For example, the absorptivity of a surface may differ from its absorptivity to radiation emitted by the walls of a furnace.

In many engineering problems (a notable exception being problems involving solar radiation or radiation from other very high temperature sources), liquids can be considered opaque, and gases can be considered transparent, to radiation heat transfer (Siegel and Howell [17] and Incropera et al. [1]). Solids can be opaque (as is the case for metals) or *semi-transparent* (as is the case for thin sheets of some polymers and some semiconducting materials). A special case that occurs frequently involves radiation exchange between a small surface at T_s and a much larger, isothermal surface that completely surrounds the smaller one (Figure 1.11). The surroundings could, for example, be the walls of a room or a furnace, whose temperature T_{sur} differs from that of an enclosed surface ($T_{sur} \neq T_s$). The irradiation may be approximated by emission from a blackbody at T_{sur}, in which case $G = \sigma T_{sur}^4$. If the surface is assumed to be one for which $\alpha = \varepsilon$ (a grey surface), the net rate of radiation heat transfer *from* the surface, expressed per unit area of the surface, is

$$q''_{rad} = \frac{q}{A} = \varepsilon E_b(T_s) - \alpha G = \varepsilon\sigma\left(T_s^4 - T_{sur}^4\right) \tag{1.135}$$

This expression provides the difference between thermal energy that is released due to radiation emission and that which is gained due to radiation absorption.

There are many applications for which it is convenient to express the net radiation heat exchange emission and that which is gained due to radiation absorption.

$$q_{rad} = h_r A(T_s - T_{sur}) \tag{1.136}$$

where, from Equation 1.135, the radiation heat transfer coefficient h_r is

$$h_r \equiv \varepsilon\sigma(T_s + T_{sur})\left(T_s^2 + T_{sur}^2\right) \tag{1.137}$$

Here we have modelled the radiation mode in a manner similar to convection. In this sense we have linearized the radiation rate equation, making the heat rate proportional to a temperature difference rather than to the difference between two temperatures to the fourth power. Note, however, that h_r depends strongly on temperature, while the temperature dependence of the convection heat transfer coefficient h is generally weak.

The surfaces of Figure 1.11 may also simultaneously transfer heat by convection to an adjoining gas. For the conditions of Figure 1.11, the total rate of heat transfer from the surface is then

$$q = q_{conv} + q_{rad} = hA(T_s - T_\infty) + \varepsilon A\sigma\left(T_s^4 - T_{sur}^4\right) \tag{1.138}$$

1.7.1 Radiation – Fundamental Concepts

A representation of the exchanges of energy between the source (the Sun) are, the Earth's surface, the Earth's atmosphere, and the ultimate sink outer space is called the 'greenhouse effect'. The ability of the atmosphere to capture and recycle energy emitted by

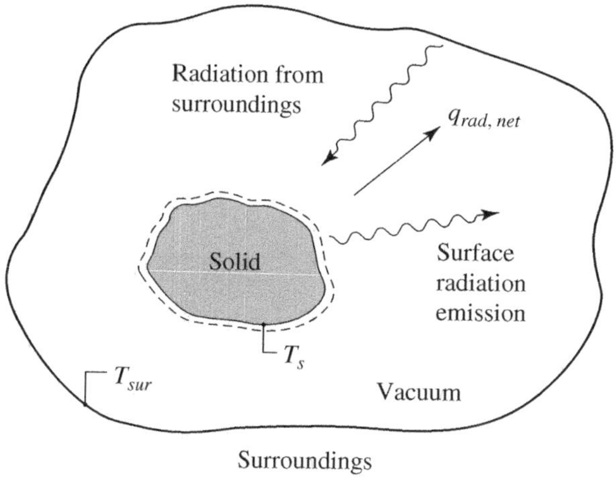

Figure 1.11 Radiation cooling of a heated solid.

the Earth surface is the defining characteristic of the greenhouse effect. The greenhouse effect is a process by which thermal radiation from a planetary surface is absorbed by atmospheric greenhouse gases, and is re-radiated in all directions. Since part of this re-radiation is back towards the surface and the lower atmosphere, it results in an elevation of the average surface temperature above what it would be in the absence of those gases.

Consider a solid that is initially at a higher temperature T_s than that of its surroundings T_{sur}, but around which there exists a vacuum (Figure 1.11). The presence of the vacuum precludes energy loss from the surface of the solid by conduction or convection. However, our intuition tells us that the solid will cool and eventually achieve thermal equilibrium with its surroundings. This cooling is associated with a reduction in the internal energy stored by the solid and is a direct consequence of the emission of thermal radiation from the surface. In turn, the surface will intercept and absorb radiation originating from the surroundings. However, if $T_s > T_{sur}$ the net heat transfer rate by radiation $q_{rad,\ net}$ is from the surface, and the surface will cool until T_s reaches T_{sur}.

We associated thermal radiation with the rate at which energy is emitted by matter as a result of its finite temperature. At this moment, thermal radiations is being emitted by all the matter that surrounds you: by the furniture and walls of the room, if you are indoors, or by the ground, the buildings and the atmosphere and sun if you are outdoors. The mechanism of emission is related to energy released as a result of oscillations or transitions of the many electrons that constitute matter. These oscillations are, in turn, sustained by the internal energy, and therefore the temperature, of the matter. Hence we associated the emission of thermal radiation with thermally excited conditions within the matter.

All forms of matter emit radiation. For gases and for semi-transparent solids, such as glass and salt crystals at elevated temperatures, emission is a volumetric phenomenon, as illustrated in Figure 1.12. That is, radiation emerging from a finite volume of matter is the integrated effect of local emission through the volume. However, in this text we concentrate on situations for which radiation is a surface phenomenon. In most solids and

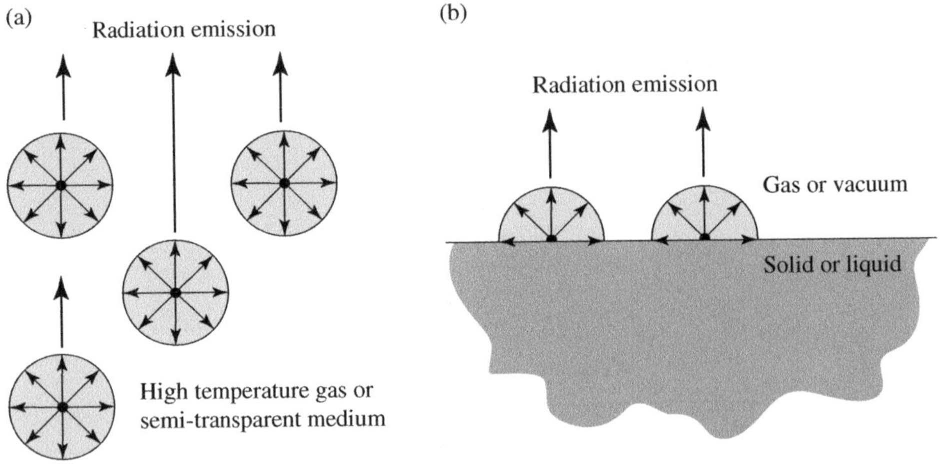

Figure 1.12 The emission process: (a) as a volumetric phenomenon, (b) as a surface phenomenon.

liquids, radiation emitted from interior molecules is strongly absorbed by adjoining molecules. Accordingly, radiation that is emitted from a solid or a liquid originates from molecules that are within a distance of approximately 1 μm from the exposed surface. It is for this reason that emission from a solid or a liquid into an adjoining gas or a vacuum can be viewed as a surface phenomenon, except in situations involving nano-scale or micro-scale devices.

We know that radiation is due to emission by matter and that its subsequent transport does not require the presence of any matter. But what is the nature of this transport? One theory views radiation as the propagation of a collection of particles termed photons or quanta. Alternatively, radiation may be viewed as the propagation of electromagnetic waves. In any case, we wish to attribute to radiation the standard wave properties of frequency ν and wavelength λ. For radiation propagating in a particular medium, the two properties are related by

$$\lambda = \frac{c}{\nu} \tag{1.139}$$

where c is the speed of light in the medium. For propagation in a vacuum, $c_0 = 2.998 \times 10^8$ m/s. The unit of wavelength is commonly the micrometer (μm), where $1\mu m = 10^{-6} m$.

The complete electromagnetic spectrum is delineated in Figure 1.13. The short wavelength gamma rays, X-rays, and ultraviolet (UV) radiation are primarily of interest to the high-energy physicist and the nuclear engineer, while the long wavelength microwaves and radio waves ($\lambda > 10^5 \mu m$) are of concern to the electrical engineer. It is the intermediate portion of the spectrum, which extends from approximately 0.1 to 100 μm and includes a portion of the UV and all of the visible and infrared (IR), that is termed thermal radiation because it is both caused by and affects the thermal state or temperature of matter. For this reason, thermal radiation is pertinent to heat transfer.

Thermal radiation emitted by a surface encompasses a range of wavelengths. As shown in Figure 1.14a, the magnitude of the radiation varies with wavelength, and the term

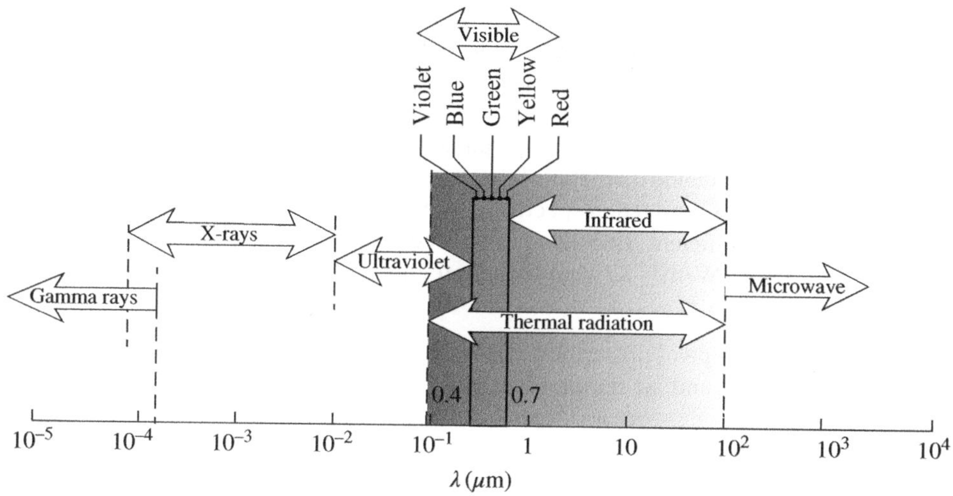

Figure 1.13 Spectrum of electromagnetic radiation.

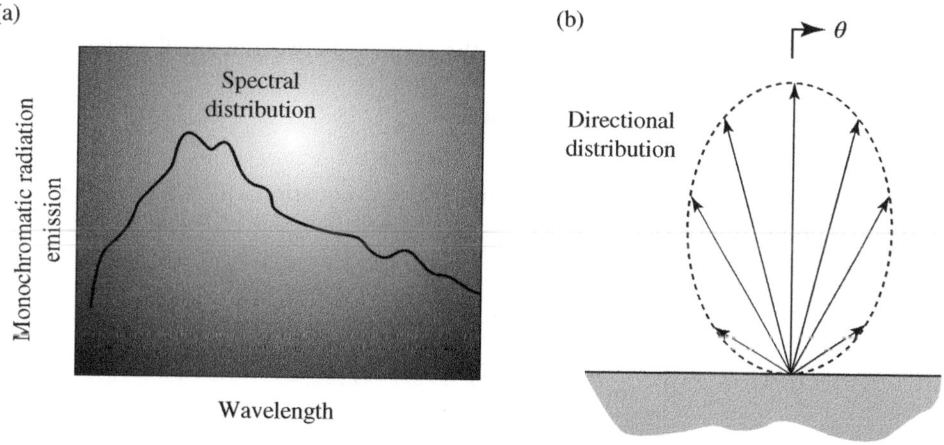

Figure 1.14 Radiation emitted by a surface: (a) spectral distribution, (b) directional distribution.

spectral is used to refer to the nature of this dependence. As we will find, both the magnitude of the radiation at any wavelength and the spectral distribution vary with the nature and temperature of the emitting surface.

The spectral nature of thermal radiation is one of two features that complicate its description. The second feature relates to its directionality. As shown in Figure 1.14b, a surface may emit preferentially in certain directions, creating a directional distribution of the emitted radiation. To quantify the emission, absorption, reflection and transmission concepts, we must be able to consider both spectral and directional effects.

Radiative cooling is the process by which a body loses heat by radiation. Outgoing energy is an important effect in the Earth's energy budget. In the case of the Earth-atmosphere system, it refers to the process by which long-wave (infrared) radiation is emitted to balance the absorption of short-wave (visible) energy from the Sun. Convective transport of heat and evaporative transport of latent heat both remove heat from the surface and redistribute it in the atmosphere.

Example 1.2 An un-insulated steam pipe passes through a room in which the air and walls are at 25°C. The outside diameter of the pipe is 70 mm, and its surface temperature and emissivity are 200°C and 0.8, respectively. What are the surface emissive power and irradiation? If the coefficient associated with free convection heat transfer from the surface to the air is 15 W/m^2 · K, what is the rate of heat loss from the surface per unit length of pipe?

Known: Un-insulated pipe of prescribed diameter, emissivity and surface temperature in a room with fixed wall and air temperatures

Find:

Surface emissive power and irradiation
Pipe heat loss per unit length, q'

Schematic: Figure 1.15

Assumptions:

Steady-state conditions
Radiation exchange between the pipe and the room is between a small surface and a much larger enclosure
Surface emissivity and absorptivity are equal

Analysis:

The surface emissive power may be evaluated from Equation 1.133, while the irradiation corresponds to $G = \varepsilon\sigma T_{sur}^4$, hence

$$E = \varepsilon\sigma T_s^4 = 0.8\left(5.67 \times 10^{-8}\frac{\text{W}}{\text{m}^2}\cdot\text{K}^4\right)(473\,\text{K})^4 = 2270\,\text{W/m}^2$$

$$G = \sigma T_{sur}^4 = 5.67 \times 10^{-8}\frac{\text{W}}{\text{m}^2}\cdot\text{K}^4\,(298\,\text{K})^4 = 447\,\text{W/m}^2$$

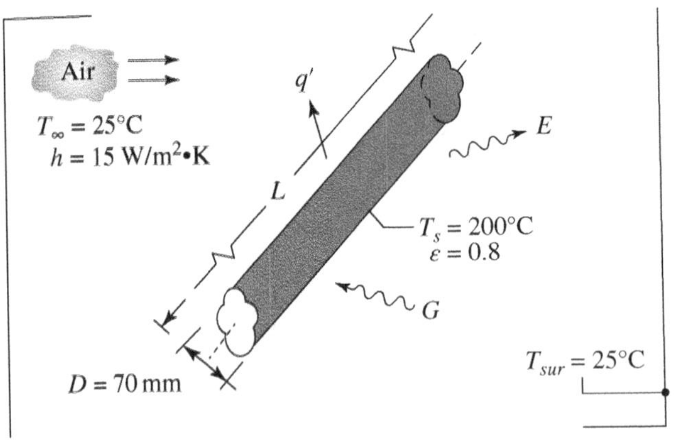

Figure 1.15 Schematic view for Example 1.2.

Heat loss from the pipe is by convection to the room air and by radiation exchange with the walls. Hence, $q = q_{conv} + q_{rad}$ and from Equation 1.138, with $A = \pi DL$,

$$q = h(\pi DL)(T_s - T_\infty) + \varepsilon(\pi DL)\sigma(T_s^4 - T_{sur}^4)$$

The heat loss per unit length of pipe is then

$$q' = \frac{q}{L} = 15\,\text{W/m}^2\cdot\text{K}(\pi \times 0.07\,\text{m})\,(200-25)^\circ\text{C}$$

$$+0.8(\pi \times 0.07\,\text{m})\,5.67 \times 10^{-8}\frac{\text{W}}{\text{m}^2}\cdot\text{K}^4\left(473^4 - 298^4\right)^\circ\text{C}$$

$$q' = 577\,\text{W/m} + 421\,\text{W/m} = 998\,\text{W/m}$$

Comments:

Note that temperature may be expressed in units of °C or K when evaluating the temperature difference for a convection (or conduction) heat transfer rate. However, temperature must be expressed in kelvins (K) when evaluating a radiation transfer rate.

The net rate of radiation heat transfer from the pipe may be expressed as

$$q'_{rad} = \pi D(E - \alpha G)$$

$$q'_{rad} = \pi \times 0.07\,\text{m}\,(2270 - 0.8 \times 447)\,\text{W/m}^2 = 421\,\text{W/m}$$

In this situation the radiation and convection heat transfer rates are comparable because T_s is large compared to T_{sur}, and the coefficient associated with free convection is small. For more moderate values of T_s and the larger values of h associated with forced convection, the effect of radiation may often be neglected. The radiation heat transfer coefficient may be computed from Equation 1.137, and for the conditions of this problem its value is

$$h_r = 11\frac{\text{W}}{\text{m}^2}\cdot\text{K}.$$

1.8 Boiling Heat Transfer

Boiling is the liquid-to-vapour phase change process that occurs at a solid–liquid interface when the surface is heated above the saturation temperature of the liquid. The formation and rise of the bubbles and the liquid entrainment coupled with the large amount of heat absorbed during liquid–vapour phase change at essentially constant temperature are responsible for the very high heat transfer coefficients associated with nucleate boiling. Boiling is classified as pool boiling or flow boiling, depending on the presence of bulk fluid motion. Boiling is called pool boiling in the absence of bulk fluid flow and flow boiling (or forced convection boiling) in the presence of it. In pool boiling, the fluid is stagnant, and any motion of the fluid is due to natural convection currents, and the motion of the bubbles is due to influence of buoyancy. In flow boiling, the fluid has velocity relative to heating surface and is forced to move in a heated pipe or over a surface by external means such as a pump. Therefore, flow boiling is always accompanied by other convection effects.

There are three different boiling heat transfer mechanisms. The first is nucleate boiling, in which heat is transferred by means of vapour bubbles nucleating, growing and finally detaching from the surface. In nucleate boiling of water, the boiling point of a substance is the temperature at which the vapour pressure of the liquid equals the pressure surrounding the liquid Kandlikar et al. [18] and the liquid evaporates, resulting in an abrupt change in vapour volume. The saturation temperature is the temperature for a corresponding saturation pressure at which a liquid boils into its vapour phase. The liquid can be said to be saturated with thermal energy. Any addition of thermal energy results in a phase transition. At standard atmospheric pressure and low temperatures, no boiling occurs and the heat transfer rate is controlled by the usual single-phase mechanisms. As the surface temperature is increased, local boiling occurs and vapour bubbles nucleate, grow into the surrounding cooler fluid and collapse. This is sub-cooled nucleate boiling, and is a very efficient heat transfer mechanism. At high bubble generation rates, the bubbles begin to interfere, and the heat flux no longer increases rapidly with surface temperature (this is the departure from nucleate boiling, or DNB).

At similar standard atmospheric pressure and high temperatures, the hydrodynamically quieter regime of film boiling is reached. Heat fluxes across the stable vapour layers are low, but rise slowly with temperature. Any contact between fluid and the surface that may be seen probably leads to the extremely rapid nucleation of a fresh vapour layer ("spontaneous nucleation"). At higher temperatures still, a maximum in the heat flux is reached (the critical heat flux, or CHF). The Leiden frost effect demonstrates how nucleate boiling slows heat transfer due to gas bubbles on the heater's surface. As mentioned, gas-phase thermal conductivity is much lower than liquid-phase thermal conductivity, so the outcome is a kind of "gas thermal barrier."

The second boiling heat transfer mechanism is convective boiling, in which heat is conducted through the liquid, which evaporates at the liquid–vapour interface and there isn't any bubble formation.

Finally, there is film boiling in which the heat is transferred by conduction and radiation through a film of vapour that covers the heated surface, and the liquid vaporizes at the vapour liquid interface. Nucleate boiling and film boiling may occur in both pool boiling and flow boiling, whereas forced convective boiling occurs only in flow boiling. Pool boiling and flow boiling further be classified into sub-cooled boiling and saturated boiling. Boiling is called sub-cooled when the temperature of the liquid is below the saturation temperature and it is called saturated when the liquid temperature is equal to saturation temperature of the liquid.

1.8.1 Flow Boiling

In flow boiling, the fluid is forced to move by an external source, as it undergoes a phase change process from liquid to vapour. The boiling in this case exhibits the combined effects of convection and pool boiling. The different stages encountered in flow boiling in a heated tube are illustrated in Figure 1.16, together with the variation of the heat transfer coefficient, wall temperature, flow patterns and liquid temperature along the tube. Initially, the liquid is sub-cooled, and heat transfer to the liquid is by single-phase forced convection. Then bubbles start forming on the inner surfaces of the tube, and the detached bubbles are drifted into the mainstream. This gives the fluid flow a bubbly appearance, and thus the name *bubbly flow regime*. As the fluid is heated further, the bubbles grow in size and eventually coalesce into slugs of vapour. Up to half of the volume in the tube in this slug flow regime is occupied by vapour. After a while the core of the flow consists of vapour only, and the liquid is confined only in the annular space

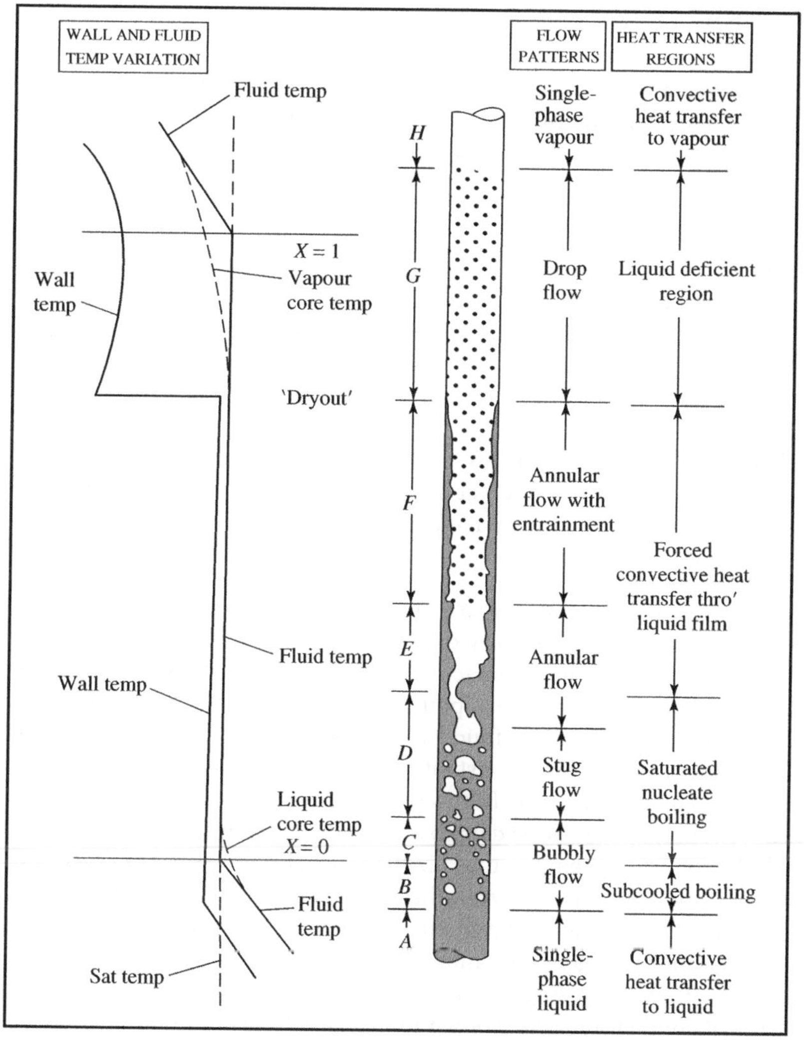

Figure 1.16 Flow boiling in a uniformly heated circular tube. (Collier and Thome [19])

between the vapour core and the tube walls. This is the annular-flow regime, and very high heat transfer coefficients are realized in this regime. As the heating continues, this annular liquid layer gets thinner and thinner, and eventually dry spots start to appear on the inner surfaces of the tube. The appearance of dry spots is accompanied by a sharp decrease in the heat transfer coefficient. This transition regime continues until the inner surface of the tube is completely dry. Any liquid at this moment is in the form of droplets suspended in the vapour core, which resembles a mist, and we have a *mist-flow regime* until all the liquid droplets are vaporized. At the end of the mist-flow regime there is saturated vapour, which becomes superheated with any further heat transfer.

With regard to details of flow boiling, refer to Chapter 5 for experimental evaluation of flow boiling correlations on different types of compact heat exchanger fins and refer to Chapter 6 for worked example for estimation of flow boiling heat transfer coefficient and pressure drop.

1.9 Condensation

Condensation occurs when a vapour is cooled and changes its phase to a liquid. During condensation, the latent heat of vaporization must be released. The amount of the heat is the same as that absorbed during vaporization at the same fluid pressure (Butterworth [20]).

There are several types of condensation:

- homogeneous condensation, as during a formation of fog
- condensation in direct contact with sub-cooled liquid
- condensation on direct contact with a cooling wall of a heat exchanger – the most common mode used in industry
- film condensation, when a liquid film is formed on the sub-cooled surface – usually occurs when the liquid wets the surface
- drop-wise condensation, when liquid drops are formed on the sub-cooled surface – usually occurs when the liquid does not wet the surface. Drop-wise condensation is difficult to sustain consistently, so industrial equipment is normally designed to operate in film condensation mode.

If a vapour is exposed to a surface at a temperature below saturation temperature then condensation occurs on the surface in the form of a liquid film or individual droplets. It is usually done by bringing the vapour into contact with a solid surface whose temperature is below the saturation temperature of the vapour. This is considered to be a form of convection heat transfer, since it involves fluid motion such as the flow of condensate to the bottom. The temperature remains constant under equilibrium conditions during a phase-change process at a fixed pressure. In this process, large amounts of heat (due to the large latent heat of vaporization released or absorbed) can be transferred during condensation, particularly at constant temperature (Collier and Thome [19]). However, it is necessary to maintain some difference between the surface temperature and the saturation temperature for an effective heat transfer mechanism. Heat transfer coefficients associated with condensation are generally much higher than those in other forms of convection processes that involve a single phase.

Two distinct forms of condensation are shown in Figure 1.17. They are film condensation and drop-wise condensation. In film condensation, the condensate wets the

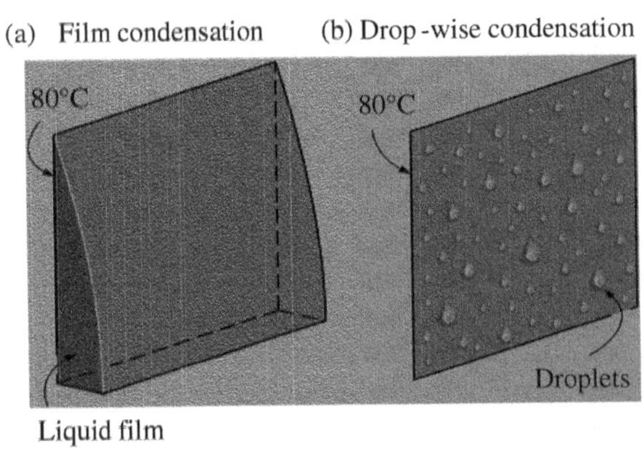

Figure 1.17 Types of condensation. (Collier and Thome [19])

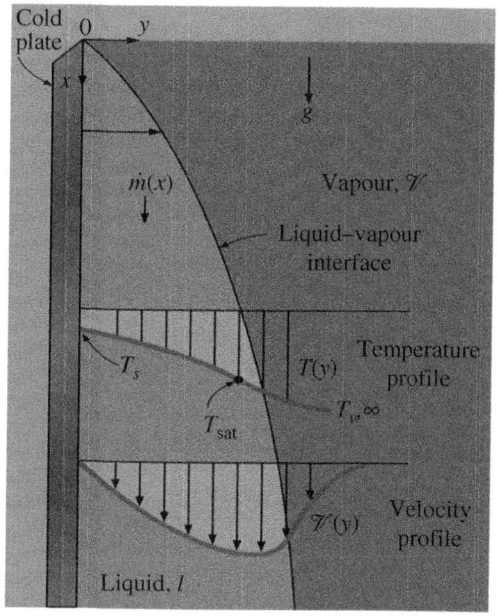

Figure 1.18 Typical velocity and temperature profile of the condensate. (Collier and Thome [19])

surface and forms a liquid film on the surface that slides down under the influence of gravity. The liquid film thickness increases in the flow direction as more vapour condenses on the film. In Drop-wise condensation, the condensed vapour forms droplets on the surface instead of a continuous film, and the surface is covered by large numbers of droplets of varying diameters.

1.9.1 Film Condensation

In film condensation, the surface is covered by a liquid film of increasing thickness, and this "liquid wall" between the solid surface and the vapour serves as a resistance to heat transfer. The velocity and temperature profiles of the condensate are also given in Figure 1.18. Due to no-slip condition, the velocity of the condensate at the wall is zero and it reaches a maximum at the liquid–vapour interface. At the interface, the temperature of the condensate is saturation temperature and it decreases gradually to surface temperature at the wall. In forced convection, involving a single phase, heat transfer in condensation also depends on whether the condensate flow is laminar or turbulent. Again, the Reynolds number provides the criterion for the flow regime.

1.9.2 Drop-wise Condensation

It is characterized by countless droplets of varying diameters on the condensing surface instead of a continuous liquid film. This is one of the most effective mechanisms of heat transfer, and extremely large heat transfer coefficients can be achieved with this mechanism and is shown in Figure 1.19.

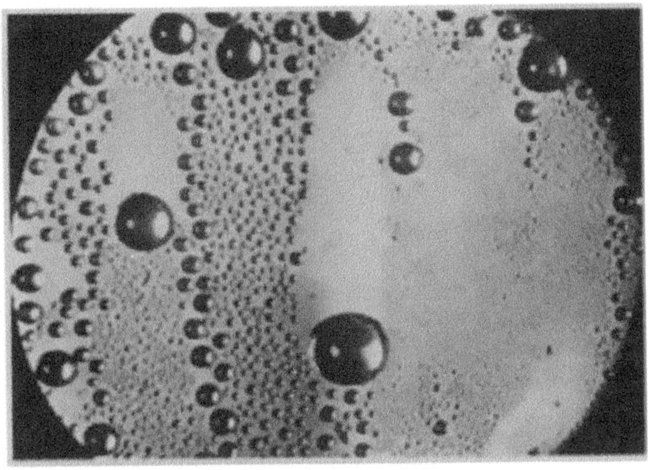

Figure 1.19 Drop-wise condensation. (Collier and Thome [19])

In this mechanism, the small droplets that form at the nucleation sites on the surface grow as a result of continued condensation, coalesce into large droplets, and slide down when they reach a certain size. As a result, with drop-wise condensation, heat transfer coefficients can be achieved that are 10 times larger than those associated with film condensation. Therefore, drop-wise condensation is the preferred mode of condensation in heat transfer applications and is achieved by adding a promoting chemical into the vapour or coating the surface with a polymer such as Teflon or a noble metal such as gold, silver, rhodium, palladium or platinum. The drop-wise condensation achieved does not last long and converts to film condensation after some time, so it is common practice to be conservative and assume film condensation in the design of heat transfer equipment (Collier and Thome [19]).

For details of the condensation data, refer to Chapter 5 for condensation experiments on different types of compact heat exchanger fins and Chapter 6 for worked example for estimation of condensation heat transfer coefficient and pressure drop estimation.

Nomenclature

A	cross-sectional area or heat transfer area, m^2
A_w	wall area, m^2
b	length, m
$C_{f,x}$	local skin friction coefficient, dimensionless
C_p	specific heat at constant pressure, J/kg.K
D	spacing, diameter, m
D_h	hydraulic diameter, m
$\Delta x, \Delta y, \Delta z$	differential distance, m

∂t	time
E	heat flux, W/m^2
f	friction factor, dimensionless
G	emission from block body, irradiation
G_z	Graetz number, dimensionless
H	height, m
h	heat transfer coefficient, $W/m^2 \cdot K$
	specific bulk enthalpy, J/kg
k	thermal conductivity, W/m.K
L	flow/ heat transfer length, m
$\dot{m}$	mass flow rate, kg/s
m', n'	direction cosines
Nu	Nusselt number, dimensionless
Nu_x	local Nusselt number, dimensionless
P	pressure, Pa
ΔP	pressure difference, Pa
Pr	Prandtl number, dimensionless
Pr_t	turbulent Prandtl number, dimensionless
p	perimeter, m
Q	Heat flow (W)
q'' or q	heat flux, W/m^2
r	radial position, m
r_h	hydraulic radius, m
r_0	tube radius, m
Ra	Rayleigh number, dimensionless
Re_D	Reynolds number based on D, dimensionless
Re_{Dh}	Reynolds numbers based on D_h, dimensionless
Re_L	Reynolds number based on L, dimensionless
S_t	Stanton number, dimensionless
Sh_x	Sherwood number, dimensionless
Sc	Schmidt number
S_{opt}	optimum distance, m
t	plate thickness, m
T	temperature, K
T_{in}	inlet temperature, K

T_{out}	outlet temperature, K
T_m	mean temperature, K
ΔT_{avg}	average temperature difference, K
ΔT_{lm}	log mean temperature difference, K
u	longitudinal velocity, m/s
u_*	friction velocity, m/s
U	mean velocity, m/s
V	volume, m^3
v	transversal velocity, m/s
w	width, m
x,y,z	positions, m
X	flow entrance length, m
X_T	thermal entrance length, m
y	transversal position, m
y_{VSL}	viscous sub-layer thickness, m

Greek Symbols

α	thermal diffusivity, m^2/s
ε	emissivity
λ	wave length
σ	Stefan–Boltzmann constant
θ_{*m}	bulk temperature, dimensionless
Γ	boundary
μ	viscosity, kg/s-m
ν	kinematic visocity, m^2/s
ρ	density, kg/m^3
τ_{app}	apparent sheer stress, Pa

Subscripts

a	atmosphere
r,θ,z	coordinates
lm	log mean
w	wall

References

1 Incropera F.P., Dewitt D.P., Bergman T.L., Lavine A.S., Seetharamu K.N. and Setharam T.R., *Fundamentals of Heat and Mass Transfer*, Wiley publications, New Delhi, 2014.
2 Bejan A. *Heat Transfer*, John Wiley & Sons, New York, 1993.
3 Holman, *Heat Transfer*, McGraw-Hill, New York, 1989.
4 Incropera F.P.,and Dewitt D.P. *Fundamentals of Heat and Mass Transfer*, John Wiley & Sons, New York, 1990.
5 Sukhatme S.P. *A Text Book on Heat Transfer*, Third Edition, Orient Longman Publishers, 1992.
6 Ozisik M.N. *Boundary Value Problems of Heat Conduction*, International Text Book Company, Scranton, P.A., 1968.
7 Carslaw H.S. and Jaeger J.C. *Conduction of Heat in Solids*, Second Edition, Oxford University Press, Fairlawn, N.J., 1959.
8 Roland W. Lewis, Perumal Nithiarasu and Kankanhally N. Seetharamu, Fundamentals of the Finite Element Method for Heat and Fluid Flow, John Wiley & Sons Ltd., 2004.
9 Jaluria Y., *Natural Convection Heat and Mass Transfer*, Pergamon Press, 1986.
10 Massarotti N., Nithiarasu P. and Zienkiewicz O.C. Characteristic – based – split (CBS) algorithm for incompressible flow problems with heat transfer, *International Journal of Numerical Methods for Heat and Fluid Flow*, 8, 969–990, 1998.
11 Minkowycz W.J., Sparrow E.M., Schneider G.E. and Pletcher R.H. *Handbook of Numerical Heat Transfer*, John Wiley & Sons, 1988.
12 Aung W. and Worku G. B., Theory of fully developed, combined convection including flow reversal, *Trans. of ASME J. Heat Transfer*,108, 485–488, 1986.
13 Nithiarasu P., Seetharamu K.N. and Sundararajan T. Finite element analysis of transient natural convection in an odd-shaped enclosure, *International Journal of Numerical Methods for Heat and Fluid Flow*, 8, 199–216, 1998.
14 Blasius H.Z., Math. Phys., 56, 1, English translation in National Advisory Committee for Aeronautics Technical Memo no. 1256, 1908.
15 Chilton T.H. and A.P. Colburn, *Ind. Eng. Chem.*, 26, 1183, 1934.
16 Rohsenow W.M., James P. Hartnett and Young I. Cho, *Handbook of Heat Transfer*, McGraw-Hill Education, U.S., 1998.
17 Siegel R. and Howell J.R. *Thermal Radiation Heat Transfer*, Third Edition, Hemisphere Publishing Corporation, 1992.
18 Kandlikar S.G. and A. Mariani, in S.G. Kandlikar, M. Shoji and V.K. Dhir, Eds., *Handbook of phase change: Boiling and Condensation*, Chp 15, Taylor & Francis, New York, 1999.
19 Collier J.G. and J.R. Thome, *Convective Boiling and Condensation*, 3rd Ed., Oxford University Press, New York, 1996.
20 Butterworth, D., in D. Butterworth and G.F. Hewitt, Eds., *Two-phase Flow and Heat Transfer*, Oxford University press, London, pp. 426–462, 1977.

2

Compact Heat Exchangers

2.1 Introduction

The process of heat exchange between two fluids that are at different temperatures and separated by a solid wall occurs in many engineering applications. The device used to implement this exchange is termed a heat exchanger. Decreasing size and increasing heat load is the typical feature of the modern-day heat exchanger industry. In forced-convection heat transfer between a gas and a liquid, the heat transfer coefficient of the gas may be 10 to 50 times smaller than that of the liquid. The use of specially configured surfaces can reduce the gas-side thermal resistance. For heat transfer between two gases, the difficulty in inducing the desired heat exchange is even more pronounced. In this case especially, the use of enhanced surfaces can substantially reduce heat exchanger size. This is the motivation behind the design of a category of heat exchangers with reduced size and greatly enhanced gas-side heat transfer, which are referred to as "compact." A compact heat exchanger is generally defined as one which incorporates a heat transfer surface having a high "area density." In other words, it possesses a high ratio of heat transfer surface area to volume. Quantitatively, Shah [1] arbitrarily defines a compact heat exchange surface as one that has an area density greater than 700 m^2/m^3. Figure 2.1 shows a spectrum of surface area density for heat exchangers. The range of surface area density (and hydraulic diameter) is given for various types of heat exchange surfaces, with the dividing line for compactness clearly marked.

While traditional areas of compact heat exchanger applications, such as automotive, aerospace and cryogenics, continue to demand even higher heat transfer with further shrinking of available space, there are many new areas coming up in the use of compact heat exchangers. These include areas such as cooling of electronic equipment, cooling of laser and related technologies and cooling technology for fuel cells. A number of traditional industries have also turned towards compact heat exchangers, including chemical process industry, power industry and the food and beverages industry. The use of compact heat exchangers for multiphase flow has been an area of particular focus in the recent years. For a comprehensive picture of these developments, this chapter aims to discuss at length the analysis, types of heat exchangers, their new uses, and future directions in research and development. Various techniques can be used to make heat exchangers more compact, with three main general types of extended surface geometries used to increase gas-side heat transfer coefficients. These include: (a) a finned-tube heat

Compact Heat Exchangers – Analysis, Design and Optimization using FEM and CFD Approach,
First Edition. C. Ranganayakulu and K.N. Seetharamu.
© 2018 John Wiley & Sons Ltd. Published 2018 by John Wiley & Sons Ltd.

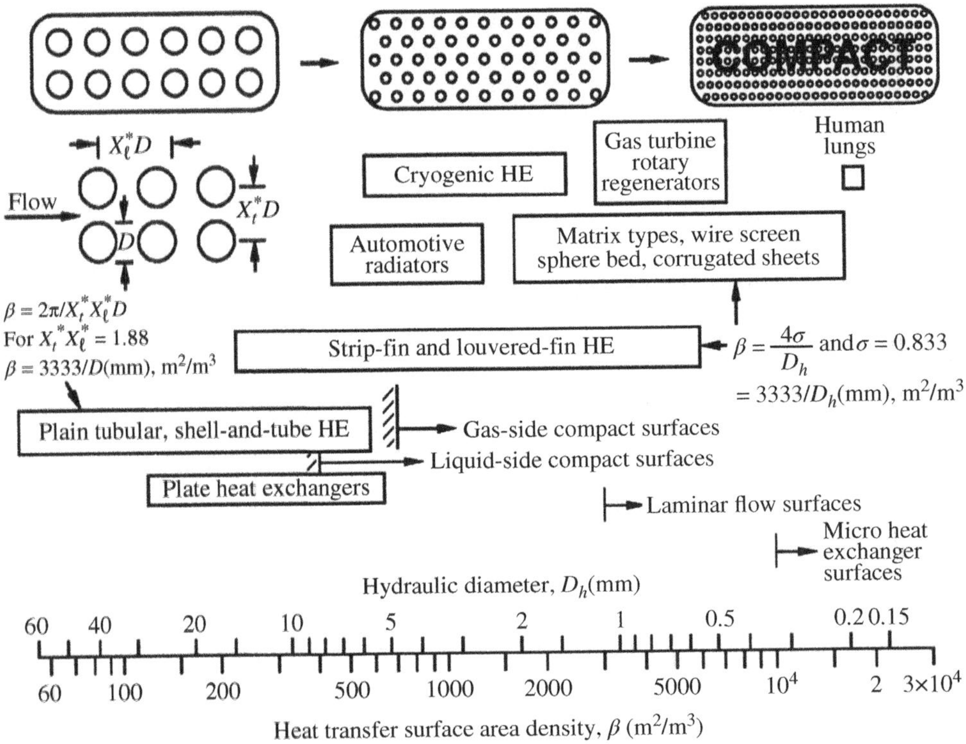

Figure 2.1 A spectrum of surface area density for heat exchangers.

exchanger with flat fins, (b) a finned tube heat exchanger with individually finned tubes and (c) a plate-fin heat exchanger. A wide variety of plate-fin geometries have been used to obtain enhanced heat transfer, and engineers are constantly working to develop new and more effective surfaces.

2.2 Motivation for Heat Transfer Enhancement

The study of enhanced heat transfer has gained more momentum during recent years, especially due to increased demands by industry for heat exchange equipment that is less expensive to build and operate than standard heat exchange devices. Savings in materials and energy also provide strong motivation for the development of improved methods of enhancement. When designing cooling systems for automobiles and aircraft/spacecraft, it is imperative that the heat exchangers are especially compact and lightweight. Also, enhancement devices are necessary for the high heat duty exchangers found in power plants (e.g. air-cooled condensers, nuclear fuel rods). These applications, as well as numerous others, have led to the development of various enhanced heat transfer surfaces.

In general, enhanced heat transfer surfaces can be used for three purposes: (1) to make heat exchangers more compact in order to reduce their overall volume, and possibly their cost, (2) to reduce the pumping power required for a given heat transfer process, and (3) to increase the overall UA value of the heat exchanger. A higher UA value can be exploited in two ways: (1) to obtain an increased heat exchange rate for fixed fluid inlet temperatures, or (2) to reduce the mean temperature difference for the heat exchange; this increases the thermodynamic process efficiency, which can reduce operating costs. Enhancement techniques can be separated into two categories: passive and active. Passive methods require no direct application of external power. Instead, passive techniques employ special surface geometries or fluid additives that cause heat transfer enhancement. On the other hand, active schemes such as electromagnetic fields and surface vibration require external power for operation, Bergles et al. [2]. The majority of commercially interesting enhancement techniques are passive ones. Active techniques have attracted little commercial interest because of the costs involved, and the problems that are associated with vibration or acoustic noise as per Webb [3]. In case of gas-side heat transfer enhancement, using special surface geometries, it can be established with higher hA per unit base surface area. Clearly, there are three basic ways of accomplishing this as per Webb [3].

1) Increase the effective heat transfer surface area (A) per unit volume without appreciably changing the heat transfer coefficient (h). Plain fin surfaces enhance heat transfer in this manner.
2) Increase h without appreciably changing A. This is accomplished by using a special channel shape, such as a wavy or corrugated channel, which provides mixing due to secondary flows and boundary-layer separation within the channel. Vortex generators also increase h without a significant area increase by creating longitudinally spiralling vortices exchange fluid between the wall and core regions of the flow, resulting in increased heat transfer as per Brockmeier et al. [4].
3) Increase both h and A. Interrupted fins (i.e. offset strip and louvred fins) act in this way. These surfaces increase the effective surface area, and enhance heat transfer through repeated growth and destruction of the boundary layers.

When looking for a special surface geometry to enhance heat transfer in an industrial heat exchange application, there are a large number of options. We can judge the relative heat transfer enhancement for selected geometries by comparing the heat transfer coefficients or dimensionless heat transfer parameters (i.e. Nusselt number, Stanton number, etc.) yielded by each enhanced surface. But this will only give a partial indication of performance. Enhanced surfaces do provide a greater heat transfer coefficient, but they also lead to increased fluid flow friction and pressure drop. Sometimes the benefits gained from heat transfer enhancement are not great enough to offset the increased friction losses. Clearly then, the performance goal is to gain maximum enhancement of heat transfer with minimum penalty on pumping power. However, this balance is difficult to quantify in a manner that allows straightforward comparisons between various enhanced surface geometries. Numerous methods have been suggested to accurately evaluate the performance improvement provided by enhancement techniques.

2.3 Comparison of Shell and Tube Heat Exchanger

High compactness can be achieved in compact heat exchangers (CHEs) when compared to shell and tube heat exchangers. CHEs can be classified by the kinds of compact elements that they employ. The compact elements usually fall into five classes:

1) Circular and flattened circular tubes: These are the simplest form of CHE surfaces. The designation ST indicates flow inside straight tubes (e.g. ST-1), FT indicates flow inside straight flattened tubes (e.g. FT-1) and FTD indicates flow inside straight flattened dimpled tubes. Dimpling interrupts the boundary layer, which tends to increase the heat transfer coefficient without increasing the flow velocity.
2) Tubular surfaces: These are arrays of tubes of small diameter, from 10 mm down to 9 mm, used in service where the ruggedness and clean-ability of the conventional shell-and-tube exchanger are not required. Usually, tube sheets are comparatively thin, and soldering or brazing a tube to a tube sheet provides an adequate seal against inter-leakage and differential thermal expansion.
3) Surfaces with flow normal to banks of smooth tubes: Unlike the radial low fin tubes, smooth round tubes are expanded into fins that can accept a number of tube rows, as shown in Figure 2.2a. Holes may be stamped in the fin with a drawn hub or foot to improve contact resistance or as a spacer between successive fins, or brazed directly to the fin with or without a hub. Other types reduce the flow resistance outside the tubes by using flattened tubes and brazing, as indicated in Figure 2.2b and c below. Flat tubing is made from strips similar to the manufacture of welded circular tubing but is much thinner and is joined by soldering or brazing rather than welding.
4) Plate fin surfaces: A typical cross-flow heat exchanger with plate-fin surface is shown in Figure 2.3. A plate-fin heat exchanger with six fins per cm provides approximately 1300 m^2 of surface per m^3 of volume. This heat exchanger would be approximately 10% of the volume of an equivalent shell and tube heat exchanger with 19 mm tubes. A rule-of-thumb calculation suggests that, for a given duty, a shell and tube unit will be 5–10 times heavier. The weight benefit is coupled with significant volume reductions.
5) Finned-tube surfaces: Circular tubes with spiral radial fins are designated by the letters CF followed by one or two numerals. The first numeral indicates the number of fins per inch, and the second (if one is used) refers to the nominal tube size. With circular tubes with continuous fins, no letter prefix is employed and the two numerals

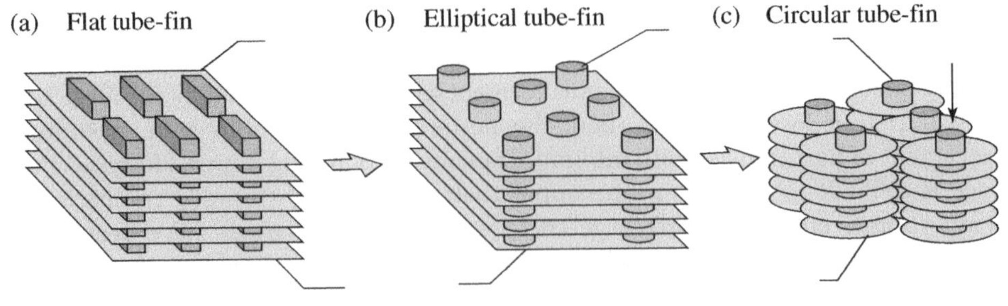

Figure 2.2 Tube-fin heat exchangers.

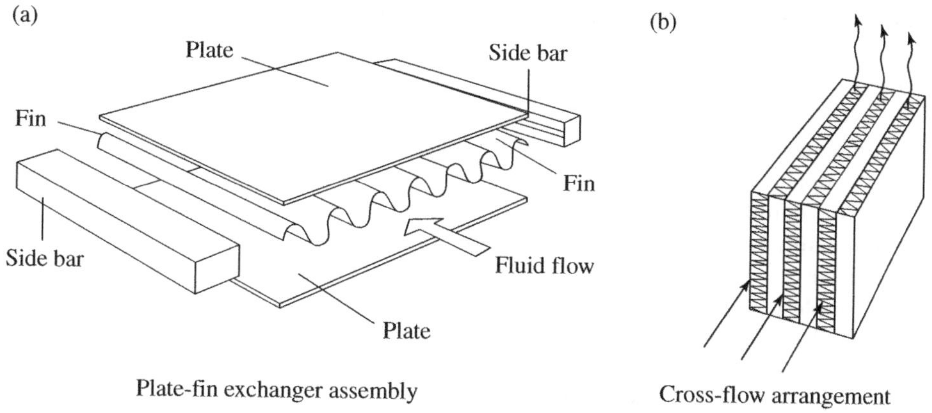

Figure 2.3 Typical cross-flow heat exchanger with plate-fins.

have the same meaning as those used for circular tubes with spiral radial fins. For finned flat tubes, no letter prefix is used; the first numeral indicates the fins per inch and the second numeral indicates the largest tube dimension. When CF does not appear in the designation of the circular tube with spiral radial fins, the surface may be presumed to have continuous fins.

6) Matrix surfaces: These are surfaces that are used in rotating, regenerative equipment such as combustion flue gas–air preheaters for conventional fossil furnaces. In this application, metal is deployed for its ability to absorb heat with minimal fluid friction while exposed to hot flue gas and to give up this heat to incoming cold combustion air when it is rotated into the incoming cold airstream. No designation is employed.

2.4 Classification of Heat Exchangers

The classification of heat exchangers is defined on number of passes, transfer processes, surface compactness, flow pattern, heat transfer mechanisms etc. as follows:

1) **Classification according to transfer process**
   a) Indirect contact type
      i) Direct transfer type
         1) Single phase
         2) Multiphase
      ii) Storage type
      iii) Fluidized Bed
   b) Direct contact type
      i) Immiscible fluids
      ii) Gas–liquid
      iii) Liquid–vapour
2) **Classification according to number of fluids**
   a) Two fluids
   b) Three fluids
   c) N fluids (N > 3)

3) **Classification according to surface compactness**
   a) Gas-to-fluid
      i) Compact ($\beta \geq 700\ m^2/m^3$)
      ii) Non-compact ($\beta < 700\ m^2/m^3$)
   b) Liquid-to-liquid and phase change
      i) Compact ($\beta \geq 400\ m^2/m^3$)
      ii) Non-compact ($\beta < 400\ m^2/m^3$)
4) **Classification according to construction**
   a) **Tubular**
      i) Double-pipe
      ii) Shell-and-tube
         1) Cross flow to tubes
         2) Parallel flow to tubes
      iii) Spiral Tube
      iv) Pipe coils
   b) **Plate type**
      i) PHE (plate heat exchanger)
         1) Gasket
         2) Welded
         3) Brazed
      ii) Spiral
      iii) Plate coil
      iv) Printed Circuit
   c) **Extended surface**
      i) Plate-fin
      ii) Tube-fin
         1) Ordinary separating wall
         2) Heat-pipe wall
   d) **Regenerative**
      i) Rotary
      ii) Fixed-matrix
      iii) Rotary hoods
5) **Classification according to flow arrangements**
   a) Single-pass
      i) Counter flow
      ii) Parallel flow
      iii) Cross flow
      iv) Split-flow
      v) Divided-flow
   b) Multi-pass
      i) Extended surface
         1) Cross- counter flow
         2) Cross- parallel flow
         3) Compound flow
   c) Shell-and-tube
      i) Parallel counter flow
         1) m-shell passes
         2) n-tube passes

    ii) Split-flow
    iii) Divided-flow
  d) Plate
    i) Fluid 1 m passes
    ii) Fluid 2 n passes
6) **Classification according to heat transfer mechanisms**
  a) Single-phase convection on both sides
  b) Single-phase convection on one side, two-phase convection on other side
  c) Two-phase convection on both sides
  d) Combined convection and radiative heat transfer

2.5 Heat Transfer Surfaces

A wide variety of plate-fin geometries, as shown in Figure 2.4, have been used to obtain enhanced heat transfer, and engineers are constantly working to develop new and more effective enhanced surfaces. Typical fin spacings are 400 to 1000 fins per metre. Due to their small hydraulic diameter and the low density of gases, these surfaces are usually operated in the Reynolds number range 500–10,000. As a result, plate-fin enhancement geometries must be effective in the low Reynolds number regime. A large number of plate-fin geometries have been proposed for use in compact heat exchangers, and more are still being developed. A high-performance surface will enhance the heat transfer that takes place within the heat exchanger, without incurring penalties on friction

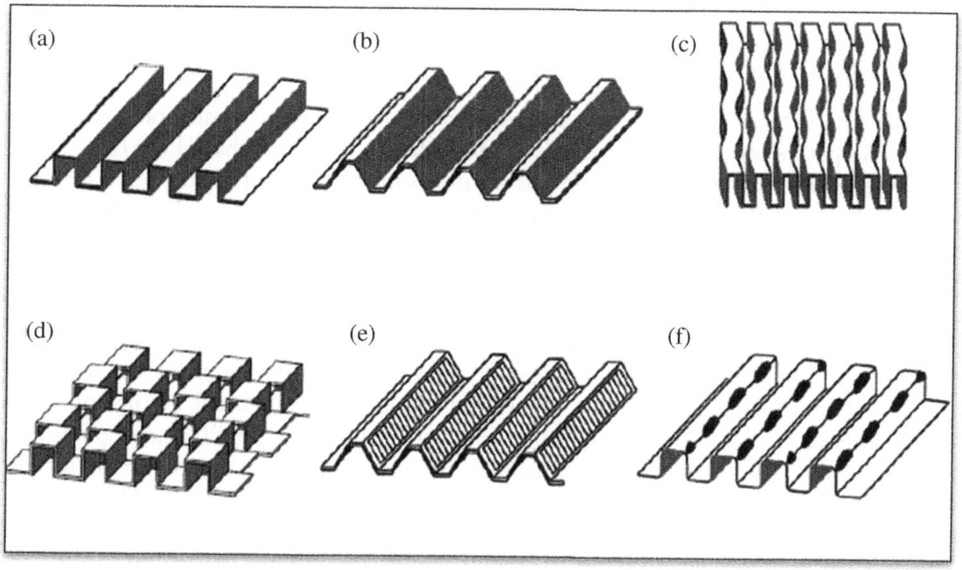

Figure 2.4 Types of plate fin surfaces: (a) plain rectangular (b) plain trapezoidal (c) wavy fin (d) offset strip fin (e) louvred (f) perforated.

and pressure drop that are severe enough to negate the benefits of heat transfer augmentation. In this section, the following types of plate-fin geometries are examined: plain fins, wavy and corrugated channels, offset-strip fins, louvred fins and vortex generators.

2.5.1 Rectangular Plain Fin

Plain fins are the most common of all compact cores or surfaces used in CHEs. The plain fin surfaces are characterized by long uninterrupted flow passages, with performance similar to that obtained inside long circular tubes – Kays & London [5]. Although passages of triangular and rectangular cross-section are more common, any desired shape can be given to the fins, considering only manufacturing constraints. Straight fins in triangular arrangement can be manufactured at high speeds and hence are less expensive than rectangular fins. But generally they are structurally weaker than rectangular fins for the same passage size and fin thickness. They also have lower heat transfer performance compared to rectangular fins, particularly in laminar flow. Plain fins are used in those applications where core pressure drop is critical. A heat exchanger with rectangular plain fin, as shown in Figure 2.4a, requires a smaller flow frontal area than that with interrupted fins for specified pressure drop, heat transfer and mass flow rate. The heat transfer enhancement achieved with plain fins results mainly from increased area density, rather than any substantial rise in the heat transfer coefficient. The plain-fin surfaces are characterized by long uninterrupted flow passages, with performance similar to that obtained inside long circular tubes. Note that the plain-fin surfaces include rectangular passages, triangular passages and passages with rounded and re-entry corners, so that there is considerably more variation.

2.5.2 Louvred-Fin

Louvred fin surfaces (Figure 2.4e) are commonly used in automobile radiators. The louvred fin geometry consists of an interrupted surface similar to that of the offset-strip fin. However, the slit strips of louvred fins are not completely offset. Instead, the slit fin is rotated between 20° and 60° relative to the direction of the airflow. For equal strip width, the louvred fin geometry provides enhancement comparable to that of offset strip fins. Moreover, louvred fins are less expensive than offset strip fins for large-quantity production, because of their ease of manufacture using high-speed mass production technology provided by O'Brien and Sparrow [6].

The louvred-fin surfaces are characterized by fins that have been cut and bent out into the flow stream at frequent intervals. The purpose of louvring is to break up the boundary layers so as to yield higher heat transfer coefficients than are possible with plain fins under the same flow conditions. As a general rule, the more frequent the interruption, the higher the conductance, although the friction factor is also increased. The louvred-fin surfaces are designed by two numbers. The first refers to the length of the louvred fin in the flow direction; the second refers to the fins per inch transverse to the flow. Thus surface ¾ – 11.1 has ¾ louvres and 11.1 fins per inch.

2.5.3 Strip-Fin or Lance and Offset Fin

The offset strip fin geometry (Figure 2.4d) is one of the most widely used enhanced surfaces in CHEs. Basically, a laminar boundary layer develops on the short strip length, and is then dissipated in the wake region between strips. Typical strip lengths are 3–6 mm, and offset strip fins are usually employed in the laminar regime as per Webb [3]. The enhancement provided by offset strip fins results from an increase in both the effective surface area and the heat transfer coefficient – Brockmeier et al. [4].

The strip-fin surfaces are similar in principle to the louvred-fin surfaces, the only difference being that the short sections of fins are aligned entirely with the flow direction. With the strip-fin configuration it is feasible to have very short flow-length fins and thus very high heat transfer coefficients. Fins of this type are generally constructed by a machine-cutting process that inevitably leaves a slightly bent and scarped fin edge that differs depending upon the fin material and the character of the cutting tool. Since a few ten-thousandths of an inch of scarring can have a considerable effect, it is difficult to either dimensionally describe this effect or to exactly duplicate one of the test surfaces. Most of the fins of the test surfaces are of aluminium construction and although this by no means precludes the use of the performance data for steel fins, it does suggest some conservatism in design. The designation scheme for the strip-fin surfaces is essentially the same as that used for the louvred surfaces.

2.5.4 Wavy-Fin

Several studies have been conducted to evaluate the heat transfer enhancement provided by using wavy passages in CHEs. Wavy and corrugated channels both enhance heat transfer by promoting mixing due to complex re-circulatory flows and boundary layer separation. However, less friction is expected in wavy channels because the sharp corners of the corrugated channel are not present. The wavy-fin surfaces (Figure 2.4c) are also high-performance surfaces with performance quite similar to the louvred and strip-fin surfaces. The change in flow direction induced by the fins causes boundary-layer separation with effects similar to complete fin interruption. The wavy-fin surfaces are designed by two figures, giving the number of fins per inch and the wavelength, followed by the letter W. Thus surface 11.5 – ¾ W has 11.5 fins per inch and a complete wave every ¾ in.

2.5.5 Pin-Fin

Pin-fin surfaces (Figure 2.5) are another example of the plate-fin system, where the purpose is to achieve very high heat transfer coefficients by maintaining thin boundary layers on the fins. By constructing the fins from small-diameter wire, the effective flow length of the fins can be very small indeed. The pin-fin surfaces are, however, characterized by quite high friction factors attributable primarily to a form drag associated with the boundary-layer. The designation scheme for the pin-fin surfaces is not descriptive.

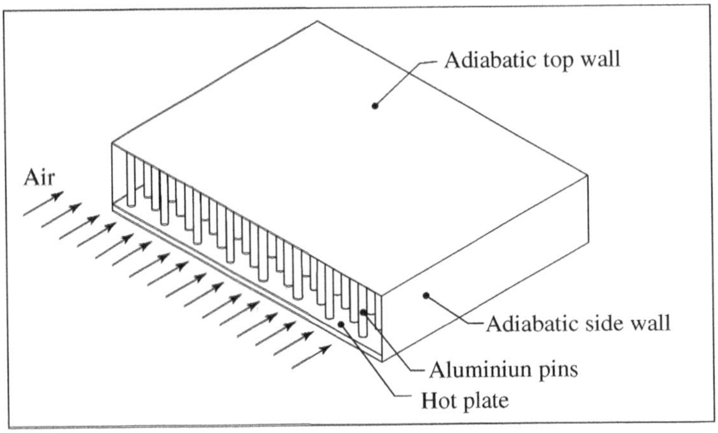

Figure 2.5 Cooling of hot plate with pin-fin surface.

2.5.6 Rectangular Perforated Fin

This surface geometry is made by punching a pattern of spaced holes in the fin material before it is formed into flow channels, as shown in Figure 2.4f. The channels may be triangular or rectangular in shape with either round or rectangular perforations. The thermal performance of the perforated fin is less than that of a good offset strip fin. Perforated fins are now used only in limited number of applications such as tabulators in oil coolers. Even though a negligible quantity of material is removed for making holes or perforations, there is an improvement in performance of the heat exchanger.

Perforated fins are used in the distributor sections of exchangers and also for boiling applications. The perforations in distributors allow for lateral migration of flow (including the vapour phase), without the pressure drop penalty of the offset strip fin. The interruptions given by them give a slight performance improvement per unit surface area, but this is offset by loss of surface area, and there is generally a higher friction factor.

2.5.7 Triangular Plain Fin

The triangular plain fin is similar to rectangular plain fin, but the fin flow path is a triangular shape. The model of the triangular plain fin is shown in Figure 2.6. The enhancement in heat transfer achieved with plain fins is due mainly to increased area density, rather than any increase in the heat transfer coefficient Brockmeier et al. [4]. Plain fins require a smaller flow frontal area than interrupted surfaces (i. e. offset strip fins and louvred fins) for given values of heat duty, pressure drop, and flow rate, but the flow length with plain fins will be greater, resulting in a higher overall heat exchanger volume as per O'Brien and Sparrow [6].

2.5.8 Triangular Perforated Fin

The perforated triangular fin is similar to the plain triangular fin, the only difference being that the perforated fins have pierced holes along the sides of the fin; this is used to study the effects on heat transfer rate, but it also has its own limitations because the

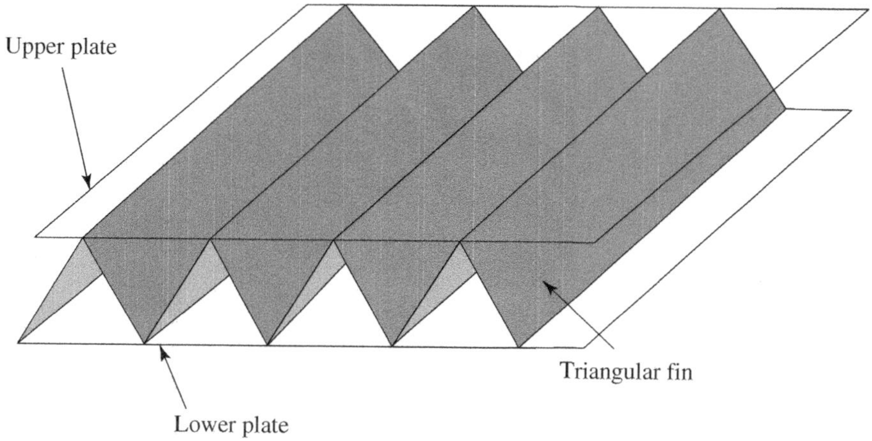

Figure 2.6 Triangular plain-fin.

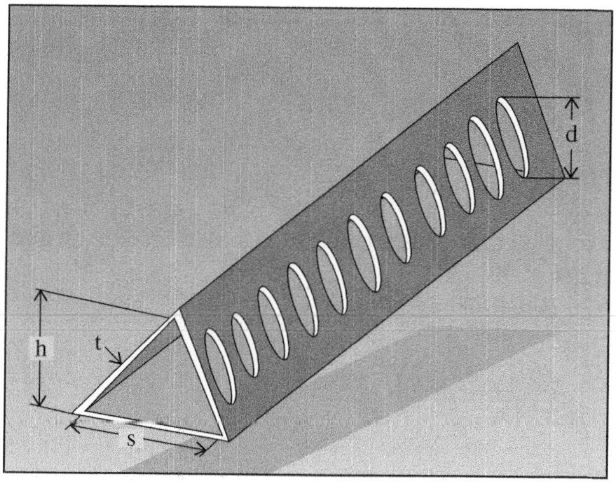

Figure 2.7 Model of triangular perforated fin.

diverged fluid path increases the pumping power needed at the inlet. The channel is in a triangular shape with round perforations, as shown in Figure 2.7; this surface geometry is made by punching a pattern of spaced holes in the fin material before it is formed into flow channels.

2.5.9 Vortex Generator

A wing-type vortex generator plate is shown in Figure 2.8. Vortex generators do not significantly change the effective heat transfer surface area of the plate, but they do increase the heat transfer coefficient by creating longitudinally spiralling vortices that promote mixing between the wall and core regions of the flow – Brockmeier et al. [4]. Vortex generators are a relatively new type of enhancement device, and an optimum geometry has

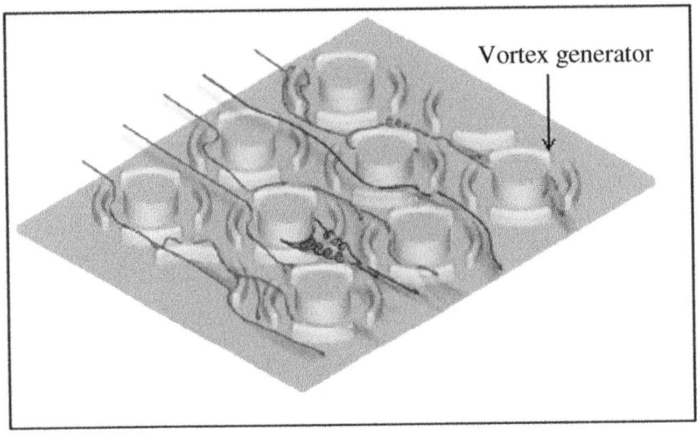

Figure 2.8 A pin-fin surface with vortex generators.

not yet been found There are many possibilities for different vortex generator surfaces, since we can vary the size, angle of attack, aspect ratio and/or arrangement of the vortex generators.

2.6 Heat Exchanger Analysis

In this analysis, basic types of heat exchanger analysis are provided, based on of Shah and Sekulic [7], Webb [8] and Incropera et al. [9].

The heat exchanger effectiveness, ε is defined as

$$\varepsilon = \frac{C_h}{C_{min}}\left(\frac{(T_{h_{in}} - T_{h_{out}})}{(T_{h_{in}} - T_{c_{in}})}\right) = \frac{C_c}{C_{min}}\left(\frac{(T_{c_{out}} - T_{c_{in}})}{(T_{h_{in}} - T_{c_{in}})}\right) \tag{2.1}$$

where C_{min} is the smaller of C_c and C_h.

To define the effectiveness of a heat exchanger, we must first determine the maximum possible heat transfer rate, q_{max}, for the exchanger. This heat transfer rate could, in principle, be achieved in a counter-flow heat exchanger (Figure 2.9) of infinite length. In such an exchanger, one of the fluids would experience the maximum possible temperature difference, $T_{h,i} - T_{c,i}$. To illustrate this point, consider a situation for which $C_c > C_h$, in which case, from Equation 2.1, $|dT_c| > |dT_h|$. The cold fluid would then experience the larger temperature change, and since $L \to \infty$, it would be heated to the inlet temperature of the hot fluid ($T_{c,o} = T_{h,i}$). Accordingly, from Equation 2.1,

$$C_c > C_h: \quad q_{max} = C_c(T_{h,i} - T_{c,i}) \tag{2.2}$$

Similarly, if $C_h > C_c$, the hot fluid would experience the larger temperature change and would be cooled to the inlet temperature of the cold fluid ($T_{h,o} = T_{c,i}$). From Equation 2.1, we then obtain

$$C_h > C_c: \quad q_{max} = C_h(T_{h,i} - T_{c,i}) \tag{2.3}$$

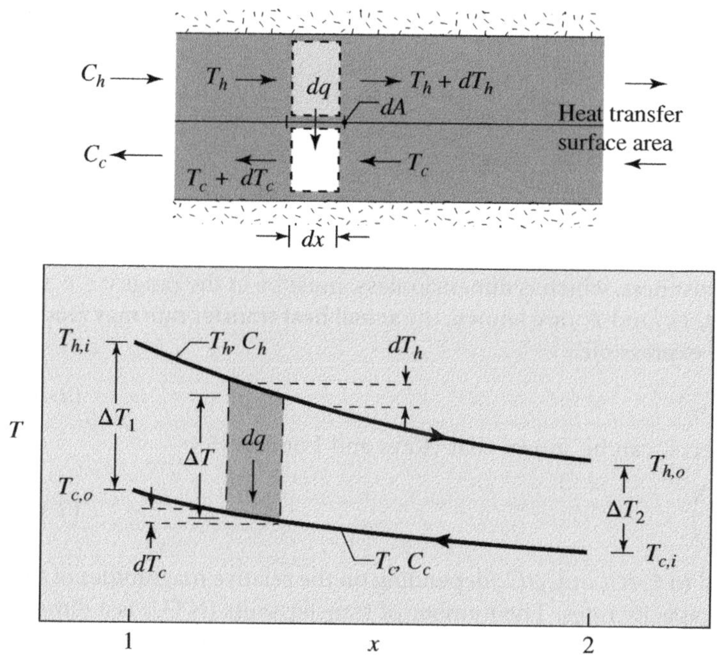

Figure 2.9 Temperature distribution for a counter-flow heat exchanger.

From the foregoing results we are then prompted to write the general expression

$$q_{max} = C_{min}\,(T_{h,i} - T_{c,i}) \tag{2.4}$$

where C_{min} is equal to C_c or C_h, whichever is smaller. For prescribed hot and cold fluid inlet temperatures, Equation 2.4 provides the maximum heat transfer rate that could possibly be delivered by the exchanger. A quick mental exercise should convince the reader that the maximum possible heat transfer rate is not equal to $C_{max}\,(T_{h,i} - T_{c,i})$. If the fluid having the larger heat capacity rate are to experience the maximum possible temperature change, conservation of energy in the form of

$$C_c(T_{c,o} - T_{c,i}) = C_h(T_{h,i} - T_{h,o}) \tag{2.5}$$

would require that the other fluid experience a yet larger temperature change. For example, if $C_{max} = C_c$ and we could argue that it is possible for $T_{c,o}$ to be equal to $T_{h,i}$, it follows that as

$$(T_{h,i} - T_{h,o}) = (C_c/C_h)(T_{h,i} - T_{c,i}) \tag{2.6}$$

in which case $(T_{h,i} - T_{h,o}) > (T_{h,i} - T_{c,i})$. Such a condition is clearly impossible.

It is now logical to define the effectiveness, ε, as the ratio of the actual heat transfer rate for a heat exchanger to the maximum possible heat transfer rate:

$$\varepsilon \equiv \frac{q}{q_{max}} \tag{2.7}$$

From Equations 2.1 and 2.4, it follows that

$$\varepsilon = \frac{C_h(T_{h,i} - T_{h,o})}{C_{min}(T_{h,i} - T_{c,i})} \tag{2.8}$$

or

$$\varepsilon = \frac{C_c(T_{c,o} - T_{c,i})}{C_{min}(T_{h,i} - T_{c,i})} \tag{2.9}$$

By definition, the effectiveness, which is dimensionless, must be in the range $0 \le \varepsilon \le 1$. This is useful because if ε, $T_{h,i}$ and $T_{c,i}$ are known, the actual heat transfer rate may readily be determined from the expression

$$q = \varepsilon C_{min}(T_{h,i} - T_{c,i}) \tag{2.10}$$

For any heat exchanger, it can be shown that (Kays and London [5])

$$\varepsilon = f\left(NTU, \frac{C_{min}}{C_{max}}\right) \tag{2.11}$$

where C_{min}/C_{max} is equal to C_c/C_h or C_h/C_c, depending on the relative magnitudes of the hot and cold fluid heat capacity rates. The number of transfer units (NTU) is a dimensionless parameter that is widely used for heat exchanger analysis and is defined as

$$NTU \equiv \frac{UA}{C_{min}} \tag{2.12}$$

2.6.1 Use of the Log Mean Temperature Difference

To design or to predict the performance of a heat exchanger, it is essential to relate the total heat transfer rate to quantities such as the inlet and outlet fluid temperatures, the overall heat transfer coefficient and the total surface area for heat transfer. Two such relations may readily be obtained by applying overall energy balances to the hot and cold fluids, as shown in Figure 2.10. In particular, if q is the total rate of heat transfer between the hot and cold fluids and there is negligible heat transfer between the exchanger and its surroundings, as well as negligible potential and kinetic energy changes, application of the steady flow energy equation gives

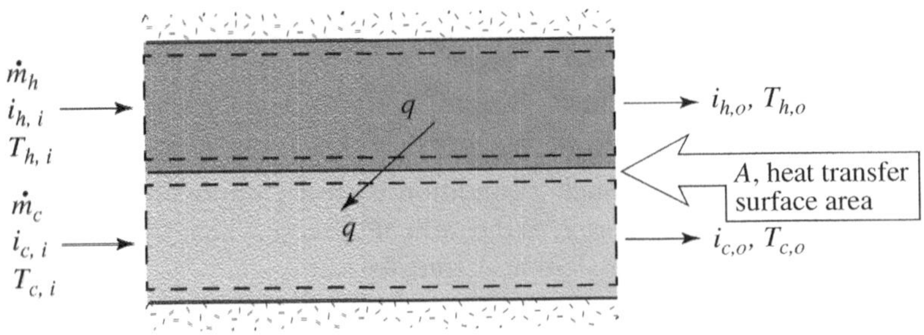

Figure 2.10 Overall energy balances for the hot and cold fluids of a two-fluid heat exchanger.

$$q = \dot{m}_h(i_{h,i} - i_{h,o}) \tag{2.13a}$$

and

$$q = \dot{m}_c(i_{c,o} - i_{c,i}) \tag{2.13b}$$

where i is the fluid enthalpy. The subscripts h and c refer to the hot and cold fluids, whereas i and o designate the fluid inlet and outlet conditions. If the fluids are not undergoing a phase change and constant specific heats are assumed, these expressions reduce to

$$q = \dot{m}_h c_{p,h}(T_{h,i} - T_{h,o}) \tag{2.14a}$$

and

$$q = \dot{m}_c c_{p,c}(T_{c,o} - T_{c,i}) \tag{2.14b}$$

where the temperatures appearing in the expressions refer to the mean fluid temperatures at the designated locations. Note that Equations 2.13 and 2.14 are independent of the flow arrangement and heat exchanger type.

Another useful expression may be obtained by relating the total heat transfer rate q to the temperature difference ΔT between the hot and cold fluids, where

$$\Delta T \equiv T_h - T_c \tag{2.15}$$

Such an expression would be an extension of Newton's law of cooling, with the overall heat transfer coefficient U used in place of the single convection coefficient h. However, since ΔT varies with position in the heat exchanger, it is necessary to work with a rate equation of the form

$$q = UA\Delta T_m \tag{2.16}$$

where ΔT_m is an appropriate mean temperature difference. Equation 2.16 may be used with Equations 2.13 and 2.14 to perform a heat exchanger analysis. Before this can be done, however, the specific form of ΔT_m must be established.

2.6.1.1 Parallel-Flow Heat Exchanger

The hot and cold fluid temperature distributions associated with a parallel-flow heat exchanger are shown in Figure 2.11. The temperature difference ΔT is initially large but decays with increasing x, approaching zero asymptotically. It is important to note that, for such an exchanger, the outlet temperature of the cold fluid never exceeds that of the hot fluid. In Figure 2.11 the subscripts 1 and 2 designate opposite ends of the heat exchanger. This convention is used for all types of heat exchangers considered. For parallel flow, it follows that $T_{h,i} = T_{h,1}$, $T_{h,o} = T_{h,2}$, $T_{c,i} = T_{c,1}$, $T_{c,o} = T_{c,2}$.

The form of ΔT_m may be determined by applying an energy balance to differential elements in the hot and cold fluids. Each element is of length dx and heat transfer surface area dA, as shown in Figure 2.11. The energy balances and the subsequent analysis are subject to the following assumptions.

1) The heat exchanger is insulated from its surroundings, in which case the only heat exchange is between the hot and cold fluids.
2) Axial conduction along the tubes is negligible.

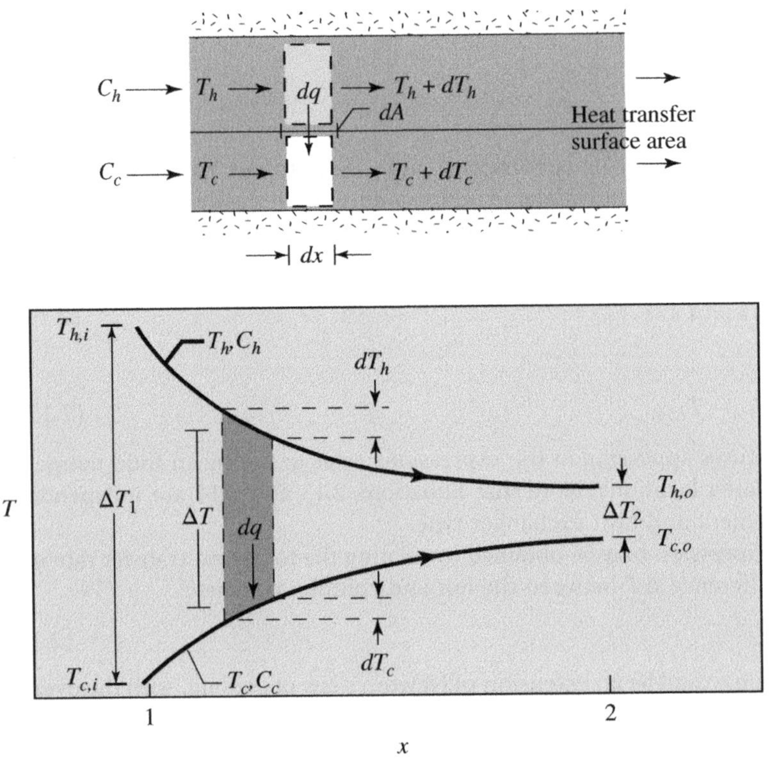

Figure 2.11 Temperature distribution for a parallel-flow heat exchanger.

3) Potential and kinetic energy changes are negligible.
4) The fluid specific heats are constant.
5) The overall heat transfer coefficient is constant.

The specific heats may of course change as a result of temperature variations, and the overall heat transfer coefficient may change because of variations in fluid properties and flow conditions. However, in many applications such variations are not significant and it is reasonable to work with average values of $C_{p,c}$, $C_{p,h}$, and U for the heat exchanger.

Applying an energy balance to each of the differential elements of Figure 2.11, it follows that

$$dq = -\dot{m}_h C_{p,h} dT_h \equiv -C_h dT_h \tag{2.17}$$

and

$$dq = \dot{m}_c C_{p,c} dT_c \equiv C_c dT_c \tag{2.18}$$

where C_h and C_c are the hot and cold fluid heat capacity rates, respectively. These expressions may be integrated across the heat exchanger to obtain the overall energy balances given by Equations 2.13b and 2.14b. The heat transfer across the surface area dA may also be expressed as

$$dq = U\,\Delta T\,dA \tag{2.19}$$

where $\Delta T = T_h - T_c$ is the local temperature difference between the hot and cold fluids.

To determine the integrated form of Equation 2.19, we begin by substituting Equations 2.17 and 2.18 into the differential form of Equation 2.15:

$$d(\Delta T) = dT_h - dT_c \tag{2.20}$$

to obtain

$$d(\Delta T) = -dq\left(\frac{1}{C_h} + \frac{1}{C_c}\right) \tag{2.21}$$

Substituting for dq from Equation 2.19 and integrating across the heat exchanger, we obtain

$$\int_1^2 \frac{d(\Delta T)}{\Delta T} = -U\left(\frac{1}{C_h} + \frac{1}{C_c}\right)\int_1^2 dA \tag{2.22}$$

or

$$\ln\left(\frac{\Delta T_2}{\Delta T_1}\right) = -UA\left(\frac{1}{C_h} + \frac{1}{C_c}\right) \tag{2.23}$$

Substituting for C_h and C_c from Equations 2.13b and 2.14b, respectively, it follows that

$$\begin{aligned}\ln\left(\frac{\Delta T_2}{\Delta T_1}\right) &= -UA\left(\frac{T_{h,i} - T_{h,o}}{q} + \frac{T_{c,o} - T_{c,i}}{q}\right)\\ &= -\frac{UA}{q}\left[(T_{h,i} - T_{c,i}) - (T_{h,o} - T_{c,o})\right]\end{aligned} \tag{2.24}$$

Recognizing that for the parallel-flow heat exchanger of Figure 2.11, $\Delta T_1 = (T_{h,i} - T_{c,i})$ and $\Delta T_2 = (T_{h,o} - T_{c,o})$, we then obtain

$$q = UA\frac{\Delta T_2 - \Delta T_1}{In\,(\Delta T_2/\Delta T_1)} \tag{2.25}$$

Comparing the above expression with Equation 2.16, we conclude that the appropriate average temperature difference is a log mean temperature difference, ΔT_{lm}. Accordingly, we may write

$$q = UA\Delta T_{lm} \tag{2.26}$$

where

$$\Delta T_{lm} = \frac{\Delta T_2 - \Delta T_1}{In\,(\Delta T_2/\Delta T_1)} = \frac{\Delta T_1 - \Delta T_2}{In\,(\Delta T_1/\Delta T_2)} \tag{2.27}$$

Remember that, for the parallel-flow exchanger,

$$\begin{bmatrix}\Delta T_1 \equiv T_{h,1} - T_{c,1} = T_{h,i} - T_{c,i}\\ \Delta T_2 \equiv T_{h,2} - T_{c,2} = T_{h,o} - T_{c,o}\end{bmatrix} \tag{2.28}$$

Example 2.1 Consider the following parallel-flow heat exchanger specification: cold flow enters at 40 °C:

$$C_c = 20{,}000\,\text{W/K}$$

hot flow enters at 150 °C:

$$C_h = 10{,}000\,\text{W/K}$$

$$A = 30\,\text{m}^2 \qquad U = 500\,\text{W/m}^2\cdot\text{K}.$$

Determine the heat transfer and the exit temperatures.

Solution

In this case, we do not know the exit temperatures, so it is not possible to calculate the LMTD. However, using parallel flow effectiveness equation,

$$\varepsilon = \frac{1-\exp\left(-NTU\left(1+\frac{C_{min}}{C_{max}}\right)\right.}{1+\frac{C_{min}}{C_{max}}} \text{ (refer Equation 2.38)}$$

$$\text{and with } NTU = UA/C_{min} = \frac{500(30)}{10{,}000} = 1{,}5$$

and $\frac{C_{min}}{C_{max}} = 0.5$ we obtain $\varepsilon = 0.596$. Now from following equation, we find that

$$Q = \epsilon\, C_{min}(T_{h_{in}} - T_{c_{in}}) = 0.596(10{,}000)(110) = 655{,}600\,\text{W} = 655.6\,\text{kW}$$

Finally, from energy balances, we get

$$T_{h_{out}} = T_{h_{in}} - \frac{Q}{C_h} = 150 - \frac{655{,}600}{10000} = 84.44°\text{C}$$

$$T_{c_{out}} = T_{c_{in}} + \frac{Q}{C_c} = 40 + \frac{655{,}600}{20000} = 72.78°\text{C}$$

2.6.1.2 Counter-Flow Heat Exchanger

The hot and cold fluid temperature distributions associated with a counter-flow heat exchanger are shown in Figure 2.9. In contrast to the parallel-flow exchanger, this configuration provides for heat transfer between the hotter portions of the two fluids at one end, and between the colder portions at the other. For this reason, the change in the temperature difference, $\Delta T = T_h - T_c$, with respect to x is nowhere as large as it is for the inlet region of the parallel-flow exchanger. Note that the outlet temperature of the cold fluid may now exceed the outlet temperature of the hot fluid.

Equations 2.13b and 2.14b apply to any heat exchanger and hence may be used for the counter-flow arrangement. Moreover, from an analysis like that performed in

Section 2.6.1.1, it may be shown that Equations 2.25 and 2.26 also apply. However, for the counter-flow exchanger the endpoint temperature differences must now be defined as

$$\begin{bmatrix} \Delta T_1 \equiv T_{h,1} - T_{c,1} = T_{h,i} - T_{c,o} \\ \Delta T_2 \equiv T_{h,2} - T_{c,2} = T_{h,o} - T_{c,i} \end{bmatrix} \tag{2.29}$$

Note that, for the same inlet and outlet temperatures, the log mean temperature difference for counter flow exceeds that for parallel flow, $\Delta T_{lm,CF} > \Delta T_{lm,PF}$. Hence the surface area required to effect a prescribed heat transfer rate q is smaller for the counter-flow than for the parallel-flow arrangement, assuming the same value of U. Also note that $T_{c,o}$ can exceed $T_{h,o}$ for counter-flow but not parallel-flow.

Example 2.2 The counter-flow, concentric tube heat exchanger of Example 11.1 from Incropera et al. [9] is replaced with a compact, plate-type heat exchanger that consists of a stack of thin metal sheets, separated by N gaps of width a. The oil and water flows are subdivided into $N/2$ individual flow streams, with the oil and water moving in opposite directions within alternating gaps. It is desirable for the stack to be of a cubical geometry, within a characteristic exterior dimension L. Determine the exterior dimensions of the heat exchanger as a function of the number of gaps if the flow rates, inlet temperatures and desired oil outlet temperature are the same. Compare the pressure drops of the water and oil streams within the plate-type heat exchanger to the pressure drops of the flow streams, if 60 gaps are specified.

Solution

Known: Configuration of a plate-type heat exchanger with given dimensions (length = breath = 0.131 m and gap between plates, a = 0.00218 m), fluid flow rates, inlet temperatures and desired oil outlet temperature.

Find: Pressure drops within the given dimensions of plate-type heat exchanger, and concentric tube heat exchanger.

Schematic: Figure 2.12

Assumptions:

1) Negligible heat loss to the surroundings
2) Negligible kinetic and potential energy changes
3) Constant properties
4) Negligible plate thermal resistance and fouling factors
5) Fully developed conditions for the water and oil
6) Identical gap-to-gap heat transfer coefficients
7) Heat exchanger exterior dimension large compared to the gap width

Properties: See Example 11.1, and Table A.5 in [9], unused engine oil $(\bar{T}_h = 353\,\text{K}) : \rho = 852.1\ \text{kg/m}^3$. Table A.6, water $(\bar{T}_c \approx 35°\text{C}) : \rho = v_f^{-1} = 994\,\text{kg/m}^3$.

Analysis:

$$Nu_D = \frac{hD_h}{k} = 7.54$$

Assume the hydraulic diameter as $D_h = 2a$.

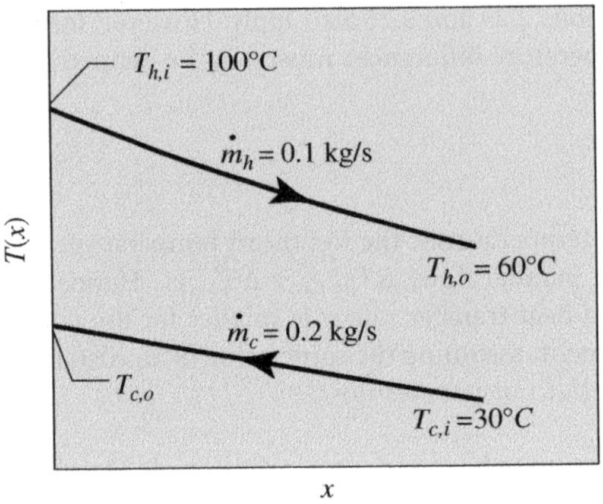

Figure 2.12 Inlet and outlet temperatures of counter-flow heat exchanger.

The hydraulic diameter is $D_h = 0.00436$ m, and the mean velocity in each water-filled gap is

$$u_m = \frac{\dot{m}}{\rho L^2/2} = \frac{2 \times 0.2\,\text{kg/s}}{994\dfrac{\text{kg}}{\text{m}^3} \times 0.131^2\,\text{m}^2} = 0.0235\,\text{m/s}$$

Providing a Reynolds number of

$$Re_D = \frac{\rho u_{mD_h}}{\mu} = \frac{994\dfrac{\text{kg}}{\text{m}^3} \times 0.0235\dfrac{\text{m}}{\text{s}} \times 0.00436\,\text{m}}{725 \times 10^{-6}\,\text{N.s/m}^2} = 141$$

for the oil-filled gaps

$$u_m = \frac{\dot{m}}{\rho L^2/2} = \frac{2 \times 0.1\,\text{kg/s}}{852.1\dfrac{\text{kg}}{\text{m}^3} \times 0.131^2\text{m}^2} = 0.0137\,\text{m/s}$$

yielding a Reynolds number of

$$Re_D = \frac{\rho u_{mD_h}}{\mu} = \frac{852.1\dfrac{\text{kg}}{\text{m}^3} \times 0.0137\dfrac{\text{m}}{\text{s}} \times 0.00436\,\text{m}}{3.25 \times 10^{-2}\,\text{N.s/m}^2} = 1.57$$

Therefore, the flow is laminar for both fluids, the pressure drop for the water:

$$\Delta p = \frac{64}{Re_D} \cdot \rho \frac{u_m^2}{2D_h} . \text{L} = \frac{64}{141} \times \frac{994\dfrac{\text{kg}}{\text{m}^3} \times 0.0235^2\,\text{m}^2/\text{s}^2}{2 \times 0.00436\,\text{m}} \times 0.131\,\text{m} = 3.76\,\text{N/m}^2$$

Similarly, for the oil

$$\Delta p = \frac{64}{Re_D} \cdot \rho \frac{u_m^2}{2D_h} . L = \frac{64}{1.57} \times \frac{852.1\dfrac{\text{kg}}{\text{m}^3} \times 0.0137^2\,\text{m}^2/\text{s}^2}{2 \times 0.00436\,\text{m}} \times 0.131\,\text{m} = 98.2\,\text{N/m}^2$$

The friction factor associated with the water flow can be calculated as follows (Incropera et al.[9]), for a smooth surface condition,

$$f = (0.790\ \ln(14{,}050) - 1.64)^{-2} = 0.0287.$$

The mean velocity is

$$u_m = 4\dot{m}/\left[\rho\pi D^2_i = 4 \times 0.2\,\text{kg/s}/\left(994\,\text{kg/m}^3 \times \pi \times 0.025^2\,\text{m}^2\right)\right] = 0.410\,\text{m/s},$$

and the pressure drop is

$$\Delta p = \frac{64}{Re_D}\cdot\frac{\rho\, u_m^2}{2D_h}.\text{L} = \frac{64}{56} \times \frac{852.1\frac{\text{kg}}{\text{m}^3} \times 0.107^2\,\text{m}^2/\text{s}^2}{2 \times 0.020\,\text{m}} \times 65.9\,\text{m} = 18{,}300\,\text{N/m}^2$$

Comments:

1) The volume occupied by the concentric tube heat exchanger is $V = \pi\frac{D^2_o L}{4} = \pi \times 0.045^2\,\text{m}^2 \times 65.9\,\text{m}/4 = 0.10\,\text{m}^3$, while the volume of the compact plate-type exchanger is $V = L^3 = 0.131^3\,\text{m}^3 = 0.0022\,\text{m}^3$. Use of the plate-type heat exchanger results in a 97.8% reduction in volume relative to the conventional, concentric tube heat exchanger.
2) The pressure drops associated with use of the compact heat exchanger are significantly less than for a conventional concentric tube configuration. Pressure drops are reduced by 99.9% and 99.5% for the water and oil flows, respectively.

2.6.2 Effectiveness-NTU Method

Two general types of heat exchanger problems are commonly encountered by the practising engineer. In the heat exchanger design problem, the fluid inlet temperatures and flow rates, as well as a desired hot or cold fluid outlet temperature, are prescribed. The design problem is then one of specifying a specific heat exchanger type and determining its size – that is, the heat transfer surface area A – required to achieve the desired outlet temperature. The design problem is commonly encountered when a heat exchanger is to be custom-built for a specific application. Alternatively, in a heat exchanger performance calculation, an existing heat exchanger is analysed to determine the heat transfer rate and the fluid outlet temperatures for prescribed flow rates and inlet temperatures. The performance calculation is commonly associated with the use of off-the-shelf heat exchanger types and sizes available from a vendor.

For heat exchanger design problems, the NTU method may be used by first calculating ε and (C_{min}/C_{max}). The appropriate equation (or chart) may then be used to obtain the NTU value, which in turn may be used to determine A. For a performance calculation, the NTU and (C_{min}/C_{max}) values may be computed and ε may then be determined from the appropriate equation (or chart) for a particular exchanger type. Since q_{max} may also be computed from Equation 2.4, it is a simple matter to determine the actual heat transfer rate from the requirement that $q = \varepsilon q_{max}$. Both fluid outlet temperatures may then be determined from Equations 2.13b and 2.14b.

To determine the exit temperatures of the two fluids for a given exchanger, the LMTD method requires solving two simultaneous equations for the case of pure parallel-flow and counter-flow arrangements and an iterative solution for all other flow arrangements. This iterative procedure can be avoided by using the effectiveness-NTU method. The limitations of the LMTD method are illustrated in Example 2.3.

Example 2.3 In an oil-to-water heat exchanger, the oil $\left[C_p = 2000\,\mathrm{J}/(\mathrm{kg}\cdot\mathrm{K})\right]$ enters the exchanger at 100 °C with a heat capacity rate of 3700 W/K. Water $\left(C_p = 4190\frac{\mathrm{J}}{\mathrm{kg}}\cdot\mathrm{K}\right)$ enters the exchanger at 15 °C with a mass flow rate of 0.6 kg/s. Determine the exit temperatures of the two fluids for (i) parallel flow arrangement, (ii) counter flow arrangement and (iii) one shell pass arrangement. Assume the total surface area of the heat exchanger is 10 m^2 and the overall heat transfer coefficient is 500 W/(m^2 · K).

Solution

Known: Fluid flow rates, inlet temperatures, overall heat transfer coefficients and heat transfer area, mass flow rates and specific heats of the two fluids

Find: Exit temperatures of the two fluids

Schematic: Shown in each case separately

Assumptions:

1) Negligible heat loss to the surroundings
2) Negligible kinetic and potential energy changes
3) Constant fluid properties
4) Fully developed conditions for both fluids

Analysis:

Case (i) Solution for parallel-flow arrangement: From Equation 2.23 we have

$$\Delta T_2 = \Delta T_1 \exp\left\{-UA\left(\frac{1}{C_h} + \frac{1}{C_c}\right)\right\}$$

or

$$T_{h_o} - T_{c_o} = (T_{h_i} - T_{c_i})\exp\left\{-UA\left(\frac{1}{C_h} - \frac{1}{C_c}\right)\right\} \tag{a}$$

From heat balance equation we have

$$\left(\frac{C_h}{C_c}\right)T_{h_o} + T_{c_o} = \left(\frac{C_h}{C_c}\right)T_{h_i} + T_{c_i} \tag{b}$$

The RHS of both Equations a and b can be determined from the given information and hence these two equations can be solved simultaneously for $T_{h,o}$ and $T_{c,o}$.

Case (ii) Solution for counter-flow arrangement: For this arrangement Equation 2.23 reduces to

$$T_{h_o} - T_{c_i} = (T_{h_i} - T_{c_o})\exp\left\{-UA\left(\frac{1}{C_h} - \frac{1}{C_c}\right)\right\} \tag{c}$$

$$T_{h_o} + \exp\left\{-UA\left(\frac{1}{C_h} - \frac{1}{C_c}\right)\right\}T_{c_o} = \exp\left\{-UA\left(\frac{1}{C_h} - \frac{1}{C_c}\right)\right\}T_{h_i} + T_{c_i} \tag{d}$$

Equation b holds good for counter-flow arrangement also. Hence Equations b and d can be solved simultaneously to get $T_{h,o}$ and $T_{c,o}$.

Case (iii) Solution for cross-flow and multipass arrangements: Unfortunately the solutions for these types of heat exchangers are not as straightforward as the above two types as the mean temperature difference involves the correction factor F, which requires a knowledge of the exit temperatures. Hence the problem has to be solved by trial and error, as shown below.

Calculate R. Assume a value for F and read the value of P from the LMTD correction factor chart for the particular arrangement. Knowing P, calculate $T_{h,o}$, $T_{c,o}$, and Q. Then calculate ΔT_m for counter-flow arrangement and check for the assumed value of F by calculating F from the equation

$$F = \frac{Q}{UA(\Delta T_m)_{counter\ flow}} \tag{e}$$

The procedure is repeated until the solution converges. This method for case (iii) is illustrated by Shah and Sekulic [7] with an example, and is reproduced here.

Given: Hot fluid is oil with: $T_{h,i} = 100°C$; $C_h = 3700$ W/K

Cold fluid is water with: $T_{c,i} = 15°C$; $m_c = 0.6$ kg/s; $C_{p,c} = 4190$ J/(kg.K)

$U = 500$ W/(m^2.K); $A = 10$ m^2

a) **Solution for parallel-flow arrangement:** $C_c = m_c C_{pc} = 0.6 \times 4190 = 2514$ W/K

From Equation a we have

$$T_{h_o} - T_{c_o} = (100 - 15)\exp\left\{-500 \times 10\left(\frac{1}{3700} + \frac{1}{2514}\right)\right\}$$

Or

$$T_{h_o} - T_{c_o} = 3.01°C \tag{i}$$

From Equation b we have

$$\left(\frac{3700}{2514}\right)T_{h_o} + T_{c_o} = \left(\frac{3700}{2514}\right) \times 100 + 15$$

or

$$1.47T_{h_o} + T_{c_o} = 162.20 \tag{ii}$$

Solving for $T_{h,o}$ and $T_{c,o}$ from Equations i and ii we get

$$T_{h,o} = 66.89°C, T_{c,o} = 63.88°C, \text{ and } Q = 3700 \times (100 - 66.89) = 122{,}507\,W$$

b) **Solution for counter-flow arrangement:** From Equation d we have

$$T_{h_o} + \exp\left\{-500 \times 10 \times \left(\frac{1}{3700}\right) - \left(\frac{1}{2514}\right)\right\}T_{c_o}$$

$$= \exp\left\{-500 \times 10 \times \left(\frac{1}{3700}\right) - \left(\frac{1}{2514}\right)\right\} \times 100 + 15$$

or

$$T_{h_o} + 1.982 T_{c_o} = 204.2°\text{C} \qquad \text{(iii)}$$

Solving for $T_{h,o}$ and $T_{c,o}$ from Equations ii and iii we get

$$T_{h,o} = 57.6\,°\text{C}$$

$$T_{c,o} = 77.47\,°\text{C}$$

and

$$Q = 2514 \times (77.47 - 15) = 157{,}050\,\text{W}$$

c) **Solution for one shell pass two tube pass arrangement**
Trial 1: Assume $F = 0.8$ (this generally is a minimum value for good design)

$$R = \frac{T_1 - T_2}{t_2 - t_1} = \frac{T_{co} - T_{ci}}{T_{hi} - T_{ho}} = \frac{C_h}{C_c} = \frac{3771}{2514} = 1.50$$

From chart, corresponding to $F = 0.8$ and $R = 1.50$, $P = 0.40$. Hence

$$\frac{(t_2 - t_1)}{(T_1 - t_1)} = 0.40$$

$$t_2 = t_1 + 0.40(T_1 - t_1) = 100 + 0.40 \times (15 - 100)$$

or

$$t_2 = 66.0°\text{C} = T_{h,o}$$

$$\text{Q} = 3771\tilde{\text{n}}(10066.0) = 128{,}214\ \text{W}$$

Hence

$$T_{c_o} = T_{c_1} + \frac{Q}{C_c} = 15 + \frac{128214}{2514} = 66.00°\text{C}$$

$$(\Delta T_m)_{\text{counter flow}} = \frac{(100 - 66) - (66 - 15)}{\ln\left\{\dfrac{(100 - 66)}{(66 - 15)}\right\}} = 41.93°\text{C}$$

Hence

$$F = \frac{Q}{UA(\Delta T_m)_{\text{counterflow}}} = \frac{128214}{500 \times 10 \times 41.93} = 0.612$$

The calculated value of F is about 31% lower than the assumed value. Hence further iterations have to be carried out. Table 2.1 shows the difference between the assumed value of F and calculated value in % for first four iterations.

It can be seen from the above table that the convergence of the iterations is slow and F continues to fluctuate, but it will converge. This rating problem can be guaranteed convergent using the Newton–Raphson method (i.e. the solution of $f(x) = 0$ where function f is non-linear in the present case).

The above example illustrates the limitations of the mean temperature difference method for solving the rating problem of a given heat exchanger. Though the method

Table 2.1 Iterative solution for 1-2 heat exchanger. (Incropera et al. [9])

Trial	*F* (assumed)	*P*	$T_{h,o}$	$T_{c,o}$	*F* (calculated)	Δ*F* in %
1	0.80	0.4052	65.56	65.69	0.6080	+31.58
2	0.6080	0.4513	61.64	71.46	0.7701	−21.05
3	0.7701	0.4163	64.61	67.09	0.6435	+19.67
4	0.6435	0.4456	62.13	70.75	0.7477	−13.94

is straightforward for parallel-flow and counter-flow arrangements, but it requires solving two simultaneous equations to find $T_{h,o}$ and $T_{c,o}$. For cross-flow and multipass type heat exchangers, the method involves a trial and error solution wherein the solution converges very slowly iterating on F. This problem can be solved relatively easily by effectiveness-NTU method.

2.6.3 Effectiveness-NTU Relations

To determine a specific form of the effectiveness-NTU relation, Equation 2.24, consider a parallel-flow heat exchanger for which $C_{\min} = C_h$. From Equation 2.21 we then obtain

$$\varepsilon = \frac{T_{h,i} - T_{h,o}}{T_{h,i} - T_{c,i}} \tag{2.30}$$

and from Equations 2.13b and 2.14b it follows that

$$\frac{C_{min}}{C_{max}} = \frac{m_h c_{p,h}}{m_c c_{p,c}} = \frac{T_{c,o} = T_{c,i}}{T_{h,i} - T_{h,o}} \tag{2.31}$$

Now consider Equation 2.24, which may be expressed as

$$\ln\left(\frac{T_{h,o} - T_{c,o}}{T_{h,i} - T_{c,i}}\right) = -\frac{UA}{C_{min}}\left(1 + \frac{C_{min}}{C_{max}}\right) \tag{2.32}$$

Or from Equation 2.25

$$\frac{T_{h,o} - T_{c,o}}{T_{h,i} - T_{c,i}} = \exp\left[-NTU\left(1 + \frac{C_{min}}{C_{max}}\right)\right] \tag{2.33}$$

Rearranging the left-hand side of this expression as

$$\frac{T_{h,o} - T_{c,o}}{T_{h,i} - T_{c,i}} = \frac{T_{h,o} - T_{h,i} + T_{h,i} - T_{c,o}}{T_{h,i} - T_{c,i}} \tag{2.34}$$

and substituting for $T_{c,o}$ from Equation 2.31, it follows that

$$\frac{T_{h,o} - T_{c,o}}{T_{h,i} - T_{c,i}} = \frac{(T_{h,o} - T_{h,i}) + (T_{h,i} - T_{c,i}) - (C_{min}/C_{max})(T_{h,i} - T_{h,o})}{T_{h,i} - T_{c,i}} \tag{2.35}$$

or from Equation 2.30

$$\frac{T_{h,o} - T_{c,o}}{T_{h,i} - T_{c,i}} = -\varepsilon + 1 - \left(\frac{C_{min}}{C_{max}}\right)\varepsilon = 1 - \varepsilon\left(1 + \frac{C_{min}}{C_{max}}\right) \tag{2.36}$$

Substituting the above expression into Equation 2.33 and solving for ε, we obtain for the parallel-flow heat exchanger

$$\varepsilon = \frac{1 - \exp\{-NTU[1 + (C_{min}/C_{max})]\}}{1 + (C_{min}/C_{max})} \tag{2.37}$$

since precisely the same result may be obtained for $C_{min} = C_c$, Equation 2.37 applies for any parallel-flow heat exchanger, irrespective of whether the minimum heat capacity rate is associated with the hot or cold fluid.

Similar expressions have been developed for a variety of heat exchangers, Kays and London [5], and representative results are summarized in Table 2.2, where C_r is the capacity ratio $C_r \equiv C_{min}/C_{max}$. In deriving Equation 2.41 for a shell-and-tube heat exchanger with multiple shell passes, it is assumed that the total NTU is equally distributed between shell passes of the same arrangement, $NTU = n(NTU)_1$. In order to determine ε, $(NTU)_1$ would first be calculated using the heat transfer area for one shell, ε_1 would then be calculated from Equation 2.36, and ε would finally be calculated from Equation 2.41. Note that for $C_r = 0$, as in a boiler or condenser, ε is given by Equation 2.45 for all flow arrangements. Hence, for this special case, it follows that heat exchanger behaviour is independent of the flow arrangement. For the cross-flow heat exchanger with both fluids unmixed, Equation 2.42 is exact only for $C_r = 1$. However, it may be used to a good approximation for all $0 < C_r \leq 1$. For $C_r = 0$, Equation 2.45 must be used.

Explicit relations as a function of ε and C_r are provided in Table 2.3 for NTU. Note that Equation 2.42 may not be manipulated to yield a direct relationship for NTU as a function of ε and C_r, Note also that to determine the NTU for a shell-and-tube heat exchanger with multiple shell passes, ε would first be calculated for the entire heat exchanger. The variables F and ε_1 would then be calculated using Equations 2.53c and 2.53b, respectively. The parameter E would subsequently be determined from Equation 2.52 and substituted into Equation 2.51 to find $(NTU)_1$. Finally, this result would be multiplied by n to obtain the NTU for the entire exchanger, as indicated in Equation 2.53d.

Note that for $C_r = 0$, all heat exchangers have the same effectiveness, which may be computed from Equation 2.56. Moreover, if $NTU \lesssim 0.25$, all heat exchangers have approximately the same effectiveness, regardless of the value of C_r, and ε may again be computed from Equation 2.56. More generally, for $C_r > 0$ and $NTU \gtrsim 0.25$, the counter flow exchanger is the most effective. For any exchanger, maximum and minimum values of the effectiveness are associated with $C_r = 0$ and $C_r = 1$, respectively.

As noted previously, in the context of cross-flow heat exchangers, the terms mixed and unmixed are idealizations representing limiting cases of actual flow conditions. That is, most flows are neither completely mixed nor unmixed, but exhibit partial degrees of mixing. This issue has been addressed by DiGiovanni and Webb [11], and algebraic expressions have been developed to determine the ε-NTU relationship for arbitrary values of partial mixing.

We also note that both the LMTD and ε-NTU methods approach heat exchanger analysis from a global perspective and provide no information concerning conditions within the exchanger. Although flow and temperature variations within a heat exchanger may

Table 2.2 Heat exchanger effectiveness relations. (Hesselgreaves [10])

Flow arrangement	Relation	
Concentric tube		
Parallel flow	$\varepsilon = \dfrac{1-\exp[-NTU(1+C_r)]}{1+C_r}$	(2.38)
Counter flow	$\varepsilon = \dfrac{1-\exp[-NTU(1-C_r)]}{1-C_r\exp[-NTU(1-C_r)]} \quad (C_r < 1)$ $\varepsilon = \dfrac{NTU}{1+NTU} \quad (C_r = 1)$	(2.39)
Shell-and-tube		
One shell pass	$\varepsilon_1 = 2\left\{1 + C_r + \left(1+C_r^2\right)^{1/2} \times \dfrac{1+\exp\left[-(NTU)_1\left(1+C_r^2\right)^{1/2}\right]}{1-\exp\left[-(NTU)_1\left(1+C_r^2\right)^{1/2}\right]}\right\}^{-1}$ (2,4, ...tube passes)	(2.40)
N Shell passes (2n, 4n, ...tube passes)	$\varepsilon = \left[\left(\dfrac{1-\varepsilon_1 C_r}{1-\varepsilon_1}\right)^n - 1\right]\left[\left(\dfrac{1-\varepsilon_1 C_r}{1-\varepsilon_1}\right)^n - C_r\right]^{-1}$	(2.41)
Cross-flow (single pass)		
Both fluids unmixed	$\varepsilon = 1 - \exp\left[\left(\dfrac{1}{C_r}\right)(NTU)^{0.22}\left\{\exp\left[-C_r(NTU)^{0.78}\right]-1\right\}\right]$	(2.42)
$\dfrac{C_{max}(\text{mixed})}{C_{min}(\text{unmixed})}$	$\varepsilon = \left(\dfrac{1}{C_r}\right)(1-\exp\{-C_r[1-\exp(-NTU)]\})$	(2.43)
$\dfrac{C_{min}(\text{mixed})}{C_{max}(\text{unmixed})}$	$\varepsilon = 1-\exp(-C_r^{-1}\{1-\exp[-C_r(NTU)]\})$	(2.44)
All exchangers ($C_r = 0$)	$\varepsilon = 1-\exp(-NTU)$	(2.45)
Multipass overall counter-flow, fluids mixed between passes $\varepsilon = \dfrac{\left(\frac{1-\varepsilon\varepsilon_p C^*}{1-\varepsilon_p}\right)^n - 1}{\left(\frac{1-\varepsilon\varepsilon_p C^*}{1-\varepsilon\varepsilon_p}\right)^n - C^*}$		(2.46)

(Continued)

Table 2.2 (Continued)

Flow arrangement	Relation	
with	n = number of identical passes (i.e. each pass having the same ε_p)	
	ε_p = effectiveness of each pass (as a function of $NTU_p = \dfrac{NTU}{n}$	
and	$\varepsilon_p = \dfrac{\left(\frac{1-\varepsilon C^*}{1-\varepsilon}\right)^{1/n} - 1}{\left(\frac{1-\varepsilon C^*}{1-\varepsilon}\right)^{1/n} - C^*}$	(2.47)

(limiting value = $\varepsilon_{\text{counter flow}}$, as $n \to \infty$).

Multipass overall parallel flow fluids mixed between passes:

$$\varepsilon = \frac{1}{1+C^*}\left[1 - \{1-(1+C^*)\varepsilon_p\}^n\right] \quad (2.48)$$

Table 2.3 Heat exchanger NTU relations. (Kays and London [5] and Hesselgreaves [10])

Flow arrangement	Relation	
Concentric tube		
Parallel flow	$NTU = -\dfrac{\ln[1-\varepsilon(1+C_r)]}{1+C_r}$	(2.49)
Counter flow	$NTU = \dfrac{1}{C_r - 1} In\left(\dfrac{\varepsilon - 1}{\varepsilon C_r - 1}\right) \quad (C_r < 1)$ $\varepsilon = \dfrac{\varepsilon}{1-\varepsilon} \quad (C_r = 1)$	(2.50)
Shell-and-tube		
One shell pass	$(NTU)_1 = -\left(1+C_r^2\right)^{-1/2} \ln\left(\dfrac{E-1}{E+1}\right)$	(2.51)
(2,4, ...tube passes)	$E = \dfrac{2/\varepsilon_1 - (1+C_r)}{\left(1+C_r^2\right)^{1/2}}$	(2.52)
n Shell passes	use Equations 2.51 and 2.52 with	
(2n, 4n,...tube passes) $\varepsilon_1 = \dfrac{F-1}{F-C_r} F = \left(\dfrac{\varepsilon C_r - 1}{\varepsilon - 1}\right)^{1/n}$	$NTU = n(NTU)_1$	(2.53b, c, d)

(*Continued*)

Cross-flow (single pass)

$$C_{max}(\text{mixed}),\ C_{min}(\text{unmixed})\quad NTU = -\ln\left[1+\left(\frac{1}{C_r}\right)\ln(1-\varepsilon C_r)\right] \tag{2.54}$$

$$C_{min}(\text{mixed}),\ C_{max}(\text{unmixed})\quad NTU = -\left(\frac{1}{C_r}\right)\ln[\ln(1-\varepsilon)+1] \tag{2.55}$$

All exchangers $(C_r = 0)$

$$NTU = -\ln(1-\varepsilon) \tag{2.56}$$

be determined using commercial computational fluid dynamic (CFD) computer codes, simpler numerical procedures may be adopted. Such procedures have been applied to determine temperature variations in concentric tube and shell-and-tube heat exchangers by Lienhard [12].

2.6.4 Evaluation of Heat Transfer and Pressure Drop Data

The basic performance data for an enhanced surface is often shown as curves of the Colburn factor ($j = StPr^{2/3}$), and the Fanning friction factor (f), plotted versus Reynolds number. Kays and London [5] present j and f vs Re for a large number of compact surfaces, in one of the first comprehensive collections of data on enhanced surfaces for CHEs. Since that time, j and f curves have become a customary tool for presenting performance data for heat transfer surface geometries. However, the curves for j and f plotted vs Re tend to vary over a wide range, in magnitude as well as slope, making comparisons between different geometries difficult. This fact suggests plotting j and f as functions of a parameter other than Re. It has long been acknowledged (Kays and London [5]) that the flow length between boundary layer disturbances has an effect on the performance of a heat transfer surface. Kays and London [5] wrote: "One of the most widely used ways of increasing conductance is to interrupt the wall surfaces so that the boundary layers can never become thick." From this principle, plotting the factors j and f versus the ratio of the flow length between major boundary layer disturbances to the equivalent diameter (l/D_{eq}) is the next logical step. Kays and London [5] presented j and f vs (l/D_{eq}) plots for many of the previously mentioned compact heat transfer surfaces.

2.6.4.1 Flow Properties and Dimensionless Numbers

The hydraulic diameter is given by

$$D_h = \frac{4A_cL_2}{A_t} \tag{2.57}$$

The mass velocity G is given by

$$G = \frac{\dot{m}}{A_c} \tag{2.58}$$

The Reynolds number is given by

$$R_e = \frac{\rho u D_h}{\mu} = \frac{GD_h}{\mu} \tag{2.59}$$

The pin pitch is defined by

$$P_f = \frac{1}{N_f} \tag{2.60}$$

In CHEs, the Stanton number *St* is often used, which is defined as

$$S_t = \frac{N_u}{R_e . P_r} \tag{2.61}$$

The Nusselt number is given by

$$N_u = \frac{hD_h}{k} \tag{2.62}$$

The Reynolds number is given by

$$R_e = \frac{GD_h}{\mu} \tag{2.63}$$

With the kinematic viscosity $(\nu = \mu/\rho)$, and the thermal diffusivity $\left(\alpha = k/\rho c_p\right)$, the Prandtl number, *Pr*, can be written as

$$P_r = \frac{\nu}{\alpha} = \frac{\frac{\mu}{\rho}}{\frac{k}{\rho c_p}} = \frac{c_p \mu}{k} \tag{2.64}$$

The Stanton number, S_t, is then expressed using Equations 2.61 to 2.64 as

$$S_t = \frac{N_u}{R_e . P_r} = \frac{\frac{hD_h}{k}}{\frac{GD_h}{\mu}\frac{c_p \mu}{k}} = \frac{h}{Gc_p} \tag{2.65}$$

It is customary to use the Colburn factor, *j*, to represent the thermal characteristics of the CHEs. The Colburn factor *j* is defined and expressed using Equation 2.65 as

$$j = S_t . Pr^{\frac{2}{3}} = \frac{h}{Gc_p} Pr^{2/3} \tag{2.66}$$

This gives the convection heat transfer coefficient *h* in terms of the Colburn factor *j* as

$$h = j \frac{Gc_p}{Pr^{\frac{2}{3}}} \tag{2.67}$$

The failure of *j* and *f* curves to portray the relative performance of heat transfer surfaces in a satisfactory fashion has led to the development of other performance parameters. Two such parameters are the "area goodness" comparison and the "volume goodness" comparison – Hesselgreaves [10].

The **area goodness method** actually makes a direct comparison between *j* and *f* values, since it consists of plotting

$$\frac{j}{f}\ vs\ Re \tag{2.68}$$

where the flow area goodness factor (j/f) is given by

$$\frac{j}{f} = (S_t Pr^{\frac{2}{3}})/f \tag{2.69}$$

For fully developed laminar flow of a specified fluid, j/f is constant for a given surface, regardless of Reynolds number. A higher j/f is considered desirable. Because j and f are both dimensionless quantities, they are independent of the hydraulic diameter (D_h). Therefore, when the area goodness factor j/f is compared for different surfaces, it reveals the influence of the cross-sectional shape regardless of the scale of the geometry.

For the **volume goodness factor**, a comparison of surface geometries in terms of core volume is provided by plotting

$$h_{std}\ \text{vs}\ E_{std} \tag{2.70}$$

where:

h_{std} is the heat transfer coefficient at a standard set of fluid properties, $W/(m^2 \cdot K)$,
E_{std} is flow-friction power per unit area for a standard set of fluid properties, W/m^2,
h_{std} is referred to as the volume goodness factor. Expressions for h_{std} and for E_{std} are given below

$$h_{std} = \frac{k}{D_h} N_u = \frac{c_p \mu}{Pr^{2/3}} \frac{1}{D_h} jRe \tag{2.71}$$

$$E_{std} = \frac{W \Delta p}{\rho A} = \frac{\mu^3}{2\rho^2 g_c} \frac{1}{D_h^3} fRe^3 = u_m r_h \frac{\Delta p}{L} \tag{2.72}$$

The dimensionless heat transfer in a heat exchanger is measured by the exchanger effectiveness (ε), which is dependent on the number of transfer units (NTU) for fixed flow rates. In a "balanced" heat exchanger, the thermal resistances of both sides of the heat exchanger are of the same order of magnitude. This means that NTU is proportional to hA (or $h_{std} A$) for the side of the heat exchanger that is under examination. As a result, a higher h_{std} for a given E_{std} will yield a lower heat transfer area (A) for the specified exchanger effectiveness – Hesselgreaves [10]. Consequently, a high position on the h_{std} vs E_{std} plot signifies a desirable surface geometry. And, since

$$A = \beta V = \left(\frac{4\sigma}{D_h}\right) V \tag{2.73}$$

where σ is the dimensionless ratio of free flow area to frontal flow area in a heat exchanger, a high h_{std} will yield a smaller heat exchanger volume at a given E_{std} for constant σ and D_h.

2.6.4.2 Data Curves for *j* and *f*

Many different tubular and plate configurations have been considered, where differences are primarily in fin design and arrangement in CHEs. Heat transfer and flow characteristics have been determined for specific configurations and are typically presented in the format of Re vs j and f curves (Figures 11.16 and 11.17 of Kays and London [5]).

Heat transfer results are correlated in terms of the Colburn j factor $j_H = St\,Pr^{2/3}$ and the Reynolds number, where both the Stanton $(St = \frac{h}{GC_p})$ and Reynolds $(Re = GD_h/\mu)$ numbers are based on the maximum mass velocity

$$G \equiv \rho u_{max} = \frac{\rho u A_{fr}}{A_{ff}} = \frac{\dot{m}}{A_{ff}} = \frac{\dot{m}}{\sigma A_{fr}} \tag{2.74}$$

The quantity σ is the minimum free-flow area of the finned passages (cross-sectional area perpendicular to flow direction), A_{ff}, to the frontal area, A_{fr}, of the exchanger. Values of σ, D_h (the hydraulic diameter of the flow passage), α (the heat transfer surface area per total heat exchanger volume), A_f/A (the ratio of fin to total heat transfer surface area), and other geometrical parameters are listed for each configuration. The ratio A_f/A is used in $\eta_o = 1 - \frac{A_f}{A}(1-\eta_f)$ to evaluate the temperature effectiveness η_o. In a design calculation, α would be used to determine the required heat exchanger volume, after the total heat transfer surface area has been found; in a performance calculation, it would be used to determine the surface area from knowledge of the heat exchanger volume.

In a CHE calculation, empirical information, such as that provided in Figures 10.75 to 10.84 of Kays and London [5], would first be used to determine the average convection coefficient of the finned surfaces. The overall heat transfer coefficient would then be determined, and using the ε-NTU method, the heat exchanger design or performance calculations would be performed. A detailed alternative procedure based on CFD is presented in Chapter 5 for generation of j and f data for different types of CHE surfaces.

The pressure drop associated with flow across finned-tube banks may be computed from the expression

$$\Delta p = \frac{G^2 v_i}{2}\left[\left(1+\sigma^2\right)\left(\frac{v_o}{v_i} - 1\right) + f\frac{A}{A_{ff}}\frac{v_m}{v_i}\right] \tag{2.75}$$

where v_i and v_o are the fluid inlet and outlet specific volumes and $v_m = (v_i + v_o)/2$. The first term on the right-hand side of Equation 2.75 accounts for the cumulative effects of pressure change due to inviscid fluid acceleration and deceleration at the exchanger inlet and outlet, respectively. The effects are reversible, and if fluid density variations may be neglected $(v_o = v_i)$, the term is negligible. The second term accounts for losses due to fluid friction in the heat exchanger core, with fully developed conditions presumed to exist throughout the core. For a prescribed core configuration, the friction factor is known as a function of Reynolds number for a prescribed heat exchanger size. The area ratio may be evaluated from the relation $(A/A_{ff}) = (\alpha V/\sigma A_{fr})$, where V is the total heat exchanger volume. Equation 2.75 does not account for irreversible losses due to viscous effects at the inlet and outlet of the heat exchanger. The losses depend on the nature of the ductwork used to transport fluids to and from the heat exchanger core. If the transition between the duct and the core occurs with little flow separation, the losses are small. However, if there are abrupt changes between the duct cross-sectional area and the free flow area of the heat exchanger, separation is pronounced and the attendant losses are large. Inlet and exit losses may be estimated from empirical contraction and expansion coefficients obtained for a variety of core geometries. In addition to the classic work of Kays and London [5], other sources of information are provided by Webb [3], Brockmeier [4], Incropera et. al. [9] and Hesselgreaves [10].

2.7 Plate-Fin Heat Exchanger

2.7.1 Description

In this book, FEM and CFD analysis carried out mainly for plate-fin type CHEs are presented in Chapters 4 and 5 respectively. Plate-fin exchangers have various geometries, in order to compensate for the high thermal resistance, particularly if one of the fluids is a gas. This type of exchanger is corrugated with either triangular fins, rectangular fins, wavy fins, offset strip fins (OSF), louvred fins or perforated fins. Plate-fin heat exchangers have been used since the 1910s in the auto industry, and since the 1940s in the aerospace industry. They are now widely used in many applications: cryogenics, gas turbines, nuclear, aerospace and fuel cells.

Plate-fin heat exchangers are generally designed for moderate operating pressures, less than 700 kPa (gauge pressure) and have been built with a surface area density of up to 5900 m^2/m^3. Common fin thicknesses range from 0.05 to 0.25 mm, while fin heights may range from 2 to 25 mm. Although typical fin densities are 120–700 fins/m, applications exist for as much as 2100 fins/m – Bejan and Pfister [13].

Plate-fin heat exchangers are made by the stacking of corrugated sheets (fins) separated by planar sheets and closed on the sides by lateral bars. The separation sheet usually has a clad alloy that will melt at a lower temperature than the parent aluminium during brazing to bond the various components. The gaps between constitute a fluid layer. A core is made of a great number of layers. The exchanger can be made of one or more cores. The number of plate and fin layers, the size of the plates and fin, the height of the fin and the type of fin are engineered for optimum performance. The core is assembled (stacked) and typically held together by tack welding a weld rod to the top and bottom layer of the core. The stacked core is then placed within a fixture that exerts force on the individual pieces to keep them in contact. The part is then vacuum brazed in an environmentally controlled room to ensure high quality and reliability. After brazing, the core is typically heat treated or aged in order to increase its strength. Manifold ducting and mounting brackets are then welded in place as required, and any required paint or coating can be added.

The plate fin heat exchanger is an extended surface type of CHE where the heat transfer surface area is enhanced by providing the extended metal surface (fin) interface between the two fluids. Out of the various CHEs, plate-fin heat exchangers are unique due to their construction and performance. They are characterized by high effectiveness, compactness, low weight and moderate cost. A schematic view of such a heat exchanger is shown in Figure 2.13. Appropriate headers are welded to provide the necessary interface with the inlet and exit streams. Plate-fin heat exchangers can be made in a variety of materials. Aluminium is preferred in cryogenic and aerospace applications because of its low density, high thermal conductivity and high strength at low temperature. The maximum design pressure for brazed aluminium plate-fin heat exchanger is around 100 bars. At temperatures above ambient, most aluminium alloys lose mechanical strength. Stainless steel, nickel and copper alloys have been used at temperatures up to 550 °C.

Figure 2.14 shows the exploded view of two layers of a plate-fin heat exchanger. Such layers are arranged together in a monolithic block to form a heat exchanger. The corrugations serve both as secondary heat transfer surface and as mechanical support against the internal pressure between layers. The side bars prevent one fluid spilling over and

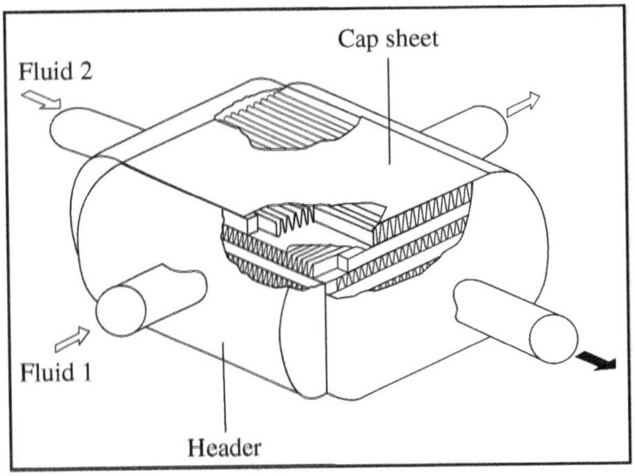

Figure 2.13 Compact heat exchanger with plate-fins.

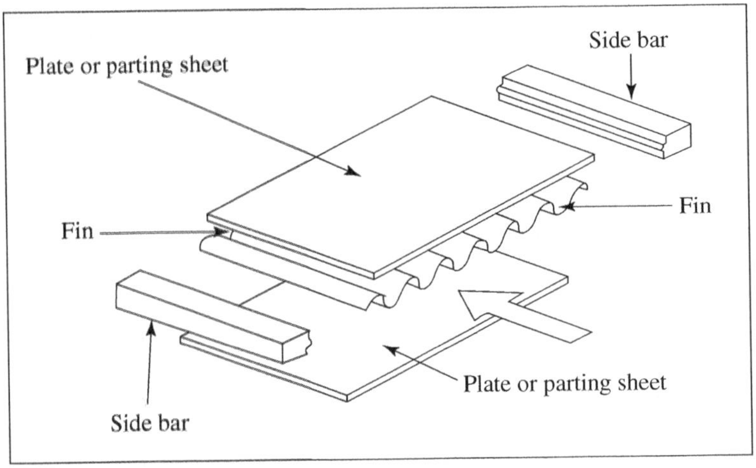

Figure 2.14 Basic components of plate-fin heat exchanger.

mixing with the other fluid. The fins and side bars are brazed with the parting sheet to ensure a good thermal link and to provide the mechanical stability.

2.7.2 Geometric Characteristics

A schematic of a single-pass cross-flow plate-fin heat exchanger employing offset strip fins is shown in Figure 2.15. The idealized fin geometry is shown in Figure 5.41 in Chapter 5. Formulating the total heat transfer area for each fluid is an important task in the analysis, similar to the *Plate-Fin Heat Exchangers Guide to their Specification and Use*, Editor: Taylor [14]. The total heat transfer area consists of the primary area and the fin area. The primary area consists of the plate area except the fin base area, multi-passage side walls, and multi-passage front and back walls. Usually, the numbers

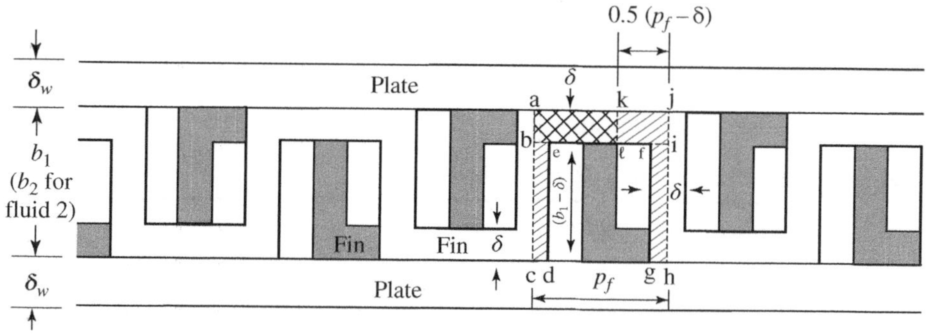

Figure 2.15 Schematic of offset strip fin geometry.

of passages for the hot fluid side and the cold fluid side are N_p and $N_p + 1$ to minimize the heat loss to the ambient. The top and bottom passages in Figure 2.15 are designated to be cold fluid. The number of passages can be obtained from an expression for L_3 as

$$L_3 = \underbrace{N_p b_1}_{\textit{Length of fluid } 1} + \underbrace{(N_p + 1) b_2}_{\textit{Length of fluid } 2} + \underbrace{2(N_p + 1)\delta_w}_{\textit{Thickness of total plates}} \tag{2.76}$$

Solving for N_p gives the number of passages for hot fluid

$$N_p = \frac{L_3 - b_2 - 2\delta_\omega}{b_1 + b_2 + 2\delta_\omega} \tag{2.77}$$

By definition, the number of passages counts the number based on one flow passage between two plates, and not all the individual channels between the plates. The total number of fins for fluid 1 (hot) is calculated by

$$nf_1 = \frac{L_1}{pf_1} N_p \tag{2.78}$$

For fluid 2 (cold),

$$nf_2 = \frac{L_2}{pf_2} (N_p + 1) \tag{2.79}$$

where p_{f1} is the fin pitch, which is usually obtained by taking the inverse of the fin density. The total number of fins n_{f1} is based on the hatched area (a-c-d-e-f-g-h-j-a in Figure 2.15 counting as a unit fin.

Since total primary area = Total plate areas – Fin base areas + Passage side wall areas + Passage front and back wall areas, the primary area for fluid 1 is expressed by

$$A_{p1} = \underbrace{2L_1L_2N_p}_{\textit{Total plate area}} - \underbrace{2\delta L_2 n_{f1}}_{\textit{Fin base area}} + \underbrace{2b_1L_2N_p}_{\textit{Passage side wall area}} + \underbrace{2(b_2 + 2\delta_w)L_1(N_p + 1)}_{\textit{Passage front back wall area}} \tag{2.80}$$

The number of offset strip fin n_{off1} per the number of fins is obtained from

$$n_{off2} = \frac{L_1}{\lambda_2} \tag{2.81}$$

$$n_{off1} = \frac{L_2}{\lambda_1} \tag{2.82}$$

where $\lambda_{1,2}$ is the offset strip fin length for fluids 1 and 2. The total fin area A_{f1} consists of the fin area and offset-strip edge areas.

$$A_{f1} = \underbrace{2(b_1-\delta)L_2 n_{f1}}_{\substack{\text{Fin surface areas} \\ (d-e \text{ and } g-f)}} + \underbrace{2(b_1-\delta)\delta n_{off1} n_{f1}}_{\substack{\text{Offset-strip edge areas} \\ 2(b-c-d-e)}} + \underbrace{(p_{f1}-\delta)\delta(n_{off1}-1)n_{f1}}_{\substack{\text{Internal offset-strip edge area} \\ 2(i-j-k-l-i)}} + \underbrace{2p_{f1}\delta n_{f1}}_{\substack{\text{First and last offset-strip edge area} \\ 2(a-b-i-j-a)}} \tag{2.83}$$

Note that the cross-hatched area (a-b – l-k-a) was not included in the offset-strip edge area because the area is blocked by the next strip fin, as shown in Figure 2.15, so that no heat transfer in the area is expected. The total heat transfer area A_{t1} is the sum of the primary area and the fin area.

$$A_{t1} = A_{p1} + A_{f1} \tag{2.84}$$

The free-flow (cross-sectional) area A_{c1} is obtained from

$$A_{c1} = (b_1 - \delta)(p_{f1} - \delta)n_{f1} \tag{2.85}$$

It is assumed in Equation 2.83 that there is a small gap between the offset strip fin, whereby the next strip fin shown in the unit fin in Figure 2.15 is not considered as an obstructing structure to the free flow. The frontal area A_{fr1} for fluid 1 where fluid 1 is entering is defined by

$$A_{fr1} = L_1 L_3 \tag{2.86}$$

The hydraulic diameter for fluid 1 is generally defined by

$$D_{h1} = \frac{4A_{c1}L_2}{A_{t1}} \tag{2.87}$$

For the fin efficiency η_f of the offset strip fin, it is assumed that the heat flow from both plates is uniform and that the adiabatic plane occurs at the middle of the plate spacing b_1. Hence, the fin profile length L_{f1} is defined by

$$L_{f1} = \frac{b_1}{2} - \delta \tag{2.88}$$

The m value is obtained from the fin equation as

$$m_1 = \sqrt{\frac{2h}{k_f \delta}} \tag{2.89}$$

The single fin efficiency η_f is obtained using Equation $\eta_f = \dfrac{\tanh(mb)}{mb}$ as

$$\eta_f = \frac{\tanh(m_1 L_{f1})}{m_1 L_{f1}} \tag{2.90}$$

The overall surface (fin) efficiency η_o is then obtained using Equations ($\eta_{01} = 1 - n\frac{A_f}{A_t}(1-\eta_f)$, 2.83 and 2.84 as (where n = number of fins)

$$\eta_{01} = 1 - \frac{A_{f1}}{A_{t1}}(1-\eta_f) \tag{2.91}$$

2.7.3 Correlations for Offset Strip Fin (OSF) Geometry

This geometry has one of the highest heat transfer performances relative to the friction factors. Extensive analytical, numerical and experimental investigations have been conducted over the past 50 years. The most comprehensive correlations for j and f factors for the laminar, transition, and turbulent regions are provided by Manglik and Bergles [15]:

$$j = 0.6522Re^{-0.5403}\left(\frac{p_f-\delta}{b-\delta}\right)^{-0.1541}\left(\frac{\delta}{\lambda}\right)^{0.1499}\left(\frac{\delta}{p_f-\delta}\right)^{-0.0678} \times \left[1 + 5.269 \times 10^{-5}Re^{1.34}\left(\frac{p_f-\delta}{b-\delta}\right)^{0.504}\left(\frac{\delta}{\lambda}\right)^{0.456}\left(\frac{\delta}{p_f-\delta}\right)^{-1.055}\right]^{0.1} \tag{2.92}$$

$$f = 9.6243Re^{-0.7422}\left(\frac{p_f-\delta}{b-\delta}\right)^{-0.1856}\left(\frac{\delta}{\lambda}\right)^{0.3053}\left(\frac{\delta}{p_f-\delta}\right)^{-0.2659} \times \left[1 + 7.669 \times 10^{-8}Re^{4.429}\left(\frac{p_f-\delta}{b-\delta}\right)^{0.92}\left(\frac{\delta}{\lambda}\right)^{3.767}\left(\frac{\delta}{p_f-\delta}\right)^{0.236}\right]^{0.1} \tag{2.93}$$

where the hydraulic diameter is defined by them as

$$D_{h\_MB} = \frac{4(p_f-\delta)(b-\delta)\lambda}{2[(p_f-\delta)\lambda + (b-\delta)\delta] + (p_f-\delta)\delta} \tag{2.94}$$

Equation 2.60 is an approximation of Equation 2.87. However, Equation 2.94 is in good agreement with Equation 2.87, which indicates that the general definition of hydraulic diameter, Equation 2.87, can be used in place of Equation 2.94. Similar correlations are developed using CFD and are presented in Chapter 5 for OSF and other types of fins.

2.8 Finned-Tube Heat Exchanger

Finned-tube heat exchangers have been used for heat exchange between gases and liquids for many years. In a gas-to-liquid exchanger, the heat transfer coefficient on the liquid side is generally one order of magnitude higher than that on the gas side. Hence, to have balanced thermal conductances on both sides for a minimum-size heat exchanger, fins are used on the gas side to increase the surface area. Figure 2.16 shows three important finned-tube heat exchanger construction types as given in Incropera et al. [9]. Figure 2.16a shows circular finned-tube geometry, Figure 2.16b shows the plate finned-tube geometry, and Figure 2.16c shows the plate-fin flat-tube geometry. These exchangers are widely used in the air conditioning, refrigerators and the automotive industry. Some examples are cooling towers, evaporators, condensers and radiators.

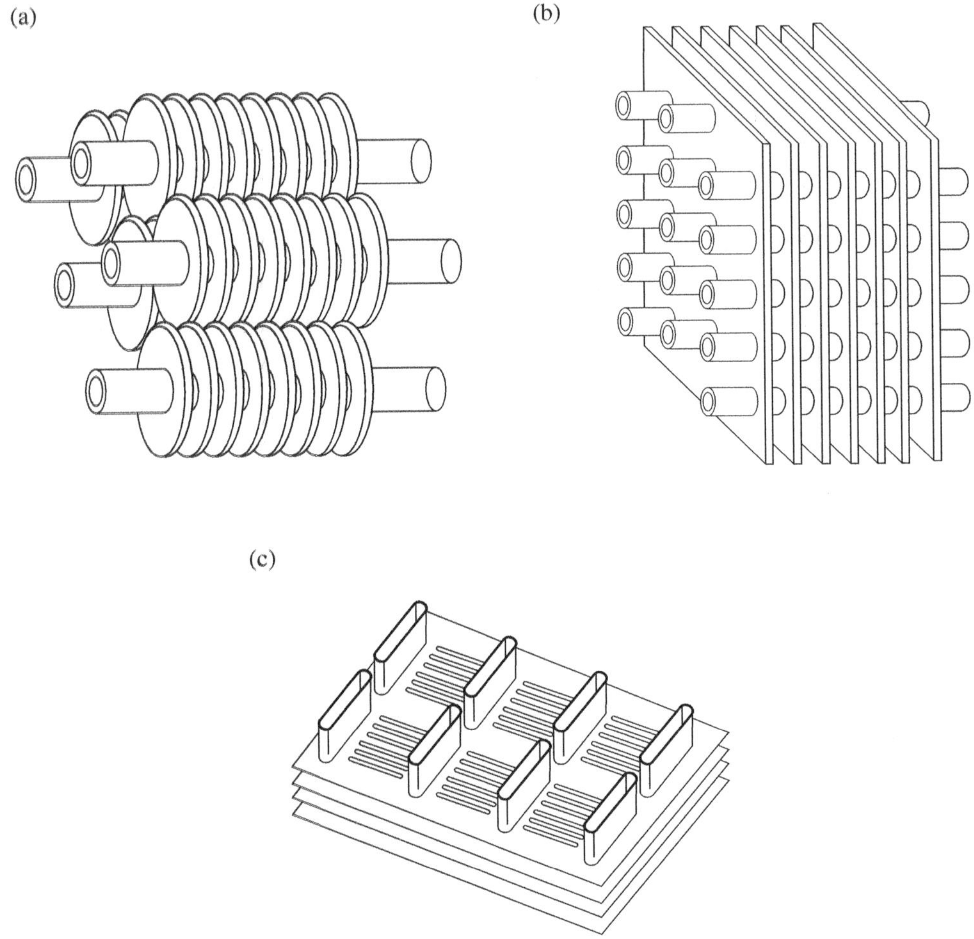

Figure 2.16 Finned heat exchangers: (a) circular finned-tube type, (b) plate finned-tube type, (c) louvred plate-fin flat-tube type.

Finned-tube exchangers can withstand high pressure on the tube side. The highest temperature is again limited by the type of bonding, materials employed and material thickness. Finned-tube exchangers are usually less compact than plate-fin exchangers. Finned-tube exchangers with a surface area density of about 3300 m^2/m^3 are available commercially. On the fin side, the desired surface area can be achieved through the proper fin density and fin geometry. Typical fin densities for plate fins vary from 250 to 800 fins per meter (6 to 20 fins per inch), fin thickness 0.08–0.25 mm, and fin flow lengths 25–250 mm. A plate finned-tube exchanger with 400 fins per meter has a surface area density of about 720 m^2/m^3 – Bejan and Pfister [13].

2.8.1 Geometrical Characteristics

The basic core geometry for an idealized single-pass cross-flow circular finned-tube exchanger with a staggered tube arrangement is shown in Figure 2.17. We assume that the staggered tube arrangement has an equilateral triangular shape.

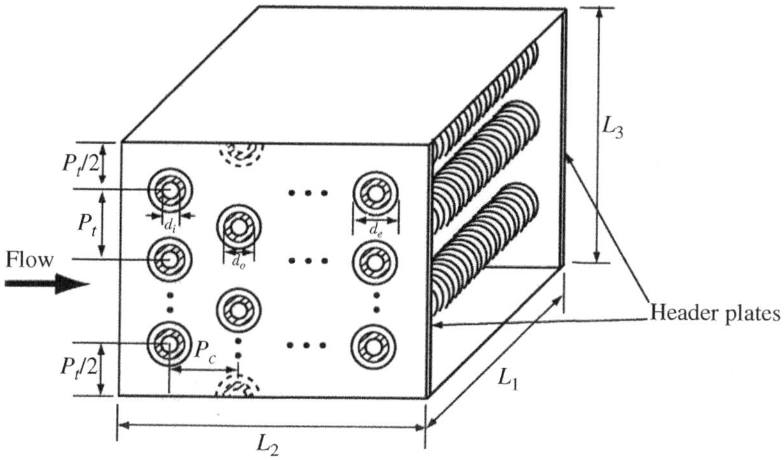

Figure 2.17 Circular finned-tube heat exchanger.

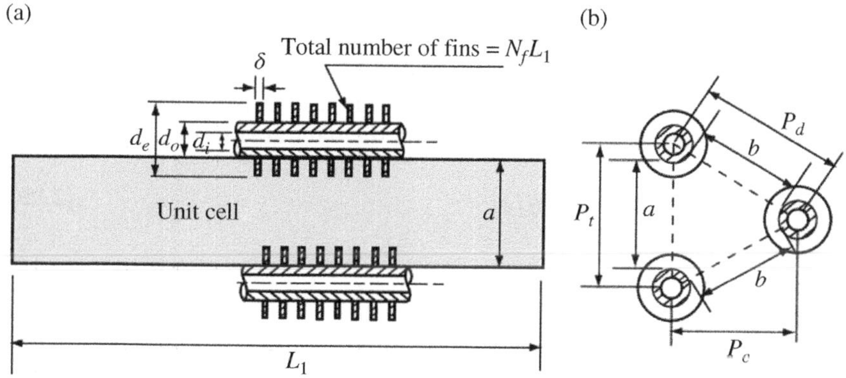

Figure 2.18 Unit cell of a staggered finned-tube arrangement.

The geometrical characteristics are derived for the staggered arrangement in Figures 2.17 and 2.18. The total number of tubes can be calculated in terms of tube pitches, P_t and P_c, and the dimensions of the exchanger, where P_t is the transverse tube pitch and P_c is the longitudinal tube pitch. Note that, in Figure 2.17, the number of tubes in the first row perpendicular to the flow direction is L_3/P_t and the number of tubes in the second row is $(L_3/P_t - 1)$. Also, note that the number of rows with the same pattern as the first row is $(L_2/P_c + 1)/2$ and the number of rows with the same pattern as the second row is $(L_2/P_c - 1)/2$. Therefore, the total number of tubes is

$$N_t = \frac{L_3}{P_t}\frac{\frac{L_2}{P_c}+1}{2} + \left(\frac{L_3}{P_t} - 1\right)\frac{\frac{L_2}{P_c}-1}{2} \tag{2.95}$$

The total heat transfer area A_t is composed of the primary surface area A_p and the fin surface area A_f. The primary surface area consists of the tube surface area, except the fin base area (Figure 2.18b) and two header plate surface areas, except the tube outer diameters (Figure 2.17). N_f is the number of fins per unit length, as shown in Figure 2.18b. The primary surface area is

$$A_p = \pi d_o (L_1 - \delta N_f L_1) N_t + 2\left(L_2 L_3 - \frac{\pi d_o^2}{4} N_t\right) \tag{2.96}$$

The fin surface area is

$$A_f = \left(\frac{2\pi \left(d_e^2 - d_o^2\right)}{4} + \pi d_e \delta\right) N_f L_1 N_t \tag{2.97}$$

The total heat transfer area is then

$$A_t = A_p + A_f \tag{2.98}$$

We consider a unit cell of a staggered finned-tube exchanger, as shown in Figure 2.18. The total minimum free-flow area can then be calculated by multiplying the unit cell area by the number of unit cells (L_3/P_t). The minimum free-flow area occurs either at a plane through 'a' or at two planes through 'b' in Figure 2.18b, and whichever is the smaller area takes the minimum free-flow area. However, since two planes through b are always greater than one plane through a for the equilateral triangular arrangement, the minimum free-flow area must occur at the plane through a. In other words, the flow through a is not restricted by the diagonal planes through b because 2b > a. Note that a = $(P_t - d_o)$, and the minimum free-flow area is calculated using the dark area in Figure 2.18(a) as

$$A_c = \frac{L_3}{P_t}\left[(P_t - d_o)L_1 - (d_e - d_o)\delta N_f L_1\right] \tag{2.99}$$

2.8.2 Correlations for Circular-Finned-Tube Geometry

The fin geometry, helically wrapped (or extruded) circular fins on a tube similar to the one shown in Figure 2.16a, are commonly used in the process and waste heat industry. Briggs and Young [16] recommended a correlation for the Colburn factor j.

$$j = 0.134 Re^{-0.319}\left(\frac{P_f - \delta}{d_e - d_o}\right)^{0.2}\left(\frac{P_f - \delta}{\delta}\right)^{0.11} \tag{2.100}$$

where P_f is the fin pitch, and the symbols are found in Figures 2.19 and 2.20. From the Colburn factor j, the convection heat transfer coefficient h can be calculated.

For friction factors, Robinson and Briggs [17] recommended the following correlation:

$$f_{RB} = 9.464 Re^{-316}\left(\frac{P_t}{P_d}\right)^{-0.927}\left(\frac{P_t}{P_d}\right)^{0.515} \tag{2.101}$$

The friction factor f_{RB} used by Robinson and Briggs [17] is given by

$$f_{RB} = \frac{\Delta P_\rho}{N_r G^2} = \frac{\Delta P_\rho}{\frac{L_2}{P_c} G^2} = \frac{\Delta P_\rho P_c}{G^2\ L_2} \tag{2.102}$$

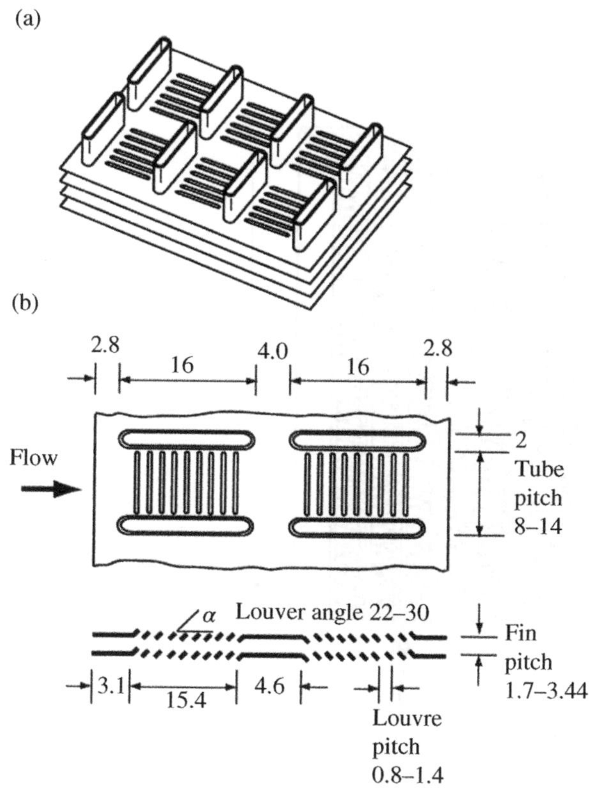

Figure 2.19 (a) Flat tubes and (b) louvred plate fin geometry (unit: mm).

where N_r is the number of rows that obtained by dividing the flow length L_2 by the longitudinal pitch P_c in Figure 2.17. The general friction factor f can be expressed in terms of ΔP for the geometry

$$f = \frac{\Delta P_\rho}{G^2} \frac{D_h}{L} \tag{2.103}$$

The relationship between f and f_{RB} is given by

$$f = \frac{1}{2} \frac{D_h}{P_c} f_{RB} \tag{2.104}$$

2.8.3 Pressure Drop

For a circular finned-tube heat exchanger, the tube outside flow in each tube row experiences a contraction and an expansion. Thus the pressure losses associated with a tube row within the core are of the same order of magnitude as those at the entrance to the first tube row and those at the exit with the last tube row. Consequently, the entrance and exit pressure drops are not calculated separately, but they are generally lumped into the friction factor for individually finned tubes.

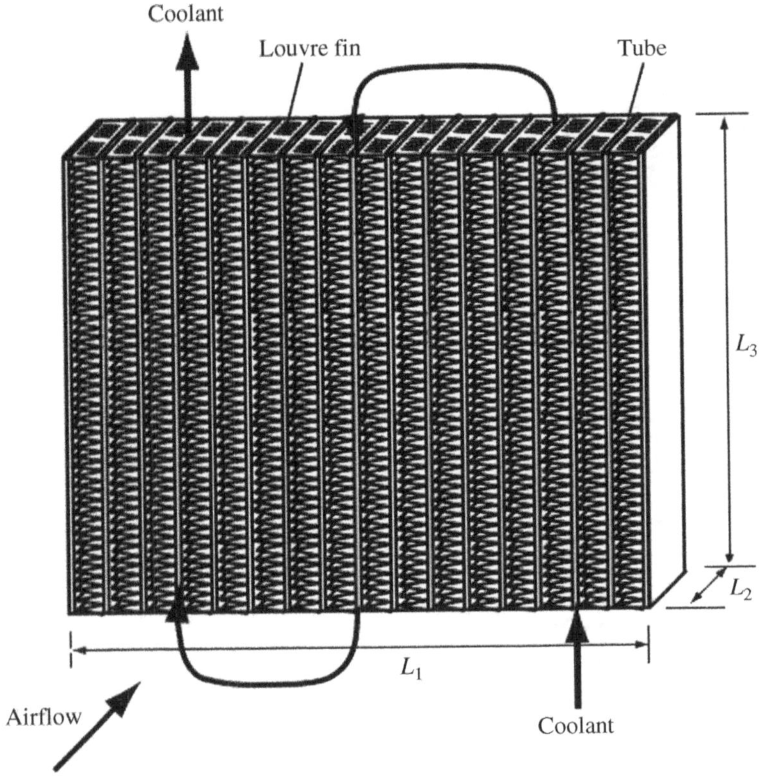

Figure 2.20 Louvre fin type plate-fin heat exchanger (1-pass for air and 3-pass for coolant).

The core pressure drop is generally obtained by Kays and London [5]

$$\Delta P = \frac{G^2}{2\rho_i}\left[1-\sigma^2+K_c+2\left(\frac{\rho_i}{\rho_o}-1\right)+\frac{4fL}{D_h}\rho_i\left(\frac{1}{\rho}\right)_m-\left(1-\sigma^2-K_e\right)\frac{\rho_i}{\rho_o}\right] \tag{2.105}$$

By eliminating the entrance and exit terms from above equation, the total pressure drop associated with the core becomes [5]

$$\Delta P = \frac{G^2}{2\rho_i}\left[2\left(\frac{\rho_i}{\rho_o}-1\right)+\frac{4fL}{D_h}\rho_i\left(\frac{1}{\rho}\right)_m\right] \tag{2.106}$$

where $(1/\rho)_m$ was defined in Equation 2.105 and is rewritten here for convenience.

$$\left(\frac{1}{\rho}\right)_m = \frac{1}{\rho_m} = \frac{1}{2}\left(\frac{1}{\rho_i}+\frac{1}{\rho_o}\right) \tag{2.107}$$

It should be emphasized that the friction factor f in Equation 2.104 is based on the hydraulic diameter.

2.8.4 Correlations for Louvred Plate-Fin Flat-Tube Geometry

Figure 2.19 shows a widely used automotive radiator geometry, having louvred plate fins on flat tubes. Achaichia and Cowell [18] developed a correlation for the inline louvre fin geometry. Their test data is in the range $1.7 \le P_f \le 3.44\text{mm}$, $8 \le P_t \le 14\text{mm}$, $1.7 \le P_f \le 3.44\text{mm}$ and $22 \le \alpha \le 30°$. The Reynolds number limits are $150 \le R_e \le 3{,}000$.

The Colburn factor j is given by

$$j = 1.234\gamma\,\mathrm{Re}^{-0.59}\left(\frac{P_t}{P_L}\right)^{-0.09}\left(\frac{P_f}{P_L}\right)^{-0.04} \tag{2.108}$$

where

$$\gamma = \frac{1}{\alpha}\left(0.936 - \frac{243}{R_e} - 1.76\frac{P_f}{P_L} + 0.995\alpha\right) \tag{2.109}$$

where P_L is the louver pitch, P_t the tube pitch, P_f the fin pitch, and α the louvre angle (must be in degrees).

The friction factor f is given by

$$f = 834.CF.P_f^{-0.22}P_L^{0.25}P_t^{0.26}H_L^{0.33}\left[Re^{(0.318\log_{10}Re^{-2.25})}\right]^{1.07} \tag{2.110}$$

where CF is the correction factor (see Equation 2.110). H_L is the louvre height and can be calculated by

$$H_L = P_L \;\sin(\alpha \text{ deg}) \tag{2.111}$$

Compare the correlations for louvred plate-fin flat-tube geometry developed by Achaichia and Cowell [18] and the experimental data from a smooth plate-fin flat-tube geometry obtained by Kays and London [5].

A conversion factor (CF) is inserted by Achaichia and Cowell [18] into Equation 2.108 in order to correct the units in dimensions because, in the original equation, all dimensions are in millimetres. Hence, we can enter metres in Equation 2.100.

$$CF = \left(\frac{1000}{m}\right)^{(-0.22+0.25+0.26+0.33)} \tag{2.112}$$

where m is in metres.

Example 2.4 Consider a finned-tube CHE having the core configuration of Figure 2.13 of Incropera et al. [9]. The core is fabricated from aluminium, and the tubes have an inside diameter of 13.8 mm. In a waste heat recovery application, water flows through the tubes provides an inside convection coefficient of $h_i = 1500\,\text{W/m}^2\cdot\text{K}$, while combustion gases at 1 atm and 825 K are in cross flow over the tubes. If the gas flow rate is 1.25 kg/s and the frontal area is 0.20 m^2, what is the gas-side overall heat transfer coefficient? If a water flow rate of 1 kg/s is to be heated from 290 to 370 K, what is the required heat exchanger volume?

Solution

Known: Compact heat exchanger geometry, gas-side flow rate and temperature and water-side convection coefficient. Water flow rate and inlet and outlet temperatures.

Find: Gas-side overall heat transfer coefficient. Heat exchanger volume.

Schematic:

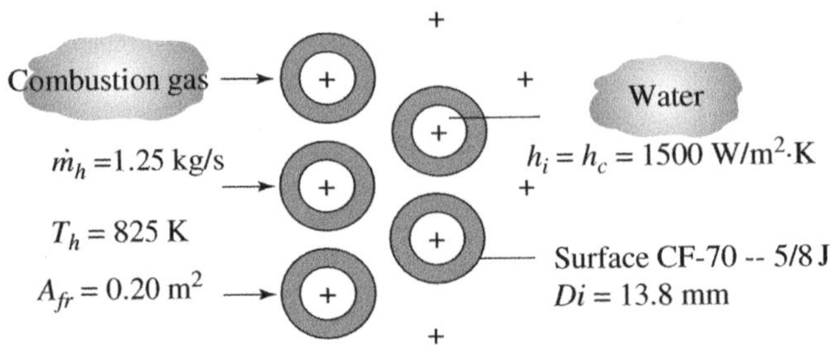

Assumptions:

1) Gas has properties of atmospheric air at an assumed mean temperature of 700 K.
2) Fouling is negligible.

Properties: Table A.1, aluminium $(T \approx 300\,\text{K}): k = 237\,\text{W/m}\cdot\text{K}$. Table A.4, air ($p = 1$ atm, $\bar{T} = 700$ K): $c_p = 1075\,\text{J/kg}\cdot\text{K}, \mu = 338.8 \times 10^{-7}\,\text{N}\cdot\text{s/m}^2, Pr = 0.695$. Table A.6, water ($\bar{T} = 330$ K): $c_p = 4184\,\text{J/kg}\cdot\text{K}$.

Analysis: Referring to Incropera et al [9], the combustion gas and the water are the hot and cold fluids, respectively. Hence, neglecting fouling effects, and acknowledging that the tube inner surface is not finned $(\eta_{o,c} = 1)$, the overall heat transfer coefficient based on the gas (hot) side surface area is given by

$$\frac{1}{U_h} = \frac{1}{h_c(A_c/A_h)} + A_h R_w + \frac{1}{\eta_{o,h} h_h}$$

where A_h and A_c are the total gas-side (hot) and water-side areas, respectively. If the fin thickness is assumed to be negligible, it is readily shown that

$$\frac{A_c}{A_h} = \frac{D_i}{D_o}\left(1 - \frac{A_{f,h}}{A_h}\right)$$

where $A_{f,h}$ is that portion of the total gas-side area associated with the fins. The approximation is valid to within 10%, and for the heat exchanger core conditions

$$\frac{A_c}{A_h} \approx \frac{13.8}{16.4}(1 - 0.830) = 0.143$$

Obtaining the wall conduction resistance of a cylindrical wall, it follows that

$$A_h R_w = \frac{In(D_o/D_i)}{2\pi Lk/A_h} = \frac{D_i In(D_o/D_i)}{2k(A_c/A_h)}$$

Hence

$$A_h R_w = \frac{(0.0138\text{m})\ln(16.4/13.8)}{2\left(237\frac{\text{W}}{\text{m}}\cdot\text{K}\right)(0.143)} = 3.51 \times 10^{-5}\,\text{m}^2\cdot\text{K/W}$$

The gas-side convection coefficient may be obtained by the equation below,

$$G = \frac{\dot{m}}{\sigma A_{fr}} = \frac{1.25\,\text{kg/s}}{0.449 \times 0.20\,\text{m}^2} = 13.9\frac{\text{kg}}{\text{s}} \cdot \text{m}^2$$

Hence

$$Re = \frac{13.9\,\text{kg/s} \cdot \text{m}^2 \times 6.68 \times 10^{-3}\text{m}}{338.8 \times 10^{-7}\text{kg/s} \cdot \text{m}} = 2740$$

and from Figure 11.16 of Incropera et al. [9]), $j_H \approx 0.010$. Hence

$$h_h \approx 0.010\frac{Gc_p}{Pr^{2/3}} = 0.010\frac{(13.9\,\text{kg/s} \cdot \text{m}^2)(1075\,\text{J/kg} \cdot \text{K})}{(0.695)^{2/3}}$$

$$= 190\,\text{W/m}^2 \cdot \text{K}$$

To obtain the hot-side temperature effectiveness from Equation $\eta_o = 1 - \frac{A_f}{A}\left(1 - \eta_f\right)$, the fin efficiency must first be determined. With $r_{2c} = 14.38\,\text{mm}$, $r_{2c}/r_1 = 1.75$, $L_c = 6.18\,\text{mm}$, $A_p = 1.57 \times 10^{-6}\,\text{m}^2$ and $L_c^{3/2}\left(h_h/kA_p\right)^{1/2} = 0.34$, it follows that $\eta_f \approx 0.89$.

Hence

$$\eta_{o,h} = 1 - \frac{A_f}{A}(1 - \eta_f) = 1 - 0.830\,(1 - 0.89) = 0.91$$

We then obtain

$$\frac{1}{U_h} = \left(\frac{1}{1500 \times 0.143} + 3.51 \times 10^{-5} + \frac{1}{0.91 \times 190}\right)\text{m}^2 \cdot \text{K/W}$$

$$\frac{1}{U_h} = \left(4.66 \times 10^{-3} + 3.51 \times 10^{-5} + 5.78 \times 10^{-3}\right) = 0.010\,\text{m}^2 \cdot \text{K/W}$$

or

$$U_h = 100\,\text{W/m}^2 \cdot \text{K}$$

With $C_c = \dot{m}_c\, c_{p,c} = 1\frac{\text{kg}}{\text{s}} \times 4184\frac{\text{J}}{\text{kg}} \cdot \text{K} = 4184\,\text{W/K}$, the heat exchanger must be large enough to transfer heat

$$q = C_c(T_{c,o} - T_{c,i}) = 4184\,\text{W/K}(370 - 290)\,\text{K} = 3.35 \times 10^5\,\text{W}$$

with $C_h = \dot{m}_h\, c_{p,h} = 1.25\,\text{kg/s} \times 1075\,\text{J/kg} \cdot \text{K} = 1344\,\text{W/K}$, the minimum heat capacity rate corresponds to the hot fluid and the maximum possible heat transfer rate is

$$q_{max} = C_{min}(T_{h,i} - T_{c,i}) = 1344\frac{\text{W}}{\text{K}}(825 - 290)\text{K} = 7.19 \times 10^5\,\text{W}$$

It follows that

$$\varepsilon = \frac{q}{q_{max}} = \frac{3.35 \times 10^5\,\text{W}}{7.19 \times 10^5\,\text{W}} = 0.466$$

Hence, with $(C_{min}/C_{max}) = 0.321$, for a cross-flow heat exchanger with both fluids unmixed yields

$$NTU = \frac{U_h A_h}{C_{min}} \approx 0.65$$

The required gas-side heat transfer surface area is then

$$A_h = \frac{0.65 \times 1344\,\mathrm{W/K}}{100\frac{\mathrm{W}}{\mathrm{m}^2}\cdot\mathrm{K}\cdot} = 8.7\mathrm{m}^2$$

with the gas-side surface area per unit heat exchanger volume corresponding to $\alpha = 269\,\mathrm{m}^2/\mathrm{m}^3$ (Figure 2.13 of Incropera et al. [9]), the required heat exchanger volume is

$$V = \frac{A_h}{\alpha} = \frac{8.7\,\mathrm{m}^2}{269\,\mathrm{m}^2/\mathrm{m}^3} = 0.032\,\mathrm{m}^3$$

Comments:

1) The effect of the tube wall thermal conduction resistance is negligible, while contributions due to the cold- and hot-side convection resistances are comparable.
2) Knowledge of the heat exchanger volume yields the heat exchanger length in the gas-flow direction, $L = \frac{V}{A_{fr}} = 0.032\frac{\mathrm{m}^3}{0.20}\mathrm{m}^2 = 0.16\,\mathrm{m}$, from which the number of tube rows in the flow direction may be determined.

$$N_L \approx \frac{L - D_f}{S_L} + 1 = \frac{(160 - 28.5)\mathrm{mm}}{34.3\,\mathrm{mm}} + 1 = 4.8 \approx 5$$

3) The temperature of the gas leaving the heat exchanger is

$$T_{h,o} = T_{h,i} - \frac{q}{C_h} = 825\mathrm{K} - \frac{3.35 \times 10^5\,\mathrm{W}}{1344\,\mathrm{W/K}} = 576\,\mathrm{K}$$

Hence the assumption of $T_h = 700\,\mathrm{K}$ is good.

4) From Figure 11.16 of Incropera et al. [9]), the friction factor is $f \approx 0.033$. With $(A/A_{ff}) = (\alpha V/\sigma A_{fr}) = (269 \times 0.032/0.449 \times 0.20) = 96$, $v_i(825\,\mathrm{K}) = 2.37\,\mathrm{m}^3/\mathrm{kg}$, $v_o(576\,\mathrm{K}) = 1.65\,\mathrm{m}^3/\mathrm{kg}$, and $v_m = 2.10\,\mathrm{m}^3/\mathrm{kg}$, Equation 2.44 yields a pressure drop of

$$\Delta p = \frac{(13.9\,\mathrm{kg/s\cdot m^2})^2(2.37\,\mathrm{m}^3/\mathrm{kg})}{2}[(1 + 0.202)(0.696 - 1) + 0.033 \times 96 \times 0.848]$$

$$\Delta p = 530\frac{\mathrm{kg}}{\mathrm{s}^2}\cdot\mathrm{m} = 530\,\mathrm{N/m}^2$$

2.8.5 Louvre-Fin-Type Flat-Tube Plate-Fin Heat Exchangers

Louvre surface is the standard geometry for automotive radiators and is also often used in aircraft heat exchangers. A core of the louvre-fin-type plate-fin heat exchanger is shown in Figure 2.20. The louvre-fin geometry provides heat transfer coefficients comparable to those of the offset-strip-fin (OSF) geometry.

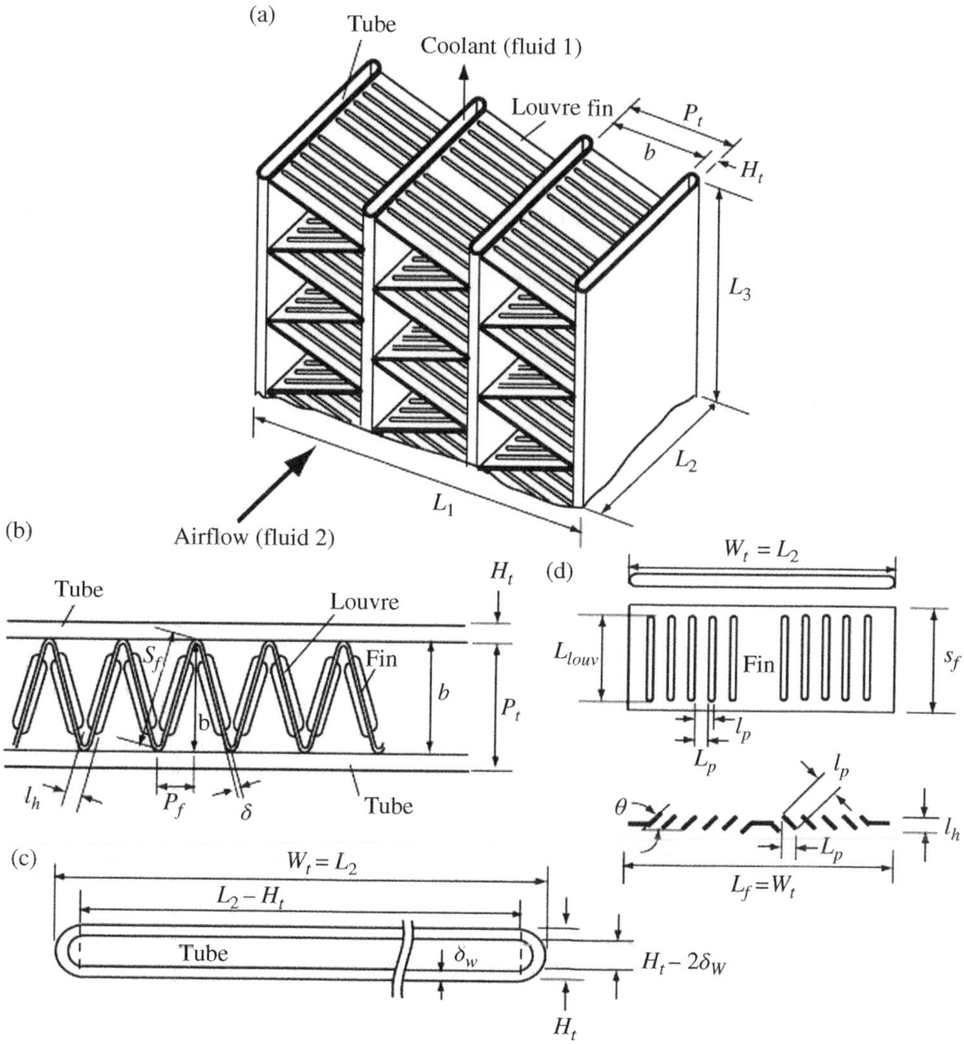

Figure 2.21 Definition of geometrical parameters of corrugated louvre fin.

2.8.5.1 Geometric Characteristics

A louvre fin geometry is illustrated in Figure 2.21. For convenience, coolant and air are referred to as fluid 1 and fluid 2, respectively. We consider first the air side.

The number of passages N_{pg} on the air side is defined as the air flow passages between the flat tubes (Shah and Webb [19]). The core width L_1 is expressed in terms of the number of passages N_{pg} as

$$L_1 = N_{pg} b + \left(N_{pg} + 1\right) H_1 \tag{2.113}$$

where b is the tube spacing and H_t the tube height. Solving for the number of passages N_{pg} gives

$$N_{pg} = \frac{L_1 - H_t}{b + H_t} \tag{2.114}$$

The total number of fins n_f are

$$n_f = \frac{L_3}{p_f} N_{pg} \tag{2.115}$$

where p_f is the fin pitch, as shown in Figure 2.21b and L_3 the core height. The total heat transfer area A_t is generally obtained from the sum of the primary area A_p and the fin area A_f. The primary area A_p is calculated by subtracting the fin base areas from the tube outer surface areas, considering the circular front and end of the tubes (Figure 2.21c).

$$A_p = [2(L_2 - H_t) + \pi H_t] L_3 (N_{pg} + 1) - 2\delta L_2 n_f \tag{2.116}$$

The total number of louvres in the core is obtained from Figure 2.21d as

$$n_{louv} = \left(\frac{L_f}{l_p} \; 1 \right) n_f \tag{2.117}$$

where L_f is the fin length in the flow direction and lp the louvre pitch. Note that when the louvres change the direction of opening we allow one louvre missing in the middle of the fin. The total fin area A_f is the sum of the fin area and the louvre edge area as

$$A_f = \underbrace{2(s_f L_2 + s_f \delta) n_f}_{\textit{fin area}} + \underbrace{2L_{louv} \delta n_{louv}}_{\textit{Louver edge area}} \tag{2.118}$$

where s_f is the fin width and L_{louv} the louvre length. The total heat transfer area A_t is obtained by

$$A_t = A_p + A_f \tag{2.119}$$

The minimum free-flow area A_c is expressed by

$$A_c = \underbrace{bL_3 N_{pg}}_{\textit{Spacing between tubes}} - \underbrace{[\delta(\delta_f - L_{louv}) + L_{louv} l_h] n_f}_{\textit{fin and louver edge area}} \tag{2.120}$$

Now we consider coolant-side geometry in Figure 2.21a–c. The total number of tubes N_t is obtained by

$$N_t = N_{pg} + 1 \tag{2.121}$$

Considering the circular shapes at both ends, the total heat transfer area A_{t1} on the coolant side is obtained by

$$A_{t1} = \left[\underbrace{2(L_2 - H_t)}_{\textit{Tube straight length}} + \underbrace{\pi (H_t - 2\delta_w)}_{\textit{Circular shape at the end}} \right] L_3 N_t \tag{2.122}$$

The free-flow area A_{c1} on the coolant side is obtained by

$$A_{c1} = \left[\underbrace{(L_2 - H_t)(H_t - 2\delta_w)}_{\textit{Rectangle of tube}} + \underbrace{\frac{\pi}{4}(H_t - 2\delta_w)^2}_{\textit{Circular parts at the end}} \right] \frac{N_t}{N_p} \tag{2.123}$$

2.8.5.2 Correlations for Louvre Fin Geometry

Louvre fins are used extensively in the auto industry due to their mass production manufacturability and, hence, their lower cost. They have generally higher j and f factors than those for the offset strip fin geometry, and also the increase in the friction factors is, in general, higher than the increase in j factors [5]. Davenport [20] tested 32 one-row louvre fin geometries. Davenport systematically varied the louvre dimensions for two louvre heights (12.7 and 7.8 mm) and developed correlations for j and f vs Re. Chang and Wang [21] provided a correlation for the Colburn factor j based on an extensive database for air flow over louvre fin.

$$j = Re^{-0.49}\left(\frac{\theta}{90}\right)^{0.27}\left(\frac{p_f}{l_p}\right)^{-0.14}\left(\frac{b}{l_p}\right)^{-0.29}\left(\frac{W_t}{l_p}\right)^{-0.23}\left(\frac{L_{louv}}{l_p}\right)^{0.68}\left(\frac{p_t}{l_p}\right)^{-0.28}\left(\frac{\delta}{l_p}\right)^{-0.05} \tag{2.124}$$

where l_p is the louvre pitch, θ the louvre angle, p_f the fin pitch, p_t the tube pitch, b the tube spacing, W_t the tube outside width, L_{louv} the louvre cut length and δ the fin thickness. Figure 2.20 shows the dimensions of the geometrical parameters of louvre fin.

Equation 2.122 is valid for the following ranges of the parameters: $0.82 \le D_h \le 5.02$ mm, $0.51 \le p_f \le 3.33$ mm, $0.5 \le l_p \le 3$ mm, $2.84 \le b \le 20$ mm, $15.6 \le W_t \le 57.4$ mm, $2.13 \le L_l \le 18.5$ mm, $7.51 \le p_t \le 25$ mm, $0.0254 \le \delta \le 0.16$ mm, $1 \le N_r \le 2$, and $8.4 \le \theta \le 35°$. This correlation predicts j factors within ±15% for $30 < Re < 5000$. The correlation for the Fanning friction factor based on the same database by Changa et al. [22] is

$$f = f_a f_b f_c \tag{2.125}$$

where

$$f_a = \begin{cases} 14.39\,Re^{-0.805\,pf/b}\left[In\left(1.0 + p_f/l_p\right)\right]^{3.04} & Re < 150 \\ 4.97\,Re^{\left(0.6049 - 1.064/\theta^{0.2}\right)}\left\{In\left[\left(\delta/p_f\right)^{0.5} + 0.9\right]\right\}^{-0.527} & 150 \le Re < 5{,}000 \end{cases} \tag{2.126}$$

$$f_b = \begin{cases} \left\{\ln\left[\left(\delta/p_f\right)^{0.48} ı\, 0.9\right]\right\}^{-1.453}\left(D_h/l_p\right)^{-3.01}\left[lıı\left(0.5\,Re\right)\right]^{-3.01} & Re < 150 \\ \left[\left(D_h/l_p\right)\ln\left(0.3\,Re\right)\right]^{-2.966}\left(p_f/L_{louv}\right)^{-0.7931\,pf/b} & 150 \le Re < 5{,}000 \end{cases} \tag{2.127}$$

$$f_c = \begin{cases} \left(p_f/L_{louv}\right)^{-0.308}\left(L_f/L_{louv}\right)^{-0.308}\left[\exp\left(-0.1167\,p_t/H_t\right)\right]\theta^{0.35} & Re < 150 \\ \left(p_t/H_t\right)^{-0.0446}\left\{\ln\left[1.2 + \left(l_p/p_f\right)^{1.4}\right]\right\}^{-3.553}\theta^{-0.477} & 150 \le Re < 5{,}000 \end{cases} \tag{2.128}$$

where H_t is the tube outside height and L_f the fin length.

The detailed information for the louvre fin geometry of the experiments is not provided.

2.9 Plate-Fin Exchangers Operating Limits

The maximum operating temperature of a plate-fin heat exchanger is a function of its construction materials. Aluminium brazed plate-fin heat exchangers can be used from cryogenic temperatures (−270 °C) up to 220 °C, depending on the pipe and header alloys.

Table 2.4 Features of fin-type applications: Relative ΔP and relative heat transfer.

		Features	
Fin type	**Application**	**Relative ΔP**	**Relative heat transfer**
Plain	General	Lowest	Lowest
Perforated	Boiling streams	Low	Low
Herringbone	Gas stream with low allowable ΔP High pressure streams Gas streams for hydrocarbon and natural gas applications	High	High
Serrated	Gas streams in air separation applications general	Highest	Highest

Stainless steel plate-fin heat exchangers are able to operate at up to 580 °C, while titanium units can tolerate temperatures approaching 530 °C. Aluminium brazed units can operate at up to 80 bar, depending on the physical size and the maximum operating temperature. Stainless steel plate-fin heat exchangers are currently limited to 50 bar, with developments expected that will extend the capability to 80 bar. Higher pressures can be tolerated by using a diffusion-bonded structure.

The size of a plate-fin heat exchanger is a function of the procedure used to assemble the core. In the case of aluminium vacuum-brazed units, modules of 6.25 × 2.4 × 1.2 m are available. When selecting brazed aluminium plate-fin exchangers, the engineer should ensure that

- All fluids must be clean and dry. Filtration must be used to remove particulate matter over 0.3 mm.
- Fluids must be non-corrosive to aluminium. Water is suitable if it is in a closed loop and contains corrosion inhibitors.
- Fluids must be in the temperature range −270 to +220 °C.
- The maximum design pressure is less than 80 bar.
- Features are listed in Table 2.4 for fin-type CHEs with respect to relative Δp and relative heat transfer.

2.10 Plate-Fin Exchangers – Monitoring and Maintenance

Condition monitoring of heat exchanger tubes may be conducted through nondestructive methods such as eddy current testing. The mechanics of water flow and deposits are often simulated by computational fluid dynamics (CFD). Fouling is a serious problem in some heat exchangers. River water is often used as cooling water, which results in biological debris entering the heat exchanger and building layers, thus decreasing the heat transfer coefficient. Another common problem is scale, which is made up of deposited layers of chemicals such as calcium carbonate or magnesium carbonate. Plate heat exchangers need to be disassembled and cleaned periodically. Tubular heat exchangers can be cleaned by such methods as acid cleaning, sandblasting, high-pressure water jet, bullet cleaning or drill rods. In large-scale cooling water systems for heat exchangers, water treatment such as purification, addition of chemicals and testing is used to minimize fouling of the heat exchanger equipment. Other water treatment is also used in steam systems for power plants etc. to minimize fouling and corrosion of the heat exchanger and other equipment.

A variety of companies have started using water-borne oscillation technology to prevent bio-fouling. Without the use of chemicals, this type of technology has helped in providing a low-pressure drop in heat exchangers.

2.10.1 Advantage

As mentioned earlier, plate-fin heat exchangers are extremely efficient and lightweight, making them an ideal solution for applications where weight (and therefore performance) is critical, such as aircraft applications. They are also very reliable parts under uniform operating pressures. A quality braze joint is often stronger than the parent material being brazed.

2.10.2 Disadvantages

Normally the plate-fin heat exchanger fins cannot be easily cleaned with high density in aerospace applications. The biggest drawback of PFHEs is the price. Owing to the enormous amount of labour involved in stacking the layers of the core, and the various processes (fin manufacturing, brazing, heat treating and welding), a small core can cost more when compared to other constructions.

Nomenclature

A	exchanger total heat transfer area on one side (m^2)
A_f	exchanger total fin area on one side (m^2)
A_{fr}	exchanger total frontal area (m^2)
$2A$	wavy fin amplitude (m)
A_s	surface area (m^2)
A_c	flow area (m^2)
A_{fr}	frontal area (m^2)
A_t	total area of plate and fin (m^2)
A_m	arithmetic mean diameter, m
a	plate thickness, m
b	effective plate gap (m)
c_p	specific heat at constant pressure ($J/kg \cdot K$)
C	heat capacity rate (W/K)

C^*,Cr	ratio of heat capacity rates, C_{min}/C_{max}
d	hole diameter, m
d_h	hydraulic diameter (m)
d_i	tube inside diameter (m)
d_o	tube outside diameter (m)
f	fanning friction factor
F	LMTD correction factor
FL	fin length (m)
F_H	fin height of louvred surface (m)
FT	fin thickness of louvred surface (m)
FN	flow non-uniformity
g_c	gravitational constant, (m/s^2)
G	mass velocity (kg/m$^2 \cdot$ s)
h	passage height (m), convective heat transfer coefficient, W/m$^2 \cdot$ K
j	Colburn factor
K_c	contraction loss coefficient
K_e	expansion loss coefficient
K_d	momentum flux correction factor, dimensionless
k	thermal conductivity, W/m $\cdot$ K
l	length (height) of fin in efficiency formula (m)
l_f	length of strip (fin) (m)
L	flow length, louvre pitch (m)
$LMTD$	log mean temperature difference
L_t	tube length (m)
m	fin efficiency parameter (m^{-1})
M	molecular weight (kg/kmol)
$N_{c,h}$	NTU for cold or hot side
N_f	number of fins/unit length (m^{-1})
N_t	number of tubes in exchanger
NTU	number of transfer units
Nu	Nusselt number
P	pressure (Pa)
p_c	critical pressure (Pa)
p_f	fin pitch (m)
p_m	design pressure (Pa)

p_{sat}	saturation pressure (Pa)
P_r	Prandtl number
q	heat flow (W)
R	ideal gas constant (J/kgK)
R	capacity rate ratio
R_c	ideal gas constant
Re	Reynolds number
R_f	fouling resistance (m^2K/W)
r_h	hydraulic radius (m)
R_w	wall resistance (m^2K/W)
s	fin spacing (i.e. space between fins) (m)
S	nucleate boiling suppression factor
S_t	Stanton number
t	thickness (m)
TN	temperature non-uniformity
t	separation plate thickness (m)
t_f	fin thickness (m)
T	absolute temperature (K)
U	overall heat transfer coefficient (w/m^2 K)
u	mean velocity (m/s)
V	volume (m^3)
v	specific volume (m^3/kg)
$\dot{m}$	mass flow rate, kg/s

Greek Symbols

α	ratio of heat transfer surface area per total exchanger volume
β	ratio of total transfer area on one side of heat exchanger to the volume between the plates
Δ	denotes difference
ε	exchanger effectiveness, dimensionless
ε_p	effectiveness of one pass of a multipass heat exchanger, dimensionless
λ	longitudinal conduction parameter, dimensionless
η_f	fin temperature effectiveness, dimensionless
η_o	total overall surface temperature effectiveness

μ	dynamic viscosity (Ns/m^2 or kg/ms or Pas)
η_f	fin efficiency
η_o	surface effectiveness
ρ	density (kg/m^3)
σ	porosity
σ	ratio of free flow area to frontal area, A_c/A_{fr}
ϕ_f^2	two-phase frictional multiplier

Subscripts

1	cold stream
2	hot stream
i	inlet
o	outlet
c	cold stream
cb	convective boiling
f	fin
g	gas/vapour
gr	gravity
h	hot stream
l	liquid
lm	log mean
m	mean
min	minimum
max	maximum
n	number
p	pass
T	tube
w	wall

References

1 R.K. Shah., Classification of Heat Exchangers, *Low Reynolds Number Flow Heat Exchangers*. S. Kakac, R.K. Shah, and A.E. Bergles ed. pp. 9–14 (Hemisphere Publishing Corporation, Washington, 1983).

2 A.E. Bergles, R.L. Webb and G.H. Junkan, Energy Conservation Via Heat Transfer Enhancement, *Energy*, vol. 4, pp. 193–200, 1979.

3 R.L. Webb., Enhancement of Single-phase Heat Transfer, Enhancement of Single-phase Heat Transfer. S. Kakac, R.K. Shah, and A.E. Bergles ed. pp. 17: 1–62, Wiley, New York, 1987.

4 U. Brockmeier, T. Guentermann and M. Fiebig, Performance Evaluation of a Vortex Generator Heat Transfer Surface and Comparison with Different High Performance Surfaces, *International Journal of Heat and Mass Transfer*, vol. 36, pp. 2575–2587, 1993.

5 W.M. Kays and A.L. London, Compact Heat Exchangers, The National Press, Palo Alto, CA, 1964.

6 J.E. O'Brien and E.M. Sparrow, Corrugated-Duct Heat Transfer, Pressure Drop, and Flow Visualization, Transactions of the ASME, Journal of Heat Transfer, vol. 104, pp. 410–416, 1982.

7 Shah, R.K. and Dusan P. Sekulic, *Fundamentals of Heat Exchanger Design*, John Wiley & Sons, 2003.

8 R.L. Webb, Performance Evaluation Criteria for Use of Enhanced Heat Transfer Surfaces in Heat Exchanger Design, *International Journal of Heat and Mass Transfer*, vol. 24, pp. 715–726, 1981.

9 Incropera F.P., Dewitt D.P., Bergman T.L., Lavine A.S., Seetharamu K.N. and Setharam T. R., 2014, *Fundamentals of Heat and Mass Transfer*, Wiley publications, New Delhi.

10 J.E. Hesselgreaves, Compact Heat Exchangers, Selection, Design and Operation, Pergamon, UK, 2001.

11 Di Giovanni M.A. and Webb R.L, Uncertainty in effectiveness – NTU calculations for cross flow heat exchangers, *Heat Transfer Eng.*, 10(3):61–70, 1989.

12 John H. Lienhard IV and John H Lienhard V, *A Heat Transfer Text Book*, 3rd Edition, Phlogiston Press, Cambridge Massachusetts, 2004.

13 A. Bejan and Peter A. Pfister, Jr., Evaluation of Heat Transfer Augmentation Techniques Based on Their Impact on Entropy Generation, *Letters in Heat and Mass Transfer*, vol. 7, pp. 97–106, 1980.

14 *Plate Fin Heat Exchangers Guide to their Specification and Use*, Editor: Taylor, H.T.F.S., 1st Edition, 1987.

15 Manglik, R.M. and Bergles, A.E, Heat Transfer and Pressure Drop Correlations for the Rectangular Offset Strip Fin Compact Heat Exchanger, *Experimental Thermal and Fluid Science*, vol. 10, no. 2: p. 171–180. 1995.

16 Briggs, D.E. and Young, E.H., Convection Heat Transfer and Pressure Drop of Air Flowing Across Triangular Pitch Banks of Finned Tubes, *IT Chem. Eng. Progress Symposium Series – Heat Transfer*, vol. 59, no. 41, 1963, pp. 1–10.

17 Robinson, K.K. and Briggs, D.E., Pressure drop of air flowing across triangular pitch banks of finned tubes, *Chem. Eng. Prog. Symp. Ser.*, vol. 62, no. 64, pp. 177–184. 1966.

18 Achaichia, A. and Cowell, T.A., Heat Transfer and Pressure Drop Characteristics of Flat Tube and Louvered Plate Fin Surfaces, *Exp. Therm. Fluid Sci.*, 1, pp. 147–157. 1988.

19 R.K. Shah and R.L. Webb., *Compact and Enhanced Heat Exchangers, Heat Exchangers: Theory and Practice.* J. Taborek G.F. Hewitt, and N. Afgan ed. pp. 425–468, Hemisphere Publishing Corporation, Washington, 1983.

20 Davenport, C.J., Correlation for Heat Transfer and Flow Friction Characteristics of Louvered Fin, *AIChE Symp. Ser.*, 79, pp. 19–27. 1983.

21 Chang, Y.J. and Wang, C.C., A Generalized Heat Transfer Correlation for Louver Fin Geometry, *Int. J. Heat Mass Transfer*, 40, pp. 533–544. 1997.

22 Yu-Juei Changa, Kuei-Chang Hsub, Yur-Tsai Linb, Chi-Chuan Wang, A generalized friction correlation for louver fin geometry, Technical Note, *International Journal of Heat and Mass Transfer* 43, pp: 2237–2243, 2000.

3

Fundamentals of Finite Element and Finite Volume Methods

3.1 Introduction

There are three distinct methods of numerical solution techniques: finite difference, finite volume and finite element methods. The purpose in each is to convert the differential equations into algebraic equations. The main differences between the three methods are associated with the way the differential equations are converted to algebraic equations.

Finite difference methods (FDM) describe the unknown variable of the flow problem by means of point samples at the node points of a grid of coordinate lines.

In the finite element method (FEM), the continuous variable in the domain is replaced by piecewise continuous functions defined in sub-regions (elements). The approximating function is substituted in differential equations resulting in residuals which are minimized by weighted residuals (Lewis et al. [1]). This process determines the unknown coefficients in the assumed profile.

In the finite volume method (FVM), discretization of the domain is carried out into number of cells. Here the flux entering the face and leaving is used by integrating the differential equations in the cell. This produces algebraic equations. Incidentally, the sub-domain methods (with weighting function of 1) gives the same algebraic equations as the finite volume method.

3.2 Finite Element Method

The finite element method is a numerical tool for determining approximate solutions to a large class of engineering problems. The method was originally developed to study the stresses in complex airframe structures (Clough [2]) and is later extended to the general field of continuum mechanics (Zienkiewicz and Cheung [3]). Recently, heat transfer engineers have sensed that finite element methodology has considerable power and versatility for solving many types of problems encountered in the field of heat transfer and have used this method very successfully to solve many heat transfer problems including heat exchangers. Among the various numerical methods that have evolved over the years, the most commonly used techniques are the finite difference, finite volume and finite element methods. The finite difference is a well-established and conceptually

Compact Heat Exchangers – Analysis, Design and Optimization using FEM and CFD Approach, First Edition. C. Ranganayakulu and K.N. Seetharamu.
© 2018 John Wiley & Sons Ltd. Published 2018 by John Wiley & Sons Ltd.

simple method that requires a point-wise approximation to the governing equations. The model, formed by writing the difference equations for an array of grid points, can be improved by increasing the number of points. Although many heat transfer problems may be solved using finite difference methods (Ozisik and Czisik [4]), as soon as irregular geometries or an unusual specification of boundary conditions are encountered, the finite difference technique becomes difficult to use.

The finite element method (Baker [5]; Bathe [6]; Chandrupatla and Belegundu [7]; Huebner and Thornton [8]; Hughes [9]; Lewis et al. [10]; Rao [11]; Reddy [12]; Segerlind [13]; Zienkiewicz and Morgan [14]; Zienkiewicz and Taylor [15]) considers that the solution region comprises many small, interconnected, sub-regions or elements and gives a piecewise approximation to the governing equations; that is, the complex partial differential equations are reduced to either linear or non-linear simultaneous equations. Thus, the finite element discretization procedure (i.e. dividing the region into a number of smaller regions) reduces the continuum problem, which has an infinite number of unknowns, to one with a finite number of unknowns at specified points referred to as nodes. Since the finite element method allows us to form the elements, or sub-regions, in an arbitrary sense, a close representation of the boundaries of complicated domains is possible. Therefore, in this chapter we intend to present a brief procedure for the finite element method as applied to heat transfer/heat exchanger problems. The solution of a continuum problem by the finite element method is approximated by the following step-by-step process.

1) Discretize the continuum
   Divide the solution region into non-overlapping elements or sub-regions. The finite element discretization allows a variety of element shapes, for example, triangles, quadrilaterals. Each element is formed by the connection of a certain number of nodes. The number of nodes employed to form an element depends on the type of element (or interpolation function).
2) Select interpolation or shape functions
   The next step is to choose the type of interpolation function that represents the variation of the field variable over an element. A number of nodes form an element; the nature and number of unknowns at each node decide the variation of a field variable within the element.
3) Form element equations (formulation)
   Next, we have to determine the matrix equations that express the properties of the individual elements by forming an element left hand side (LHS) matrix and load vector.
4) Assemble the element equations to obtain a system of simultaneous equations
   To find the properties of the overall system, we must assemble all the individual element equations, that is, to combine the matrix equations of each element in an appropriate way such that the resulting matrix represents the behaviour of the entire solution region of the problem.
5) Solve the system of equations
   The resulting set of algebraic equations may now be solved to obtain the nodal values of the field variable, for example, temperature.
6) Calculate the secondary quantities
   From the nodal values of the field variable, for example, temperatures, we can then calculate the secondary quantities, for example, space heat fluxes.

3.2.1 Finite Element Form of the Conduction Equation

The finite element form of the conduction equation is illustrated below for one-dimensional steady-state conduction in a composite slab, as shown in Figure 3.1. If the height and depth of the slab are very large in comparison to the thickness of the slab, then the total heat flow will be constant at any section perpendicular to the heat flow direction, leading to one-dimensional heat flow. Then the energy balance equation for different nodes of the composite slab shown in Figure 3.1 can be written as follows:

For node 1:

$$q''A = k_1 A\frac{(T_1 - T_2)}{L_1} \tag{3.1}$$

For node 2:

$$k_1 A\frac{(T_1 - T_2)}{L_1} = k_2 A\frac{(T_2 - T_3)}{L_2} \tag{3.2}$$

For node 3:

$$k_2 A\frac{(T_2 - T_3)}{L_2} = hA(T_3 - T_\infty) \tag{3.3}$$

In the above equations, q'' represents the heat flux on the left face of the composite slab and h is the heat transfer coefficient between the right face, and the ambient at a uniform temperature T_∞. Equations 3.1 through 3.3 can be rearranged as follows:

$$\frac{k_1 A}{L_1}T_1 - \frac{k_1 A}{L_1}T_2 = q''A \tag{3.4a}$$

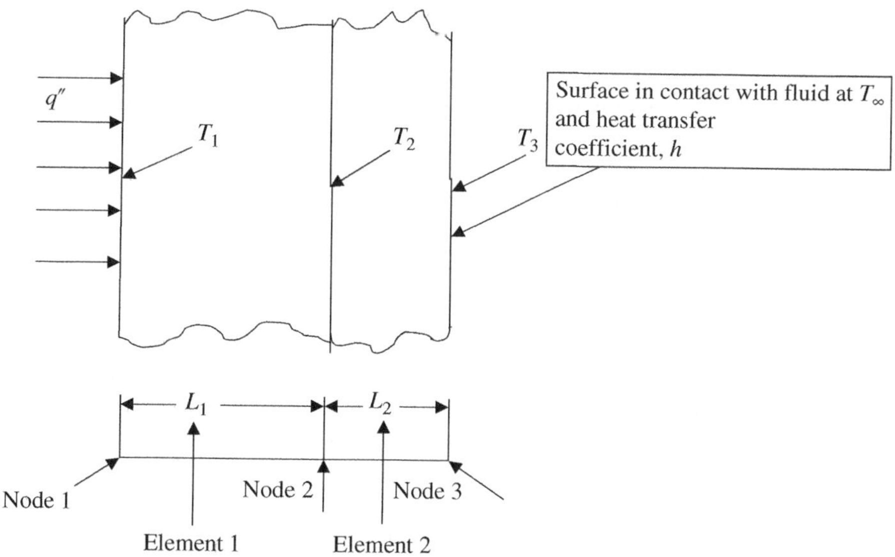

Figure 3.1 Nodes and elements for one-dimensional conduction in a composite slab.

$$-\frac{k_1 A}{L_1}T_1 + \left\{\frac{k_1 A}{L_1} + \frac{k_2 A}{L_2}\right\}T_2 - \frac{k_2 A}{L_2}T_3 = 0 \tag{3.4b}$$

$$-\frac{k_2 A}{L_2}T_2 + \left\{\frac{k_2 A}{L_2} + hA\right\}T_3 = hAT_\infty \tag{3.4c}$$

Equations 3.4a through 3.4c can be recast in the matrix form as

$$\begin{bmatrix} \frac{k_1 A}{L_1} & -\frac{k_1 A}{L_1} & 0 \\ -\frac{k_1 A}{L_1} & \left\{\frac{k_1 A}{L_1} + \frac{k_2 A}{L_2}\right\} & -\frac{k_2 A}{L_2} \\ 0 & -\frac{k_2 A}{L_2} & \left\{\frac{k_2 A}{L_2} + hA\right\} \end{bmatrix} \begin{Bmatrix} T_1 \\ T_2 \\ T_3 \end{Bmatrix} = \begin{Bmatrix} q''A \\ 0 \\ hAT_\infty \end{Bmatrix} \tag{3.5a}$$

Equation 3.5a can be written in a more compact form as

$$[K]\{T\} = \{f\} \tag{3.5b}$$

where $[K]$ is called the global stiffness matrix, $\{T\}$ the vector of unknown temperatures, and $\{f\}$ the global load vector (thermal force vector). Solution of Equation 3.5b gives the temperatures T_1, T_2, and T_3. In the case of heat conduction there is only one degree of freedom at each node as temperature is a scalar quantity. The following important features of Equation 3.5b should be observed.

1) The characteristics of each layer of the composite slab for heat conduction with linear temperature variation in each layer can be written as

$$\frac{kA}{L}\begin{bmatrix} 1 & -1 \\ -1 & 1 \end{bmatrix} \begin{Bmatrix} T_i \\ T_j \end{Bmatrix} = \begin{Bmatrix} Q \\ Q \end{Bmatrix} \tag{3.6}$$

   where Q is the total heat flux and is constant for steady-state conduction.
2) The global stiffness matrix $[K]$ can be obtained by assembling the stiffness matrix of each layer, and the result is a symmetric and positive definite matrix.
3) The load vector consists of known values. The method of assembly can be extended to more than two layers.
4) The effect of heat flux boundary conditions appears only in the loading vector $\{f\}$.
5) The convection effect appears both in the stiffness matrix and the loading vector.
6) The effect of the natural boundary conditions (heat flux and convective conditions) is taken care at the formulation stage itself.

To summarize, if $[K]$ and $\{f\}$ can be formed, then the temperature distribution can be determined by any standard matrix solution procedure. This methodology can be extended to fluid flow network (water flow distribution, electronic cabinets, gas flow distribution in gas turbine blade passages, etc.) and also in electrical network.

3.2.2 Elements and Shape Functions

The fundamental concept in the finite element method is that a continuous function such as temperature (or velocity, pressure, displacement, etc.) in a given domain can be replaced by piecewise continuous functions defined in each of the sub-regions called "elements."

Consider a fin of uniform cross-section whose base is maintained at a uniform temperature T_1. The actual temperature distribution in the fin is replaced by piecewise continuous functions defined in each of the sub-regions (elements) by means of linearly varying temperatures, as shown in Figure 3.2. Since the domain of the fin is discretized into elements, each element can have a linear or quadratic temperature variation. Divide the solution region into non-overlapping elements or sub-regions. The finite element discretization allows a variety of element shapes, for example, triangles, quadrilaterals. Each element is formed by the connection of a certain number of nodes, as shown in Figure 3.3.

The variation of temperature for a typical linear element with end nodes i and j with corresponding temperatures T_i and T_j (Figure 3.4) can be represented by the equation

$$T(x) = \alpha_1 + \alpha_2 x \tag{3.7}$$

where $T(x)$ is the temperature at any location x, and α_1 and α_2 are constants. These can be calculated from the following equations:

$$T_i = \alpha_1 + \alpha_2 x_i \tag{3.8a}$$

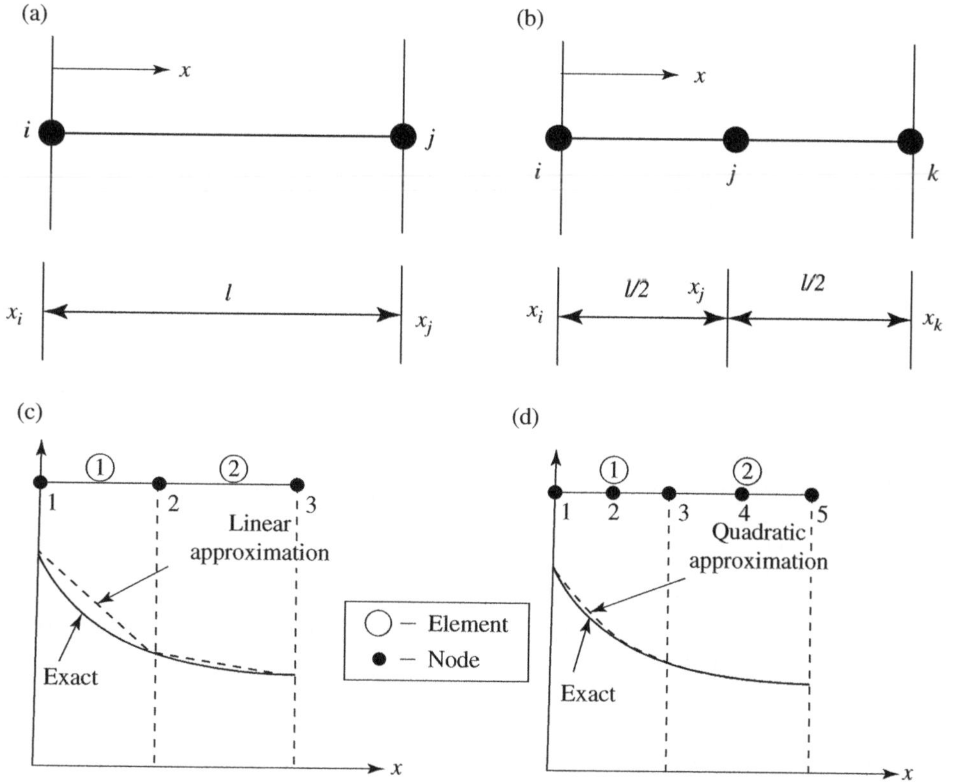

Figure 3.2 Approximation of temperature distribution in a fin by linear elements.

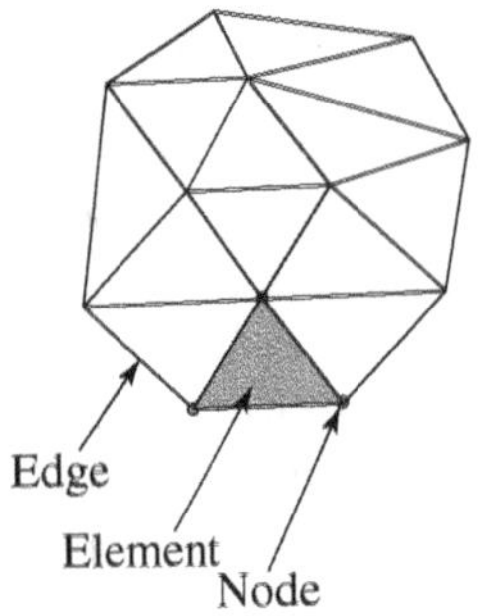

Figure 3.3 Two-dimensional discretization with linear triangular elements.

Figure 3.4 One-dimensional linear elements.

$$T_j = \alpha_1 + \alpha_2 x_j \tag{3.8b}$$

Solving Equations 3.8a and 3.8b we get the values of α_1 and α_2 as

$$\alpha_1 = \frac{(T_i x_j - T_j x_i)}{(x_j - x_i)} \tag{3.9a}$$

$$\alpha_2 = \frac{(T_j - T_i)}{(x_j - x_i)} \tag{3.9b}$$

Substituting these expressions for α_1 and α_2 in Equation 3.7 we get

$$T = T_i \frac{(x_j - x)}{(x_j - x_i)} + T_j \frac{(x - x_i)}{(x_j - x_i)} \tag{3.10}$$

or

$$T = N_i T_i + N_j T_j = [N_i N_j] \begin{pmatrix} T_i \\ T_j \end{pmatrix} \tag{3.11}$$

where N_i and N_j are given by

$$N_i = \frac{(x_j - x)}{(x_j - x_i)}; N_j = \frac{(x - x_i)}{(x_j - x_i)} \tag{3.12}$$

N_i and N_j are called "shape functions" or interpolation functions or basis functions. Equation 3.11 can be written as

$$T = [N]\{T\} \tag{3.13}$$

where T is the temperature anywhere in the element, $[N]$ is the shape function matrix and $\{T\}$ is the vector of unknown temperatures. Equation 3.13 shows that the temperature T at any location within the element can be calculated using the shape functions N_i and N_j

Table 3.1 Properties of linear shape functions.

Item	Node *i*	Node *j*	Arbitrary *x*
Ni	1	0	Between 0 and 1
N_j	0	1	Between 0 and 1
$N_i + N_j$	1	1	1

evaluated at x. The shape functions at different locations within the element are tabulated in Table 3.1.

The shape function assumes the value of unity at the designated node and zeros at all other nodes. It is also observed that the sum of the shape functions in an element is equal to unity anywhere within the element including the boundaries. These are the two essential requirements for the shape functions of any element in one, two or three dimensions. Figure 3.5 shows the variation of the shape functions and their derivatives within a linear element. A typical linear variation of temperature within the element is also shown in the figure. It can be seen that the derivatives of the shape functions are constant within the element.

The temperature gradient is calculated from Equation 3.1 as

$$\frac{dT}{dx} = \frac{dN_i}{dx}T_i + \frac{dN_j}{dx}T_j = \frac{-1}{(x_j - x_i)}T_i + \frac{1}{(x_j - x_i)}T_j$$

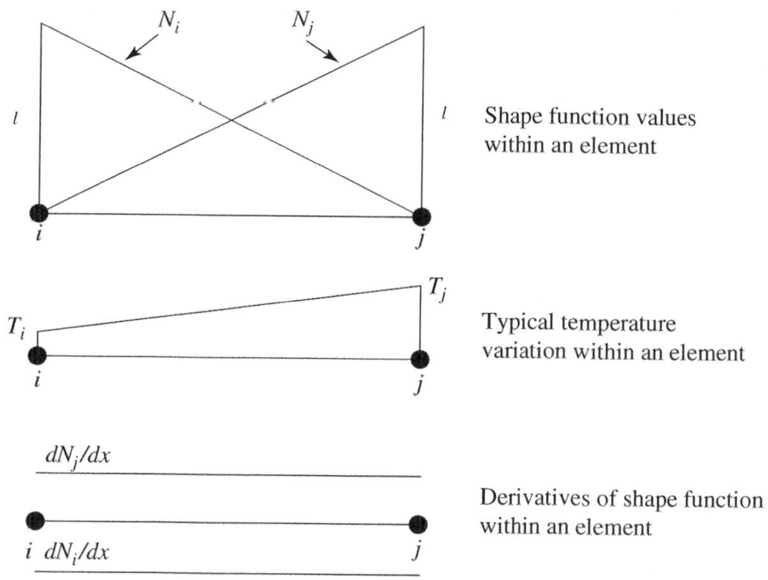

Figure 3.5 Variation of shape function, temperature and derivatives of shape function with a linear element.

or

$$\frac{dT}{dx} = \left[-\frac{1}{l}\ \frac{1}{l}\right]\begin{pmatrix} T_i \\ T_j \end{pmatrix} \tag{3.14}$$

where l is the length of the element and equal to $(x_j - x_i)$. Thus it is observed that the temperature gradient is constant within the element as the temperature variation is linear. Equation 3.14 can be written as

$$g = [B]\{T\} \tag{3.15}$$

where g is the gradient of the field variable T, $[B]$ is the derivative matrix (or strain matrix in structural mechanics) which relates the gradient of the field variable to the nodal values and $\{T\}$ is the temperature vector. The shape function matrix $[N]$ and the derivative matrix $[B]$ are the two important matrices used in the determination of the element properties as we shall see later in this section.

Example 3.1 Calculate the temperature in a 10 cm long bar at a distance of 7 cm from one end where the temperature is 100 °C, with the other end at a temperature of 200 °C. Assume the temperature variation between the two ends of the rod to be linear.

Solution

Known: Nodal temperatures and geometry
Find: Temperature at a location of 7 cm from one end
Schematic:

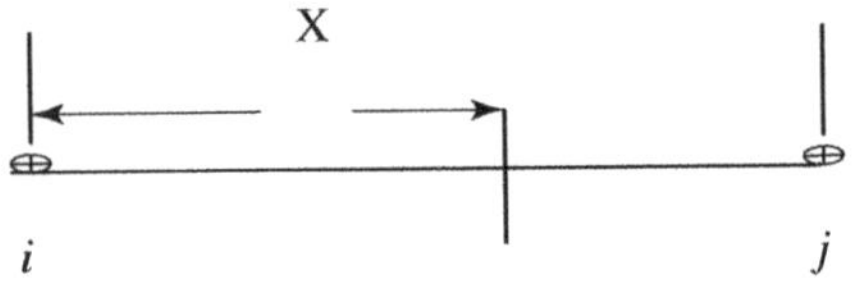

$T_i = 100°C$ $\qquad$ $T = ?$ $T_j = 200°C$

Assumption: Linear variation of temperature in the element
Analysis: From Equation 3.11 the temperature distribution in the rod, considering it as an element, can be written as

$$T = N_i T_i + N_j T_j$$

At $x = 7$ cm,

$$N_i = \frac{(x_j - x)}{(x_j - x_i)} = \frac{(10-7)}{(10-0)} = \frac{3}{10};\quad N_j = \frac{(x - x_i)}{(x_j - x_i)} = \frac{(7-0)}{(10-0)} = \frac{7}{10}$$

It can be verified that

$$N_i + N_j = \frac{3}{10} + \frac{7}{10} = 1$$

Substitution in Equation 3.11 gives

$$T = \frac{3}{10}(100) + \frac{7}{10}(200) = 170°\text{C}$$

Comment: The answer is correct, as the temperature varies linearly from 100 °C at x = 0 to 200 °C at x = 10 cm.

3.2.3 Two-Dimensional Linear Triangular Elements

When one-dimensional approximations are not sufficient, a multidimensional solution procedure needs to be employed. In this section a two-dimensional element is introduced for the first time. The simplest geometric shape that can be employed to approximate irregular surfaces is the triangle, and it is one of the popular elements currently used in finite element calculations. This is partly due to the advances made on unstructured and adaptive mesh generation techniques in recent times (Thompson et al. [16]).

The two-dimensional linear triangular element, also known as a simplex element, is represented by

$$T(x,y) = \alpha_1 + \alpha_2 x + \alpha_3 y \tag{3.16}$$

where the polynomial is linear in x and y and contains three coefficients. Since a linear triangle has three nodes (Figure 3.6), the values of α_1, α_2, and α_3 are determined from the following equations:

$$T_i = \alpha_1 + \alpha_2 x_i + \alpha_3 y_i \tag{3.17a}$$

$$T_j = \alpha_1 + \alpha_2 x_j + \alpha_3 y_j \tag{3.17b}$$

$$T_k = \alpha_1 + \alpha_2 x_k + \alpha_3 y_k \tag{3.17c}$$

The expressions for α_1, α_2, and α_3 are determined from Equation 3.17 as

$$\alpha_1 = \frac{1}{2A}\left[\left(x_j y_k - x_k y_j\right)T_i + \left(x_k y_i - x_i y_k\right)T_j + \left(x_i y_j - x_j y_i\right)T_k\right] \tag{3.18a}$$

$$\alpha_2 = \frac{1}{2A}\left[\left(y_j - y_k\right)T_i + \left(y_k - y_i\right)T_j + \left(y_i - y_j\right)T_k\right] \tag{3.18b}$$

Figure 3.6 A linear triangular element.

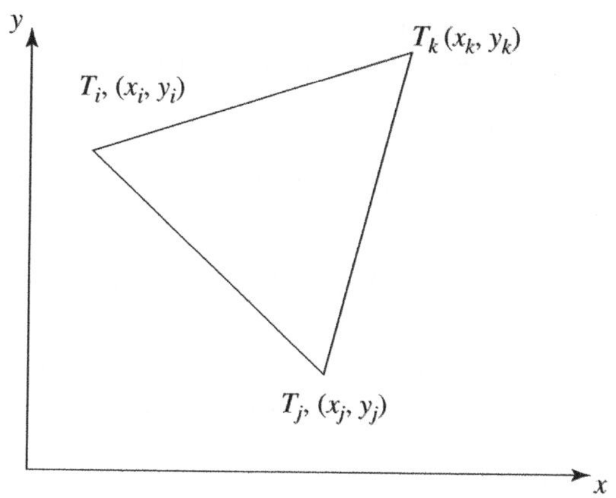

$$\alpha_3 = \frac{1}{2A}\left[\left(x_k - x_j\right)T_i + \left(x_i - x_k\right)T_j + \left(x_j - x_i\right)T_k\right] \tag{3.18c}$$

where A is the area of the triangle given by

$$2A = \begin{vmatrix} 1 & x_i & y_i \\ 1 & x_j & y_j \\ 1 & x_k & y_k \end{vmatrix} = \left(x_i y_j - x_j y_i\right) + \left(x_k y_i - x_i y_k\right) + \left(x_j y_k - x_k y_j\right) = a_k + a_j + a_i \tag{3.19}$$

Substituting the values of α_1, α_2, and α_3 into Equation 3.17 and collating the coefficients of T_i, T_j, and T_k, we get

$$T = N_i T_i + N_j T_j + N_k T_k = \left[N_i\, N_j\, N_k\right]\begin{Bmatrix} T_i \\ T_j \\ T_k \end{Bmatrix} \tag{3.20}$$

where

$$N_i = \frac{1}{2A}\left(a_i + b_i x + c_i y\right) \tag{3.21a}$$

$$N_j = \frac{1}{2A}\left(a_j + b_j x + c_j y\right) \tag{3.21b}$$

$$N_k = \frac{1}{2A}\left(a_k + b_k x + c_k y\right) \tag{3.21c}$$

and

$$a_i = x_j y_k - x_k y_j;\ b_i = y_j - y_k;\ c_i = x_k - x_j \tag{3.22a}$$

$$a_j = x_k y_i - x_i y_k;\ b_j = y_k - y_i;\ c_j = x_i - x_k \tag{3.22b}$$

$$a_k = x_i y_j - x_j y_i;\ b_k = y_i - y_j;\ c_k = x_j - x_i \tag{3.22c}$$

If N_i at node i where the coordinates are (x_i, y_i) is evaluated, then we get

$$\left(N_i\right)_i = \frac{1}{2A}\left[\left(x_j y_k - x_k y_j\right) + \left(y_j - y_k\right)x_i + \left(x_k - x_j\right)y_i\right] = \frac{2A}{2A} = 1 \tag{3.23}$$

Similarly, it can be readily verified that $(N_j)_i = (N_k)_i = 0$. Thus we see that the shape functions have a value of unity at the designated vertex, and zero at other vertices of the triangular element. It can be verified that

$$N_i + N_j + N_k = 1 \tag{3.24}$$

everywhere in the element, including the boundaries. The gradients of the temperature T are given by

$$\frac{\partial T}{\partial x} = \frac{\partial N_i}{\partial x}T_i + \frac{\partial N_j}{\partial x}T_j + \frac{\partial N_k}{\partial x}T_k = \frac{b_i}{2A}T_i + \frac{b_j}{2A}T_j + \frac{b_k}{2A}T_k \tag{3.25a}$$

$$\frac{\partial T}{\partial y} = \frac{\partial N_i}{\partial y}T_i + \frac{\partial N_j}{\partial y}T_j + \frac{\partial N_k}{\partial y}T_k = \frac{c_i}{2A}T_i + \frac{c_j}{2A}T_j + \frac{c_k}{2A}T_k \tag{3.25b}$$

or

$$\{g\} = \begin{Bmatrix} \dfrac{\partial T}{\partial x} \\ \dfrac{\partial T}{\partial y} \end{Bmatrix} = \frac{1}{2A}\begin{bmatrix} b_i & b_j & b_k \\ c_i & c_j & c_k \end{bmatrix}\begin{Bmatrix} T_i \\ T_j \\ T_k \end{Bmatrix} = [B]\{T\} \tag{3.25c}$$

It should be noted that both $\partial T/\partial x$ and $\partial T/\partial y$ are constants within an element as b_i, b_j, b_k and c_i, c_j and c_k are constants for a given triangle. Hence the heat fluxes q_x and q_y are also constants within a linear triangular element. Since the temperature varies linearly within an element, it is possible to draw the isotherms within a linear triangle and this is illustrated in the following example.

Example 3.2 Calculate the temperature T and the heat fluxes q_x and q_y at the location (2.0, 1.0) within an element for the data given below if the thermal conductivity of material is 2 W/(m·K). Draw the isothermal line for 60 °C in the triangle.

Node	*x* (cm)	*y* (cm)	*T* (°C)
i	0.0	0.0	50.0
j	4.0	0.0	70.0
k	0.0	2.5	100.0

Solution:

Known: Nodal coordinates and corresponding temperatures

Find: Temperature and heat fluxes q_x and q_y at location (2.0,1.0)

Schematic: Isotherm within a linear triangular element (Figure 3.7)

Assumptions: Linear variation of temperature in the triangle

Analysis: The temperature at any location within the triangle is given by Equation 3.20. The shape functions are calculated using Equation 3.21 with the x- and y-coordinates as given in the problem. The result is

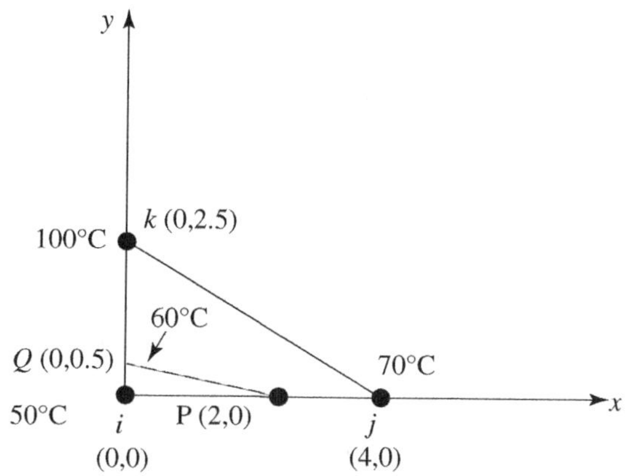

Figure 3.7 Isotherm within a linear triangular element.

$$N_i = \frac{1}{10}; N_j = \frac{5}{10}; N_k = \frac{4}{10}$$

The substitution of the nodal temperatures and the above shape function values into Equation 3.13 results in the temperature of the point (2.0, 1.0) being

$$T = N_i T_i + N_j T_j + N_k T_k = \frac{1}{10}(50) + \frac{5}{10}(70) + \frac{4}{10}(100) = 80°\text{C}$$

The components of heat flux in the x and y directions are calculated as

$$\begin{Bmatrix} q_x \\ q_y \end{Bmatrix} = -\frac{k}{2A}\begin{bmatrix} b_i & b_j & b_k \\ c_i & c_j & c_k \end{bmatrix}\begin{Bmatrix} T_i \\ T_j \\ T_k \end{Bmatrix} = -\frac{2}{10}\begin{Bmatrix} 50 \\ 200 \end{Bmatrix}$$

The position of the 60 °C isotherm may be obtained from Figure 3.7. From the given temperature values, it is clear that one 60 °C point lies on the side ij (point P) and another lies on the side ik (point Q). It should be noted that the temperature varies linearly along these sides, that is, temperature is directly proportional to distance.

In order to determine the location of P on ij, we have the following linear relation between the distances and temperature values:

$$\frac{60.0 - 50.0}{70.0 - 50.0} = \frac{\sqrt{(x_p - x_i)^2 + (y_p - y_i)^2}}{\sqrt{(x_j - x_i)^2 + (y_j - y_i)^2}}$$

From the data given, it is clear that the y-coordinates on the ij side are equal to zero and thus the above equation is simplified to

$$\frac{10.0}{20.0} = \frac{(x_p - x_i)}{(x_j - x_i)}$$

which results in $x_p = 2.0$ cm. The location of Q along ik can be determined in a similar fashion as

$$\frac{60.0 - 50.0}{100.0 - 50.0} = \frac{(y_Q - y_i)}{(y_j - y_i)}$$

which gives $y_Q = 0.5$ cm. The x-coordinate of this point is zero.

The line joining P and Q will be the 60 °C isothermal (Figure 3.7). It should be noted that the same principle can be used for arbitrary triangles.

3.2.3.1 Area Coordinates

An area, or natural, coordinate system will now be introduced for triangular elements, in order to simplify the solution process. Let us consider a point P within a triangle at any location as shown in Figure 3.8 The local coordinates L_i, L_j, and L_k of this point can be established by calculating appropriate non-dimensional distances or areas. For example, L_i is defined as the ratio of the perpendicular distance from point P to the side jk (OP) to the perpendicular distance of point i from the side jk (QR). Thus,

Figure 3.8 Area coordinates of a triangular element.

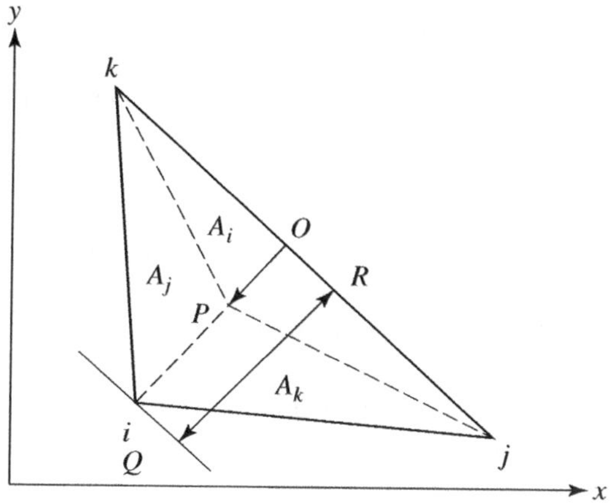

$$L_i = OP/QR \tag{3.26}$$

Similarly, L_j and L_k are also defined. The value of L_i is also equal to the ratio of the area A_i (opposite to node i) to the total area of the triangle, that is,

$$L_i = \frac{A_i}{A} = \frac{0.5\,(OP)(jk)}{0.5\,(QR)(jk)} = \frac{OP}{QR} \tag{3.27}$$

Thus, the local coordinate L_i varies from 0 on the side jk to 1 at the node i. From Figure 3.8 it is obvious that

$$A_i + A_j + A_k = A \tag{3.28}$$

or

$$\frac{A_i}{A} + \frac{A_j}{A} + \frac{A_k}{A} = 1 \tag{3.29}$$

Therefore

$$L_i + L_j + L_k = 1 \tag{3.30}$$

The relationships between the (x, y) coordinates and the natural, or area coordinates are given by

$$x = L_i x_i + L_j x_j + L_k x_k \tag{3.31}$$

and

$$y = L_i y_i + L_j y_j + L_k y_k \tag{3.32}$$

From Equations 3.30–3.32, the following relations for the local coordinates can be derived:

$$L_i = \frac{1}{2A}\left(a_i + b_i x + c_i y\right) \tag{3.33a}$$

$$L_j = \frac{1}{2A}\left(a_j + b_j x + c_j y\right) \tag{3.33b}$$

$$L_k = \frac{1}{2A}(a_k + b_k x + c_k y) \tag{3.33c}$$

where the constants a, b and c are defined in Equation 3.22. Comparing with Equation 3.21 it is clear that

$$L_i = N_i,\ L_j = N_j,\ L_k = N_k \tag{3.34}$$

Thus, the local or area coordinates in a triangle are the same as the shape functions for a linear triangular element. In general, the local coordinates and shape functions are the same for linear elements, irrespective of whether they are of one, two or three dimensions.

For a two-dimensional linear triangular element, with local coordinates L_i, L_j, and L_k, we have a simple formula for integration over the triangle:

$$\int_0^A N_i^a N_j^b N_k^c \, dA = \int_0^A L_i^a L_j^b L_k^c \, dA = \frac{a!b!c!}{(a+b+c+2)!} 2A \tag{3.35}$$

where A is the area of a triangle. Note that L_i, L_j and L_k happen to be the shape functions for a linear triangular element. Example 3.2 can also be solved using the local coordinates via Equations 3.30 and 3.33, that is, on substituting the x and y coordinates of the three points of the triangle into Equation 3.33, we obtain

$$L_j = \frac{x}{4} \tag{3.36}$$

$$L_k = \frac{y}{2.5} \tag{3.37}$$

From Equation 3.30, we get

$$L_i = 1 - \left(\frac{x}{4}\right) - \left(\frac{y}{2.5}\right) \tag{3.38}$$

At (x, y) = (2, 1), we have

$$L_i = 0.1 = N_i \tag{3.39a}$$

$$L_j = 0.5 = N_j \tag{3.39b}$$

$$L_k = 0.4 = N_k \tag{3.39c}$$

Note that these local coordinates are exactly the same as the shape function values calculated in Example 3.2.

3.2.4 Formulation for the Heat Conduction Equation

In many practical situations, finding the temperature in a solid body is of vital importance in terms of the maximum allowable temperature, for example, the maximum allowable displacement in semiconductor devices, the maximum allowable thermal stress in steam and gas turbines or the maximum number of repeated thermal cycles in fatigue-dominated problems. In this section, we shall give the derivation of the finite element equations, both by the variational method and the Galerkin method (Segerlind [13]) for the three-dimensional heat conduction equation of stationary systems under steady-state conditions.

The governing differential equation is

$$\frac{\partial}{\partial x}\left(k_x\frac{\partial T}{\partial x}\right)+\frac{\partial}{\partial y}\left(k_y\frac{\partial T}{\partial y}\right)+\frac{\partial}{\partial z}\left(k_z\frac{\partial T}{\partial z}\right)+G=0 \tag{3.40}$$

with the following boundary conditions

$$T=T_b \text{ on surface } S_1 \tag{3.41a}$$

$$\left(k_x\frac{\partial T}{\partial x}\tilde{l}\right)+\left(k_y\frac{\partial T}{\partial y}\tilde{m}\right)+\left(k_z\frac{\partial T}{\partial z}\tilde{n}\right)+q=0 \text{ on surface } S_2 \tag{3.41b}$$

$$\left(k_x\frac{\partial T}{\partial x}\tilde{l}\right)+\left(k_y\frac{\partial T}{\partial y}\tilde{m}\right)+\left(k_z\frac{\partial T}{\partial z}\tilde{n}\right)+h(T-T_a)=0 \text{ on surface } S_3 \tag{3.41c}$$

where $\tilde{l}$, $\tilde{m}$, and $\tilde{n}$ are surface normals, h is the heat transfer coefficient, k is the thermal conductivity and q is the heat flux.

3.2.4.1 Variational Approach

The variational integral, I, corresponding to the above differential equation with its boundary conditions is given by

$$I(T)=\frac{1}{2}\int_{\Omega}\left[k_x\left(\frac{\partial T}{\partial x}\right)^2+k_y\left(\frac{\partial T}{\partial y}\right)^2+k_z\left(\frac{\partial T}{\partial z}\right)^2-2GT\right]d\Omega+\int_{S_2}q\,T\,dS+\int_{S_3}\frac{1}{2}h(T-T_a)^2dS \tag{3.42}$$

The given domain is divided into n number of finite elements with each element having r nodes. The temperature is expressed in each element by

$$T^e=\sum_{i=1}^{r}N_iT_i=[N]\{T\} \tag{3.43}$$

where $[N]=[N_i,N_j,\ldots,N_r]$ = shape function matrix and

$$\{T\}^T=\{T_i\ T_j\ldots\ldots\ldots\ldots\ldots T_r\} \tag{3.44}$$

is the transpose of vector of nodal temperatures.

The finite element solution to the problem involves selecting the nodal values of T so as to make the function $I(T)$ stationary. In order to make $I(T)$ stationary with respect to the nodal values of T, we require that

$$\delta I(T)=\sum_{i=1}^{n}\frac{\partial I}{\partial T_i}=0 \tag{3.45}$$

where n is the total number of discrete values of T assigned to the solution domain. Since T_i are arbitrary, Equation 3.45 holds good only if

$$\frac{\partial I}{\partial T_i}=0 \text{ for } i=1,2,\ldots,n \tag{3.46}$$

The functional $I(T)$ can be written as a sum of individual functions, defined for the assembly of elements, only if the shape functions giving piecewise representation of T obey certain continuity and compatibility conditions. These conditions will be discussed later in the text. Now

$$I(T) = \sum_{e=1}^{n} I^e T^e \tag{3.47}$$

Thus, instead of working with a functional defined over the whole solution region, our attention is now focused on a functional defined for the individual elements. Hence,

$$\delta I = \sum_{e=1}^{n} \delta I^e = 0 \tag{3.48}$$

where the variation in I^e is taken only with respect to the r nodal values associated with the element e, that is,

$$\left\{\frac{\delta I^e}{\delta T}\right\} = \frac{\delta I^e}{\delta T_j} = 0 \text{ with } j = 1,2,\ldots,r \tag{3.49}$$

Equation 3.43 comprises a set of r equations that characterize the behaviour of the element e. The fact that we can represent the functional for the assembly of elements as a sum of the functional for all individual elements provides the key to formulating individual element equations from a variational principle. The complete set of assembled finite element equations for the problem is obtained by adding all the derivatives of I, as given by Equation 3.49, for all the elements. We can write the complete set of equations as

$$\frac{\partial I}{\partial T_i} = \sum_{e=1}^{n} \frac{\partial I^e}{\partial T_i} = 0 \text{ with } i = 1,2,\ldots,m \tag{3.50}$$

The problem is complete when the M set of equations are solved simultaneously for the M nodal values of T. We now give the details for formulating the individual finite element equations from a variational principle:

$$\frac{1}{2}\int_{\Omega}\left[k_x\left(\frac{\partial T^e}{\partial x}\right)^2 + k_y\left(\frac{\partial T^e}{\partial y}\right)^2 + k_z\left(\frac{\partial T^e}{\partial z}\right)^2 - 2GT^e\right]d\Omega + \int_{S_{2e}} q\,T^e dS + \int_{S_{3e}} \frac{1}{2}h(T^e - T_a)^2 dS = I^e \tag{3.51}$$

with

$$\begin{aligned} T^e &= [N]\{T\} = [N_1, N_2, N_3, \ldots, N_r][T_1, T_2, \ldots, T_r]^T \\ &= N_1T_1 + N_2T_2 + N_3T_3 + \ldots + N_rT_r \end{aligned} \tag{3.52}$$

and

$$\frac{\partial T^e}{\partial T_1} = N_1; \quad \frac{\partial T^e}{\partial T_2} = N_2; \quad \frac{\partial T^e}{\partial T_3} = N_3; \quad \ldots \quad \frac{\partial T^e}{\partial T_r} = N_r \tag{3.53}$$

or

$$\frac{\partial T^e}{\partial \{T\}} = \begin{Bmatrix} N_1 \\ N_2 \\ \vdots \end{Bmatrix} = \{N\}^T = [N]^T \tag{3.54}$$

The gradient matrix is written as

$$\{g\} = \begin{Bmatrix} \dfrac{\partial T^e}{\partial x} \\ \dfrac{\partial T^e}{\partial y} \\ \dfrac{\partial T^e}{\partial z} \end{Bmatrix} = \begin{bmatrix} \dfrac{\partial N_1}{\partial x} & \dfrac{\partial N_2}{\partial x} & \dfrac{\partial N_3}{\partial x} & & \dfrac{\partial N_r}{\partial x} \\ \dfrac{\partial N_1}{\partial y} & \dfrac{\partial N_2}{\partial y} & \dfrac{\partial N_3}{\partial y} & \cdots & \dfrac{\partial r}{\partial y} \\ \dfrac{\partial N_1}{\partial z} & \dfrac{\partial N_2}{\partial z} & \dfrac{\partial N_3}{\partial z} & & \dfrac{\partial N_r}{\partial z} \end{bmatrix} \begin{Bmatrix} T_1 \\ T_2 \\ T_3 \\ \vdots \\ T_r \end{Bmatrix} [B]\{T\} \tag{3.55}$$

Consider

$$\{g\}^T [D]\{g\} = \left\{ \frac{\partial T^e}{\partial x} \; \frac{\partial T^e}{\partial y} \; \frac{\partial T^e}{\partial z} \right\} \begin{bmatrix} k_x & 0 & 0 \\ 0 & k_y & 0 \\ 0 & 0 & k_z \end{bmatrix} \begin{Bmatrix} \dfrac{\partial T^e}{\partial x} \\ \dfrac{\partial T^e}{\partial y} \\ \dfrac{\partial T^e}{\partial z} \end{Bmatrix}$$
$$= k_x \left(\frac{\partial T^e}{\partial x} \right)^2 + k_y \left(\frac{\partial T^e}{\partial y} \right)^2 + k_z \left(\frac{\partial T^e}{\partial z} \right)^2 \tag{3.56}$$

Substituting into Equation 3.45, we have

$$I^e = \frac{1}{2} \int_{\Omega} \left[\{g\}^T [D] \{g\} - 2GT^e \right] d\Omega + \int_{S_{2e}} q\, T^e dS + \int_{S_{3e}} \frac{1}{2} h(T^e - T_a)^2 dS \tag{3.57}$$

From Equation 3.55 we can substitute

$$\{g\}^T [D]\{g\} = [T]^T [B]^T [D][B]\{T\} \tag{3.58}$$

and minimizing the integral, we have (employing Equation 3.54)

$$\frac{\partial I^e}{\partial \{T\}} = \int \frac{1}{2}\, 2\,[B]^T [D][B]\{T\}\, d\Omega - \int \frac{1}{2}\, 2\,G[N]^T \{T\}\, d\Omega$$
$$+ \int_{S_{2e}} q[N]^T \{T\}\, dS + \int_{S_{3e}} h[N]^T \{T\}\, dS - \int_{S_{3e}} h[N]^T T_e dS = 0 \tag{3.59}$$

The above equation can be written in a compact form as

$$[K]\{T\} = \{f\} \tag{3.60}$$

where

$$[K] = \int_{\Omega} [B]^T [D][B] d\Omega + \int_{S_{3e}} h[N]^T \{N\} dS$$

and

$$\{f\} = \int G[N]^T d\Omega - \int_{S_2} q[N]^T dS + \int_{S_3} hT_a[N]^T dS \tag{3.61}$$

Equation 3.59 forms the backbone of the calculation method for finite element analysis of heat conduction problems. It can be easily noted that when there is no heat generation within an element ($G = 0$), the corresponding term disappears. Similarly, for an insulated boundary (i.e. $q = 0$ or $h = 0$) the corresponding term again disappears. Thus, for an insulated boundary, we do not have to specify any contribution, but leave it unattended. In this respect, this is a great deal more convenient as compared to the finite-difference method, where nodal equations have to be written for insulated boundaries.

3.2.4.2 Galerkin Method

The method requires that the following expression be satisfied:

$$\int w_k \boldsymbol{L}(\bar{T}) d\Omega = 0 \tag{3.62}$$

where the weight w_k is replaced by the shape functions at nodes, $N_k(x)$

$$\int N_k \left\{ \frac{\partial}{\partial x}\left(k_x \frac{\partial \bar{T}}{\partial x}\right) + \frac{\partial}{\partial y}\left(k_y \frac{\partial \bar{T}}{\partial y}\right) + \frac{\partial}{\partial z}\left(k_z \frac{\partial \bar{T}}{\partial z}\right) + G \right\} d\Omega = 0 \tag{3.63}$$

Integration by parts is often essential when dealing with second-order derivatives. Using Green's lemma we can rewrite the second derivatives in two parts as

$$N_k \int \frac{\partial}{\partial x}\left(k_x \frac{\partial \bar{T}}{\partial x}\right) d\Omega = \int_S N_k \left(k_x \frac{\partial \bar{T}}{\partial x}\right) dS - \int_\Omega \frac{\partial N_k}{\partial x} k_x \frac{\partial N_m}{\partial x} \{\overline{T_m}\} d\Omega \tag{3.64}$$

where m represents nodes. With the boundary conditions (Equation 3.41), we can rewrite Equation 3.63 as

$$\begin{aligned} &-\int_\Omega \left(k_x \frac{\partial N_k}{\partial x}\frac{\partial N_m}{\partial x} + k_y \frac{\partial N_k}{\partial y}\frac{\partial N_m}{\partial y} + k_z \frac{\partial N_k}{\partial z}\frac{\partial N_m}{\partial z} \right) \{\overline{T_m}\} d\Omega \\ &+ \int_\Omega G N_k d\Omega - \int_S N_k q \, dS + \int_S h N_k N_m \{\overline{T_m}\} dS + \int_S h T_a N_k \{\overline{T_m}\} dS = 0 \end{aligned} \tag{3.65}$$

Now collecting the coefficients of the nodal variables $\{T_m\}$, we get

$$[K]\{T\} = \{f\} \tag{3.66}$$

or

$$[K_{km}]\{T_m\} = \{f_k\}$$

where

$$K_{km} = \int_\Omega \left(k_x \frac{\partial N_k}{\partial x}\frac{\partial N_m}{\partial x} + k_y \frac{\partial N_k}{\partial y}\frac{\partial N_m}{\partial y} + k_z \frac{\partial N_k}{\partial z}\frac{\partial N_m}{\partial z} \right) d\Omega + \int_S h N_k N_m \, dS \tag{3.67}$$

and

$$f_k = \int_{\Omega} G N_k d\Omega - \int_{S} N_k \, q \, dS + \int_{S} h T_a N_k dS \tag{3.68}$$

It may be observed that Equations 3.59 and 3.66 are identical, which substantiates the fact that both variational and Galerkin methods give the same result because there exist a classical variational integral for the heat conduction equation.

3.2.5 Requirements for Interpolation Functions

The procedure for formulating the individual element equations from a variational (Rao [11]) principle and the assemblage of these equations relies on the assumption that the interpolation functions satisfy the following requirements. This arises from the need to ensure that Equation 3.53 holds and that our approximate solution converges to the correct solution when we use an increasing number of elements, that is, when we refine the mesh.

Compatibility: At element interfaces, the field variable T and any of its partial derivatives up to one order less than the highest-order derivative appearing in $I(T)$ must be continuous.

Completeness: All uniform states of T and its partial derivatives up to the highest order appearing in $I(T)$ should have representation in T, when in the limit the element size decreases to zero.

If the field variables are continuous at the element interfaces, then we have C° continuity. If, in addition, the first derivatives are continuous, we have C^1 continuity, and if the second derivatives are continuous, then we have C^2 continuity, and so on. If the functions appearing in the integrals of the element equations contain derivatives up to the $(r + 1)$th order, then to have a rigorous assurance of convergence as the element size decreases, we must satisfy the following requirements.

For compatibility: At the element interfaces, we must have C^r continuity.

For completeness: Within an element, we must have C^{r+1}continuity.

These requirements will hold regardless of whether the element equations (integral expressions) were derived using the variation method, the Galerkin method, the energy balance methods, or any other method yet to be devised. These requirements govern the selection of proper interpolation functions depending on the order of the differential equation. Thus, for a conduction heat transfer problem, the highest derivative in I is of the first order. Thus, the shape function selected should provide for the continuity of temperature at the interface between two elements and should also ensure the continuity of temperature and heat flux within each element.

In addition to the requirements of continuity of the field variable and convergence to the correct solution as the element size reduces, we require that the field variable representation (polynomials used) within an element remain unchanged under a linear transformation from one Cartesian coordinate system to another. Polynomials that exhibit this invariance property are said to possess "geometric isotropy." Clearly, we cannot expect a realistic approximation if our field variable representation changes with respect to a movement in origin, or in the orientation of the coordinate system. Hence, the need to ensure geometric isotropy in our polynomial interpolation functions is apparent. Fortunately, we have two simple guidelines that allow us to construct polynomial series with geometric isotropy. These are as follows:

1) Polynomials of order n that are complete, that is, those that contain all terms have geometric isotropy. The triangle family satisfies this condition whether it be a linear, quadratic or cubic form.
2) Polynomials of order n that are incomplete yet contain the appropriate terms to preserve "symmetry" have geometric isotropy. We neglect only these terms that occur in symmetric pairs that is (x^3, y^3), (x^2y, xy^2), and so on.

Example: For an eight-node element, the following polynomial, P, satisfies geometric isotropy, that is,

$$P(x,y) = \alpha_1 + \alpha_2 x + \alpha_3 y + \alpha_4 x^2 + \alpha_5 xy + \alpha_6 y^2 \tag{3.69}$$

with either $\alpha_7 x^3 + \alpha_8 y^3$ or $\alpha_7 x^2 y + \alpha_8 y^2 x$ added to it.

Example 3.3 It is important to consider a numerical problem for illustrating the theory presented. For this purpose, we consider again a fin problem as shown in Figure 3.9. The linear variation for the temperature within each finite element is assumed. We shall derive the element equations from the most general formulation given earlier and determine the temperature distribution, heat dissipation and the efficiency of the fin assuming that the tip is insulated.

Solution

Known:

Thermal conductivity of the material, geometry of the fin and the boundary conditions

Find:

1) Steady state temperature distribution in the fin
2) Heat dissipation capability of the fin
3) Efficiency of the fin

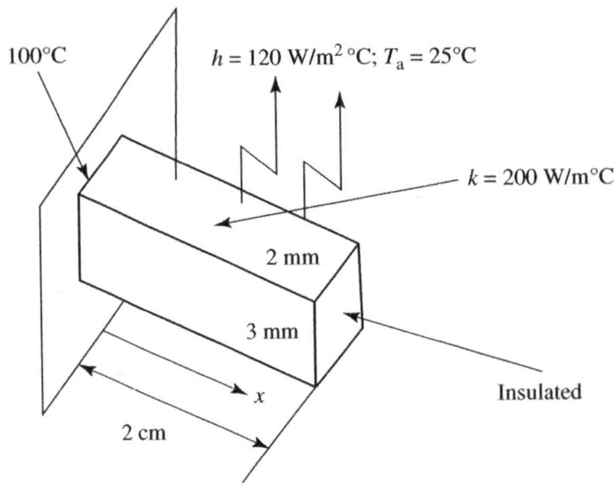

Figure 3.9 Heat transfer from a rectangular fin.

Table 3.2 Element and node numbers of linear one-dimensional elements.

Element No.	Node *i*	Node *j*
1	1	2
2	2	3
e	i	j
n	n	$n+1$

Schematic: Figure 3.9
Assumptions: Fin tip is insulated
Analysis: (See Table 3.2)

Since we are using linear elements, the element will only have two nodes. First, we divide the given length of the fin into number of divisions – say n elements. Therefore, we will have $(n + 1)$ nodes to represent the fin as shown in Table 3.2. The variation of temperature in the elements is linear. Hence,

$$T = N_i T_i + N_j T_j \tag{3.70}$$

and the first derivative is given by

$$\begin{aligned} \frac{dT}{dx} &= \frac{dN_i}{dx}T_i + \frac{dN_j}{dx}T_j \\ &= -\frac{1}{l}T_i + \frac{1}{l}T_j \end{aligned} \tag{3.71}$$

that is, the gradient matrix is

$$g = \frac{dT}{dx} = \left[\frac{1}{-l}\ \frac{-1}{l}\right]\begin{pmatrix} T_i \\ T_j \end{pmatrix} = [B]\{T\} \tag{3.72}$$

With the above relationships, we can write the relevant element matrices as follows:

$$[K]_e = \int_0^l \frac{1}{l}\begin{pmatrix} -1 \\ 1 \end{pmatrix}[k_x]\frac{1}{l}[-1 \quad 1]A\,dx + \int h \begin{pmatrix} N_i \\ N_j \end{pmatrix}[N_i \quad N_j]P dx \tag{3.73}$$

where A is the cross-sectional area of the fin and P is the perimeter of the fin from which convection takes place. Note that $[D] = k_x$ for one-dimensional problems. Rearranging Equation 3.73, we have

$$[K]_e = \int_0^l \frac{Ak_x}{l^2}\begin{bmatrix} 1 & -1 \\ -1 & 1 \end{bmatrix} dx + \int hP \begin{bmatrix} N_i^2 & N_iN_j \\ N_iN_j & N_j^2 \end{bmatrix} dx \tag{3.74}$$

Here $N_i = L_i$ and $N_j = L_j$, which is generally true for all linear elements. Hence, we can make use of the formula

$$\int_0^l L_i^a L_j^b \, dl = \frac{a!b!l}{(a+b+1)!} \tag{3.75}$$

For example,

$$\int_0^l N_i^2 \, dl = \int_0^l L_i^2 \, dl = \frac{2!0!}{(2+0+1)!} = \frac{l}{3} \tag{3.76}$$

and other terms can be similarly integrated.

If A, k_x, P and h are all assumed to be constant throughout the element (Figure 3.9), we obtain the following $[K]$ matrix:

$$[K]_e = \frac{AK_x}{l}\begin{bmatrix} 1 & -1 \\ -1 & 1 \end{bmatrix} + \frac{hPl}{6}\begin{bmatrix} 2 & 1 \\ 1 & 2 \end{bmatrix} \tag{3.77}$$

Let us next consider the thermal loading. From Equation 3.68, we can write

$$\{f\}_e = \frac{GAl}{2}\begin{Bmatrix} 1 \\ 1 \end{Bmatrix} - \frac{qPl}{2}\begin{Bmatrix} 1 \\ 1 \end{Bmatrix} + \frac{hT_aPl}{2}\begin{Bmatrix} 1 \\ 1 \end{Bmatrix} \tag{3.78}$$

In this case, the loads are distributed equally between the two nodes, which is a general characteristic of linear elements.

The solution of the given problem may be found by substitution of the numerical values.

First let us consider a one-element solution for the case where $l = 2$ cm, as shown in Figure 3.10. The element stiffness matrix is

$$[K]_e = \frac{Ak_x}{l}\begin{bmatrix} 1 & -1 \\ -1 & 1 \end{bmatrix} + \frac{hPl}{6}\begin{bmatrix} 2 & 1 \\ 1 & 2 \end{bmatrix}$$

$$= \begin{bmatrix} 0.06 & -0.06 \\ -0.06 & 0.06 \end{bmatrix} + \begin{bmatrix} 0.008 & 0.004 \\ 0.004 & 0.008 \end{bmatrix} = \begin{bmatrix} 0.068 & -0.056 \\ -0.056 & 0.068 \end{bmatrix}$$

and the loading term is given by

$$\{f\} = \frac{hPlT_a}{2}\begin{Bmatrix} 1 \\ 1 \end{Bmatrix} = \begin{Bmatrix} 0.3 \\ 0.3 \end{Bmatrix}$$

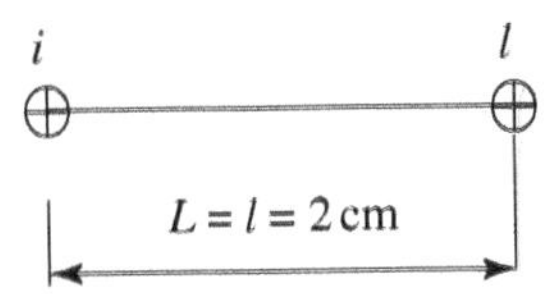

Figure 3.10 Heat transfer from a rectangular fin. One linear element.

Note that T_a is in °C as h is expressed in W/m^2°C.

Since only one element is employed, no assemblage of element contribution is necessary. Thus, the simultaneous equation system may be written as

$$\begin{bmatrix} 0.068 & -0.056 \\ -0.056 & 0.068 \end{bmatrix} \begin{Bmatrix} T_1 \\ T_2 \end{Bmatrix} = \begin{Bmatrix} 0.3 \\ 0.3 \end{Bmatrix}$$

We now incorporate the known base temperature of 100 °C at node 1. It is done in such a way that the symmetry of the [K] matrix is retained. This is essential if a symmetric matrix solution procedure is employed in the solution of the simultaneous equations. The following steps give a typical implementation procedure for the temperature boundary condition:

1) The diagonal element of the first row is assigned a value of 1 and the remaining elements on that row are zero.
2) Replace the first row value of the loading vector by the known value of T_1, that is, 100.
3) In order to retain the symmetry, the first term of the second row in the [K] matrix is transferred to the right-hand side and replaced with a zero value as given below:

$$\begin{bmatrix} 1.0 & 0.0 \\ 0.0 & 0.068 \end{bmatrix} \begin{Bmatrix} T_1 \\ T_2 \end{Bmatrix} = \begin{Bmatrix} 100.0 \\ 0.30 + 0.056 \times 100 \end{Bmatrix}$$

The equation to be solved is

$$0.068T_2 = 0.3 + 0.056(100)$$

Therefore, the solution is $T_1 = 100$ °C and $T_2 = 86.765$ °C. Heat dissipated is

$$Q = \frac{kA}{l}(T_1 - T_2) = 0.7941\ \text{W}$$

The above answer is very approximate. However, a more accurate value can be determined by using the following convection condition, that is,

$$Q = \sum_{e=1}^{M} hPl\left(\frac{T_1 + T_2}{2} - T_a\right) = 1.64\ \text{W}$$

where M is the total number of elements. The maximum theoretically possible heat transfer is

$$Q_{max} = hPl(T_1 - T_a) = 1.8\text{W}$$

The efficiency is defined as

$$\eta_f = \frac{Q}{Q_{max}} = \frac{1.64}{1.80} = 91.11\%$$

The exact solution for this problem is

$$Q_{exact} = kAm(T_b - T_a)\tanh(ml) = 1.593\ \text{W}$$

where $m = \sqrt{hP/kA} = 31.62$. Therefore, the exact fin efficiency is

$$\left(\eta_f\right)_{exact} = \frac{Q_{exact}}{Q_{max}} = \frac{1.593}{1.80} = 88.48\ \%$$

Let us consider a two-element solution of the same problem (three nodes). The length of the fin is divided equally into two elements, that is, $l = 1.0$ cm. The stiffness matrix calculation is similar to the one for the single-element case, that is,

$$[K_1] = [K_2] = \begin{bmatrix} 0.124 & -0.118 \\ -0.118 & 0.124 \end{bmatrix}$$

and the loading vectors are

$$\{f_1\} = \{f_2\} = \begin{Bmatrix} 0.15 \\ 0.15 \end{Bmatrix}$$

On assembly we obtain

$$\begin{bmatrix} 0.124 & -0.118 & 0.0 \\ -0.118 & 0.124+0.124 & -0.118 \\ 0.0 & -0.118 & 0.124 \end{bmatrix} \begin{Bmatrix} T_1 \\ T_2 \\ T_3 \end{Bmatrix} = \begin{Bmatrix} 0.15 \\ 0.15+0.15 \\ 0.15 \end{Bmatrix}$$

Now we have to incorporate the known value of base temperature, that is, $T_1 = 100\,°\text{C}$.

$$\begin{bmatrix} 1.0 & 0.0 & 0.0 \\ 0.0 & 0.248 & -0.118 \\ 0.0 & -0.118 & 0.124 \end{bmatrix} \begin{Bmatrix} T_1 \\ T_2 \\ T_3 \end{Bmatrix} = \begin{Bmatrix} 100 \\ 0.3+0.118\times 100 \\ 0.15 \end{Bmatrix}$$

Therefore, the two equations to be solved are:

$$0.248T_2 - 0.118T_3 = 12.1 \text{ and}$$
$$-0.118T_2 + 0.124T_3 = 0.15$$

Solving these equations, we get $T_2 = 90.209\,°\text{C}$, $T_3 = 87.057\,°\text{C}$. Results that have been generated using different number of elements are tabulated in Tables 3.3 and 3.4.

Comments: As can be seen, the two-element solution is very good and is further improved with the use of four elements. As a first idealization, even the one-element solution is reasonably good, considering the small effort involved.

Table 3.3 Summary of results: temperatures.

x, mm	Exact	1 element	2 elements	4 elements
0.0	100.00	100.00	100.0	100
5.0	94.28	—	—	94.26
10.0	90.28	—	90.209	90.25
15.0	87.93	—	—	87.908
20.0	87.15	86.77	87.07	87.128

Table 3.4 Summary of results: heat dissipated and efficiency.

Case	Q_w (Watts)	η_f (Per cent)
1 Element	1.640	91.11
2 Elements	1.604	89.11
4 Elements	1.596	88.65
Exact Solution	1.590	88.48

Example 3.4 Determine the temperature distribution in a plane wall of thickness 60 mm, which has an internal source of 0.3 MW/m^3 and the thermal conductivity of the material is 21 W/(m · K). Assume that the surface temperature of the wall is 40 °C.

Solution

Known:

Thermal conductivity of the material, thickness of the wall, internal heat generation and the surface temperature

Find:

Steady state temperature distribution in the wall

Schematic: Figure 3.11

Assumptions: Steady state conduction

Analysis:

Because of symmetry, we can consider only one half of the plane wall, as shown in Figure 3.11. Let us consider four elements, each of length 7.5 mm. Let the area for heat flow, $A = 1\ m^2$.

The element stiffness matrix is

$$[K]_e = \frac{kA}{L}\begin{bmatrix} 1 & -1 \\ -1 & 1 \end{bmatrix} = \begin{bmatrix} 2800 & -2800 \\ -2800 & 2800 \end{bmatrix}$$

which is identical for every element and

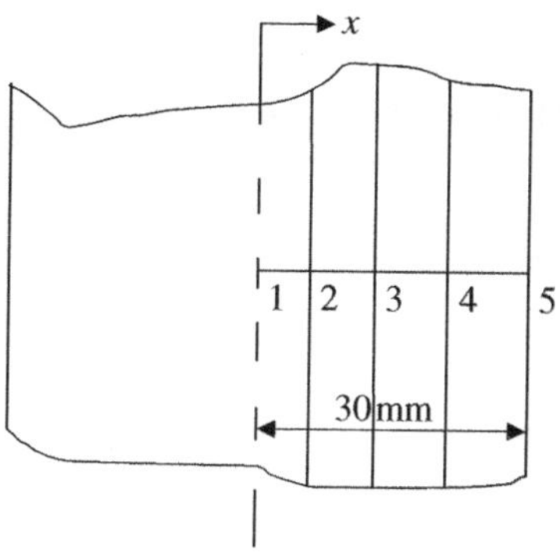

Figure 3.11 Finite element discretization.

$$\{f\}_e = \frac{GAL}{2}\begin{Bmatrix} 1 \\ 1 \end{Bmatrix} = \begin{Bmatrix} 1125 \\ 1125 \end{Bmatrix}$$

which is also identical for all elements. Assembly gives

$$\begin{bmatrix} 2800 & -2800 & 0.0 & 0.0 & 0.0 \\ -2800 & 5600 & 2800 & 0.0 & 0.0 \\ 0.0 & -2800 & 5600 & -2800 & 0.0 \\ 0.0 & 0.0 & -2800 & 5600 & -2800 \\ 0.0 & 0.0 & 0.0 & -2800 & 2800 \end{bmatrix}\begin{Bmatrix} T_1 \\ T_2 \\ T_3 \\ T_4 \\ T_5 \end{Bmatrix} = \begin{Bmatrix} 1125 \\ 2250 \\ 2250 \\ 2250 \\ 1125 \end{Bmatrix}$$

Applying the boundary condition, $T_5 = 40\,°\text{C}$, the modifications are necessary to retain the symmetry of the stiffness matrix:

$$\begin{bmatrix} 2800 & -2800 & 0.0 & 0.0 & 0.0 \\ -2800 & 5600 & 2800 & 0.0 & 0.0 \\ 0.0 & -2800 & 5600 & -2800 & 0.0 \\ 0.0 & 0.0 & -2800 & 5600 & 0.0 \\ 0.0 & 0.0 & 0.0 & -2800 & 1.0 \end{bmatrix}\begin{Bmatrix} T_1 \\ T_2 \\ T_3 \\ T_4 \\ T_5 \end{Bmatrix} = \begin{Bmatrix} 1125 \\ 2250 \\ 2250 \\ 2250 + 2800 \times 40 \\ 1125 \end{Bmatrix}$$

Solving the above system of equations, we obtain the temperature distribution as shown in Table 3.5.

Comment: We observe that the finite element method results are either very close to or equal to the exact solution. The method can be extended for the case of either a known wall heat flux condition or a convective boundary condition at the wall as shown in Example 3.5.

Table 3.5 Summary of results – temperatures for Example 3.4.

Temperature	FEM solution (°C)	Exact solution (°C)
T_1	46.43	46.43
T_2	46.03	46.03
T_3	44.83	44.82
T_4	42.82	42.81
T_5	40.00	40.00

Example 3.5 If in Example 3.4 the left-hand face of the wall is insulated and the right-hand face is in contact with a fluid at 93 °C with a surface heat transfer coefficient of 570 W/m^2K, determine the temperature distribution within the wall.

Solution

Known: Thermal conductivity of the material, thickness of the wall, internal heat generation and the surface heat transfer coefficient

Find:

Steady state temperature distribution in the wall

Schematic: Figure 3.12

Assumptions: Steady state conduction

Analysis:

Since there is no symmetry in the wall with respect to the boundary conditions, we have to consider the entire domain. Let us subdivide the domain into eight elements (Figure 3.12) each of 7.5 mm width.

Now

$$[K]_1 = [K]_2 = [K]_3 = \dots = [K]_7 = \begin{bmatrix} 2800 & -2800 \\ -2800 & 2800 \end{bmatrix}$$

$$[K]_8 = \begin{bmatrix} 2800 & -2800 \\ -2800 & 2800 \end{bmatrix} + 570 \begin{bmatrix} 0 & 0 \\ 0 & 1 \end{bmatrix} = \begin{bmatrix} 2800 & -2800 \\ -2800 & 3370 \end{bmatrix}$$

The elemental forcing vectors are the same as for Example 3.4, except for the last element, for which,

$$\{f\}_8 = \begin{Bmatrix} 1125 \\ 1125 \end{Bmatrix} + hAT_a \begin{Bmatrix} 0 \\ 1 \end{Bmatrix} = \begin{Bmatrix} 1125 \\ 54135 \end{Bmatrix}$$

Assembly may be carried out as in Example 3.4. The solution of the assembled equation results in the temperature distribution within the wall. The FEM solution is compared with the analytical results as shown in Table 3.6.

Comment: The FEM solution compares very favourably with the exact solution.

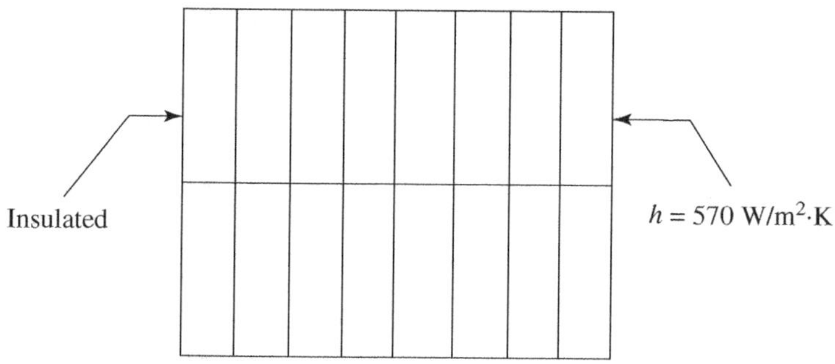

Figure 3.12 Finite element discretization for Example 3.5.

Table 3.6 Summary of results: temperatures.

Location	Temperature	FEM solution (°C)	Exact solution (°C)
1	T_1	150.28	150.20
2	T_2	149.88	149.89
3	T_3	148.68	148.68
4	T_4	146.67	146.67
5	T_5	143.86	143.86
6	T_6	140.24	140.24
7	T_7	135.82	135.83
8	T_8	130.60	130.60
9	T_9	124.59	123.59

3.2.6 Plane Wall with a Heat Source – Solution by Quadratic Element

The analytical solution to the problem of a plane wall with a heat source gives a quadratic temperature distribution. Thus it is appropriate to solve such a problem using quadratic elements. Let us consider Example 3.6. We require three nodes for each element in order to represent a quadratic variation, that is

$$T = N_i T_i + N_j T_j + N_k T_k \tag{3.79}$$

with

$$\begin{aligned} N_i &= \left[1 - \frac{3x}{l} + \frac{2x^2}{l^2}\right] \\ N_j &= \left[\frac{x}{l} - \frac{4x^2}{l^2}\right] \\ N_k &= \left[\frac{2x^2}{l^2} - \frac{x}{l}\right] \end{aligned} \tag{3.80}$$

The stiffness matrix is defined as

$$\begin{aligned} [K] &= \int_0^{\Omega} [B]^T [D][B] d\Omega \\ &= \frac{Ak}{6l} \begin{bmatrix} 14 & -16 & 2 \\ -16 & 32 & -16 \\ 2 & -16 & 14 \end{bmatrix} \end{aligned} \tag{3.81}$$

where

$$[B] = \left[\left(\frac{4x}{l^2} - \frac{3}{l}\right)\left(\frac{4}{l} - \frac{8x}{l^2}\right)\left(\frac{4x}{l^2} - \frac{1}{l}\right)\right) \tag{3.82}$$

The loading vector is

$$\{f\} = \int_0^l G \begin{Bmatrix} L_i(2L_i - 1) \\ 4L_iL_j \\ L_j(2L_j - 1) \end{Bmatrix} Adx = \frac{GAl}{6} \begin{Bmatrix} 1 \\ 4 \\ 1 \end{Bmatrix} \tag{3.83}$$

In the above equation, the shape functions N_i, N_j and N_k are expressed in terms of the local coordinate system L_i and L_j, the use of which will facilitate the integration process by using

$$\int_0^l N_i^a N_j^b \, dl = \frac{a!b!}{(a+b+1)!}$$

Example 3.6 Solve Example 3.4 using one quadratic element.

Solution

Known:

Thermal conductivity of the material, thickness of the wall, internal heat generation and the surface temperature

Find:

Temperature distribution in the wall

Schematic: Figure 3.13

Assumptions: Steady state conduction

Analysis:

As before, we consider only one half of the wall as shown in Figure 3.13 with $L = 30$ mm. Substitution of the given values into Equations 3.81 and 3.83 gives

$$[K]_e = \begin{bmatrix} 1633.33 & -1866.66 & 233.33 \\ -1866.66 & 3733.33 & -1866.66 \\ 233.33 & -1866.66 & 1633.33 \end{bmatrix}$$

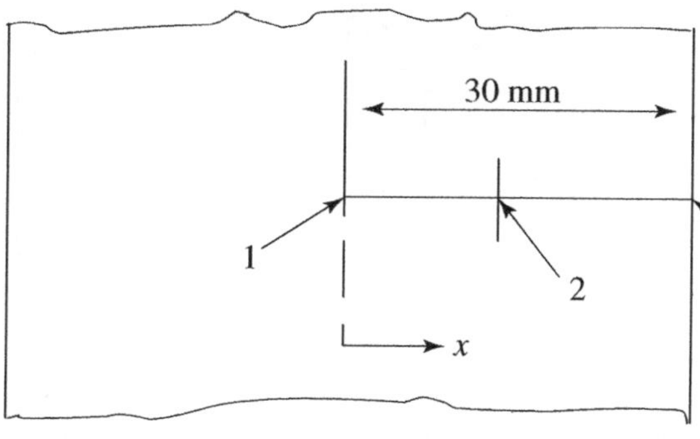

Figure 3.13 Quadratic finite element.

and

$$\{f\}_e = \begin{Bmatrix} 1500 \\ 6000 \\ 1500 \end{Bmatrix}$$

Incorporating the boundary condition that is $T_3 = 40\,°C$, results in the following set of equations:

$$\begin{bmatrix} 1633.33 & -1866.66 & 0.0 \\ -1866.66 & 3733.33 & 0.0 \\ 0.0 & 0.0 & 1.0 \end{bmatrix} \begin{Bmatrix} T_1 \\ T_2 \\ T_3 \end{Bmatrix} = \begin{Bmatrix} 1500 - 233.33x40 \\ 6000 + 1866.66x40 \\ 40.0 \end{Bmatrix}$$

The solution to the above system of equations gives $T_1 = 46.43\,°C$ and $T_2 = 44.82\,°C$, which are identical to the exact solution.

Comment:

Only one quadratic element is capable of giving exact solution which has been obtained with four linear elements.

3.2.7 Two-Dimensional Plane Problems

Both conduction heat transfer in an infinitely long, hollow rectangular tube, which is exposed to different boundary conditions inside and outside the tube (Figure 3.14a), and heat conduction in a thin plate, which has negligible heat transfer in the direction of the thickness, may be approximated as two-dimensional problems.

In certain situations, it is often difficult to simplify the problem to two dimensions without sacrificing accuracy. Most complex industrial heat transfer problems are three-dimensional in nature because of the complicated geometries involved. Heat transfer in aircraft structures and heat shields used in space vehicles are examples of

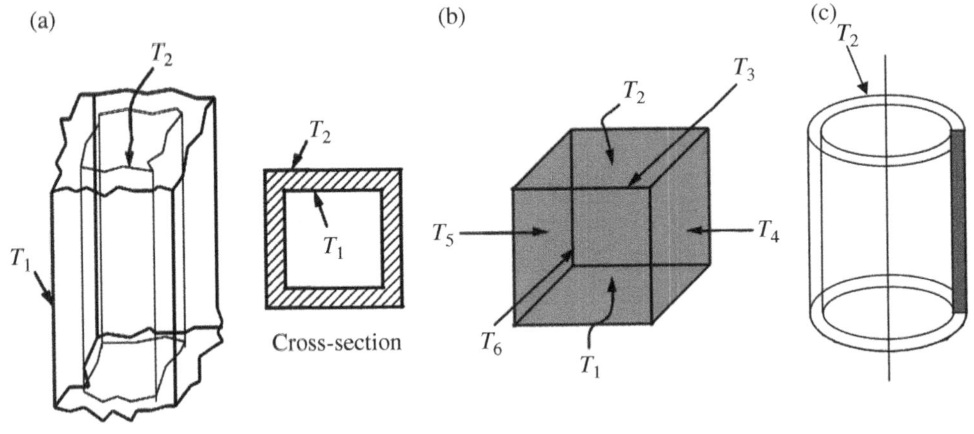

Figure 3.14 Examples of heat conduction in two-dimensional, three-dimensional and axi-symmetric geometries.

such problems. It is, however, important to note that even geometries that are simple but which have complex boundary conditions become three-dimensional in nature. For example, the same hollow, rectangular tube mentioned previously, but in this case having non-uniform conditions along the length, is a three-dimensional problem. Also, if the hollow rectangular tube is finite, again it may be necessary to treat it as a three-dimensional problem (Figure 3.14). One typical simple example of three-dimensional heat conduction is that of a solid cube subjected to different boundary conditions on all six faces, as shown in Figure 3.14b. Another approximation commonly employed in heat conduction studies is the axi-symmetric formulation. This type of problem is often considered as a two-and-a-half-dimensional case as it has the features of both a two- and a three-dimensional approximation. If geometry is generated by revolving a surface through 360° with reference to its axis then it is referred to as being axi-symmetric. For instance, the revolution of a rectangular surface through 360° with respect to a vertical axis, produces a vertical cylinder as shown in Figure 3.14c. Therefore, the heat conduction equations need to be written in three-dimensional cylindrical coordinates for such a system. However, if no significant variation in temperature is expected in the circumferential direction (θ direction), which is often the case, the problem can be reduced to two dimensions, and a solution based on the shaded rectangular plane in Figure 3.14c is sufficient.

Unlike one-dimensional problems, two- and three-dimensional situations are usually geometrically complex and expensive to solve. The complexity of the problem is increased in multidimensions by the occurrence of irregular geometry shapes and the appropriate implementation of boundary conditions on their boundaries. In the case of complicated geometries, it is often necessary to use unstructured meshes (unstructured meshes are generated employing arbitrarily generated points in a domain) to divide the domain into finite elements. Fortunately, owing to present-day computing capabilities, even complex three-dimensional problems can be solved on a standard personal computer (PC). In the following sections, we demonstrate the solution of multidimensional steady-state problems with relevant examples.

3.2.7.1 Triangular Elements

The simplest finite element discretization that can be employed in two dimensions is the use of linear triangular elements. Earlier, we discussed the use of triangular elements in detail. These principles are employed here to solve two-dimensional conduction heat transfer problems.

In order to demonstrate the use of linear triangular elements, let us consider a general problem as shown in Figure 3.15. Here the geometry is irregular, and both the flat faces of the plate are insulated. The surface in the thickness direction is exposed to various boundary conditions. This is an ideal two-dimensional heat conduction problem with no temperature variation allowed in the thickness direction. The final matrix form of the finite element equations is

$$[K]\{T\} = \{f\} \tag{3.84}$$

where

$$[K] = \int_0^{\Omega} [B]^T [D][B] d\Omega + \int_0^{\Gamma} h [N]^T [N] d\Gamma \tag{3.85}$$

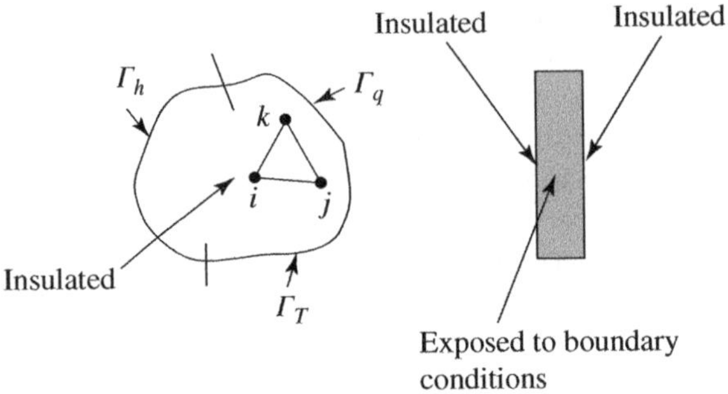

Figure 3.15 Typical two-dimensional plane geometry and triangular element.

and

$$\{f\} = \int_0^{\Omega} G\,[N]^T d\Omega - \int_0^{\Gamma} q\,[N]^T d\Gamma + \int_0^{\Gamma} hT_\infty [N]^T d\Gamma \tag{3.86}$$

For a linear triangular element, the temperature distribution can be written as

$$T = N_i T_i + N_j T_j + N_k T_k \tag{3.87}$$

The gradient matrix is given as

$$\{g\} = \begin{Bmatrix} \dfrac{\partial T}{\partial x} \\ \dfrac{\partial T}{\partial y} \end{Bmatrix} = \begin{bmatrix} \dfrac{\partial N_i}{\partial x} & \dfrac{\partial N_j}{\partial x} & \dfrac{\partial N_k}{\partial y} \\ \dfrac{\partial N_i}{\partial y} & \dfrac{\partial N_j}{\partial y} & \dfrac{\partial N_k}{\partial y} \end{bmatrix} \begin{Bmatrix} T_i \\ T_j \\ T_k \end{Bmatrix} = [B]\{T\} \tag{3.88}$$

where

$$[B] = \begin{bmatrix} \dfrac{\partial N_i}{\partial x} & \dfrac{\partial N_j}{\partial x} & \dfrac{\partial N_k}{\partial y} \\ \dfrac{\partial N_i}{\partial y} & \dfrac{\partial N_j}{\partial y} & \dfrac{\partial N_k}{\partial y} \end{bmatrix} = \frac{1}{2A} \begin{bmatrix} b_i & b_j & b_k \\ c_i & c_j & c_k \end{bmatrix} \tag{3.89}$$

Note that {g} in the above is a uniform heat source. Assuming an anisotropic material, we have

$$[D] = \begin{bmatrix} k_x & 0 \\ 0 & k_y \end{bmatrix} \tag{3.90}$$

Note that the off-diagonal terms are neglected from the above equation for the sake of simplicity. Substituting [D] and [B] into Equation 3.85, we get, for a boundary element as shown in Figure 3.16,

Figure 3.16 Typical two-dimensional triangular element with heat generation and with heat flux and convection boundary conditions.

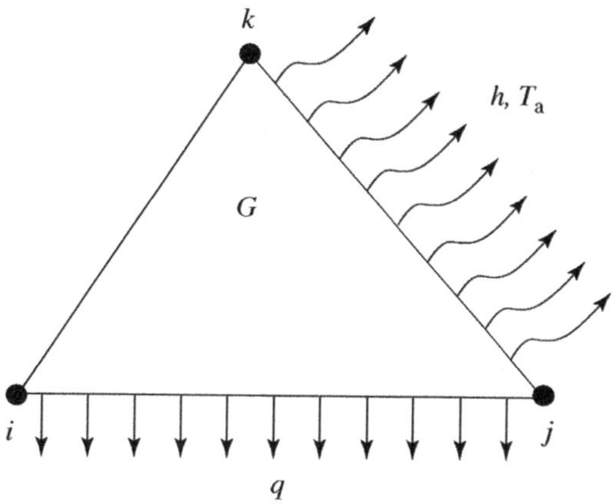

$$[K]e = \frac{t}{4A}\left\{k_x\begin{bmatrix} b_i^2 & b_ib_j & b_ib_k \\ b_ib_j & b_j^2 & b_jb_k \\ b_ib_k & b_jb_k & b_k^2\end{bmatrix} + k_y\begin{bmatrix} c_i^2 & c_ic_j & c_ic_k \\ c_ic_j & c_j^2 & c_jc_k \\ c_ic_k & c_jc_k & c_k^2\end{bmatrix}\right\} + \frac{htl_{jk}}{6}\begin{bmatrix} 0 & 0 & 0 \\ 0 & 2 & 1 \\ 0 & 1 & 2\end{bmatrix} \tag{3.91}$$

The subscript *e* in Equation 3.91 denotes a single element. It should be noted that in Equation 3.85, $d\Omega$ is equal to tdA and $d\Gamma$ is equal to tdl, where t is the thickness of the plate and l is the length of the element side on the domain boundary. In a similar manner, the forcing vector can be written as

$$\{f\}^e = \frac{GAt}{3}\begin{Bmatrix} 1 \\ 1 \\ 1 \end{Bmatrix} - \frac{qtl_{ij}}{2}\begin{Bmatrix} 1 \\ 1 \\ 0 \end{Bmatrix} + \frac{hT_a\,t\,l_{jk}}{2}\begin{Bmatrix} 0 \\ 1 \\ 1 \end{Bmatrix} \tag{3.92}$$

The integration formulae used for the above equations are simple. As seen in the previous equations, the effect of uniform heat generation contributes to all three nodes of an element, irrespective of its position. However, the convection and flux boundary conditions are applicable only on the boundaries of the domain. If we need to have a "point source" G^* instead of a "uniform source" G'', the first term in Equation 3.92 is replaced with

$$\{f\} = G^*t\begin{Bmatrix} N_i \\ N_j \\ N_k \end{Bmatrix}_{(xo,yo)} \tag{3.93}$$

where x_o and y_o are the coordinates of the point source. In the above equations, all the shape function values must be evaluated at (x_o, y_o) (note that although G^* is a point source, in two dimensions, it is a line source in the thickness direction and expressed in units of W/m). The contribution from the point source is then appropriately distributed to the three nodes of the element that contains the point source. In order to demonstrate the characteristics of two-dimensional steady-state heat transfer, the temperature distribution in a flat plate having constant temperature boundary conditions is considered in the following example.

Example 3.7 A square plate of unit thickness, size 100 cm, as shown in Figure 3.17, is subjected to isothermal boundary conditions of 100 °C on all sides except the top side, which is subjected to 500 °C. If the thermal conductivity of the material is constant and equal to 10 W/m°C, determine the temperature distribution using linear triangular finite elements.

Solution

Known:

Thermal conductivity of the material, dimensions of the plate and the surface temperatures

Find: Temperature distribution in the plate

Schematic: Figure 3.17

Assumptions: Steady state conduction

Analysis:

The square domain is first divided into eight equal-sized linear triangular elements, as shown in Figure 3.18. Two sets of elemental [*K*] matrices exist because of the orientation

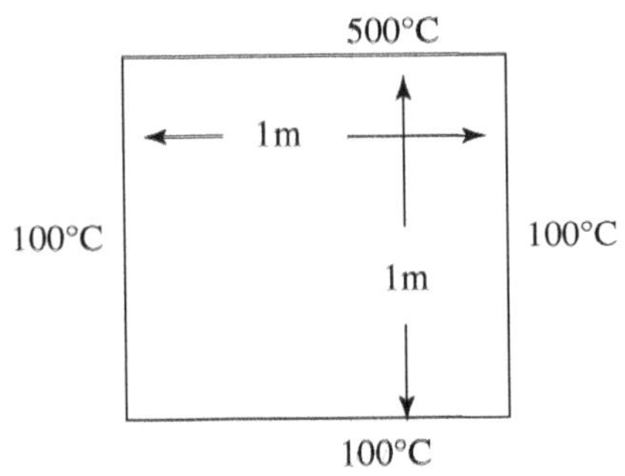

Figure 3.17 Square plate with different temperature boundary conditions.

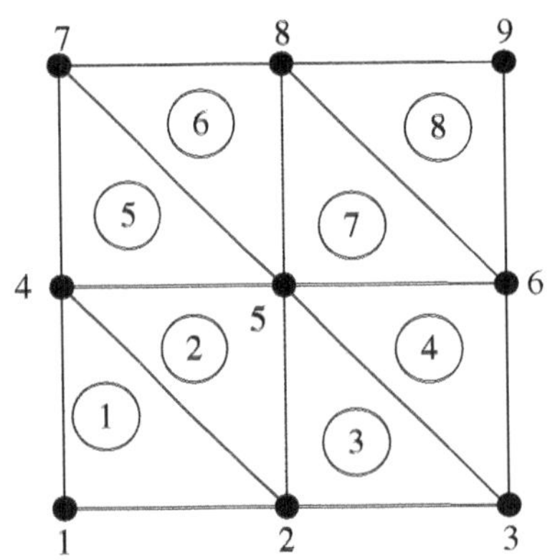

Figure 3.18 Discretization using triangular elements.

of the triangles. For elements 1, 3, 5 and 7, we have the following elements of the [K] matrix:

$$b_1 = y_2 - y_4 = -0.50; \quad c_1 = x_4 - x_2 = -0.50$$

$$b_2 = y_4 - y_1 = -0.50; \quad c_2 = x_1 - x_4 = 0.00$$

$$b_4 = y_1 - y_2 = 0.00; \quad c_4 = x_2 - x_1 = 0.50$$

The elemental [K] matrices for elements 1, 3, 5 and 7 can be written as (refer to Equation 3.91)

$$[K]_1 = [K]_3 = [K]_5 = [K]_7 = \frac{tk}{4A}\begin{bmatrix} b_1^2 + c_1^2 & b_1b_2 + c_1c_2 & b_1b_4 + c_1c_4 \\ b_1b_2 + c_1c_2 & b_2^2 + c_2^2 & b_2b_4 + c_2c_4 \\ b_1b_4 + c_1c_4 & b_2b_4 + c_2c_4 & b_3^2 + c_3^2 \end{bmatrix}$$

where the area of the elements can be written as

$$2A = \det\begin{vmatrix} 1.0 & 0.0 & 0.0 \\ 1.0 & 0.5 & 0.5 \\ 1.0 & 0.0 & 0.0 \end{vmatrix} = 0.25\,\text{m}^2$$

Substituting the area in the expressions for [K], we get the final form of the elemental equation as

$$[K]_1 = [K]_3 = [K]_5 = [K]_7 = \frac{tk}{2}\begin{bmatrix} 2.0 & -1.0 & -1.0 \\ -1.0 & 1.0 & 0.0 \\ -1.0 & 0.0 & 1.0 \end{bmatrix}$$

Similarly, we can calculate the elemental [K] matrices for elements 2, 4, 6 and 8 as

$$[K]_2 = [K]_4 = [K]_6 = [K]_8 = \frac{tk}{2}\begin{bmatrix} 2.0 & -1.0 & -1.0 \\ -1.0 & 2.0 & -1.0 \\ 0.0 & -1.0 & 1.0 \end{bmatrix}$$

The assembled equations are

$$\frac{tk}{2}\begin{bmatrix} 2.0 & -1.0 & 0.0 & -1.0 & 0.0 & 0.0 & 0.0 & 0.0 & 0.0 \\ -1.0 & 4.0 & -1.0 & 0.0 & -2.0 & 0.0 & 0.0 & 0.0 & 0.0 \\ 0.0 & -1.0 & 2.0 & 0.0 & 0.0 & -1.0 & 0.0 & 0.0 & 0.0 \\ -1.0 & 0.0 & 0.0 & 4.0 & -2.0 & 0.0 & -1.0 & 0.0 & 0.0 \\ 0.0 & -2.0 & 0.0 & -2.0 & 8.0 & -2.0 & 0.0 & -2.0 & 0.0 \\ 0.0 & 0.0 & -1.0 & 0.0 & -2.0 & 4.0 & 0.0 & 0.0 & -1.0 \\ 0.0 & 0.0 & 0.0 & -1.0 & 0.0 & 0.0 & 2.0 & -1.0 & 0.0 \\ 0.0 & 0.0 & 0.0 & 0.0 & -2.0 & 0.0 & -1.0 & 4.0 & -1.0 \\ 0.0 & 0.0 & 0.0 & 0.0 & 0.0 & -1.0 & 0.0 & -1.0 & 2.0 \end{bmatrix}\begin{Bmatrix} T_1 \\ T_2 \\ T_3 \\ T_4 \\ T_5 \\ T_6 \\ T_7 \\ T_8 \\ T_9 \end{Bmatrix} = \begin{Bmatrix} 0.0 \\ 0.0 \\ 0.0 \\ 0.0 \\ 0.0 \\ 0.0 \\ 0.0 \\ 0.0 \\ 0.0 \end{Bmatrix}$$

The only unknown quantity in the above equation is T_5, which can be calculated from the equation corresponding to the fifth node, that is, from

$$8T_5 = 2T_2 + 2T_4 + 2T_6 + 2T_8$$

Substituting $T_2 = T_4 = T_6 = 100\,°C$ and $T_8 = 500\,°C$, we get $T_5 = 200\,°C$. The analytical solution to this problem is given by Holman [17] as

$$T(x,y) = \left(T_{top} - T_{side}\right)\frac{2}{\pi}\sum_{n=1}^{\infty}\frac{(-1)^{n+1}+1}{n}\sin\left(\frac{n\pi x}{w}\right)\frac{\sinh\left(\frac{n\pi y}{w}\right)}{\sinh\left(\frac{n\pi H}{w}\right)} + T_{side}$$

where w is the width, H is the height of the plate, T_{top} is the temperature at the top side and T_{side} is the temperature at the other sides of the plate. Therefore,

$$T(0.5, 0.5) = 200.11°C$$

As seen, the finite element solution is in close agreement with the analytical solution. It is interesting to note that the finite difference solution is given by

$$T_5 = \frac{T_2 + T_4 + T_6 + T_8}{4} = 200°C$$

which is identical to the finite element solution. Figure 3.19 shows an unstructured mesh and a computer-generated solution for this problem. As shown, the temperature at the center is close to that obtained from the coarse mesh of Figure 3.18, and also to the analytical solution. However, the unstructured mesh solution is not as accurate as that of the regular mesh solution. This indicates that the accuracy of a regular structured mesh is superior to that of unstructured meshes. If a finer structured mesh, as shown in Figure 3.20, is used, the temperature at the center is 200 °C.

Using the nodal temperature values, the temperature at any other location within an element can be determined using linear interpolation. The calculation of the temperature at any arbitrary location has been demonstrated earlier. The following two-dimensional example is given in order to further illustrate this point.

Example 3.8 The temperature values at nodes 1, 2 and 3 for a linear triangular element shown in Figure 3.21 are 100 °C, 200 °C and 100 °C, respectively. The coordinates of these points are (50, 0), (50, 50) and (0, 50), respectively. All dimensions are in cm. Calculate the temperature at point 4 (40, 40) Also, calculate the heat flux in both the x and y directions. Assume a thermal conductivity value of 10 W/m°C.

Solution

Known:

Thermal conductivity of the material, coordinates and the corresponding temperatures at the nodes

Find:

1) Temperature at the location (40, 40)
2) Heat flux in both x and y directions at the same location

Schematic: Figures 3.19–3.21
Assumptions: Steady state conduction
Analysis:

The following expression can be used to describe the linear variation of temperature within the element:

$$T = N_1 T_1 + N_2 T_2 + N_3 T_3 \tag{a}$$

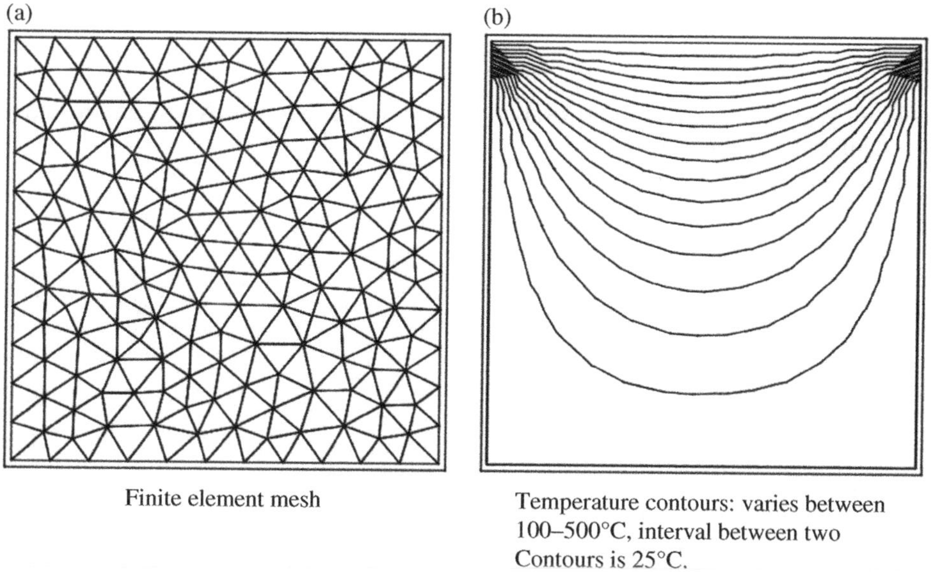

Figure 3.19 Solution for Example 3.7 on an unstructured mesh. The temperature obtained at the center of the plane is 200.42 °C.

Figure 3.20 Fine structured mesh.

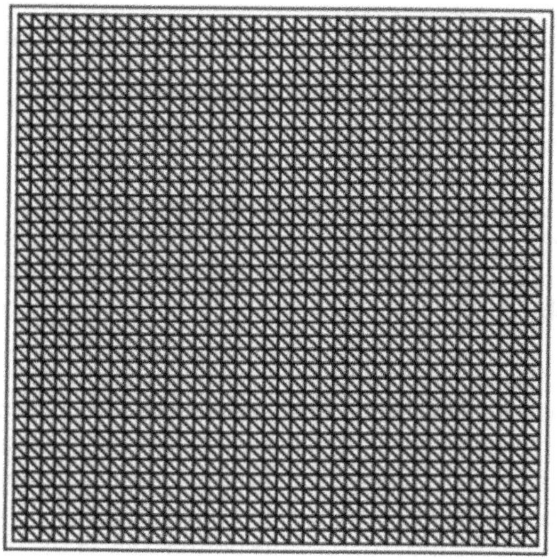

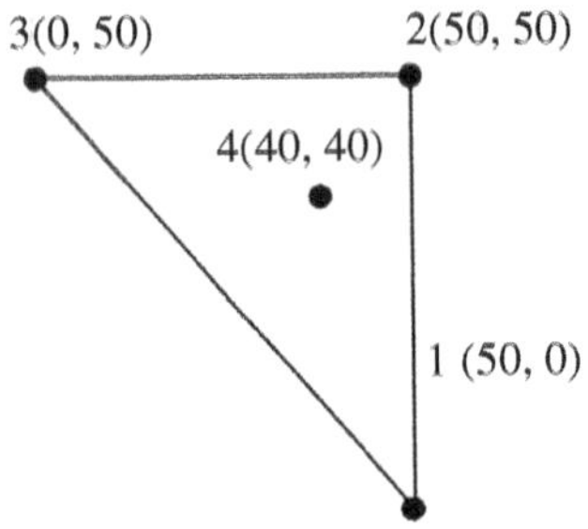

Figure 3.21 Interpolation into a triangular element.

In order to calculate the temperature at node 4, the shape functions N_1, N_2 and N_3 have to be calculated at node 4.

Therefore, for the first node

$$N_1 = \frac{1}{2A}(a_1 + b_1x_4 + c_1y_4) \tag{b}$$

where

$$a_1 = x_2y_3 - x_3y_2 = 2500.00 \tag{c}$$

$$b_1 = y_2 - y_3 = 0.0 \tag{d}$$

$$c_1 = x_3 - x_2 = -50.00 \tag{e}$$

At point 4 ($x = 40$, $y = 40$),

we get $2A = 2500$. So $N_1 = \frac{1}{5}$

Similarly, it can be verified that $N_2 = 1/5$ and $N_3 = 3/5$. Note that $N_1 + N_2 + N_3 = 1$. The substitution of these shape function values into Equation (a) above results in a value of $T_4 = 160\,°C$.

The heat flux in the x direction is calculated as

$$q_x = -k\frac{\partial T}{\partial x} = -\frac{10}{2500}(b_1T_1 + b_2T_2 + b_3T_3) = -20\,W/cm^2 \tag{f}$$

Similarly, it can be shown that heat flux in the y direction is $q_y = -20\ W/cm^2$.

Comments:

It should be noted that the flux is constant over a linear triangular element.

From Examples 3.7 and 3.8, the demonstration of problems involving constant temperature boundary conditions is clear. It is therefore essential to move on to an example with more complicated boundary conditions. Thus, in the following example, a conduction problem is considered that has mixed boundary conditions.

Example 3.9 Determine the temperature distribution in a square plate of unit thickness size 5 cm as shown in Figure 3.22. The upper triangular half has an internal heat generation of 1.2 W/cm^3, while the lower half has a point source of 5 W/cm in the thickness direction (point source on a two-dimensional plane) at the point (1,1) cm. In addition to the above heat sources, the bottom side of the plate is insulated, the right vertical

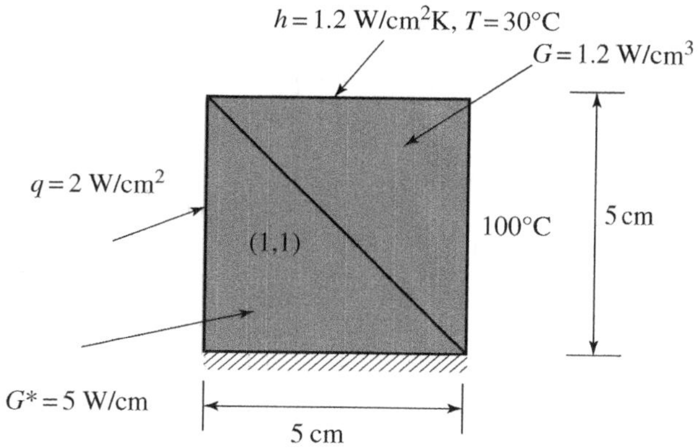

Figure 3.22 A square domain with mixed boundary conditions.

side is subjected to a temperature of 100 °C, the top side is subjected to a convective heat transfer boundary condition with a heat transfer coefficient of $h = 1.2\ \text{W/cm}^2 \cdot \text{K}$ and $T_a = 30\ °\text{C}$ and the left vertical side is subjected to a uniform heat flux of $2\ \text{W/cm}^2$. Assume a thermal conductivity of 2 W/cm°C.

Solution

Known:

Dimensions of the square plate, thermal conductivity of the plate material and boundary conditions

Find: Temperature distribution in the plate

Schematic: Figure 3.22

Assumptions: Steady state conduction

Analysis:

To make the solution procedure simple, the plate is divided into two triangular elements, as shown in Figure 3.23. The elemental equations of both elements can be set up separately, using the formulation discussed (Equations 3.91 and 3.92). For the first element, $a_1 = 25.0$, $b_1 = -5.0$, $c_1 = -5.0$, $a_2 = 0.0$, $b_2 = 5.0$, $c_2 = 0.0$, $a_3 = 0.0$, $b_3 = 0.0$, $c_3 = 5.0$.

Figure 3.23 Discretization using two triangular elements.

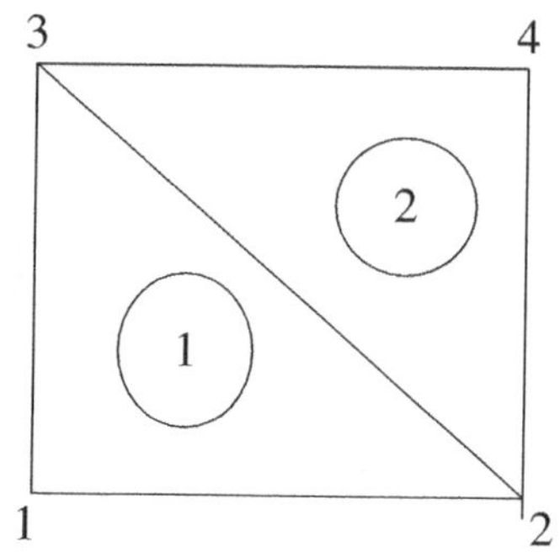

The stiffness matrix for element 1 is

$$[K]_1 = \frac{t}{4A}\left\{ k_x \begin{bmatrix} b_1^2 & b_1b_2 & b_1b_3 \\ b_1b_2 & b_2^2 & b_2b_3 \\ b_1b_3 & b_2b_3 & b_3^2 \end{bmatrix} + k_y \begin{bmatrix} c & c_1c_2 & c_1c_3 \\ c_1c_2 & c_2^2 & c_2c_3 \\ c_1c_3 & c_2c_3 & c \end{bmatrix} \right\}$$

Substituting the values for a, b and c, we obtain

$$[K]_1 = \begin{bmatrix} 2.0 & -1.0 & -1.0 \\ -1.0 & 1.0 & 0.0 \\ -1.0 & 0.0 & 1.0 \end{bmatrix}$$

The loading term for element 1 is given by

$$\{f\}_1 = -\frac{q\,l_{31}}{2}\begin{Bmatrix} 1.0 \\ 0.0 \\ 1.0 \end{Bmatrix} + G^* t \begin{Bmatrix} N_1 \\ N_2 \\ N_3 \end{Bmatrix}_{(1,1)} = \begin{Bmatrix} -2.0 \\ 1.0 \\ -4.0 \end{Bmatrix}$$

Note that the shape functions evaluated at point (1, 1) are $N_1 = 3/5$, $N_2 = 1/5$ and $N_3 = 1/5$.
In a similar way, the stiffness matrix and loading terms for the second element can be calculated. They are

$$[K]_2 = \begin{bmatrix} 1.0 & -1.0 & 0.0 \\ -1.0 & 4.0 & 0.0 \\ 0.0 & 0.0 & 3.0 \end{bmatrix}$$

and

$$\{f\}_2 = \begin{Bmatrix} 5.0 \\ 95.0 \\ 95.0 \end{Bmatrix}$$

On assembling the above contributions for the two elements, we obtain the following system of simultaneous equations

$$\begin{bmatrix} 2.0 & -1.0 & -1.0 & 0.0 \\ -1.0 & 2.0 & 0.0 & -1.0 \\ -1.0 & 0.0 & 4.0 & 0.0 \\ 0.0 & -1.0 & 0.0 & 4.0 \end{bmatrix} \begin{Bmatrix} T_1 \\ T_2 \\ T_3 \\ T_4 \end{Bmatrix} = \begin{Bmatrix} -2.0 \\ 6.0 \\ 91.0 \\ 95.0 \end{Bmatrix}$$

In the above set of equations, the temperature values T_2 and T_4 are known and are equal to 100 °C. The boundary conditions can be implemented as previously explained. Applying the boundary conditions, we get

$$\begin{bmatrix} 2.0 & 0.0 & -1.0 & 0.0 \\ 0.0 & 1.0 & 0.0 & 0.0 \\ -1.0 & 0.0 & 4.0 & 0.0 \\ 0.0 & 0.0 & 0.0 & 1.0 \end{bmatrix} \begin{Bmatrix} T_1 \\ T_2 \\ T_3 \\ T_4 \end{Bmatrix} = \begin{Bmatrix} -2.0+1(100) \\ 100.0 \\ 91.0 \\ 100.0 \end{Bmatrix}$$

Therefore, the simultaneous equations to be solved are

$$2T_1 - T_3 = 98$$

and $-T_1 + 4T_3 = 91$

The solution to these equations results in $T_1 = 69\,°\text{C}$ and $T_3 = 40\,°\text{C}$.

If, in the above example, there is a uniform heat generation of 1.2 W/cm^3 throughout the domain, then the loading term for the first element changes to (in the absence of line source)

$$\{f\}_1 = -\frac{q\,l_{31}}{2}\begin{Bmatrix} 1.0 \\ 0.0 \\ 1.0 \end{Bmatrix} + \frac{GAt}{3}\begin{Bmatrix} 1.0 \\ 1.0 \\ 1.0 \end{Bmatrix} = \begin{Bmatrix} 0.0 \\ 5.0 \\ 0.0 \end{Bmatrix}$$

The resulting simultaneous equations become $2T_1 - T_3 = 100$ and $-T_1 + 4T_3 = 95$ and the solution becomes $T_1 = 70.71\,°\text{C}$ and $T_3 = 40.42\,°\text{C}$.

Comments:

Since an extremely coarse mesh is used the solution obtained is far away from the actual values for the given problem. This problem illustrates the application of FEM to mixed boundary conditions problem. The solution can be improved by using a finer mesh.

3.2.8 Finite Element Method-Transient Heat Conduction

The time-dependent conduction equation for a stationary medium is given by

$$\frac{\partial}{\partial x}\left(k_x(T)\frac{\partial T}{\partial x}\right) + \frac{\partial}{\partial y}\left(k_y(T)\frac{\partial T}{\partial y}\right) + \frac{\partial}{\partial z}\left(k_z(T)\frac{\partial T}{\partial z}\right) + G = \rho c_p \frac{\partial T}{\partial t} \tag{3.94}$$

where $k_x(T)$ $k_y(T)$ and $k_z(T)$ are the temperature-dependent thermal conductivities in the x-, y- and z-directions, respectively. The boundary conditions for this type of problem are

$$T = T_b \text{ on } \Gamma_b \tag{3.95}$$

and

$$\left(k_x(T)\frac{\partial T}{\partial x}\right)l + \left(k_y(T)\frac{\partial T}{\partial y}\right)m + \left(k_z(T)\frac{\partial T}{\partial z}\right)n + q + h(T - T_a) + G = \text{on } \Gamma q \tag{3.96}$$

where $\Gamma_b \cup \Gamma q = \Gamma$ and $\Gamma_b \cap \Gamma_q = 0$. Γ represents the whole boundary. In Equation 3.96, l, m, and n are direction cosines, T is the atmospheric temperature, h is the heat transfer coefficient and q is the boundary heat flux.

The initial condition for the problem is

$$T = T_0 \text{ at } t = 0.0 \tag{3.97}$$

It is now possible to solve the above system, provided that appropriate spatial and temporal discretization are available. Before dealing with the temporal discretization, we introduce in the following subsection, the standard Galerkin weighted residual form for the transient conduction equation.

3.2.8.1 Galerkin Method for Transient Heat Conduction

The application of the Galerkin method for the transient conduction equations subjected to appropriate boundary and initial conditions is addressed in this section. The temperature is discretized over space as follows:

$$T(x,y,z,t) = \sum_{i=1}^{n} N(x,y,z)T\ (t) \tag{3.98}$$

where the N_i are the shape functions, n is the number of nodes in an element, and $T(t)$ are the time-dependent nodal temperatures. The Galerkin representation of Equation 3.94 is

$$\int_0^{\Omega} N_i\left[\frac{\partial}{\partial x}\left(k_x(T)\frac{\partial T}{\partial x}\right) + \frac{\partial}{\partial y}\left(k_y(T)\frac{\partial T}{\partial y}\right) + \frac{\partial}{\partial z}\left(k_z(T)\frac{\partial T}{\partial z}\right) + G - \rho c_p\frac{\partial T}{\partial t}\right]d\Omega = 0 \tag{3.99}$$

Employing integration by parts on the first three terms of Equation 3.99, we get

$$\int_0^{\Omega}\left[\left(k_x(T)\frac{\partial N_i}{\partial x}\frac{\partial T}{\partial x}\right) + \left(k_y(T)\frac{\partial N_i}{\partial y}\frac{\partial T}{\partial y}\right) + \left(k_z(T)\frac{\partial N_i}{\partial z}\frac{\partial T}{\partial z}\right) - N_i G - N_i\rho c_p\frac{\partial T}{\partial t}\right]d\Omega$$
$$+ \int_{\Gamma q} N_i\left(k_x(T)\frac{\partial T}{\partial x}\right)l\,d\Gamma q + \int_{\Gamma q} N_i\left(k_y(T)\frac{\partial T}{\partial y}\right)m\,d\Gamma q + \int_{\Gamma q} N_i\left(k_z(T)\frac{\partial T}{\partial z}\right)n\,d\Gamma q = 0 \tag{3.100}$$

But from Equation 3.96

$$+ \int_{\Gamma q} N_i\left(k_x(T)\frac{\partial T}{\partial x}\right)l\,d\Gamma q + \int_{\Gamma q} N_i\left(k_y(T)\frac{\partial T}{\partial y}\right)m\,d\Gamma q + \int_{\Gamma q} N_i\left(k_z(T)\frac{\partial T}{\partial z}\right)n\,d\Gamma q =$$
$$+ \int_{\Gamma q} N_i q\,d\Gamma q - \int_{\Gamma q} N_i h(T - T_a)\,d\Gamma q \tag{3.101}$$

On substituting the spatial approximation from Equation 3.98, Equation 3.99 finally becomes

$$-\int_0^{\Omega}\left[\left(k_x(T)\frac{\partial N_i}{\partial x}\frac{\partial N_j}{\partial x}\right)T_j(t)+\left(k_y(T)\frac{\partial N_i}{\partial y}\frac{\partial N_j}{\partial y}\right)T_j(t)+\left(k_z(T)\frac{\partial N_i}{\partial z}\frac{\partial N_j}{\partial z}\right)T_j(t)\right]d\Omega$$

$$+\int_0^{\Omega}\left[N_iG-N_i\rho c_p\frac{\partial N_j}{\partial t}T_j(t)\right]d\Omega-\int_{\Gamma q}N_iq\,d\Gamma q-\int_{\Gamma q}N_ih(T-T_a)\,d\Gamma q=0 \tag{3.102}$$

where i and j represent the nodes. Equation 3.102 can be written in a more convenient form as

$$[C]\left\{\frac{\partial T}{\partial t}\right\}+[K][T]=\{f\} \tag{3.103}$$

or

$$\left[c_{ij}\right]\left\{\frac{\partial T_j}{\partial t}\right\}+\left[k_{ij}\right]\left[T_j\right]=\{f_i\} \tag{3.104}$$

where

$$\left[c_{ij}\right]=\int_0^{\Omega}\rho c_pN_iN_jd\,\Omega \tag{3.105}$$

$$\left[k_{ij}\right]=$$

$$\int_0^{\Omega}\left[\left(k_x(T)\frac{\partial N_i}{\partial x}\frac{\partial N_j}{\partial x}\right)\{T_j\}+\left(k_y(T)\frac{\partial N_i}{\partial y}\frac{\partial N_j}{\partial y}\right)\{T_j\}+\left(k_z(T)\frac{\partial N_i}{\partial z}\frac{\partial N_j}{\partial z}\right)\{T_j\}\right]d\Omega$$

$$+\int_0^{\Gamma}hN_iN_jd\Gamma \tag{3.106}$$

$$f_i=\int_0^{\Omega}GN_id\Omega-\int_0^{\Gamma_q}qN_id\Gamma q+\int_0^{\Gamma_q}hN_iT_ad\Gamma \tag{3.107}$$

In matrix form Equations 3.105 to 3.107 can be written as

$$[C]=\int_0^{\Omega}\rho c_p\left[N^T\right][N]d\Omega \tag{3.108}$$

$$[K]=\int_0^{\Omega}\left[B^T\right][D][B]d\Omega+\int_0^{\Gamma}h\left[N^T\right][N]d\Gamma \tag{3.109}$$

$$[f] = \int_0^{\Omega} G[N^T]d\Omega - \int_0^{\Gamma_q} q[N^T]d\Gamma q + \int_0^{\Gamma} hT_a[N^T]d\Gamma \tag{3.110}$$

Since $k_x(T)$ $k_y(T)$ and $k_z(T)$ are functions of temperatures, Equation 3.103 is non-linear and requires an iterative solution. If k_x k_y and k_z are independent of temperature, then Equation 3.103 is linear.

Example 3.10 One-Dimensional Time-Dependent Conduction Problem
Equation 3.103 is employed to illustrate the application to a one-dimensional time-dependent conduction problem using a linear element as shown in Figure 3.24.

The temperature T in the element is represented by

$$T = N_i T_i + N_j T_j = [N]\{T\} \tag{3.111}$$

It should be noted that i and j in the above equation represent the nodes of the element shown in Figure 3.24.

The shape functions in Equation 3.111 are defined as

$$N_i = 1 - \frac{x}{l}$$

$$N_j = \frac{x}{l} \tag{3.112}$$

The spatial derivative of temperature is given as

$$\frac{\partial T}{\partial x} = \frac{\partial N_i}{\partial x} T_i + \frac{\partial N_j}{\partial x} T_j = -\frac{1}{l} T_i + \frac{1}{l} T_j = [B]\{T\} \tag{3.113}$$

The relevant matrices as discussed in previous section (Equation 3.108) are

$$[C] = \int_0^{\Omega} \rho c_p [N^T][N]d\Omega = \int_0^{\Omega} \rho c_p A \begin{bmatrix} N_i^2 & N_i N_j \\ N_i N_j & N_j^2 \end{bmatrix} dl \tag{3.114}$$

It should be noted that $d\Omega$ is replaced by Adl in the above equation. Here A is the uniform cross-sectional area of a one-dimensional body. The integration of Equation 3.114 gives

$$[C] = \frac{\rho c_p l A}{6} \begin{bmatrix} 2 & 1 \\ 1 & 2 \end{bmatrix} \tag{3.115}$$

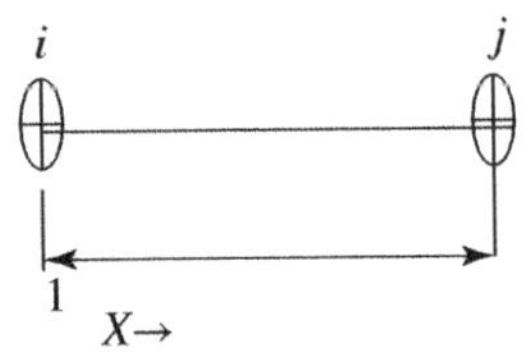

Figure 3.24 One-dimensional linear elements.

Similarly, the $[K]$ matrix and load vector $\{f\}$ can be written as

$$[K] = \frac{Ak_x}{l}\begin{bmatrix} 1 & -1 \\ -1 & 1 \end{bmatrix} + \frac{hPl}{6}\begin{bmatrix} 2 & 1 \\ 1 & 2 \end{bmatrix} \tag{3.116}$$

and

$$\{f\} = \frac{GAl}{l}\begin{Bmatrix} 1 \\ 1 \end{Bmatrix} - \frac{qPl}{2}\begin{Bmatrix} 1 \\ 1 \end{Bmatrix} + \frac{hT_a Pl}{2}\begin{Bmatrix} 1 \\ 1 \end{Bmatrix} \tag{3.117}$$

where P is the perimeter of the one-dimensional body. Substituting Equations 3.116 and 3.117 into Equation 3.113 we get

$$\begin{aligned} &\frac{\rho c_p l A}{6}\begin{bmatrix} 2 & 1 \\ 1 & 2 \end{bmatrix}\begin{Bmatrix} \frac{\partial T_i}{\partial t} \\ \frac{\partial T_j}{\partial t} \end{Bmatrix} + \left(\frac{Ak_x}{l}\begin{bmatrix} 1 & -1 \\ -1 & 1 \end{bmatrix} + \frac{hPl}{6}\begin{bmatrix} 2 & 1 \\ 1 & 2 \end{bmatrix}\right)\begin{Bmatrix} T_i \\ T_j \end{Bmatrix} \\ &= \frac{GAl}{l}\begin{Bmatrix} 1 \\ 1 \end{Bmatrix} - \frac{qPl}{2}\begin{Bmatrix} 1 \\ 1 \end{Bmatrix} + \frac{hT_a Pl}{2}\begin{Bmatrix} 1 \\ 1 \end{Bmatrix} \end{aligned} \tag{3.118}$$

Equation 3.118 is a general representation of a one-dimensional transient conduction problem with one linear element. All the terms are included irrespective of whether or not boundary fluxes and heat generation are present. This equation can be appropriately modified when solving numerical problems, and it is semi-discretized as it is discretized only in space. Now a method of discretizing the transient terms is required. The following section gives the details about the method of discretizing the transient terms.

3.2.9 Time Discretization using the Finite Element Method

In the previous section, the temporal term in the transient conduction equation has been discretized using the finite-difference method. In this section, we concentrate on the use of the finite element method to discretize the equation in time domain. In order to derive the appropriate transient relations using FEM, let us rewrite the semi-discretize one-dimensional Equation 3.103. In this equation, the temperature is discretized in the time domain as

$$T(t) = N_i(t)T_i(t) + N_j(t)T_j(t) \tag{3.119}$$

where the linear shape functions $N_i(t)$ and $N_j(t)$ are given as

$$N_i(t) = 1 - \frac{t}{\Delta t}; \quad N_j(t) = \frac{t}{\Delta t} \tag{3.120}$$

The time derivative of the temperature is thus written as follows:

$$\frac{dT(t)}{dt} = \frac{dN_i(t)}{dt}T_i(t) + \frac{dN_j(t)}{dt}T_j(t) \tag{3.121}$$

Substituting Equation 3.119 into Equation 3.121 we get

$$\frac{dT(t)}{dt} = -\frac{1}{\Delta t}T_i(t) + \frac{1}{\Delta t}T_j(t) \tag{3.122}$$

Substituting Equations 3.119 and 3.122 into Equation 3.103 and applying the weighted residual principle (Galerkin method), we obtain for a time interval of Δt,

$$\int_{\Delta t}\begin{Bmatrix} N_i \\ N_j \end{Bmatrix}\left[[C]\left(-\frac{T_i(\Delta t)}{\Delta t} + \frac{T_j(\Delta t)}{\Delta t}\right) + [K]\left(N_i(t)T_i(t) + N_j(t)T_j(t)\right) - \{f\}\right]dt = 0 \tag{3.123}$$

Using the relation

$$\int_{\Delta t} N_i^a N_j^b dt = \frac{a!b!}{(a+b+1)!}\Delta t \tag{3.124}$$

we obtain the characteristic equation for the time interval Δt as

$$\frac{[C]}{\Delta t}\begin{bmatrix} -1 & 1 \\ -1 & 1 \end{bmatrix}\begin{Bmatrix} T_i \\ T_j \end{Bmatrix} + \frac{[K]}{3}\begin{bmatrix} 2 & 1 \\ 1 & 2 \end{bmatrix}\begin{Bmatrix} T_i \\ T_j \end{Bmatrix} = \frac{1}{2}\begin{Bmatrix} f_1 \\ f_2 \end{Bmatrix} \tag{3.125}$$

The above equation involves the temperature values at the pth and $(p+1)^{th}$ time levels. A quadratic variation of temperature with respect to time may be derived in a similar manner. For convection problems using FEM, refer to the books by Lewis et al. [1,10].

3.2.10 Finite Element Method for Heat Exchangers

Application of finite element method for the analysis of heat exchangers is illustrated in this section for both pure parallel flow and counter flow arrangements. The method can be extended to other types of heat exchangers (Ravi Kumar, et al. [18–20], Ranganayakulu et al. [21–25], Quadir et al. [26–30], Jeevan et al. [31,32], Hegde et al. [33–35], Seetharamu et al. [36], Sai et al. [37], Srinivas et al. [38] Usmani et al. [39,40], Saeid et al. [41], Lee et al. [42], Beh et al. [43], Chaudhuri et al. [44], Reddy and Gartling [45], Patankar [46], Hirsch [47] and Patnaik et al. [48]).

3.2.10.1 Governing Equations

Consider the steady-state flow of hot and cold fluids as shown in Figure 3.25. The boundary conditions are given by specifying the inlet temperatures of the two fluids. Applying the law of conservation of energy to both the fluids, the following equations are derived:

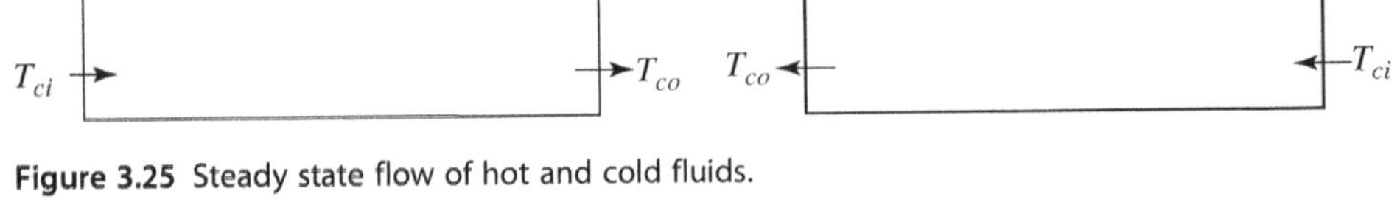

Figure 3.25 Steady state flow of hot and cold fluids.

Hot Fluid:

$$C_h \frac{dT_h}{dx} = -UP[T_h - T_c]$$

$$\frac{dT_h}{dx} = \frac{-UP}{C_h}[T_h - T_c] \tag{3.126}$$

$$\frac{dT_h}{dx} + \frac{UP}{C_h}[T_h - T_c] = 0$$

Cold Fluid:

$$C_c \frac{dT_c}{dx} = \pm UP[T_h - T_c]$$

$$\frac{dT_c}{dx} = \pm \frac{UP}{C_c}[T_h - T_c] \tag{3.127}$$

$$\frac{dT_c}{dx} \mp \frac{UP}{C_c}[T_h - T_c] = 0$$

In Equations 3.126 and 3.127, P stands for contact perimeter of the heat exchanger surface. In order to simplify and generalize the equations, the following dimensionless variables are introduced:

$$\theta = \frac{T - T_{ci}}{T_{hi} - T_{ci}} \text{ and } X = \frac{x}{L_e} \tag{3.128}$$

where L_e is the length of the element given by

$$L_e = \frac{L}{N}$$

It can be shown that the dimensionless forms of the equations are

$$\frac{d\theta_h}{dx} + \frac{UPL}{C_h}[\theta_h - \theta_c] = 0 \tag{3.129}$$

$$\frac{d\theta_c}{dx} \mp \frac{UPL}{C_c}[\theta_h - \theta_c] = 0 \tag{3.130}$$

Here the positive sign is for counter-current flow and negative sign for co-current flow. The governing parameters can be combined together in order to simplify the equations as follows:

$$Ntu = \frac{UPL_e}{C_h} \text{ and } R = \frac{C_h}{C_c} \; Ntu_e = \frac{Ntu}{N} \tag{3.131}$$

Therefore, the final forms of the dimensionless governing equations are as follows:

$$\frac{d\theta_h}{dx} + Ntu_e[\theta_h - \theta_c] = 0 \tag{3.132}$$

$$\frac{d\theta_c}{dx} \pm Ntu_e R[\theta_h - \theta_c] = 0 \tag{3.133}$$

The above equations can be solved with the specified inlet temperatures to find the outlet temperatures of the two fluids. Therefore, the boundary conditions can be defined as $\theta_{hi} = 1$ and $\theta_{ci} = 0$.

3.2.10.2 Finite Element Formulation

The heat exchanger is discretized into a number of elements. We use the method of minimizing the weighted residual to solve the equations as

$$\begin{aligned}
&\int_0^1 W\left[\frac{d\theta_h}{dx} + Ntu_e[\theta_h - \theta_c]\right]dX = 0 \\
&\int_0^1 W\left[\frac{d\theta_c}{dx} \pm Ntu_e R[\theta_h - \theta_c]\right]dX = 0
\end{aligned} \tag{3.134}$$

Assuming linear variation of the hot and cold fluid temperatures in a single element we get the following:

For co-current flow (parallel flow)

$$\left.\begin{aligned}
\theta_h &= N_1\theta_{h,i} + N_2\theta_{h,o} \\
\theta_c &= N_1\theta_{c,i} + N_2\theta_{co}
\end{aligned}\right\} \tag{3.135}$$

and for counter flow

$$\left.\begin{aligned}
\theta_h &= N_1\theta_{h,i} + N_2\theta_{h,o} \\
\theta_c &= N_2\theta_{c,i} + N_1\theta_{c,o}
\end{aligned}\right\} \tag{3.136}$$

where $N_1 = 1 - X$ and $N_2 = X$. We make use of the following results

$$\begin{aligned}
&\frac{dN_1}{dx} = -1; \quad \frac{dN_2}{dx} = 1; \\
&\int_0^1 N_1 dX = \int_0^1 N_2 dX = \frac{1}{2} \\
&\int_0^1 N_1^2 dX = \int_0^1 N_2^2 dX = \frac{1}{3} \\
&\int_0^1 N_1 N_2 dX = \frac{1}{6}
\end{aligned} \tag{3.137}$$

Substituting these approximations in Equations 3.132 and 3.133, a set of equations can be obtained if W is defined.

In the present analysis, two methods are used. The first method is the sub-domain collocation method in which $W = 1$ and the second method is the Galerkin method in which the shape functions N_1 and N_2 are taken as the weighted parameters.

The discretized equations can be written in matrix form as

$$[K]\{\theta\} = \{f\} \tag{3.138}$$

where $[K]$ is known as the stiffness matrix. Assembling the element matrices of all elements leads to the global matrix in the whole solution domain. The resultant global matrix can be solved by Gauss–Jordan elimination method for the dimensionless temperatures along the heat exchanger.

The thermal effectiveness or the overall effectiveness ε of a two-fluid heat exchanger is defined as the ratio of actual heat transfer rate in an exchanger to the maximum possible heat transfer rate, that is,

$$\varepsilon = \frac{\dot{Q}_{act}}{\dot{Q}_{max}} \tag{3.139}$$

In terms of the amount of heat lost by the hot fluid we get

$$\dot{Q}_{act} = C_h(T_{h,i} - T_{h,o}) \tag{3.140a}$$

In terms of the amount of heat gained by the cold fluid we get

$$\dot{Q}_{act} = C_c(T_{c,o} - T_{c,i}) \tag{3.140b}$$

When $C_h < C_c$

$$\dot{Q}_{max} = C_h(T_{h,i} - T_{c,i}) \tag{3.140c}$$

From Equations 3.139 and 3.140c we get

$$\varepsilon = \frac{T_{h,i} - T_{h,o}}{T_{h,i} - T_{c,i}} \tag{3.141}$$

In terms of non-dimensional parameters

$$\varepsilon = 1 - \theta_{h,o} \tag{3.142}$$

From Equations 3.140b and 3.140c we get

$$\varepsilon = \left(\frac{1}{R}\right)\left(\frac{T_{c,o} - T_{c,i}}{T_{h,i} - T_{c,i}}\right) \tag{3.143}$$

In terms of non-dimensional parameters

$$\varepsilon = \frac{\theta_{c,o}}{R} \tag{3.144}$$

When $C_c < C_h$

$$\dot{Q}_{max} = C_c(T_{h,i} - T_{c,i}) \tag{3.145}$$

From equations ($\dot{Q}_{act} = C_h(T_{h,i} - T_{h,o})$ and $\left(\varepsilon = \frac{T_{h,i} - T_{h,o}}{T_{h,i} - T_{c,i}}\right)$ we get

$$\varepsilon = R\left(\frac{T_{h,i} - T_{h,o}}{T_{h,i} - T_{c,i}}\right) \tag{3.146}$$

In terms of non-dimensional parameters

$$\varepsilon = R(1 - \theta_{h,o}) \tag{3.147}$$

From equations ($\dot{Q}_{act} = C_c(T_{c,o} - T_{c,i})$) and $\left(\varepsilon = \dfrac{T_{h,i} - T_{h,o}}{T_{h,i} - T_{c,i}}\right)$ we get

$$\varepsilon = \left(\frac{T_{c,o} - T_{c,i}}{T_{h,i} - T_{c,i}}\right) \tag{3.148}$$

In terms of non-dimensional parameters

$$\varepsilon = \theta_{c,o} \tag{3.149}$$

The analytical formula for a co-current flow heat exchanger is given by Incropera et al. [49], Ravikumar et al. [50] and Seetharamu [51]:

$$\varepsilon = \frac{1 - \exp\{-Ntu(1+R)\}}{1+R} \tag{3.150}$$

For counter-current flow we have

$$\varepsilon = \frac{1 - \exp\{-Ntu(1-R)\}}{1 - R* \exp\{-Ntu(1-R)\}}; \quad R < 1 \tag{3.151}$$

$$\varepsilon = \frac{Ntu}{1 + Ntu}; \quad R = 1 \tag{3.152}$$

Example calculations

Objective: Comparison of effectiveness values obtained from analytical method and FEM.

Data assumed: $NTU = 1$; $R = 0.5$.

i) Analytical method

For co-current flow:

$$\varepsilon = \frac{1 - \exp\{-Ntu(1+R)\}}{1+\mathrm{R}} = \frac{1 - \exp\{-1(1+0.5)\}}{1+0.5} = \frac{1 - 0.223}{1.5} = 0.5179$$

for counter current flow:

$$\varepsilon = \frac{1 - \exp\{-Ntu(1-R)\}}{1 - \mathrm{R}* \exp\{-Ntu(1-R)\}} = \frac{1 - \exp\{-1(1-0.5)\}}{1 - 0.5* \exp\{-1(1-0.5)\}} = \frac{1 - 0.6065}{0.6967} = 0.5647$$

ii) Sub-Domain Collocation Method

Here the weighted parameter is taken as 1, that is, $W = 1$.

Co-current flow

$$\begin{aligned}
&\theta_h = N_1\theta_{h,i} + N_2\theta_{h,o}\\
&\theta_c = N_1\theta_{c,i} + N_2\theta_{c,o}\\
&N_1 = 1 - X; \quad N_2 = X\\
&\frac{d\theta_h}{dX} = \frac{d(N_1\theta_{h,i} + N_2\theta_{h,o})}{dX} = -\theta_{h,i} + \theta_{h,o}\\
&\frac{d\theta_c}{dX} = \frac{d(N_1\theta_{c,i} + N_2\theta_{c,o})}{dX} = -\theta_{c,i} + \theta_{c,o}
\end{aligned} \tag{3.153}$$

Substituting for the above we get the FEM equations as follows:
Hot fluid:

$$\int_0^1 1\left[\frac{d\theta_h}{dX} + Ntu*(\theta_h - \theta_c)\right]dX = 0$$

$$\int_0^1 [(-\theta_{h,i} + \theta_{h,o}) + Ntu_e*\{(N_1\theta_{h,i} + N_2\theta_{h,o}) - (N_1\theta_{c,i} + N_2\theta_{c,o})\}]dX = 0 \tag{3.154}$$

$$\theta_{h,i}\left(\frac{Ntu_e}{2} - 1\right) + \theta_{c,i}\left(\frac{-Ntu_e}{2}\right) + \theta_{h,o}\left(\frac{Ntu_e}{2} + 1\right) + \theta_{c,o}\left(\frac{-Ntu_e}{2}\right) = 0$$

Cold fluid:

$$\int_0^1 1\left[\frac{d\theta_c}{dx} - Ntu*R*(\theta_h - \theta_c)\right]dX = 0$$

$$\int_0^1 [(-\theta_{c,i} + \theta_{c,o}) - Ntu*R*\{(N_1\theta_{h,i} + N_2\theta_{h,o}) - (N_1\theta_{c,i} + N_2\theta_{c,o})\}]dX = 0$$

$$\theta_{h,i}\left(\frac{-Ntu_e*R}{2}\right) + \theta_{c,i}\left(\frac{Ntu_e*R}{2} - 1\right) + \theta_{h,o}\left(\frac{-Ntu_e*R}{2}\right) + \theta_{c,o}\left(\frac{Ntu_e*R}{2} + 1\right) = 0 \tag{3.155}$$

Equations $(q = \varepsilon C_{min}(T_{h,i} - T_{c,i}))$ and $\left(\varepsilon = f\left(Ntu, \frac{C_{min}}{C_{max}}\right)\right)$ along with the boundary conditions $\theta_{h,l} = 1, \theta_{c,l} = 0$, can be written in matrix form as follows:

$$\begin{pmatrix} 1 & 0 & 0 & 0 \\ 0 & 1 & 0 & 0 \\ \left(\frac{Ntu_e}{2} - 1\right) & \left(\frac{-Ntu_e}{2}\right) & \left(\frac{Ntu_e}{2} + 1\right) & \left(\frac{-Ntu_e}{2}\right) \\ \left(\frac{-Ntu_e*R}{2}\right) & \left(\frac{Ntu_e*R}{2} - 1\right) & \left(\frac{-Ntu_e}{2}\right) & \left(\frac{Ntu_e*R}{2} + 1\right) \end{pmatrix} \begin{pmatrix} \theta_{h,i} \\ \theta_{c,i} \\ \theta_{h,o} \\ \theta_{c,o} \end{pmatrix} = \begin{pmatrix} 1 \\ 0 \\ 0 \\ 0 \end{pmatrix} \tag{3.156}$$

Now for one element:

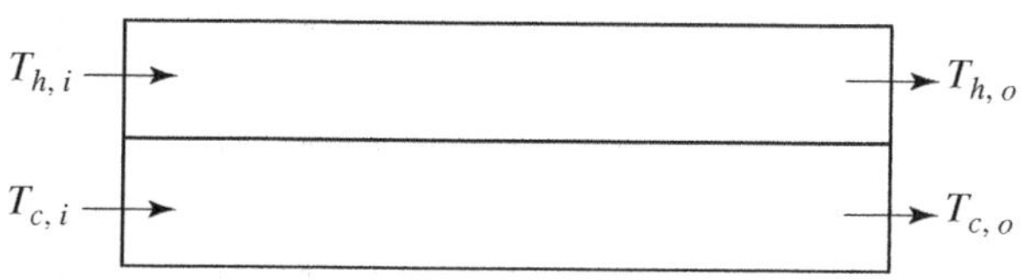

$NTU = 1$; $R = 0.5$; number of elements $= N = 1$; $Ntu_e = \dfrac{Ntu}{N}$; $NTU_e = 1$; $[K]\ \{\theta\} = \{f\}$:

$$\begin{pmatrix} 1.00 & 0.00 & 0.00 & 0.00 \\ 0.00 & 1.00 & 0.00 & 0.00 \\ -0.5 & -0.5 & 1.5 & 0.5 \\ -0.25 & -0.75 & -0.25 & 1.25 \end{pmatrix} \begin{pmatrix} \theta_{h,i} \\ \theta_{c,i} \\ \theta_{h,o} \\ \theta_{c,o} \end{pmatrix} = \begin{pmatrix} 1 \\ 0 \\ 0 \\ 0 \end{pmatrix}$$

Solving the above system of equations, we get

$$\theta_{h,i} = 1$$

$$\theta_{c,i} = 0$$

$$\theta_{h,o} = 0.4286$$

$$\theta_{c,o} = 0.2857$$

Since $R = 0.5$, $C_h < C_c$. Therefore effectiveness

$$\varepsilon = 1 - \theta_{h,o} = 1 - 0.4286 = 0.5714$$

For two elements

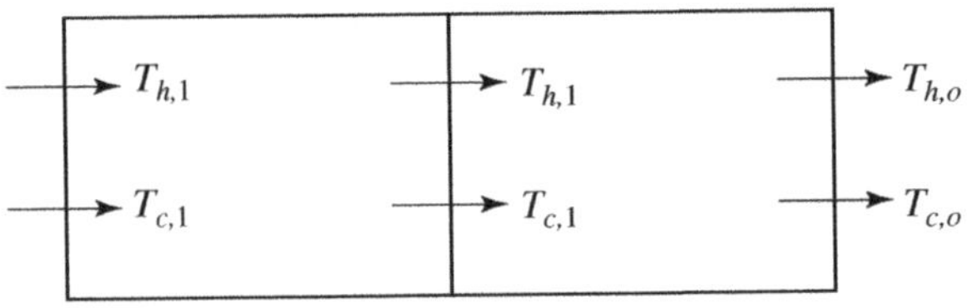

For the first element, the system of equations in matrix form is as follows:

$$\begin{pmatrix} 1 & 0 & 0 & 0 \\ 0 & 1 & 0 & 0 \\ \left(\dfrac{Ntu_e}{2} - 1\right) & \left(\dfrac{-Ntu_e}{2}\right) & \left(\dfrac{Ntu_e}{2} + 1\right) & \left(\dfrac{-Ntu_e}{2}\right) \\ \left(\dfrac{-Ntu_e * R}{2}\right) & \left(\dfrac{Ntu_e * R}{2} - 1\right) & \left(\dfrac{-Ntu_e}{2}\right) & \left(\dfrac{Ntu_e * R}{2} + 1\right) \end{pmatrix} \begin{pmatrix} \theta_{h,i} \\ \theta_{c,i} \\ \theta_{h,1} \\ \theta_{c,1} \end{pmatrix} = \begin{pmatrix} 0 \\ 0 \\ 0 \\ 0 \end{pmatrix} \tag{3.157a}$$

For the second element, system of equations in matrix form is as follows:

$$\begin{pmatrix} 1 & 0 & 0 & 0 \\ 0 & 1 & 0 & 0 \\ \left(\dfrac{Ntu_e}{2} - 1\right) & \left(\dfrac{-Ntu_e}{2}\right) & \left(\dfrac{Ntu_e}{2} + 1\right) & \left(\dfrac{-Ntu_e}{2}\right) \\ \left(\dfrac{-Ntu_e * R}{2}\right) & \left(\dfrac{Ntu_e * R}{2} - 1\right) & \left(\dfrac{-Ntu_e}{2}\right) & \left(\dfrac{Ntu_e * R}{2} + 1\right) \end{pmatrix} \begin{pmatrix} \theta_{h,1} \\ \theta_{c,1} \\ \theta_{h,o} \\ \theta_{c,o} \end{pmatrix} = \begin{pmatrix} 0 \\ 0 \\ 0 \\ 0 \end{pmatrix} \tag{3.157b}$$

Assembling the equations for the two elements and imposing the boundary conditions we get $\theta_{hi} = 1, \theta_{ci} = 0$, and the set of equations for the heat exchanger can be written as

$$\begin{pmatrix} 1 & 0 & 0 & 0 & 0 & 0 \\ 0 & 1 & 0 & 0 & 0 & 0 \\ \left(\frac{Ntu_e}{2}-1\right) & \left(\frac{-Ntu_e}{2}\right) & \left(\frac{Ntu_e}{2}+1\right) & \left(\frac{-Ntu_e}{2}\right) & 0 & 0 \\ \left(\frac{-Ntu_e*R}{2}\right) & \left(\frac{Ntu_e*R}{2}-1\right) & \left(\frac{-Ntu_e}{2}\right) & \left(\frac{Ntu_e*R}{2}+1\right) & 0 & 0 \\ 0 & 0 & \left(\frac{Ntu_e}{2}-1\right) & \left(\frac{-Ntu_e}{2}\right) & \left(\frac{Ntu_e}{2}+1\right) & \left(\frac{-Ntu_e}{2}\right) \\ 0 & 0 & \left(\frac{-Ntu_e*R}{2}\right) & \left(\frac{Ntu_e*R}{2}-1\right) & \left(\frac{-Ntu_e}{2}\right) & \left(\frac{Ntu_e*R}{2}+1\right) \end{pmatrix} \begin{pmatrix} \theta_{h,i} \\ \theta_{c,i} \\ \theta_{h,1} \\ \theta_{c,1} \\ \theta_{h,o} \\ \theta_{c,o} \end{pmatrix} = \begin{pmatrix} 1 \\ 0 \\ 0 \\ 0 \\ 0 \\ 0 \end{pmatrix} \tag{3.158a}$$

$NTU = 1$; $R = 0.5$; number of elements $= N = 2$; $Ntu_e = \frac{Ntu}{N} = \frac{1}{2}$;

$$\begin{pmatrix} 1.00 & 0.00 & 0.00 & 0.00 & 0.00 & 0.00 \\ 0.00 & 1.00 & 0.00 & 0.00 & 0.00 & 0.00 \\ 0.75 & 0.25 & 1.25 & -0.25 & 0.00 & 0.00 \\ -0.125 & -0.875 & -0.125 & 1.125 & 0.00 & 0.00 \\ 0.00 & 0.00 & -0.75 & -0.25 & 1.25 & 0.25 \\ 0.00 & 0.00 & -0.125 & -0.875 & -0.125 & 1.125 \end{pmatrix} \begin{pmatrix} \theta_{h,i} \\ \theta_{c,i} \\ \theta_{h,1} \\ \theta_{c,1} \\ \theta_{h,o} \\ \theta_{c,o} \end{pmatrix} = \begin{pmatrix} 1 \\ 0 \\ 0 \\ 0 \\ 0 \\ 0 \end{pmatrix} \tag{3.158b}$$

Solving the above system of equations, we get

$$\theta_{h,i} = 1$$
$$\theta_{c,i} = 0$$
$$\theta_{h,1} = 0.6364$$
$$\theta_{c,1} = 0.1818$$
$$\theta_{h,o} = 0.4711$$
$$\theta_{c,o} = 0.2645$$

Effectiveness $= \varepsilon = 1 - \theta_{h,o} = 1 - 0.4711 = 0.5289$

Counter-current flow

$$\theta_h = N_1\theta_{h,i} + N_2\theta_{h,o}$$
$$\theta_c = N_2\theta_{c,i} + N_1\theta_{c,o} \tag{3.159}$$
$$N_1 = 1 - X; \quad N_2 = X$$
$$\frac{d\theta_h}{dX} = \frac{d(N_1\theta_{h,i} + N_2\theta_{h,o})}{dX} = -\theta_{h,i} + \theta_{h,o}$$
$$\frac{d\theta_c}{dX} = \frac{d(N_2\theta_{c,i} + N_1\theta_{c,o})}{dX} = \theta_{c,i} - \theta_{c,o}$$

Substituting for the above we get the FEM equations as follows:
Hot fluid:

$$\int_0^1 1\left[\frac{d\theta_h}{dx} + Ntu*(\theta_h - \theta_c)\right]dX = 0$$

$$\int_0^1 [(\theta_{h,i} - \theta_{h,o}) + Ntu*\{(N_1\theta_{h,i} + N_2\theta_{h,o}) - (N_2\theta_{c,i} + N_1\theta_{c,o})\}]dX = 0 \tag{3.160}$$

$$\theta_{h,i}\left(\frac{Ntu_e}{2} - 1\right) + \theta_{c,o}\left(\frac{-Ntu_e}{2}\right) + \theta_{h,o}\left(\frac{Ntu_e}{2} + 1\right) + \theta_{c,i}\left(\frac{-Ntu_e}{2}\right) = 0$$

Cold fluid:

$$\int_0^1 1\left[\frac{d\theta_c}{dx} + Ntu*R*(\theta_h - \theta_c)\right]dX = 0$$

$$\int_0^1 [(\theta_{c,o} - \theta_{c,i}) + Ntu*R*\{(N_1\theta_{h,i} + N_2\theta_{h,o}) - (N_2\theta_{c,i} + N_1\theta_{c,o})\}]dX = 0$$

$$\theta_{h,i}\left(\frac{Ntu_e*R}{2}\right) + \theta_{c,o}\left(-\frac{Ntu_e*R}{2} - 1\right) + \theta_{h,o}\left(\frac{Ntu_e}{2}\right) + \theta_{c,i}\left(-\frac{Ntu_e*R}{2} + 1\right) = 0 \tag{3.161}$$

Equations 3.160 and 3.161 along with the boundary conditions $\theta_{h,i} = 1, \theta_{c,i} = 0$, can be written in matrix form as

$$\begin{pmatrix} 1 & 0 & 0 & 0 \\ 0 & 0 & 0 & 1 \\ \left(\frac{Ntu_e}{2} - 1\right) & \left(\frac{-Ntu_e}{2}\right) & \left(\frac{Ntu_e}{2} + 1\right) & \left(\frac{-Ntu_e}{2}\right) \\ \left(\frac{Ntu_e*R}{2}\right) & \left(-\frac{Ntu_e*R}{2} - 1\right) & \left(\frac{Ntu_e}{2}\right) & \left(-\frac{Ntu_e*R}{2} + 1\right) \end{pmatrix} \begin{pmatrix} \theta_{h,i} \\ \theta_{c,o} \\ \theta_{h,o} \\ \theta_{c,i} \end{pmatrix} = \begin{pmatrix} 1 \\ 0 \\ 0 \\ 0 \end{pmatrix} \tag{3.162}$$

For one element

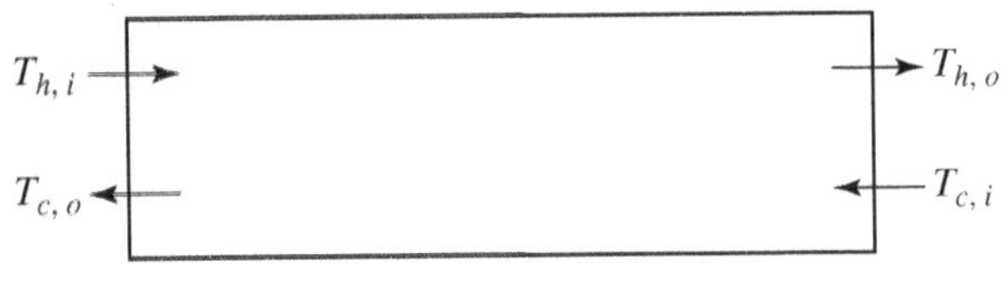

$NTU = 1$; $R = 0.5$; number of elements $= N = 1$; $Ntu_e = \dfrac{Ntu}{N} = 1$:

$$[K]\{\theta\} = \{f\}$$

$$\begin{pmatrix} 1 & 0 & 0 & 0 \\ 0 & 0 & 0 & 1 \\ -0.5 & -0.5 & 1.5 & -0.5 \\ 0.25 & -1.25 & 0.25 & 0.75 \end{pmatrix} \begin{pmatrix} \theta_{h,I} \\ \theta_{c,o} \\ \theta_{h,o} \\ \theta_{c,I} \end{pmatrix} = \begin{pmatrix} 1 \\ 0 \\ 0 \\ 0 \end{pmatrix} \tag{3.163}$$

Solving the above system of equations, we get

$$\theta_{h,i} = 1$$

$$\theta_{c,i} = 0$$

$$\theta_{h,o} = 0.4286$$

$$\theta_{c,o} = 0.2857$$

Effectiveness $= \varepsilon = 1 - \theta_{h,o} = 1 - 0.4246 = 0.5714$

For two elements

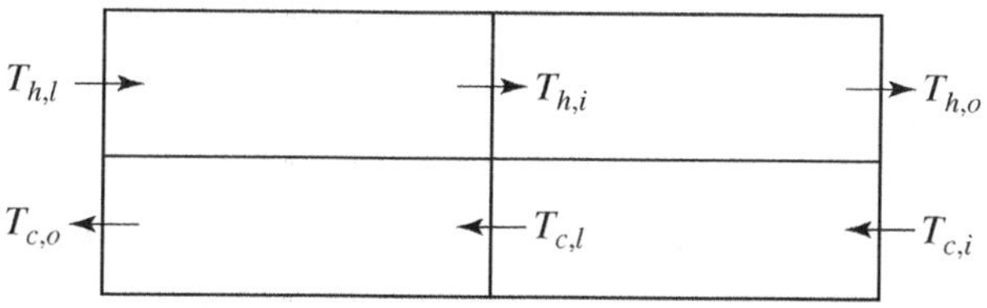

For the first element, the system of equations in matrix form are as follows:

$$\begin{pmatrix} 1 & 0 & 0 & 0 \\ 0 & 0 & 0 & 0 \\ \left(\frac{Ntu_e}{2}\;\;1\right) & \left(\frac{-Ntu_e}{2}\right) & \left(\frac{Ntu_e}{2}+1\right) & \left(\frac{-Ntu_e}{2}\right) \\ \left(\frac{Ntu_e*R}{2}\right) & \left(-\frac{Ntu_e*R}{2}-1\right) & \left(\frac{Ntu_e}{2}\right) & \left(-\frac{Ntu_e*R}{2}+1\right) \end{pmatrix} \begin{pmatrix} \theta_{h,i} \\ \theta_{c,o} \\ \theta_{h,1} \\ \theta_{c,1} \end{pmatrix} = \begin{pmatrix} 0 \\ 0 \\ 0 \\ 0 \end{pmatrix} \tag{3.164a}$$

For the second element, the system of equations in matrix form is as follows:

$$\begin{pmatrix} 0 & 0 & 0 & 0 \\ 0 & 0 & 0 & 0 \\ \left(\frac{Ntu_e}{2}-1\right) & \left(\frac{-Ntu_e}{2}\right) & \left(\frac{Ntu_e}{2}+1\right) & \left(\frac{-Ntu_e}{2}\right) \\ \left(\frac{Ntu_e*R}{2}\right) & \left(-\frac{Ntu_e*R}{2}-1\right) & \left(\frac{Ntu_e}{2}\right) & \left(-\frac{Ntu_e*R}{2}+1\right) \end{pmatrix} \begin{pmatrix} \theta_{h,1} \\ \theta_{c,1} \\ \theta_{h,o} \\ \theta_{c,i} \end{pmatrix} = \begin{pmatrix} 0 \\ 0 \\ 0 \\ 0 \end{pmatrix} \tag{3.164b}$$

Assembling the equations for the two elements and applying the boundary conditions we get $\theta_{h,i} = 1, \theta_{c,i} = 0$. The set of equations for the heat exchanger can be written as follows:

$$\begin{pmatrix} 1 & 0 & 0 & 0 & 0 & 0 \\ 0 & 0 & 0 & 0 & 0 & 1 \\ \left(\frac{Ntu_e}{2}-1\right) & \left(\frac{-Ntu_e}{2}\right) & \left(\frac{Ntu_e}{2}+1\right) & \left(\frac{-Ntu_e}{2}\right) & 0 & 0 \\ \left(\frac{Ntu_e*R}{2}\right) & \left(-\frac{Ntu_e*R}{2}-1\right) & \left(\frac{Ntu_e}{2}\right) & \left(-\frac{Ntu_e*R}{2}+1\right) & 0 & 0 \\ 0 & 0 & \left(\frac{Ntu_e}{2}-1\right) & \left(\frac{-Ntu_e}{2}\right) & \left(\frac{Ntu_e}{2}+1\right) & \left(\frac{-Ntu_e}{2}\right) \\ 0 & 0 & \left(\frac{Ntu_e*R}{2}\right) & \left(-\frac{Ntu_e*R}{2}-1\right) & \left(\frac{Ntu_e}{2}\right) & \left(-\frac{Ntu_e*R}{2}+1\right) \end{pmatrix} \begin{pmatrix} \theta_{h,i} \\ \theta_{c,o} \\ \theta_{h,1} \\ \theta_{c,1} \\ \theta_{h,o} \\ \theta_{c,i} \end{pmatrix} = \begin{pmatrix} 1 \\ 0 \\ 0 \\ 0 \\ 0 \\ 0 \end{pmatrix} \quad (3.165)$$

$NTU = 1$; $R = 0.5$; number of elements = N = 2; $Ntu_e = \dfrac{Ntu}{N} = \dfrac{1}{2}$

$$\begin{pmatrix} 1 & 0 & 0 & 0 & 0 & 0 \\ 0 & 0 & 0 & 0 & 0 & 1 \\ 00.75 & 0.25 & 1.25 & -0.25 & 0 & 0 \\ 0.125 & -1.125 & 0.125 & 0.875 & 0 & 0 \\ 0 & 0 & -0.75 & -0.25 & 1.25 & -0.25 \\ 0 & 0 & 0.125 & -1.125 & 0.125 & 0.875 \end{pmatrix} \begin{pmatrix} \theta_{h,i} \\ \theta_{c,o} \\ \theta_{h,1} \\ \theta_{c,1} \\ \theta_{h,o} \\ \theta_{c,i} \end{pmatrix} = \begin{pmatrix} 1 \\ 0 \\ 0 \\ 0 \\ 0 \\ 0 \end{pmatrix} \quad (3.166)$$

Solving the above system of equations, we get

$\theta_{h,i} = 1$

$\theta_{c,i} = 0$

$\theta_{h,1} = 0.6814$

$\theta_{c,1} = 0.1239$

$\theta_{h,o} = 0.4336$

$\theta_{c,o} = 0.2832$

Effectiveness $= \varepsilon = 1 - \theta_{h,o} = 1 - 0.4336 = 0.5664$

Table 3.7 shows the comparison of FEM solutions using sub-domain method for both co-current and counter current flow arrangements with the analytical solutions.

Table 3.7 Comparison of analytical solutions with FEM (sub-domain method) solutions.

		FEM solutions – Sub-Domain Collocation Method				
Type of flow	**Analytical value**	**1e**	**2e**	**4e**	**8e**	**16e**
Co-current	0.5179	0.5714	0.5289	0.5206	0.5186	0.5181
Counter current	0.5647	0.5714	0.5664	0.5651	0.5648	0.5648

Galerkin's Method

Here shape functions N_1 and N_2 are taken as weighted parameters and are given by

$$N_1 = 1 - X; \;\; N_2 = X$$

Co-current flow

$$\begin{aligned}
\theta_h &= N_1\theta_{h,i} + N_2\theta_{h,o} \\
\theta_c &= N_1\theta_{c,i} + N_2\theta_{c,o} \\
N_1 &= 1 - X; \;\; N_2 = X \\
\frac{d\theta_h}{dX} &= \frac{d(N_1\theta_{h,i} + N_2\theta_{h,o})}{dX} = -\theta_{h,i} + \theta_{h,o} \\
\frac{d\theta_c}{dX} &= \frac{d(N_1\theta_{c,i} + N_2\theta_{c,o})}{dX} = -\theta_{c,i} + \theta_{c,o}
\end{aligned} \tag{3.167}$$

Substituting for the above we get the FEM equations as follows:
For hot fluid:

$$\int_0^1 N_1\left[\frac{d\theta_h}{dx} + Ntu_e*(\theta_h - \theta_c)\right]dX = 0$$

$$\int_0^1 N_1[(\theta_{h,o} - \theta_{h,i}) + Ntu_e*\{(N_1\theta_{h,i} + N_2\theta_{h,o}) - (N_1\theta_{c,i} + N_2\theta_{c,o})\}]dX = 0 \tag{3.168a}$$

$$\theta_{h,i}\left(\frac{Ntu_e}{3} - 0.5\right) + \theta_{c,i}\left(\frac{-Ntu_e}{3}\right) + \theta_{h,o}\left(\frac{Ntu_e}{6} + 0.5\right) + \theta_{c,o}\left(\frac{-Ntu_e}{6}\right) = 0$$

$$\int_0^1 N_2\left[\frac{d\theta_h}{dx} + Ntu_e*(\theta_h - \theta_c)\right]dX = 0$$

$$\int_0^1 N_2[(\theta_{h,o} - \theta_{h,i}) + Ntu_e*\{(N_1\theta_{h,i} + N_2\theta_{h,o}) - (N_1\theta_{c,i} + N_2\theta_{c,o})\}]dX = 0 \tag{3.168b}$$

$$\theta_{h,i}\left(\frac{Ntu_e}{6} - 0.5\right) + \theta_{c,i}\left(\frac{-Ntu_e}{6}\right) + \theta_{h,o}\left(\frac{Ntu_e}{3} + 0.5\right) + \theta_{c,o}\left(\frac{-Ntu_e}{3}\right) = 0$$

For cold fluid:

$$\int_0^1 N_1\left[\frac{d\theta_c}{dx} - Ntu_e*R*(\theta_h - \theta_c)\right]dX = 0$$

$$\int_0^1 N_1[(\theta_{c,o} - \theta_{c,i}) - Ntu*R*\{(N_1\theta_{h,i} + N_2\theta_{h,o}) - (N_1\theta_{c,i} + N_2\theta_{c,o})\}]dX = 0$$

$$\theta_{h,i}\left(\frac{-Ntu_e*R}{3}\right) + \theta_{c,i}\left(\frac{Ntu_e*R}{3} - 0.5\right) + \theta_{h,o}\left(\frac{-Ntu_e*R}{6}\right) + \theta_{c,o}\left(\frac{Ntu_e*R}{6} + 0.5\right) = 0 \tag{3.169a}$$

$$\int_0^1 N_2\left[\frac{d\theta_c}{dx} - Ntu_e{*}R{*}(\theta_h-\theta_c)\right]dX = 0$$

$$\int_0^1 N_2[(\theta_{c,o}-\theta_{c,i}) - Ntu{*}R{*}\{(N_1\theta_{h,i}+N_2\theta_{h,o})-(N_1\theta_{c,i}+N_2\theta_{c,o})\}]dX = 0$$

$$\theta_{h,i}\left(\frac{-Ntu_e{*}R}{6}\right)+\theta_{c,i}\left(\frac{Ntu_e{*}R}{6}-0.5\right)+\theta_{h,o}\left(\frac{-Ntu_e{*}R}{3}\right)+\theta_{c,o}\left(\frac{Ntu_e{*}R}{3}+0.5\right)=0 \tag{3.169b}$$

Equations 3.168a, 3.168b, 3.169a and 3.169b can be written in matrix form as follows:

$$[K]\{\theta\} = \{f\}$$

$$\begin{pmatrix} \left(\frac{Ntu_e}{3}-0.5\right) & \left(\frac{-Ntu_e}{3}\right) & \left(\frac{Ntu_e}{6}+0.5\right) & \left(\frac{-Ntu_e}{6}\right) \\ \left(\frac{-Ntu_e{*}R}{3}\right) & \left(\frac{Ntu_e{*}R}{3}-0.5\right) & \left(\frac{-Ntu_e{*}R}{6}\right) & \left(\frac{Ntu_e{*}R}{6}+0.5\right) \\ \left(\frac{Ntu_e}{6}-0.5\right) & \left(\frac{-Ntu_e}{6}\right) & \left(\frac{Ntu_e}{3}+0.5\right) & \left(\frac{-Ntu_e}{3}\right) \\ \left(-\frac{Ntu_e{*}R}{6}\right) & \left(\frac{Ntu_e{*}R}{6}-0.5\right) & \left(-\frac{Ntu_e{*}R}{3}\right) & \left(\frac{Ntu_e{*}R}{3}+0.5\right) \end{pmatrix} \begin{pmatrix}\theta_{h,i}\\ \theta_{c,i}\\ \theta_{h,o}\\ \theta_{c,o}\end{pmatrix} = \begin{pmatrix}0\\0\\0\\0\end{pmatrix} \tag{3.170}$$

For one element

$NTU = 1$; $R = 0.5$; number of elements $= N = 1$; $Ntu_e = \frac{Ntu}{N} = 1$

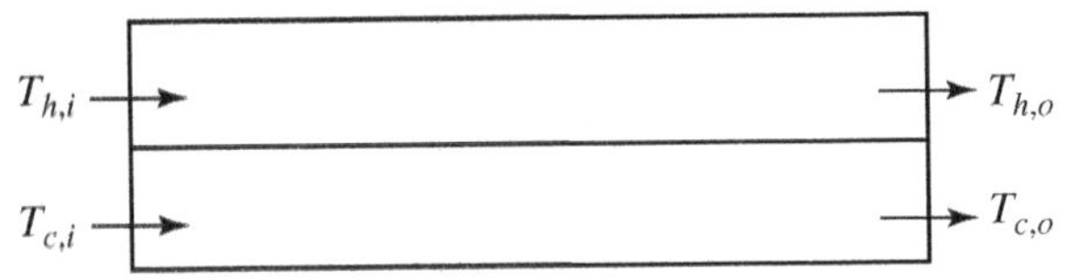

Substituting the above values in Equation 3.170, we get

$$\begin{pmatrix} -0.1667 & -0.333 & 0.6667 & -0.1667 \\ -0.1667 & -0.333 & -0.0833 & 0.5833 \\ -0.333 & -0.1667 & 0.0833 & -0.333 \\ -0.0833 & -0.4167 & -0.1667 & 0.6667 \end{pmatrix} \begin{pmatrix}\theta_{h,i}\\ \theta_{c,i}\\ \theta_{h,o}\\ \theta_{c,o}\end{pmatrix} = \begin{pmatrix}0\\0\\0\\0\end{pmatrix}$$

Imposing boundary conditions $\theta_{h,i} = 1, \theta_{c,i} = 0$, we get

$$\begin{pmatrix} 1 & 0 & 0 & 0 \\ 0 & 1 & 0 & 0 \\ -0.333 & -0.1667 & 0.0833 & -0.333 \\ -0.0833 & -0.4167 & -0.1667 & 0.6667 \end{pmatrix} \begin{pmatrix}\theta_{h,i}\\ \theta_{c,i}\\ \theta_{h,o}\\ \theta_{c,o}\end{pmatrix} = \begin{pmatrix}1\\0\\0\\0\end{pmatrix}$$

Solving the above system of equations, we get

$$\theta_{h,i} = 1.0$$

$$\theta_{c,i} = 0.0$$

$$\theta_{h,o} = 0.5$$

$$\theta_{c,o} = 0.25$$

Effectiveness $= \varepsilon = 1 - \theta_{h,o} = 1 - 0.5 = 0.5$

For two elements

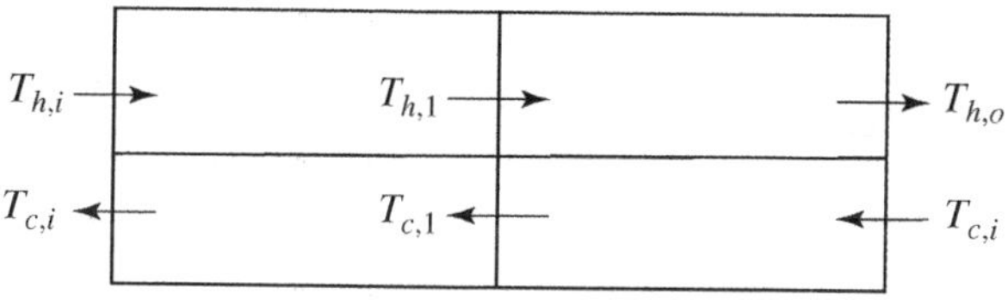

$NTU = 1$; $R = 0.5$; number of elements $= N = 2$; $Ntu_e = \dfrac{Ntu}{N} = \dfrac{1}{2}$. For the first element, the system of equations in matrix form, is

$$\begin{pmatrix} \left(\frac{Ntu_e}{3} - 0.5\right) & \left(\frac{-Ntu_e}{3}\right) & \left(\frac{Ntu_e}{6} + 0.5\right) & \left(\frac{-Ntu_e}{6}\right) \\ \left(\frac{-Ntu_e * R}{3}\right) & \left(\frac{Ntu_e * R}{3} - 0.5\right) & \left(\frac{-Ntu_e * R}{6}\right) & \left(\frac{Ntu_e * R}{6} + 0.5\right) \\ \left(\frac{Ntu_e}{6} - 0.5\right) & \left(\frac{-Ntu_e}{6}\right) & \left(\frac{Ntu_e}{3} + 0.5\right) & \left(\frac{-Ntu_e}{3}\right) \\ \left(-\frac{Ntu_e * R}{6}\right) & \left(\frac{Ntu_e * R}{2} - 0.5\right) & \left(-\frac{Ntu_e * R}{3}\right) & \left(\frac{Ntu_e * R}{3} + 0.5\right) \end{pmatrix} \begin{pmatrix} \theta_{h,i} \\ \theta_{c,i} \\ \theta_{h,1} \\ \theta_{c,1} \end{pmatrix} = \begin{pmatrix} 0 \\ 0 \\ 0 \\ 0 \end{pmatrix} \tag{3.171a}$$

Substituting for NTU_e and R, we get

$$\begin{pmatrix} -0.333 & -0.1667 & 0.5833 & -0.0833 \\ -0.0833 & -0.4167 & -0.0417 & 0.5417 \\ -0.4167 & -0.0833 & 0.6667 & -0.1667 \\ -0.0417 & -0.4583 & -0.0833 & 0.5833 \end{pmatrix} \begin{pmatrix} \theta_{h,i} \\ \theta_{c,i} \\ \theta_{h,1} \\ \theta_{c,1} \end{pmatrix} = \begin{pmatrix} 0 \\ 0 \\ 0 \\ 0 \end{pmatrix} \tag{3.171b}$$

For the second element, the system of equations in matrix form, are as follows:

$$\begin{pmatrix} \left(\frac{Ntu_e}{3} - 0.5\right) & \left(\frac{-Ntu_e}{3}\right) & \left(\frac{Ntu_e}{6} + 0.5\right) & \left(\frac{-Ntu_e}{6}\right) \\ \left(\frac{-Ntu_e * R}{3}\right) & \left(\frac{Ntu_e * R}{3} - 0.5\right) & \left(\frac{-Ntu_e * R}{6}\right) & \left(\frac{Ntu_e * R}{6} + 0.5\right) \\ \left(\frac{Ntu_e}{6} - 0.5\right) & \left(\frac{-Ntu_e}{6}\right) & \left(\frac{Ntu_e}{3} + 0.5\right) & \left(\frac{-Ntu_e}{3}\right) \\ \left(-\frac{Ntu_e * R}{6}\right) & \left(\frac{Ntu_e * R}{6} - 0.5\right) & \left(-\frac{Ntu_e * R}{3}\right) & \left(\frac{Ntu_e * R}{3} + 0.5\right) \end{pmatrix} \begin{pmatrix} \theta_{h,1} \\ \theta_{c,1} \\ \theta_{h,0} \\ \theta_{c,0} \end{pmatrix} = \begin{pmatrix} 0 \\ 0 \\ 0 \\ 0 \end{pmatrix} \tag{3.172a}$$

Substituting for NTU_e and R we get

$$\begin{pmatrix} -0.333 & -0.1667 & 0.5833 & -0.0833 \\ -0.0833 & -0.4167 & -0.0417 & 0.5417 \\ -0.4167 & -0.0833 & 0.6667 & -0.1667 \\ -0.0417 & -0.4583 & -0.0833 & 0.5833 \end{pmatrix} \begin{pmatrix} \theta_{h,1} \\ \theta_{c,1} \\ \theta_{h,o} \\ \theta_{c,o} \end{pmatrix} = \begin{pmatrix} 0 \\ 0 \\ 0 \\ 0 \end{pmatrix} \qquad (3.172b)$$

Assembling Equations 3.171b and 3.172b, we get

$$\begin{pmatrix} -0.333 & -0.1667 & 0.5833 & -0.0833 & 0 & 0 \\ -0.0833 & -0.4167 & -0.0417 & 0.5417 & 0 & 0 \\ -0.4167 & -0.0833 & (0.6666-0.3333) & (-0.1666-0.1666) & 0.5833 & -0 \\ -0.0417 & -0.4583 & (-0.0833-0.0833) & (0.5833-0.4166) & -0.0417 & 0 \\ 0 & 0 & -0.0417 & -0.0833 & 0.6667 & -0.1667 \\ 0 & 0 & -0.0417 & -0.4583 & -0.0833 & 0.5833 \end{pmatrix} \begin{pmatrix} \theta_{h,i} \\ \theta_{c,i} \\ \theta_{h,1} \\ \theta_{c,1} \\ \theta_{h,o} \\ \theta_{c,o} \end{pmatrix} = \begin{pmatrix} 0 \\ 0 \\ 0 \\ 0 \\ 0 \\ 0 \end{pmatrix}$$

Imposing boundary conditions $\theta_{h,i} = 1, \theta_{c,i} = 0$, we get

$$\begin{pmatrix} 1 & 0 & 0 & 0 & 0 & 0 \\ 0 & 1 & 0 & 0 & 0 & 0 \\ -0.4167 & -0.0833 & 0.3333 & -0.3333 & 0.5833 & -0.083 \\ -0.0417 & -0.4583 & 0.1667 & 0.1667 & -0.0417 & 0.5417 \\ 0 & 0 & -0.4167 & -0.0833 & 0.6667 & -0.1667 \\ 0 & 0 & -0.0417 & -0.4583 & -0.0833 & 0.5833 \end{pmatrix} \begin{pmatrix} \theta_{h,I} \\ \theta_{c,i} \\ \theta_{h,1} \\ \theta_{c,1} \\ \theta_{h,o} \\ \theta_{c,o} \end{pmatrix} = \begin{pmatrix} 1 \\ 0 \\ 0 \\ 0 \\ 0 \\ 0 \end{pmatrix} \qquad (3.172c)$$

By solving the above system of equations, we get

$$\theta_{h,i} = 1$$
$$\theta_{c,i} = 0$$
$$\theta_{h,1} = 0.6410$$
$$\theta_{c,1} = 0.1795$$
$$\theta_{h,o} = 0.4872$$
$$\theta_{c,o} = 0.2564$$

Effectiveness = $\varepsilon = 1 - \theta_{h,o} = 1 - 0.4872 = 0.5128$

Counter current flow

$$\begin{aligned} \theta_h &= N_1\theta_{h,i} + N_2\theta_{h,o} \\ \theta_c &= N_2\theta_{c,i} + N_1\theta_{c,o} \\ N_1 &= 1-X; \quad N_2 = X \\ \frac{d\theta_h}{dX} &= \frac{d(N_1\theta_{h,i} + N_2\theta_{h,o})}{dX} = -\theta_{h,i} + \theta_{h,o} \\ \frac{d\theta_c}{dX} &= \frac{d(N_2\theta_{c,i} + N_1\theta_{c,o})}{dX} = \theta_{c,i} - \theta_{c,o} \end{aligned} \qquad (3.173)$$

Substituting for the above we get the FEM equations as follows:
Hot fluid:

$$\int_0^1 N_1\left[\frac{d\theta_h}{dx} + Ntu*(\theta_h - \theta_c)\right]dX = 0$$

$$\int_0^1 N_1[(\theta_{h,i} - \theta_{h,o}) + Ntu*\{(N_1\theta_{h,i} + N_2\theta_{h,o}) - (N_2\theta_{c,i} + N_1\theta_{c,o})\}]dX = 0$$

$$\theta_{h,i}\left(\frac{Ntu_e}{3} - 0.5\right) + \theta_{c,i}\left(\frac{-Ntu_e}{3}\right) + \theta_{h,o}\left(\frac{Ntu_e}{6} + 0.5\right) + \theta_{c,o}\left(\frac{-Ntu_e}{6}\right) = 0 \tag{3.174a}$$

$$\int_0^1 N_2\left[\frac{d\theta_h}{dx} + Ntu*(\theta_h - \theta_c)\right]dX = 0$$

$$\int_0^1 N_2[(\theta_{h,i} - \theta_{h,o}) + Ntu*\{(N_1\theta_{h,i} + N_2\theta_{h,o}) - (N_2\theta_{c,i} + N_1\theta_{c,o})\}]dX = 0$$

$$\theta_{h,i}\left(\frac{Ntu_e}{6} - 0.5\right) + \theta_{c,i}\left(\frac{-Ntu_e}{6}\right) + \theta_{h,o}\left(\frac{Ntu_e}{3} + 0.5\right) + \theta_{c,o}\left(\frac{-Ntu_e}{3}\right) = 0 \tag{3.174b}$$

Cold fluid:

$$\int_0^1 N_1\left[\frac{d\theta_c}{dx} + Ntu*R*(\theta_h - \theta_c)\right]dX = 0$$

$$\int_0^1 N_1[(\theta_{c,i} - \theta_{c,o}) + Ntu*R*\{(N_1\theta_{h,i} + N_2\theta_{h,o}) - (N_2\theta_{c,i} + N_1\theta_{c,o})\}]dX = 0$$

$$\theta_{h,i}\left(\frac{Ntu_e*R}{3}\right) + \theta_{c,o}\left(-\frac{Ntu_e*R}{3} - 0.5\right) + \theta_{h,o}\left(\frac{Ntu_e*R}{6}\right) + \theta_{c,i}\left(-\frac{Ntu_e*R}{6} + 0.5\right) = 0 \tag{3.175a}$$

$$\int_0^1 N_2\left[\frac{d\theta_c}{dx} + Ntu*R*(\theta_h - \theta_c)\right]dX = 0$$

$$\int_0^1 N_2[(\theta_{c,i} - \theta_{c,o}) + Ntu*R*\{(N_1\theta_{h,i} + N_2\theta_{h,o}) - (N_2\theta_{c,i} + N_1\theta_{c,o})\}]dX = 0 \tag{3.175b}$$

$$\theta_{h,i}\left(\frac{Ntu_e*R}{6}\right) + \theta_{c,o}\left(-\frac{Ntu_e*R}{6} - 0.5\right) + \theta_{h,o}\left(\frac{Ntu_e*R}{3}\right)$$

$$+\theta_{c,i}\left(-\frac{Ntu_e*R}{3} + 0.5\right) = 0$$

Equations 3.174a, 3.174b, 3.175a, and 3.175b can be written in matrix form as follows:

$$\begin{pmatrix} \left(\frac{Ntu_e}{3}-0.5\right) & \left(\frac{-Ntu_e}{3}\right) & \left(\frac{Ntu_e}{6}+0.5\right) & \left(\frac{-Ntu_e}{6}\right) \\ \left(\frac{Ntu_e*R}{3}\right) & \left(-\frac{Ntu_e*R}{3}-0.5\right) & \left(\frac{Ntu_e*R}{6}\right) & \left(\frac{-Ntu_e*R}{6}+0.5\right) \\ \left(\frac{Ntu_e}{6}-0.5\right) & \left(\frac{-Ntu_e}{6}\right) & \left(\frac{Ntu_e}{3}+0.5\right) & \left(\frac{-Ntu_e}{3}\right) \\ \left(\frac{Ntu_e*R}{6}\right) & \left(-\frac{Ntu_e*R}{2}-0.5\right) & \left(\frac{Ntu_e*R}{3}\right) & \left(\frac{-Ntu_e*R}{3}+0.5\right) \end{pmatrix} \begin{pmatrix} \theta_{h,i} \\ \theta_{c,1} \\ \theta_{h,0} \\ \theta_{c,i} \end{pmatrix} = \begin{pmatrix} 0 \\ 0 \\ 0 \\ 0 \end{pmatrix} \tag{3.176}$$

For one element

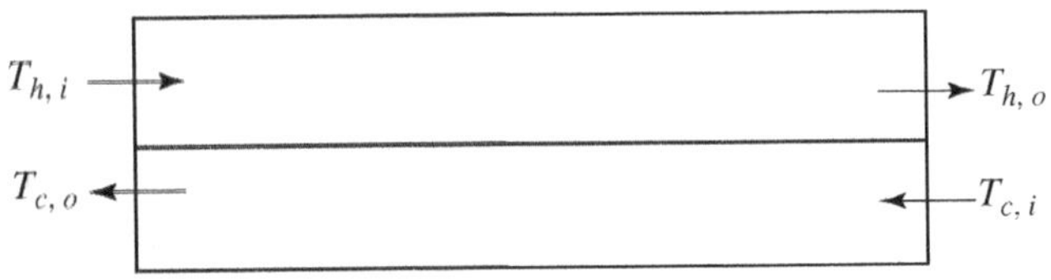

$NTU = 1$; $R = 0.5$; number of elements $= N = 1$; $Ntu_e = \frac{Ntu}{N} = 1$. Substituting above values in Equation 3.176 we get

$$\begin{pmatrix} -0.1667 & -0.3333 & 0.6667 & -0.1667 \\ 0.1667 & -0.6667 & 0.0833 & 0.4167 \\ -0.3333 & -0.1667 & 0.8333 & -0.3333 \\ 0.0833 & -0.5833 & 0.1667 & 0.3333 \end{pmatrix} \begin{pmatrix} \theta_{h,i} \\ \theta_{c,o} \\ \theta_{h,o} \\ \theta_{c,i} \end{pmatrix} = \begin{pmatrix} 0 \\ 0 \\ 0 \\ 0 \end{pmatrix}$$

Imposing boundary conditions $\theta_{h,i} = 1, \theta_{c,i} = 0$, we get

$$\begin{pmatrix} 1.0000 & 0 & 0 & 0 \\ 0.1667 & -0.6667 & 0.0833 & 0.4167 \\ -0.3333 & -0.1667 & 0.8333 & -0.3333 \\ 0 & 0 & 0 & 1.0000 \end{pmatrix} \begin{pmatrix} \theta_{h,i} \\ \theta_{c,o} \\ \theta_{h,o} \\ \theta_{c,i} \end{pmatrix} = \begin{pmatrix} 1 \\ 0 \\ 0 \\ 0 \end{pmatrix}$$

Solving the above system of equations we get

$\theta_{h,i} = 1$

$\theta_{c,i} = 0$

$\theta_{h,o} = 0.4615$

$\theta_{c,o} = 0.3077$

Effectiveness $= \varepsilon = 1 - \theta_{h,o} = 1 - 0.4615 = 0.5385$

For two elements

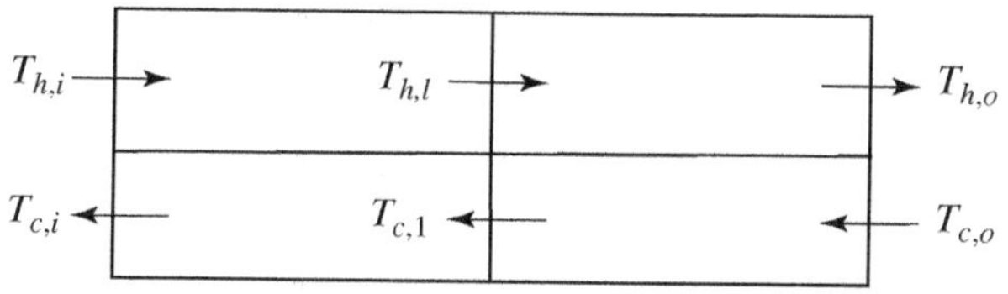

$NTU = 1$; $R = 0.5$; number of elements $= N = 2$; $Ntu_e = \dfrac{Ntu}{N} = 0.5$. For the first element, the system of equations in matrix form is as follows:

$$\begin{pmatrix} \left(\dfrac{Ntu_e}{3}-0.5\right) & \left(\dfrac{-Ntu_e}{3}\right) & \left(\dfrac{Ntu_e}{6}+0.5\right) & \left(\dfrac{-Ntu_e}{6}\right) \\ \left(\dfrac{Ntu_e*R}{3}\right) & \left(-\dfrac{Ntu_e*R}{3}-0.5\right) & \left(\dfrac{Ntu_e*R}{6}\right) & \left(\dfrac{-Ntu_e*R}{6}+0.5\right) \\ \left(\dfrac{Ntu_e}{6}-0.5\right) & \left(\dfrac{-Ntu_e}{6}\right) & \left(\dfrac{Ntu_e}{3}+0.5\right) & \left(\dfrac{-Ntu_e}{3}\right) \\ \left(\dfrac{Ntu_e*R}{6}\right) & \left(-\dfrac{Ntu_e*R}{6}-0.5\right) & \left(\dfrac{Ntu_e*R}{3}\right) & \left(\dfrac{-Ntu_e*R}{3}+0.5\right) \end{pmatrix} \begin{pmatrix} \theta_{h,i} \\ \theta_{c,0} \\ \theta_{h,1} \\ \theta_{c,1} \end{pmatrix} = \begin{pmatrix} 0 \\ 0 \\ 0 \\ 0 \end{pmatrix} \tag{3.177a}$$

Substituting for NTU_e and R, we get

$$\begin{pmatrix} -0.3333 & -0.1667 & 0.5833 & -0.0833 \\ 0.8333 & -0.5833 & 0.0417 & 0.4583 \\ -0.4167 & -0.0833 & 0.6667 & -0.1667 \\ 0.0417 & -0.5417 & 0.0833 & 0.4167 \end{pmatrix} \begin{pmatrix} \theta_{h,i} \\ \theta_{c,o} \\ \theta_{h,1} \\ \theta_{c,1} \end{pmatrix} = \begin{pmatrix} 0 \\ 0 \\ 0 \\ 0 \end{pmatrix} \tag{3.177b}$$

For the second element, the system of equations in matrix form is as follows:

$$\begin{pmatrix} \left(\dfrac{Ntu_e}{3}-0.5\right) & \left(\dfrac{-Ntu_e}{3}\right) & \left(\dfrac{Ntu_e}{6}+0.5\right) & \left(\dfrac{-Ntu_e}{6}\right) \\ \left(\dfrac{Ntu_e*R}{3}\right) & \left(-\dfrac{Ntu_e*R}{3}-0.5\right) & \left(\dfrac{Ntu_e*R}{6}\right) & \left(\dfrac{-Ntu_e*R}{6}+0.5\right) \\ \left(\dfrac{Ntu_e}{6}-0.5\right) & \left(\dfrac{-Ntu_e}{6}\right) & \left(\dfrac{Ntu_e}{3}+0.5\right) & \left(\dfrac{-Ntu_e}{3}\right) \\ \left(\dfrac{Ntu_e*R}{6}\right) & \left(-\dfrac{Ntu_e*R}{6}-0.5\right) & \left(\dfrac{Ntu_e*R}{3}\right) & \left(\dfrac{-Ntu_e*R}{3}+0.5\right) \end{pmatrix} \begin{pmatrix} \theta_{h,1} \\ \theta_{c,1} \\ \theta_{h,0} \\ \theta_{c,i} \end{pmatrix} = \begin{pmatrix} 0 \\ 0 \\ 0 \\ 0 \end{pmatrix} \tag{3.178a}$$

Substituting for NTU_e and R, we get

$$\begin{pmatrix} -0.3333 & -0.1667 & 0.5833 & -0.0833 \\ 0.8333 & -0.5833 & 0.0417 & 0.4583 \\ -0.4167 & -0.0833 & 0.6667 & -0.1667 \\ 0.0417 & -0.5417 & 0.0833 & 0.4167 \end{pmatrix} \begin{pmatrix} \theta_{h,1} \\ \theta_{c,1} \\ \theta_{h,o} \\ \theta_{c,o} \end{pmatrix} = \begin{pmatrix} 0 \\ 0 \\ 0 \\ 0 \end{pmatrix} \tag{3.178b}$$

Table 3.8 Comparison of analytical solutions with FEM (Galerkin's method) solutions.

		FEM solution –Galerkin's method				
Type of flow	**Analytical value**	**1e**	**2e**	**4e**	**8e**	**16e**
Co-current	0.5179	0.5000	0.5128	0.5165	0.5175	0.5178
Counter-current	0.5647	0.5385	0.5643	0.5646	0.5647	0.5647

Assembling Equations 3.177b and 3.178b, we get

$$\begin{pmatrix} -0.333 & -0.1667 & 0.5833 & -0.0833 & 0 & 0 \\ 0.0833 & -0.5833 & 0.0417 & 0.4583 & 0 & 0 \\ -0.4167 & -0.0833 & (0.6666-0.3333) & (-0.1666-0.1666) & 0.5833 & -0.0833 \\ 0.0417 & -0.5417 & (0.0833+0.0833) & (0.4165-0.5833) & 0.0417 & 0.4583 \\ 0 & 0 & -0.4167 & -0.0833 & 0.6667 & -0.1667 \\ 0 & 0 & 0.0417 & -0.5417 & 0.0833 & 0.4167 \end{pmatrix} \begin{pmatrix} \theta_{h,i} \\ \theta_{c,o} \\ \theta_{h,1} \\ \theta_{c,1} \\ \theta_{h,o} \\ \theta_{c,i} \end{pmatrix} = \begin{pmatrix} 0 \\ 0 \\ 0 \\ 0 \\ 0 \\ 0 \end{pmatrix}$$

Imposing the boundary conditions $\theta_{h,i} = 1, \theta_{c,i} = 0$, we get

$$\begin{pmatrix} 1.0000 & 0 & 0 & 0 & 0 & 0 \\ 0.0833 & -0.5833 & 0.0417 & 0.4583 & 0 & 0 \\ -0.4167 & -0.0833 & 0.3333 & -0.3333 & 0.5833 & -0.0833 \\ 0.0417 & -0.5417 & 0.1667 & -0.1667 & 0.0417 & 0.4583 \\ 0 & 0 & -0.4167 & -0.0833 & 0.6667 & -0.1667 \\ 0 & 0 & 0 & 0 & 0 & 1.0000 \end{pmatrix} \begin{pmatrix} \theta_{h,I} \\ \theta_{c,o} \\ \theta_{h,1} \\ \theta_{c,1} \\ \theta_{h,o} \\ \theta_{c,I} \end{pmatrix} = \begin{pmatrix} 1 \\ 0 \\ 0 \\ 0 \\ 0 \\ 0 \end{pmatrix}$$

Solving the above system of equations we get:

$\theta_{h,i} = 1$

$\theta_{c,i} = 0$

$\theta_{h,1} = 0.6740 \quad \theta_{c,1} = 0.1160$

$\theta_{h,o} = 0.4357 \quad \theta c,o = 0.2821$

Effectiveness $= \varepsilon = 1 - \theta_{h,o} = 1 - 0.4357 = 0.5643$

Table 3.8 shows the comparison of FEM solutions using Galerkin's method for both co-current and counter current flow arrangement with the analytical solutions.

3.3 Finite Volume Method

The first step, the control volume integration, distinguishes the finite volume method from all other CFD techniques. The resulting statements express the (exact) conservation of relevant properties for each finite size cell. This clear relationship between the numerical algorithm and the underlying physical conservation principle forms one of the main attractions of the finite volume method and makes its concepts more simple to understand by engineers than finite element and spectral methods. The conservation of a general flow variable, for example a velocity component or enthalpy, within a finite control volume can be expressed as a balance between the various processes tending to increase or decrease it.

Discretization techniques suitable for the treatment of the key transport phenomena, convection (transport due to fluid flow) and diffusion (transport due to variations of ø from point to point) as well as for the source terms (associated with the creation or destruction of ø) and the rate of change with respect to time. The most popular solution procedures are the TDMA line-by-line solver of the algebraic equations and the SIMPLE algorithm to ensure correct linkage between pressure and velocity.

All flow processes contain effects due to convection and diffusion. In diffusive phenomena, such as heat conduction, a change of temperature at one location affects the temperature in more or less equal measure in all directions around it. Convective phenomena involve influencing exclusively in the flow direction so that a point only experiences effects due to changes at upstream locations. Finite volume schemes with the transportiveness property must account for the directionality of influencing in terms of the relative strength of diffusion to convection.

3.3.1 Navier–Stokes Equations

The mathematical model of any fundamental fluid dynamics problem is governed by the Navier–Stokes equations. These equations are important and represent the fluid as a continuum. The equations conserve mass, momentum and energy, and can be derived following either an integral or a differential approach. The integral form of the equations is derived using the Reynolds transport theorem (RTT) and is discussed in many standard fluid mechanics texts (Shames [52]). The approach we follow in this book is the differential approach in which a differential control volume is considered in the fluid domain and a Taylor expansion is used to represent the variation of mass, momentum and energy.

The conservation of mass equation ensures that the total mass is conserved, or, in other words, that the total mass of a fluid system is completely accounted for. In order to derive a general conservation of the mass equation (Lewis et al. [53]) consider a example of Figure 3.26 and the differential control volume as shown in Figure 3.27. The reader can assume the control volume to be infinitesimal for a typical flow problem, such as flow in a channel (Figure 3.26), flow over a flat plate or the temperature (or density) difference driven circulation of air inside a room as shown in Figure 3.28.

Let us assume that the mass flux rate entering the control volume (Figure 3.27) is ρu_1 in the x_1 direction and ρu_2 in the x_2 direction. It is also assumed that there is no reaction or mass production within the fluid domain. The Taylor series expansion may be used to express the mass flux rate exiting the control volume as (Figure 3.27)

$$(\rho u_1)_{x_1+\Delta x_1} = (\rho u_1)_{x_1} + \frac{\Delta x_1}{1!}\frac{\partial(\rho u_1)}{\partial x_1} + \frac{\Delta x_1{}^2}{2!}\frac{\partial^2(\rho u_1)}{\partial x_1{}^2} + \dots \quad (3.179)$$

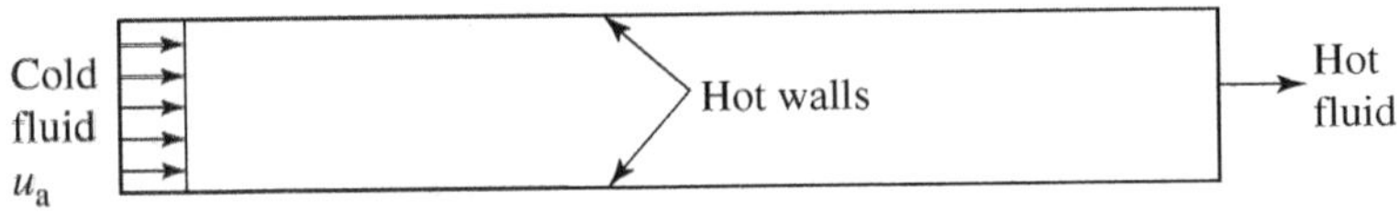

Figure 3.26 Flow and heat transport in a channel.

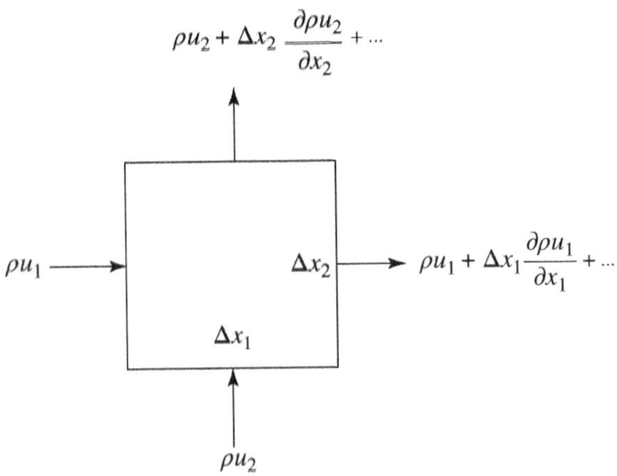

Figure 3.27 Infinitesimal control volume. (Lewis et al. [53])

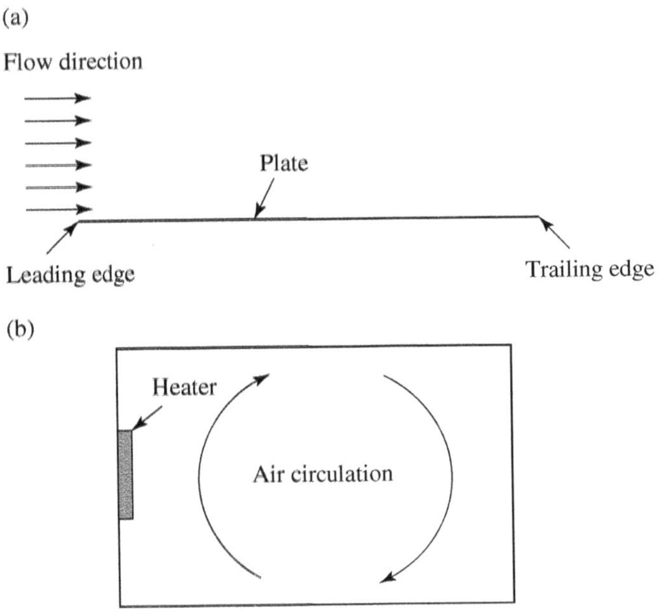

Figure 3.28 Forced flow over a flat plate and natural convection inside a room. (Lewis et al. [53])

in the x_1 direction and,

$$(\rho u_2)_{x_2+\Delta x_2} = (\rho u_2)_{x_2} + \frac{\Delta x_2}{1!}\frac{\partial(\rho u_2)}{\partial x_2} + \frac{\Delta x_2{}^2}{2!}\frac{\partial^2(\rho u_2)}{\partial x_2{}^2} + \ldots \tag{3.180}$$

in the x_2 direction. From an inspection of the control volume shown in Figure 3.27, we can write the difference between the total mass entering and exiting the control volume as

$$\Delta x_2\left[(\rho u_1)_{x_1} - (\rho u_1)_{x_1+\Delta x_1}\right] = -\Delta x_2\left[\frac{\Delta x_1}{1!}\frac{\partial(\rho u_1)}{\partial x_1} + \frac{\Delta x_1{}^2}{2!}\frac{\partial^2(\rho u_1)}{\partial x_1{}^2} + \ldots\right] \tag{3.181}$$

Similarly, in the x_2 direction

$$\Delta x_1\left[(\rho u_2)_{x_2} - (\rho u_2)_{x_2+\Delta x_2}\right] = -\Delta x_1\left[\frac{\Delta x_2}{1!}\frac{\partial(\rho u_2)}{\partial x_2} + \frac{\Delta x_2{}^2}{2!}\frac{\partial^2(\rho u_2)}{\partial x_2{}^2} + \ldots\right] \tag{3.182}$$

Note that the total mass is calculated as being the mass flux rate times the perpendicular area to the following regime. For instance, the total mass entering the control volume in the x_1 direction is $(\Delta x_2) \times (1) \times (\rho u_1)$. A unit thickness is assumed in the x_3 direction.

Adding Equations 3.181 and 3.182 gives the total mass stored inside the control volume. Neglecting the second- and higher-order terms, the total mass stored inside the control volume is

$$-\Delta x_1 \Delta x_2\left[\frac{\partial(\rho u_1)}{\partial x_1} + \frac{\partial(\rho u_2)}{\partial x_2}\right] \tag{3.183}$$

The above quantity, stored within the control volume, is equal to the rate of change of the total mass within the control volume, which is given as

$$\Delta x_1 \Delta x_2 \frac{\partial \rho}{\partial t} \tag{3.184}$$

We can therefore write

$$\Delta x_1 \Delta x_2 \frac{\partial \rho}{\partial t} = -\Delta x_1 \Delta x_2\left[\frac{\partial(\rho u_1)}{\partial x_1} + \frac{\partial(\rho u_2)}{\partial x_2}\right] \tag{3.185}$$

or

$$\frac{\partial \rho}{\partial t} + \frac{\partial(\rho u_1)}{\partial x_1} + \frac{\partial(\rho u_2)}{\partial x_2} = 0 \tag{3.186}$$

The above equation is known as the equation of conservation of mass, or the continuity equation for two-dimensional flows. In three dimensions, the continuity equation is

$$\frac{\partial \rho}{\partial t} + \frac{\partial(\rho u_1)}{\partial x_1} + \frac{\partial(\rho u_2)}{\partial x_2} + \frac{\partial(\rho u_3)}{\partial x_3} = 0 \tag{3.187}$$

If the density is assumed to be constant, then the above equation is reduced to

$$\frac{\partial u_1}{\partial x_1} + \frac{\partial u_2}{\partial x_2} + \frac{\partial u_3}{\partial x_3} = 0 \tag{3.188}$$

Using vector notation, the above equation is written as (divergence-free velocity field)

$$\nabla . u = 0 \tag{3.189}$$

or, using an indicial notation,

$$\frac{\partial u_i}{\partial x_i} = 0 \tag{3.190}$$

where $i = 1, 2$ for a two-dimensional case and $i = 1, 2, 3$ for three-dimensional flows.

3.3.1.1 Conservation of Momentum

The conservation of momentum equation can be derived in a fashion similar to the conservation of mass equation. Here, the momentum equations are derived on the basis of the conservation of momentum principle: the total force generated by the momentum transfer in each direction is balanced by the rate of change of momentum in each direction. The momentum equation has directional components and is therefore a vector equation. In order to derive the conservation of momentum equation, let us consider the control volume shown in Figure 3.29.

Since the momentum equation is a vector equation, the momentum in the x_1 direction will also have a contribution in the x_2 direction. The momentum entering the bottom face in the x_1 direction is

$$\rho u_2 \Delta x_2 u_1 = \rho u_1 u_2 \Delta x_1 \tag{3.191}$$

A Taylor expansion is employed to work out the x_1 momentum, leaving the control volume. In the x_1 direction, we have

$$\rho u_1^2 \Delta x_2 + \Delta x_2 \frac{\partial \left(\rho u_1^2 \right)}{\partial x_1} \Delta x_1 \tag{3.192}$$

Similarly, the x_1 momentum leaving the x_2 direction (top surface) is

$$\rho u_1 u_2 \Delta x_1 + \Delta x_1 \frac{\partial (\rho u_1 u_2)}{\partial x_2} \Delta x_2 \tag{3.193}$$

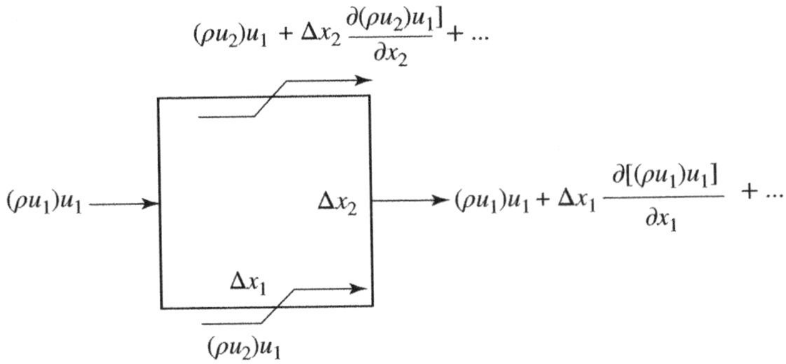

Figure 3.29 Infinitesimal control volume in a flow field: derivation of conservation of momentum in x_1 direction – rate of change of momentum. (Lewis et al. [53])

Note that the second- and higher-order terms in the previous Taylor expansion are neglected. The rate of change of momentum within the control volume due to the x_1 component is written as

$$\Delta x_1 \Delta x_2 \frac{\partial(\rho u_1)}{\partial t} \tag{3.194}$$

The net momentum of the control volume is calculated as the momentum exiting the control volume − momentum entering the control volume + rate of change of the momentum, which is

$$\Delta x_1 \Delta x_2 \left[\frac{\partial\left(\rho u_1^2\right)}{\partial x_1} + \frac{\partial(\rho u_1 u_2)}{\partial x_2} + \frac{\partial(\rho u_1)}{\partial t}\right] \tag{3.195}$$

For equilibrium, the above net momentum should be balanced by the net force acting on the control volume. In order to derive the net force acting on the control volume – Figure 3.30. From the figure, the total pressure force acting on the control volume in the x_1 direction is written as (positive in the positive x_1 direction and negative in the negative x_1 direction)

$$p\Delta x_2 - \left[p + \frac{\partial p}{\partial x_1}\Delta x_1\right]\Delta x_2 = -\frac{\partial p}{\partial x_1}\Delta x_1 \Delta x_2 \tag{3.196}$$

Similarly, the total force due to the deviatoric stress (viscosity or friction) acting on the control volume in the x_1 direction is written as (Figure 3.30)

$$\left[\tau_{11} + \frac{\partial \tau_{11}}{\partial x_1}\Delta x_1\right]\Delta x_2 - \tau_{11}\Delta x_2 + \left[\tau_{12} + \frac{\partial \tau_{12}}{\partial x_2}\Delta x_2\right]\Delta x_1 - \tau_{12}\Delta x_1 \tag{3.197}$$

Simplifying, we obtain the net force due to the deviatoric stress as

$$\frac{\partial \tau_{11}}{\partial x_1}\Delta x_1 \Delta x_2 + \frac{\partial \tau_{12}}{\partial x_2}\Delta x_1 \Delta x_2 \tag{3.198}$$

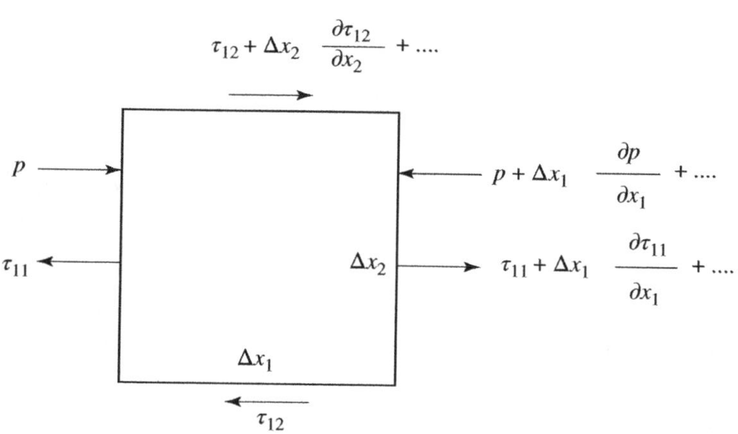

Figure 3.30 Infinitesimal control volume in a flow field: derivation of conservation of momentum in x_1 direction – viscous and pressure forces. (Lewis et al. [53])

The total force acting on the control volume in the x_1 direction is

$$\Delta x_1 \Delta x_2 \left[-\frac{\partial p}{\partial x_1} + \frac{\partial \tau_{11}}{\partial x_1} + \frac{\partial \tau_{12}}{\partial x_2} \right] \tag{3.199}$$

As mentioned before, for equilibrium, the net momentum in the x_1 direction should be equal to the total force acting on the control volume in the x_1 direction, that is,

$$\Delta x_1 \Delta x_2 \left[\frac{\partial(\rho u_1^2)}{\partial x_1} + \frac{\partial(\rho u_1 u_2)}{\partial x_2} + \frac{\partial(\rho u_1)}{\partial t} \right] = \Delta x_1 \Delta x_2 \left[-\frac{\partial p}{\partial x_1} + \frac{\partial \tau_{11}}{\partial x_1} + \frac{\partial \tau_{12}}{\partial x_2} \right] \tag{3.200}$$

Simplifying, we obtain

$$\frac{\partial(\rho u_1)}{\partial t} + \frac{\partial(\rho u_1^2)}{\partial x_1} + \frac{\partial(\rho u_1 u_2)}{\partial x_2} = -\frac{\partial p}{\partial x_1} + \frac{\partial \tau_{11}}{\partial x_1} + \frac{\partial \tau_{12}}{\partial x_2} \tag{3.201}$$

Note that the external and body forces (buoyancy) are not included in the above force balance. In the above equations, the deviatoric stresses τ_{ij} are expressed in terms of the velocity gradients and dynamic viscosity as

$$\tau_{ij} = \mu \left(\frac{\partial u_i}{\partial x_i} + \frac{\partial u_j}{\partial x_i} - \frac{2}{3} \frac{\partial u_k}{\partial x_k} \delta_{ij} \right) \tag{3.202}$$

where δ_{ij} is the Kroneker delta, which is equal to unity if $i = j$ and equal to zero if $i \neq j$. From the previous expression, τ_{11} is expressed as

$$\tau_{11} = \mu \left(\frac{\partial u_1}{\partial x_1} + \frac{\partial u_1}{\partial x_1} - \frac{2}{3} \frac{\partial u_1}{\partial x_1} - \frac{2}{3} \frac{\partial u_2}{\partial x_2} \right) \tag{3.203}$$

Note that $i = j = 1$ in the above equation and $k = 1, 2$, for a two-dimensional flow. The above equation may be simplified as

$$\tau_{11} = \mu \left(\frac{4}{3} \frac{\partial u_1}{\partial x_1} - \frac{2}{3} \frac{\partial u_2}{\partial x_2} \right) \tag{3.204}$$

Similarly, τ_{12} is

$$\tau_{12} = \mu \left(\frac{\partial u_1}{\partial x_2} + \frac{\partial u_2}{\partial x_1} \right) \tag{3.205}$$

Substituting Equations 3.204 and 3.205 into Equation 3.201, we obtain the x_1 component of the momentum equation as

$$\frac{\partial(\rho u_1)}{\partial t} + \frac{\partial(\rho u_1^2)}{\partial x_1} + \frac{\partial(\rho u_1 u_2)}{\partial x_2} = -\frac{\partial p}{\partial x_1} + \frac{\partial}{\partial x_1} \left[\mu \left(\frac{4}{3} \frac{\partial u_1}{\partial x_1} - \frac{2}{3} \frac{\partial u_2}{\partial x_2} \right) \right] + \frac{\partial}{\partial x_2} \left[\mu \left(\frac{\partial u_2}{\partial x_1} + \frac{\partial u_1}{\partial x_2} \right) \right] \tag{3.206}$$

The momentum component in the x_2 direction can be derived by the following steps, which are similar to the derivation of the x_1 component of the momentum equation. The x_2 momentum equation is

$$\frac{\partial(\rho u_2)}{\partial t} + \frac{\partial(\rho u_1 u_2)}{\partial x_1} + \frac{\partial(\rho u_2^2)}{\partial x_2} = -\frac{\partial p}{\partial x_2} + \frac{\partial}{\partial x_1} \left[\mu \left(\frac{\partial u_1}{\partial x_2} + \frac{\partial u_2}{\partial x_1} \right) \right] + \frac{\partial}{\partial x_2} \left[\mu \left(\frac{4}{3} \frac{\partial u_2}{\partial x_2} - \frac{2}{3} \frac{\partial u_1}{\partial x_1} \right) \right] \tag{3.207}$$

For a constant density flow (incompressible flow), the momentum equations can be further reduced by taking the density term out of the differential signs. In addition, substitution of the conservation of mass equation (Equation 3.188) into the momentum equation leads to a further simplification of the momentum equation. After simplification, the momentum equations are

$$\rho\left(\frac{\partial u_1}{\partial t}+u_1\frac{\partial u_1}{\partial x_1}+u_2\frac{\partial u_1}{\partial x_2}\right)=-\frac{\partial p}{\partial x_1}+\mu\left[\frac{\partial^2 u_1}{\partial x_1^2}+\frac{\partial^2 u_1}{\partial x_2^2}\right] \tag{3.208}$$

in the x_1 direction and

$$\rho\left(\frac{\partial u_2}{\partial t}+u_1\frac{\partial u_2}{\partial x_1}+u_2\frac{\partial u_2}{\partial x_2}\right)=-\frac{\partial p}{\partial x_2}+\mu\left[\frac{\partial^2 u_2}{\partial x_1^2}+\frac{\partial^2 u_2}{\partial x_2^2}\right] \tag{3.209}$$

in the x_2 direction. In vector notation, the momentum equations can be written as

$$\rho\left[\frac{\partial \mathrm{u}}{\partial t}+\nabla.(u\times u)\right]=\nabla.[-pI+\tau] \tag{3.210}$$

or, in indicial form

$$\rho\left(\frac{\partial u_i}{\partial t}+u_j\frac{\partial u_i}{\partial x_j}\right)=-\frac{\partial p}{\partial x_i}+\mu\left(\frac{\partial^2 u_i}{\partial x_i^2}\right) \tag{3.211}$$

Note that the above equation is applicable in any dimension.

3.3.1.2 Energy Equation

The energy equation can be derived by following a procedure similar to the momentum equation derivation. However, the difference here is that the temperature or energy equation, is a scalar equation. In order to derive this equation, let us consider the control volume as shown in Figure 3.31. The energy convected into the control volume in the x_1 direction is

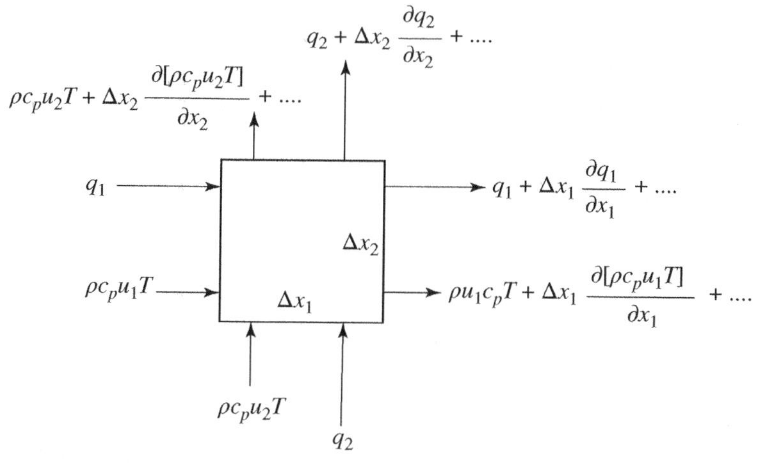

Figure 3.31 Infinitesimal control volume in a flow field: derivation of conservation of energy. (Lewis et al. [53])

$$\rho c_p u_1 T \Delta x_2 \tag{3.212}$$

Similarly, the energy convected into the control volume in the x_2 direction is

$$\rho c_p u_2 T \Delta x_1 \tag{3.213}$$

As before, a Taylor series expansion may be used to express the energy convected out of the control volume in both the x_1 and x_2 directions as

$$\rho c_p u_1 T \Delta x_2 + \rho c_p \frac{\partial (u_1 T)}{\partial x_1} \Delta x_1 \Delta x_2 \tag{3.214}$$

and

$$\rho c_p u_2 T \Delta x_1 + \rho c_p \frac{\partial (u_2 T)}{\partial x_2} \Delta x_2 \Delta x_1 \tag{3.215}$$

Note that the specific heat, *cp*, and density, ρ, are assumed to be constants in deriving the above equation. The heat diffusion into and out of the control volume is also derived using the above approach. The heat diffusing into the domain in the x_1 direction (Fourier's law of heat conduction) is

$$\Delta x_2 q_1 = -k_{x_1} \frac{\partial T}{\partial x_1} \Delta x_2 \tag{3.216}$$

and the diffusion entering the control volume in the x_2 direction is

$$\Delta x_1 q_2 = -k_{x_2} \frac{\partial T}{\partial x_2} \Delta x_1 \tag{3.217}$$

Using a Taylor series expansion, the heat diffusing out of the control volume can be written as

$$-k_{x_1} \frac{\partial T}{\partial x_1} \Delta x_2 + \frac{\partial}{\partial x_1} \left(-k_{x_1} \frac{\partial T}{\partial x_1} \right) \Delta x_2 \Delta x_1 \tag{3.218}$$

in the x_1 direction and

$$-k_{x_2} \frac{\partial T}{\partial x_2} \Delta x_1 + \frac{\partial}{\partial x_2} \left(-k_{x_2} \frac{\partial T}{\partial x_2} \right) \Delta x_1 \Delta x_2 \tag{3.219}$$

in the x_2 direction. Finally, the rate of change of energy within the control volume is

$$\Delta x_1 \Delta x_2 \rho c_p \frac{\partial T}{\partial t} \tag{3.220}$$

Now, it is a simple matter of balancing the energy entering and exiting the control volume. The energy balance can be obtained as follows:

> "heat entering the control volume by convection + heat entering the control volume by diffusion = heat exiting the control volume by convection + heat exiting the control volume by diffusion + rate of change of energy within the control volume"

Following the above heat balance approach and rearranging, we get

$$\frac{\partial T}{\partial t}+\frac{\partial (u_1 T)}{\partial x_1}+\frac{\partial (u_2 T)}{\partial x_2}=\frac{1}{\rho c_p}\left[\frac{\partial}{\partial x_1}\left(k_{x_1}\frac{\partial T}{\partial x_1}\right)+\frac{\partial}{\partial x_2}\left(k_{x_2}\frac{\partial T_2}{\partial x_2}\right)\right] \tag{3.221}$$

Differentiating the convection terms by parts and substituting Equation 3.188 (continuity) into Equation 3.221, we obtain the simplified energy equation in two dimensions as

$$\frac{\partial T}{\partial t}+u_1\frac{\partial T}{\partial x_1}+u_2\frac{\partial T}{\partial x_2}=\frac{1}{\rho c_p}\left[\frac{\partial}{\partial x_1}\left(k_{x_1}\frac{\partial T}{\partial x_1}\right)+\frac{\partial}{\partial x_2}\left(k_{x_2}\frac{\partial T_2}{\partial x_2}\right)\right] \tag{3.222}$$

If the thermal conductivity is assumed to be constant and $k = k_{x1} = k_{x2}$, the energy equation is reduced to

$$\frac{\partial T}{\partial t}+u_1\frac{\partial T}{\partial x_1}+u_2\frac{\partial T}{\partial x_2}=\alpha\left(\frac{\partial^2 T}{\partial x_1^2}+\frac{\partial^2 T}{\partial x_2^2}\right) \tag{3.223}$$

where $\alpha = k/\rho c_p$ is called the *thermal diffusivity*. The energy equation in vector form is

$$\frac{\partial T}{\partial t}+\mathrm{u}.\nabla T=\alpha\nabla^2 T \tag{3.224}$$

and in indicial form

$$\frac{\partial T}{\partial t}+u_i\frac{\partial T}{\partial x_i}=\alpha\frac{\partial^2 T}{\partial x_i^2} \tag{3.225}$$

The above equation is applicable in any space dimension.

3.3.1.3 Non-Dimensional Form of the Governing Equations

In the previous section, we discussed the derivation of the Navier–Stokes equations for an incompressible fluid. In many heat transfer applications, it is often easy to generate data by non-dimensionalizing the equations using appropriate non-dimensional scales. To demonstrate the non-dimensional form of the governing equations, let us consider the following two-dimensional incompressible flow equations in dimensional form:

Continuity equation

$$\frac{\partial u_1}{\partial x_1}+\frac{\partial u_2}{\partial x_2}=0 \tag{3.226}$$

x_1 momentum equation

$$\frac{\partial u_1}{\partial t}+u_1\frac{\partial u_1}{\partial x_1}+u_2\frac{\partial u_1}{\partial x_2}=-\frac{1}{\rho}\frac{\partial p}{\partial x_1}+\upsilon\left(\frac{\partial^2 u_1}{\partial x_1^2}+\frac{\partial^2 u_1}{\partial x_2^2}\right) \tag{3.227}$$

x_2 momentum equation

$$\frac{\partial u_2}{\partial t}+u_1\frac{\partial u_2}{\partial x_1}+u_2\frac{\partial u_2}{\partial x_2}=-\frac{1}{\rho}\frac{\partial p}{\partial x_2}+\upsilon\left(\frac{\partial^2 u_2}{\partial x_1^2}+\frac{\partial^2 u_2}{\partial x_2^2}\right) \tag{3.228}$$

Energy equation

$$\frac{\partial T}{\partial t} + u_1 \frac{\partial T}{\partial x_1} + u_2 \frac{\partial T}{\partial x_2} = \propto \left(\frac{\partial^2 T}{\partial x_1^2} + \frac{\partial^2 T}{\partial x_2^2} \right) \tag{3.229}$$

where $\nu = \mu/\rho$ is the kinematic viscosity. To obtain a set of non-dimensional equations, let us consider three different cases of convective heat transfer. We start with the forced convection problem followed by the "natural" convection problems. For each case, we discuss one set of non-dimensional scales. There are several other ways of scaling the equations. Some of these are discussed in the latter part of this chapter and others can be found in various other publications listed at the end of this chapter.

3.3.1.4 Forced Convection

In forced convection problems, the following non-dimensional scales are normally employed:

$$\begin{gathered} x_1^* = \frac{x_1}{L};\ x_2^* = \frac{x_2}{L};\ t^* = \frac{t u_a}{L}; \\ u_1^* = \frac{u_1}{u_a};\ u_2^* = \frac{u_2}{u_a};\ p^* = \frac{p}{\rho u_a^2}; \\ T^* = \frac{T - T_a}{T_w - T_a} \end{gathered} \tag{3.230}$$

where $*$ indicates a non-dimensional quantity, L is a characteristic dimension, the subscript a indicates a constant reference value and T_w is a constant reference temperature, for example, wall temperature. The density ρ and viscosity μ of the fluid are assumed to be constant everywhere and equal to the inlet value.

Substitution of the above scales into the dimensional Equations 3.226 to 3.229 leads to the following non-dimensional form of the equations:

Continuity equation

$$\frac{\partial u_1^*}{\partial x_1^*} + \frac{\partial u_2^*}{\partial x_2^*} = 0 \tag{3.231}$$

x_1 momentum equation

$$\frac{\partial u_1^*}{\partial t^*} + u_1^* \frac{\partial u_1^*}{\partial x_1^*} + u_2^* \frac{\partial u_1^*}{\partial x_2^*} = -\frac{\partial p^*}{\partial x_1^*} + \frac{1}{Re}\left(\frac{\partial^2 u_1^*}{\partial x_1^{*2}} + \frac{\partial^2 u_1^*}{\partial x_2^{*2}} \right) \tag{3.232}$$

x_2 momentum equation

$$\frac{\partial u_2^*}{\partial t^*} + u_1^* \frac{\partial u_2^*}{\partial x_1^*} + u_2^* \frac{\partial u_2^*}{\partial x_2^*} = -\frac{\partial p^*}{\partial x_2^*} + \frac{1}{Re}\left(\frac{\partial^2 u_2^*}{\partial x_1^{*2}} + \frac{\partial^2 u_2^*}{\partial x_2^{*2}} \right) \tag{3.233}$$

Energy equation

$$\frac{\partial T^*}{\partial t^*} + u_1^* \frac{\partial T^*}{\partial x_1^*} + u_2^* \frac{\partial T^*}{\partial x_2^*} = \frac{1}{RePr}\left(\frac{\partial^2 T^*}{\partial x_1^{*2}} + \frac{\partial^2 T^*}{\partial x_2^{*2}} \right) \tag{3.234}$$

where Re is the Reynolds number defined as

$$Re = \frac{u_a L}{\upsilon} \tag{3.235}$$

and Pr is the Prandtl number given as

$$Pr = \frac{\upsilon}{\alpha} \tag{3.236}$$

Once again, note that the density, kinematic viscosity and thermal conductivity are assumed to be constant in deriving the above non-dimensional equations. Appropriate changes will be necessary if an appreciable variation in these quantities occurs in a flow field. Another non-dimensional number, which is often employed in forced convection heat transfer calculations is the Peclet number and is given as $Pe = Re\ Pr = u_a L/\alpha$. For buoyancy-driven natural convection problems, a different type of non-dimensional scale is necessary if there are no reference velocity values available.

3.3.1.5 Natural Convection (Buoyancy-Driven Convection)

Natural convection is generated by the density difference induced by the temperature differences within a fluid system. Because of the small density variations present in these types of flows, a general incompressible flow approximation is adopted. In most buoyancy-driven convection problems, flow is generated by either a temperature variation or a concentration variation in the fluid system, which leads to local density differences. Therefore, in such flows, a body force term needs to be added to the momentum equations to include the effect of local density differences. For temperature-driven flows, the Boussinesq approximation is often employed, that is,

$$g(\rho - \rho_a) = g\beta(T - T_a) \tag{3.237}$$

where g is the acceleration due to gravity (9.81 m/s^2) and β is the coefficient of thermal expansion. The above body force term is added to the momentum equation in the gravity-direction. In a normal situation (Figure 3.32), the body force is added to the x_2 momentum (if the gravity-direction is negative x_2), that is,

$$\frac{\partial u_2}{\partial t} + u_1 \frac{\partial u_2}{\partial x_1} + u_2 \frac{\partial u_2}{\partial x_2} = -\frac{1}{\rho}\frac{\partial p}{\partial x_2} + \upsilon\left(\frac{\partial^2 u_2}{\partial x_1^2} + \frac{\partial^2 u_2}{\partial x_2^2}\right) + g\beta(T - T_\infty) \tag{3.238}$$

In practice, the following non-dimensional scales are adopted for natural convection in the absence of a reference velocity value:

$$\begin{aligned} x_1^* &= \frac{x_1}{L};\ x_2^* = \frac{x_2}{L};\ t^* = \frac{t\alpha}{L^2}; \\ u_1^* &= \frac{u_1 L}{\alpha};\ u_2^* = \frac{u_2 L}{\alpha};\ p^* = \frac{pL^2}{\rho\alpha^2}; \\ T^* &= \frac{T - T_a}{T_w - T_a} \end{aligned} \tag{3.239}$$

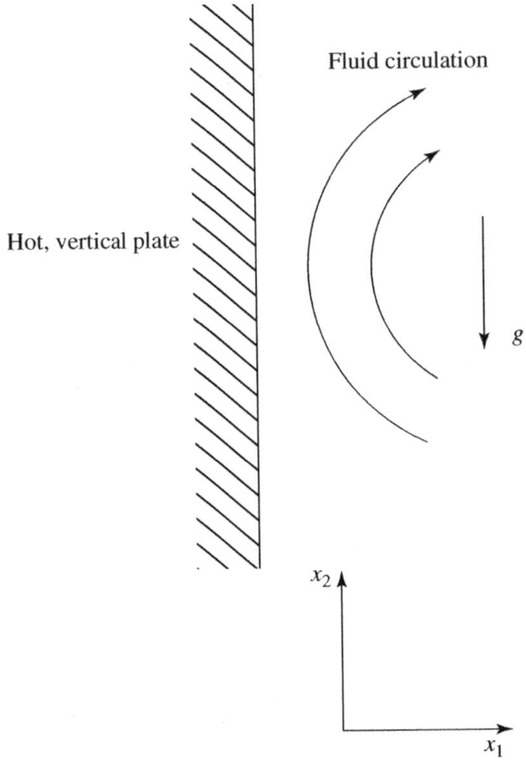

Figure 3.32 Natural convective flow near a hot, vertical plate. (Lewis et al. [53])

On introducing the above non-dimensional scales into the governing equations, we obtain the non-dimensional form of the equations as follows:

Continuity equation

$$\frac{\partial u_1^*}{\partial x_1^*} + \frac{\partial u_2^*}{\partial x_2^*} = 0 \tag{3.240}$$

x_1 momentum equation

$$\frac{\partial u_1^*}{\partial t^*} + u_1^* \frac{\partial u_1^*}{\partial x_1^*} + u_2^* \frac{\partial u_1^*}{\partial x_2^*} = -\frac{\partial p^*}{\partial x_1^*} + Pr\left(\frac{\partial^2 u_1^*}{\partial x_1^{*2}} + \frac{\partial^2 u_1^*}{\partial x_2^{*2}}\right) \tag{3.241}$$

x_2 momentum equation

$$\frac{\partial u_2^*}{\partial t^*} + u_1^* \frac{\partial u_2^*}{\partial x_1^*} + u_2^* \frac{\partial u_2^*}{\partial x_2^*} = -\frac{\partial p^*}{\partial x_2^*} + Pr\left(\frac{\partial^2 u_2^*}{\partial x_1^{*2}} + \frac{\partial^2 u_2^*}{\partial x_2^{*2}}\right) + Gr\, Pr^2 T^* \tag{3.242}$$

Energy equation

$$\frac{\partial T^*}{\partial t^*} + u_1^* \frac{\partial T^*}{\partial x_1^*} + u_2^* \frac{\partial T^*}{\partial x_2^*} = \left(\frac{\partial^2 T^*}{\partial x_1^{*2}} + \frac{\partial^2 T^*}{\partial x_2^{*2}}\right) \tag{3.243}$$

where Gr is the Grashof number given as

$$Gr = \frac{g\beta\Delta TL^3}{\upsilon^2} \tag{3.244}$$

Often, another non-dimensional number, called the *Rayleigh number*, is used in the calculations. This is given as

$$Ra = Gr\,Pr = \frac{g\beta\Delta TL^3}{\upsilon \propto} \tag{3.245}$$

On comparing the non-dimensional equations of natural and forced convection, it is easy to identify the differences. If we substitute $1/Pr$ in place of the Reynolds number for the forced convection equations, we revert to a natural convection scaling. Obviously, the extra buoyancy term needs to be added to appropriate component(s) of the momentum equation for natural convection flows.

3.3.1.6 Mixed Convection

Mixed convection involves features from both forced and natural flow conditions. The buoyancy effects become comparable to the forced flow effects at small and moderate Reynolds numbers. Since the flow is partly forced, a reference velocity value is normally known (e.g. velocity at the inlet of a channel). Therefore, non-dimensional scales of forced convection can be adopted here. However, in mixed convection problems, the buoyancy term needs to be added to the appropriate component of the momentum equation. If we replace $1/Pr$ with Re in the non-dimensional natural convection equations of the previous subsection, we obtain the non-dimensional equations for mixed convection flows. These equations are the same as for the forced convection flow problem, except for the body force term, which will be added to the momentum equation in the gravity direction. The body force term is

$$\frac{Gr}{Re^2}T^* \tag{3.246}$$

Note that sometimes a non-dimensional parameter, referred to as the *Richardson number* (Gr/Re^2), is also used in the literature.

3.3.1.7 Transient Convection – Diffusion Problem

An understanding of the fundamentals of the convection–diffusion equations is crucial in studying fluid dynamics assisted heat transfer. The equations governing the combined fluid flow and heat transfer mainly involve the convection and diffusion components. A typical scalar convection–diffusion equation may be written as

$$\frac{\partial \phi}{\partial t} + u_i\frac{\partial \phi}{\partial x_i} + \phi\frac{\partial u_i}{\partial x_i} - \frac{\partial}{\partial x_i}\left(k\frac{\partial \phi}{\partial x_i}\right) + Q = 0 \tag{3.247}$$

where ϕ is a scalar variable, k is a diffusion coefficient (thermal conductivity if $\phi = T$), u_i are the convection velocity components and Q is a source term. In the above equation, the first term is a transient term, the second and third terms are convection terms and the fourth term is the diffusion term. For a one-dimensional problem, the above equation is

$$\frac{\partial \phi}{\partial t} + u_1 \frac{\partial \phi}{\partial x_1} + \phi \frac{\partial u_1}{\partial x_1} - \frac{\partial}{\partial x_1}\left(k \frac{\partial \phi}{\partial x_1}\right) + Q = 0 \tag{3.248}$$

If the convection velocity u_1 is assumed to be constant, we can rewrite Equation 3.248 as

$$\frac{\partial \phi}{\partial t} + u_1 \frac{\partial \phi}{\partial x_1} - \frac{\partial}{\partial x_1}\left(k \frac{\partial \phi}{\partial x_1}\right) + Q = 0 \tag{3.249}$$

A one-dimensional convection equation without a source term is obtained by neglecting the diffusion and source terms as

$$\frac{\partial \phi}{\partial t} + u_1 \frac{\partial \phi}{\partial x_1} = 0 \tag{3.250}$$

Note that an appropriate solution for the above equation is valid for any similar equations such as the energy equation.

3.3.2 Boundary Conditions

All CFD problems are defined in terms of initial and boundary conditions. It is important that the user specifies these correctly and understands their role in the numerical algorithm. This section mentions of the following most common boundary conditions in the discretized equations of the finite volume method:

- inlet
- outlet
- wall
- prescribed pressure
- symmetry
- periodicity (or cyclic boundary condition)

For details of the modifications needed to implement the most common boundary conditions, see Veersteg and Malalasekera [54] and Schlichting [55]. However, the details of *j* and *f* data generated for different types of compact heat exchanger fins/surfaces are presented in Chapter 5 by applying above boundary conditions wherever necessary.

Nomenclature

A	area or heat transfer area (m^2)
C	heat capacity
f	friction factor, dimensionless
g	gradient matrix
h	heat transfer coefficient (W/m^2 K)
I	variational integral
j	Colburn j factor, dimensionless
K	global stiffness

k	thermal conductivity (W/m K)
L	element length (m)
N	shape functions, Node
NTU or Ntu	number of transfer units
p	perimeter (m), pressure (Pa)
Q	total heat flux (W/m^2)
R	heat capacity ratio
T	temperature (K)
u	velocity in x direction
v	velocity in y direction
w	velocity in z direction
x	x direction distance
y	y direction distance
z	z direction distance

Greek Symbols

$\emptyset$	flux variable
τ	Fluid shear stress
ρ	Fluid density
θ	Temperature

Subscripts

i and j	node locations
c	cold stream
h	hot stream

References

1 Lewis R.W., Morgan K., Thomas H.R. and Seetharamu K.N., *Finite Element Methods in Heat Transfer Analysis*, John Wiley & Sons, 1996.
2 Clough, R., The finite element in plane stress analysis. In *Proceedings of the Second American Society of Civil Engineers Conference on Electronic Computation*, Pittsburgh, 1960.
3 Zienkiewicz O.C. and Cheung Y.K., Finite elements in the solution of field problems, *The Engineer*, Swansea, U.K., 1965.

4 Ozisik M.N. and Czisik M.N., *Finite Difference Methods in Heat Transfer*, CRC Press, 1994.
5 Baker, A.J., *Finite Element Computational Fluid Mechanics*, Student Edition, McGraw-Hill Book Company, New York, 1995.
6 Bathe K.J., *Finite Element Procedures in Engineering Analysis*, Prentice Hill, Englewood Cliffs, N.J., 1982.
7 Chandrupatla T.R. and Belegundu A.D., *Introduction to Finite Elements in Engineering*, Prentice Hall of India Pvt Ltd. 1991.
8 Huebner, H.K. and Thornton E.A., *The Finite Element Method for Engineers, Second Edition*, John Wiley and Sons Ltd., 1982.
9 Hughes T.J.R., *The Finite Element Method: Linear Static and Dynamic Finite Element Analysis*, Dover Publications, New York, 2000.
10 Lewis R.W., Nithiarasu P. and Seetharamu K.N., *Fundamentals of Finite Element Method for Heat and Fluid Flow*, John Wiley and Sons Ltd., New Jersey, 2004.
11 Rao, S.S., The Finite Element Method in Engineering, Pergamon press., 1982.
12 Reddy J.N., *An Introduction to Finite Element Method*, Second Edition, McGraw-Hill Book Company, New York, 1993.
13 Segerlind, L.J., *Applied Finite Element Analysis*, John Wiley and Sons, Inc., 1984.
14 Zienkiewicz O.C. and Morgan K., *Finite Elements and Approximation*, Wiley-Inter Science, New York, 1983.
15 Zienkiewicz O.C. and Taylor R.L., *The Finite Element Method, Vol. 1, The Basis*, Fifth Edition, Butterworth and Heinemann, London, 2000.
16 Thompson, J.F., Soni, B.K. and Weatherhill, N.P., *Handbook of Grid Generation*, CRC Press, New York, 1999.
17 J.P. Holman, *Heat Transfer*, McGraw-Hill, New York, 1989.
18 Ravikumar S.G., Seetharamu K.N. and Aswatha Narayana P.A., Applications of finite elements in heat exchangers, *Communications in Applied Numerical Methods*, 2, 229–234, 1986.
19 Ravikumar, S.G., K.N. Seetharamu and P.A. Aswathanarayana, Finite element analysis of shell and tube heat exchanger, *Int. Commun. Heat Mass Trans.*, 15, 2, 151–163, 1988.
20 Ravikumar, S.G., K.N. Seetharamu and P.A. Aswathanarayana, Performance evaluation of Cross Flow Compact Heat Exchangers Using Finite Element Method, Int. *J. Heat Mass Trans.*,32, 5, 889–894, 1989.
21 Ranganayakulu, C., K.N. Seetharamu and K.V. Sreevatsan, Combined effects of inlet fluid flow non-uniformity on thermal performance and pressure drops inCrossflow plate-fincompact heat exchangers, *Int. J. Heat Mass Trans.*, 40, 1, 27–38, 1997.
22 Ranganayakulu, C., K.N. Seetharamu and K.V. Sreevatsan, Effects of inlet fluid flow non-uniformity on thermal performance and pressure drops in crossflow plate-fin compact heat exchangers, *Int. J. Heat Mass Trans.*, 40, 6, 1261–1277, 1997.
23 Ranganayakulu C. and K.N. Seetharamu, Combined effects of longitudinal heat conduction, flow nonuniformity and temperature nonuniformity in crossflow plate-fin heat exchangers, *Int. Commun. Heat Mass Trans.*, 26, 5, 669–678, 1999.
24 Ranganayakulu C. and K.N. Seetharamu, Combined effects of wall longitudinal heat conduction, inlet fluid flow nonuniformity and temperature nonuniformity in compact tube-fin heat exchangers: A. Finite Element Method, *Int. J. Heat Mass Trans.*, 42, 2, 263–273, 1999.

25 Ranganayakulu C. and K.N. Seetharamu, Combined effect of wall longitudinal heat conduction and inlet fluid flow maldistribution in cross flow plate-fin heat exchangers, *Int. J. Heat Mass Trans.*, 36, 3, 247–256, 2000.

26 Quadir, G.A., Anvar Mydin and K.N. Seetharamu, Analysis of microchannel heat exchangers using FEM, *Int. J. Num. Meth. Heat Fluid Flow*, 11, 1, 59–75, 2001.

27 Quadir, G.A., Krishnan G.M and K.N. Seetharamu, Modeling of wire-on-tube heat exchangers using finite element method, Finite elements in analysis and design, *Int. J. Appl. Finite Elements and Computer Aided Engineering*, 38, 5, pp. 417–434, 2002.

28 Quadir, G.A., S.L. Beh, K.N. Seetharamu and A.Y. Hassan, A transient finite element analysis of microchannel heat exchangers, *Int. J. Heat Exchangers*, 3, 1, 21–42, 2002.

29 Quadir, G.A., G.M. Krishnan and K.N. Seetharamu, Finite element Analysis of Wire-on-Tube Heat Exchangers Under Forced Convection Conditions, *Int. J. Heat Exchangers*, 3, 2, 2002.

30 Quadir, G.A. S.L. Beh, K.N. Seetharamu and A.Y. Hassan, Steady state finite element analysis of a double stack cold plate with heat losses, *Heat and Mass Transfer*, 39, 5 and 6, 519–528, 2003.

31 Jeevan, K., G.A. Quadir, K.N. Seetharamu, I.A. Azid and Z.A. Zainal, Optimization of thermal Resistance of Stacked Micro-Channel Using Genetic Algorithms, *Int. J. Num. Methods for Heat & Fluid Flow*, 15, 1, 27–42, 2005.

32 Jeevan, K., K.N. Seetharamu and I.A. Azid, Analysis and optimization of Single Layer Counter Flow Microchannel Heat Sink, *J. Microcircuits Electronic Packaging*, 2,2,110–121, 2005.

33 Hegde, P., K.N. Seetharamu, G.A. Quadir, P.A. Aswathanarayana, M.Z. Abdullah and Z.A. Zainal, Thermal analysis of micro-channel heat exchangers with two phase flow using FEM, *Int. J. Num. Methods for Heat & Fluid Flow*, 15, 1, 43–60, 2005.

34 Hegde, P., M.Z. Abdulla, K.N. Seetharamu and P.A. Aswathanarayana, Counter and parallel two-phase flow micro-channel heat sinks for electronics cooling, *Int. J. Heat Exchangers*, VII, 1, 2006.

35 Hegde, P., Mukesh Patil and K.N. Seetharamu, Numerical prediction of thermal performance of water cooled multi-stack microchannel heat sink with counter flow arrangement, *J. Microelectron. Packaging*, 8, 16–22, 2011.

36 Seetharamu, K.N., G.A. Quadir, Z.A. Zainal and G.M. Krishnan, FEM analysis of Multifluid Heat Exchangers, *Int. J. Num. Methods for Heat & Fluid Flow*, 14, 2, 242–255, 2004.

37 Sai B.V.K.S, Seetharamu K.N. and Narayana P.A.A., Solution of laminar natural convection in a square cavity by an explicit finite element scheme, *Numerical Heat Transfer Part A: Applications*, 25, 593–609, 1994.

38 Srinivas M., Ravisankar M.S., Seetharamu K.N. and Aswathanarayana P.A., Finite element analysis of internal flows with heat transfer, *Sadhana – Academy Proceedings in Engineering*, 19, 785–816, 1994.

39 Usmani A.S., Lewis R.W. and Seetharamu K.N. 1992 Finite element modelling of natural convection controlled change of phase, *Int. J. for Numerical Methods in Fluids*, 14, 1019–1036.

40 Usmani A.S., Cross J.T. and Lewis R.W., A finite element model for the simulation of mould filling in metal casting and the associated heat transfer, *Int. J. for Numerical Methods in Engineering*, 35, 787–806, 1992.

41 Saeid, N.H. and K.N. Seetharamu, Finite element analysis of co-current and counter-current parallel flow three fluid heat exchanger, *Int. J. Numer. Meth. Heat Fluid Flow*, 16, 3, 324–337, 2006.

42 Lee, H.W., Y.J. Teng, I.A. Azid and K.N. Seetharamu, Neuro-genetic optimization of micro compact heat exchanger, *Int. J. Num Meth in Heat Fluid Flow*,17, 1,20–33, 2007.
43 Beh, S.L., K.K. Tio, G.A. Quadir and K.N. Seetharamu, Fast transient solution of a two-layered counter flow micro-channel heat sink, *Int. J. Numer. Meth. Heat Flow*, vol. 19, Number 5, pp. 595–616, 2009
44 Chaudhuri, A.R., K.N. Seetharamu and T. Sundararajan, Modelling of steam Surface Condenser Using Finite Element Methods, *Communications in Numerical Methods in Engineering*, 13, 12, 909–921, 1997.
45 Reddy J.N. and Gartling G.K., *The Finite Element Method in Heat Transfer and Fluid Dynamics*, Second Edition, CRC Press, 2000.
46 Patankar, S.V., *Numerical Heat Transfer and Fluid Flow*, Hemisphere Publishing, New York, 1980.
47 Hirsch C., Numerical Computation of Internal and External Flows, vol. 1, Fundamentals of Numerical Discretization, Wiley & Sons, Chichester, 1988.
48 Patnaik B.S.V., Gowda Y.T.K., Ravisankar M.S., Narayana P.A.A. and Seetharamu K.N., Finite element simulation of internal flows with heat transfer using a velocity correction approach, *Sadhana – Academy Proceedings in Engineering Sciences*, 26, 251–283, 2001.
49 Incropera F.P., Dewitt D.P., Bergman T.L., Lavine A.S., Seetharamu K.N. and Setharam T.R., 2013, *Fundamentals of Heat and Mass Transfer*, Wiley India Pvt Ltd, New Delhi.
50 Ravikumar S.G., K.N. Seetharamu and P.A. Aswatha Narayana, Analysis of network heat exchangers using finite elements, Conference on Mathematical Modelling in Science and Technology, pp. 861–870, 1988.
51 Seetharamu K.N. Modelling of heat exchangers using FEM, *Int. Conf. on Computational Mechanics*, Tokyo, Japan, pp. VIII. 89–95, 1986.
52 Shames I.H., *Mechanics of fluids*, McGraw-Hill, 1982
53 Roland W. Lewis, Perumal Nithiarasu, Kankanhally N. Seetharamu, *Fundamentals of the Finite Element method for Heat and Mass Transfer*, Wiley Publications, 2016.
54 Versteeg & Malalasekera, An Introduction to Computational Fluid Dynamics, *The Finite Volume Method*, 1995.
55 Schlichting, H., *Boundary-layer theory*, 7th ed, McGraw-Hill, New York, 1979.

4

Finite Element Analysis of Compact Heat Exchangers

4.1 Introduction

Many types of heat transfer matrices are used in direct transfer type heat exchangers. The heat transfer matrices can generally be classified into plate-fin surfaces, tubular surfaces and micro-channels. Heat exchangers are analysed to obtain the temperature distribution in the exchanger and hence to calculate the performance variations due to longitudinal wall heat conduction, inlet flow non-uniformity and inlet temperature non-uniformity and fluid property variations. In the present section, the representation of exchangers for FEM analysis is introduced. The element models used for the analysis of various types of plate-fin, tube-fin and micro-channel heat exchangers are given. As a result, accurate prediction of the thermal performance of a heat exchanger, when the effects of longitudinal wall heat conduction, inlet flow non-uniformity and temperature non-uniformity are significant, is almost impossible before the production and testing of a prototype. Longitudinal heat conduction in the wall and the fluids is detrimental to the heat exchanger performance. Longitudinal heat conduction in the fluids is generally negligible, except for liquid metals as working fluids. However, it is important for heat exchanger to be designed with high effectiveness and having high temperature gradients in the wall.

The inlet fluid flow and temperature distribution over the heat exchanger core is usually not uniform under actual operating conditions. The flow non-uniformity through the heat exchanger is generally associated with improper exchanger entrance configuration and imperfect flow passage caused by various problems in design and manufacturing. Usually, the inlet temperatures of the fluid flowing through a great numbers of heat exchangers are not uniformly distributed under actual operating conditions, for various reasons. The flow non-uniformity through the exchanger core is known to be detrimental to its thermal performance. The performance variations due to inlet temperature non-uniformity have to be analysed for accurate prediction of thermal performance of a heat exchanger. This chapter mainly deals with the discretization of the heat exchangers, the governing equations and the finite element formulation of the governing equations, effects of longitudinal heat conduction (LHC), flow non-uniformity (FN), temperature non-uniformity (TN) and their combined effects. In addition, the heat conduction effects of evaporator tube using FEM are also presented here.

Compact Heat Exchangers – Analysis, Design and Optimization using FEM and CFD Approach, First Edition. C. Ranganayakulu and K.N. Seetharamu.
© 2018 John Wiley & Sons Ltd. Published 2018 by John Wiley & Sons Ltd.

4.2 Finite Element Discretization

A discretized model of a cross flow plate-fin heat exchanger is shown in Figure 4.1a. It is divided into a number of equal strips. The strip 1 is isolated and shown in Figure 4.1b. The exchanger may be visualized as a wall separating the two fluid streams flowing at right angles with plate-fins on both sides as shown in Figure 4.1c. Each strip consists of a number of pairs of stacks which carry hot and cold fluids. A pair of stacks is separated and shown in Figure 4.1d. It is the basic element for which the element matrices are derived. In the element, the temperatures of the fluid are assumed to vary only along their flow lengths. In the cross flow plate-fin exchanger wall, a four-noded element has been considered for studying the two-dimensional longitudinal wall heat conduction effects. Two-noded linear elements have been considered for both hot and cold fluids in the present analysis. Thus a 16-noded box element has been obtained as shown in Figure 4.1d. Similarly, the discretized exchanger for counter flow plate-fin, cross flow tube-fin and parallel-flow plate-fin type heat exchangers are shown in Figure 4.1e–h respectively. Since the wall temperature distribution in cross flow tube-fin heat exchanger is one-dimensional, a two-noded element has been considered for longitudinal wall heat conduction effects in the exchanger tube. Thus a six-noded element has been obtained for tube-fin exchanger element as shown in Figure 4.1(h). Similarly, the wall temperature distribution in counter flow plate-fin and parallel-flow plate-fin heat exchangers is one-dimensional and hence two-noded elements have been considered in the exchanger plate for longitudinal wall heat conduction effects. Thus a 10-noded element has been obtained for counter flow and parallel-flow plate-fin heat exchanger as shown in Figure 4.1e and 4.1f respectively. These are the basic elemental exchangers for which the finite element equations are formulated using Galerkin's method in Segerlind [1].

4.3 Governing Equations

In most heat exchangers, the thickness of the exchanger wall is small as compared to its other two dimensions, so conduction thermal resistance through the exchanger wall in the direction normal to the fluid flow is small enough to be neglected. The fin thickness, especially in the plate-fin heat exchanger, is much thinner than the exchanger wall (Kays and London [2]). Therefore, the temperature gradient across the fin thickness may be neglected. Other assumptions of this analysis are:

1) steady state conditions assumed
2) no phase change and no heat generation within the exchanger
3) fluids other than liquid metals considered; their mass velocities are constant with respect to time
4) exchanger where both fluids are unmixed considered; cross or transverse mixing of fluids not considered; change of flow distribution inside the exchanger neglected
5) heat transfer surface configurations and heat transfer areas on both sides per unit base area constant and uniform throughout the exchanger
6) in the elements, fluid temperatures assumed to vary only along flow lengths
7) entry length effects not considered
8) no heat transfer between the exchanger and the surrounding.

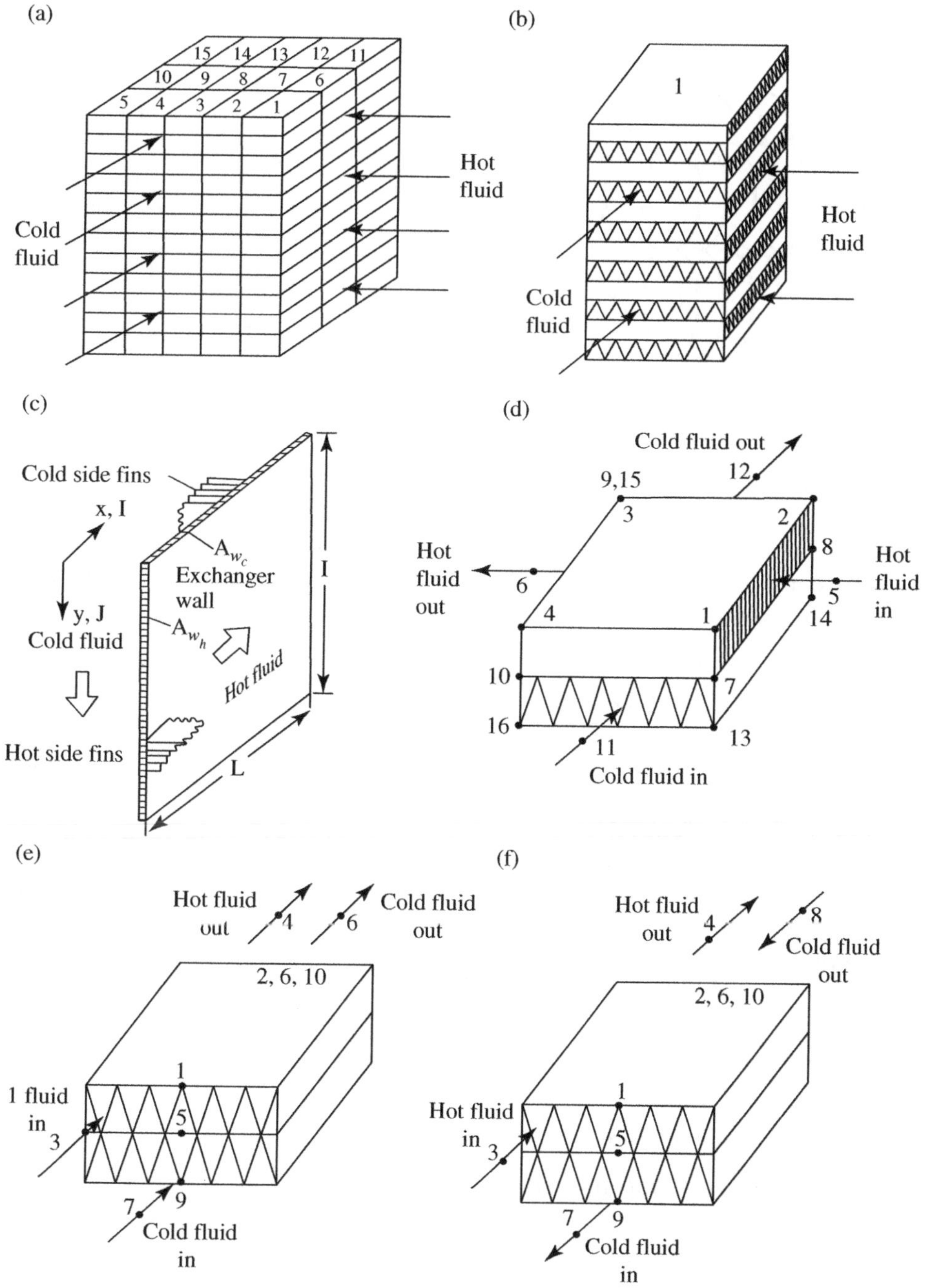

Figure 4.1 Compact Heat Exchangers: (a) discretized exchanger – cross flow heat exchanger (b) strip 1 (c) exchanger wall with fins (d) element stack – cross flow plate-fin (e) element stack – parallel flow plate-fin (f) element stack – counter flow plate-fin (g) tube-fin heat exchanger (h) element – tube-fin.

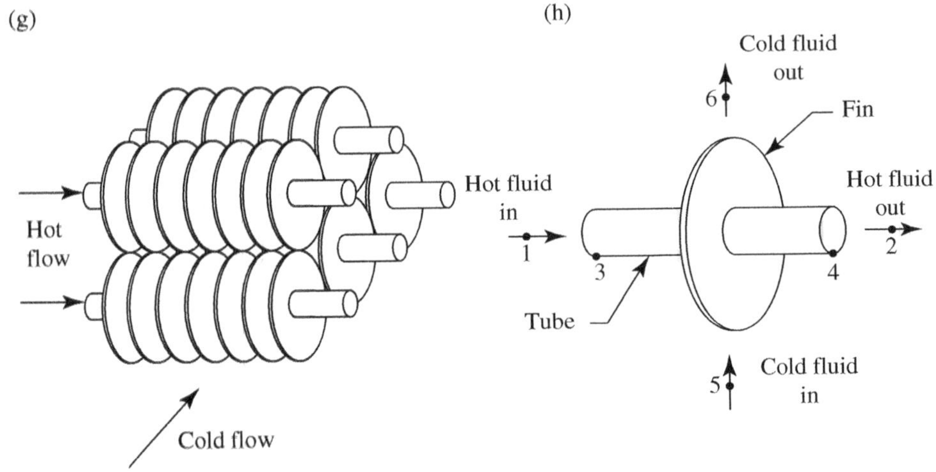

Figure 4.1 (Continued)

Based on the above assumptions, the governing energy balance equations (considering two-dimensional longitudinal heat conduction in the exchanger plate for a cross flow plate-fin exchanger) are formulated as

$$\frac{(kA_w)_h}{l}\frac{\partial^2 T_{w,t}}{\partial x^2}+\frac{(kA_w)_c}{L}\frac{\partial^2 T_{w,t}t}{\partial y^2}-(\vartheta ha')_h(T_h-T_{w,t})=q_t \quad (4.1)$$

$$\frac{-(\alpha MC_p)_h}{l}\frac{\partial T_h}{\partial x}+(\vartheta ha')_h(T_h-T_{w,t})+(\vartheta ha')_h(T_h-T_{w,m})=0 \quad (4.2)$$

$$\frac{(kA_w)_h}{l}\frac{\partial^2 T_{w,m}}{\partial x^2}+\frac{(kA_w)_c}{L}\frac{\partial^2 T_{w,m}}{\partial y^2}-(\vartheta ha')_h(T_h-T_{w,m})+(\vartheta ha')_c(T_{w,m}-T_c)=0 \quad (4.3)$$

$$\frac{(\alpha MC_p)_c}{L}\frac{\partial T_c}{\partial y}+(\vartheta ha')_c(T_{w,m}-T_c)+(\vartheta ha')_c(T_{w,b}-T_c)=0 \quad (4.4)$$

$$\frac{(kA_w)_h}{l}\frac{\partial^2 T_{w,b}}{\partial x^2}+\frac{(kA_w)_c}{L}\frac{\partial^2 T_{w,b}}{\partial y^2}+(\vartheta ha')_c(T_{w,b}-T_c)=q_b \quad (4.5)$$

The boundary conditions are

$$T_h(0,y)=\beta T_{h,in};\quad T_c(x,0)=\beta T_{c,in} \quad (4.6)$$

$$\frac{\partial T_w(0,y)}{\partial x}=\frac{\partial T_w(L,y)}{\partial x}=0;\quad \frac{\partial T_w(x,0)}{\partial y}=\frac{\partial T_w(x,l)}{\partial y}=0 \quad (4.7)$$

The temperature variation of the hot fluid (T_h), and cold fluid (T_c) in the element are approximated by a linear variation as

$$T_h=N_iT_i+N_jT_j \quad (4.8)$$

$$T_c=N_kT_k+N_1T_1 \quad (4.9)$$

The temperature variation of exchanger plate (T_w) is approximated as

$$T_w = N_m T_m + N_n T_n + N_o T_o + N_p T_p \tag{4.10}$$

where N_i, N_j, N_k, N_l, N_m, N_n, N_o and N_p are shape functions (Segerlind [1]):

$$N_i = 1 - \frac{x}{L} \tag{4.11}$$

$$N_j = \frac{x}{L} \tag{4.12}$$

$$N_k = 1 - \frac{y}{l} \tag{4.13}$$

$$N_l = \frac{y}{l} \tag{4.14}$$

$$N_m = 1 - \frac{s}{2b} - \frac{t}{2a} + \frac{st}{4ab} \tag{4.15}$$

$$N_n = \frac{s}{2b} - \frac{st}{4ab} \tag{4.16}$$

$$N_o = \frac{st}{4ab} \tag{4.17}$$

$$N_p = \frac{t}{2a} - \frac{st}{4ab} \tag{4.18}$$

The governing energy balance equations considering the one-dimensional longitudinal wall heat conduction for parallel-flow and counter flow heat exchangers are

$$\frac{(kA_w)_h \partial^2 T_{w,t}}{L\, \partial x^2} - (\vartheta h a')_h (T_h - T_{w,t}) = q_t \tag{4.19}$$

$$\frac{-(\alpha M C_p)_h \partial T_h}{l \quad \partial x} + (\vartheta h a')_h (T_h - T_{w,t}) + (\vartheta h a')_h (T_h - T_{w,m}) = 0 \tag{4.20}$$

$$\frac{(kA_w)_c \partial^2 T_{w,m}}{L\, \partial y^2} - (\vartheta h a')_h (T_h - T_{w,m}) + (\vartheta h a')_c (T_{w,m} - T_c) = 0 \tag{4.21}$$

$$\frac{(\alpha M C_p)_c \partial T_c}{L \quad \partial y} + (\vartheta h a')_c (T_{w,m} - T_c) + (\vartheta h a')_c (T_{w,b} - T_c) = 0 \tag{4.22}$$

$$\frac{(kA_w)_h \partial^2 T_{w,b}}{L\, \partial x^2} + (\vartheta h a')_c (T_{w,b} - T_c) = q_b \tag{4.23}$$

The boundary conditions are

$$T_h(0,y) = \beta T_{h,in}; \quad T_c(y,0) = \beta T_{c,in} \tag{4.24}$$

$$\frac{\partial T_w(x,0)}{\partial y} = \frac{\partial T_w(x,l)}{\partial y} = 0 \tag{4.25}$$

The temperature variation of the cold fluid (T_c), hot fluid (T_h) and exchanger plate (T_w) in the element are approximated by a linear variation as

$$T_h = N_i T_i + N_j T_j \tag{4.26}$$

$$T_c = N_k T_k + N_1 T_1 \tag{4.27}$$

$$T_w = N_m T_m + N_n T_n \tag{4.28}$$

where N_i, N_j, N_k, N_l, N_m and N_n are the shape functions

$$N_i = N_k = N_m = 1 - \frac{x}{L} \tag{4.29}$$

$$N_j = N_l = N_n = \frac{x}{L} \tag{4.30}$$

The governing energy balance equations considering the one-dimensional longitudinal wall heat conduction for cross flow tube-fin heat exchanger are

$$-(\alpha M C_p)_h \frac{\partial T_h}{\partial x} + (\vartheta h P)_h (T_h - T_w) = 0 \tag{4.31}$$

$$(kA_w)\frac{\partial^2 T_w}{\partial x^2} + (\vartheta h P)_h (T_h - T_w) - (\vartheta h P)_c (T_w - T_c) = 0 \tag{4.32}$$

$$(\alpha M C_p)_c \frac{\partial T_c}{\partial x} - (\vartheta h P)_c (T_w - T_c) = 0 \tag{4.33}$$

The boundary conditions are

$$T_h(0,y) = \beta T_{h,in}; \quad T_c(0,y) = \beta T_{c,in} \tag{4.34}$$

$$\frac{\partial T_w(x,0)}{\partial y} = \frac{\partial T_w(x,l)}{\partial y} = 0 \tag{4.35}$$

Similarly, the temperature variation of the cold fluid (T_c), hot fluid (T_h) and exchanger tube (T_w) in the cross flow tube-fin exchanger element are approximated by a linear variation as

$$T_h = N_i T_i + N_j T_j \tag{4.36}$$

$$T_c = N_k T_k + N_1 T_1 \tag{4.37}$$

$$T_w = N_m T_m + N_n T_n \tag{4.38}$$

where N_i, N_j, N_k, N_l, N_m and N_n are shape functions:

$$N_i = N_k = 1 - \frac{x}{L} \tag{4.39}$$

$$N_m = 1 - \frac{y}{l} \tag{4.40}$$

$$N_j = N_l = \frac{x}{L} \tag{4.41}$$

$$N_n = \frac{y}{l} \tag{4.42}$$

The local flow non-uniformity parameter (α_F) is defined as Chiou [3]

$$\alpha_F = \frac{\textit{actual inlet flow}}{\textit{average inlet flow if flow distribution is uniform}} \tag{4.43}$$

Similarly, the local inlet temperature non-uniformity parameter (α_T) is defined as Chiou [4]

$$\alpha_T = \frac{\textit{actua inlet temperature}}{\textit{average inlet temperature if temperature distribution is uniform}} \tag{4.44}$$

4.4 Finite Element Formulation

The finite element formulation of the above equations is carried out by Galerkin method for: cross flow plate-fin heat exchangers, counter flow/parallel flow plate-fin heat exchangers and cross flow tube-fin heat exchangers. The terms q_t and q_b get cancelled when the adjacent elements are assembled. These terms remain on the right-hand side of the global matrix only for the top and bottom pairs of stacks. When the top and bottom surfaces are insulated, then the terms q_t and q_b become zero. All the formulations of heat exchangers reduce to the form of $[K]\{T\} = \{f\}$. The objective here is to determine $[K]$ and $\{f\}$ matrices for heat exchangers.

4.4.1 Cross Flow Plate-Fin Heat Exchanger

The discretized model of a cross flow plate-fin heat exchanger is given in Figure 4.1a. The first element given in Figure 4.1d is considered for the element matrix formulation. The Galerkin weighted residual integral for equation 4.1 is written as

$$\iint [N]^T \left\{ \frac{(k_x)\partial^2 T_{w,t}}{\partial x^2} + \frac{(k_y)\partial^2 T_{w,t}}{\partial y^2} \right\} \partial x \partial y - (\vartheta h a')_h \int [N]^T (T_h - T_{w,t}) \partial x = 0 \tag{4.45}$$

Consider the two diffusion terms of the above equation. Since the assumed temperature profile doesn't have continuous first-order derivatives, the second-order derivatives have to be converted to first-order derivatives. This is carried out by Green's theorem. The integrals for the rectangular element can be evaluated using the shape functions of st-coordinate systems given by Equations 4.15–4.18. Since the st-coordinate system is parallel to the xy-coordinate system and a unit length in either s or t is the same as a unit length of x or y:

$$\int_A f(x,y)\,dx\,dy = \int_A f(s,t)\,ds\,dt \tag{4.46}$$

Equally important is the relationship between the derivatives. The chain rule gives

$$\frac{\partial N_\beta}{\partial x} = \frac{\partial N_\beta}{\partial s} \text{ and } \frac{\partial N_\beta}{\partial y} = \frac{\partial N_\beta}{\partial t} \tag{4.47}$$

Using the shape functions (Equations 4.15–4.18), the evaluation of $[K]$ matrix for Equation 4.1 is

$$\iint [N]^T \left\{ k_x \frac{\partial^2 T_{w,t}}{\partial x^2} + k_y \frac{\partial^2 T_{w,t}}{\partial y^2} \right\} dx\,dy = \int kX[N]^T \frac{\partial T_{w,t}}{\partial x} dx - \iint k_x \frac{\partial [N]^T}{\partial x} \frac{\partial [N]}{\partial x} \begin{bmatrix} T_1 \\ T_2 \\ T_3 \\ T_4 \end{bmatrix} dx\,dy$$

$$+ \int ky[N]^T \frac{\partial T_{w,t}}{\partial y} dy - \iint k_y \frac{\partial [N]^T}{\partial y} \frac{\partial [N]}{\partial y} \begin{bmatrix} T_1 \\ T_2 \\ T_3 \\ T_4 \end{bmatrix} dx\,dy \tag{4.48}$$

$$= \int k_x [N]^T \frac{\partial T_{w,t}}{\partial x} dx + \int k_y [N]^T \frac{\partial T_{w,t}}{\partial y} dy - e_x \begin{bmatrix} 2 & -2 & -1 & 1 \\ -2 & 2 & 1 & -1 \\ -1 & 1 & 2 & -2 \\ 1 & -1 & -2 & 2 \end{bmatrix} \begin{bmatrix} T_1 \\ T_2 \\ T_3 \\ T_4 \end{bmatrix}$$

$$- e_y \begin{bmatrix} 2 & 1 & -1 & -2 \\ 1 & 2 & -2 & -1 \\ -1 & -2 & 2 & 1 \\ -2 & -1 & 1 & 2 \end{bmatrix} \begin{bmatrix} T_1 \\ T_2 \\ T_3 \\ T_4 \end{bmatrix} \tag{4.49}$$

where

$$e_x = \frac{k_x l}{6L} \text{ and } e_y = \frac{k_y L}{6l} \tag{4.50}$$

We note

$$\int \eta h a' [N]^T T_h\,dx = H_h \begin{bmatrix} 2 & 1 \\ 1 & 2 \end{bmatrix} \begin{bmatrix} T_5 \\ T_6 \end{bmatrix} \tag{4.51}$$

where

$$H_h = \frac{\eta h a' L}{6} \tag{4.52}$$

$$\int (\eta h a')_h [N]^T T_{w,t} dA = \frac{H_h}{2} \begin{bmatrix} 2 & 2 & 1 & 1 \\ 1 & 1 & 2 & 2 \end{bmatrix} \begin{bmatrix} T_1 \\ T_2 \\ T_3 \\ T_4 \end{bmatrix} \tag{4.53}$$

Since we have assumed insulated conditions exist on the boundaries, the line integrals do not contribute anything to the stiffness matrix and therefore are dropped. The sum of the Equations 4.49 and 4.52 will give

$$\begin{bmatrix} 2e_x + 2e_y + H_h & -2e_x + e_y + H_h & -e_x - e_y + H_h/2 & e_x - e_y + H_h/2 & -2H_h & -H_h \\ -2e_x + e_y + H_h/2 & 2e_x + 2e_y + H_h/2 & e_x - 2e_y + H_h & -e_x - e_y + H_h & -H_h & -2H_h \\ -e_x - e_y & e_x - 2e_y & 2e_x + 2e_y & -2e_x + e_y & 0 & 0 \\ e_x - 2e_y & -e_x - e_y & -2e_x + e_y & 2e_x + 2e_y & 0 & 0 \end{bmatrix} \begin{bmatrix} T_1 \\ T_2 \\ T_3 \\ T_4 \\ T_5 \\ T_6 \end{bmatrix} = \begin{bmatrix} 0 \\ 0 \\ 0 \\ 0 \\ 0 \\ 0 \end{bmatrix} \tag{4.54}$$

The integral for Equation 4.2 is written as

$$\int_0^L [N]^T \frac{(-\alpha M C_p)_h}{l} \frac{\partial T_h}{\partial x} \partial x + \int_0^L [N]^T (\vartheta h a')_h (Th - T_{w,t}) \partial x + \int_0^L [N]^T (\vartheta h a')(T_h - T_{w,m}) \partial x = 0 \tag{4.55}$$

We note

$$\int [N]^T \frac{(-\alpha M C_p)_h}{l} \frac{\partial T_h}{\partial x} \partial x = \left(\frac{MC_P}{2l}\right) \begin{bmatrix} -1 & 1 \\ -1 & 1 \end{bmatrix} \begin{bmatrix} T_5 \\ T_6 \end{bmatrix} \tag{4.56}$$

$$\int (\eta h a')_h [N]^T T_h dx = \frac{\eta h a' L}{6} \begin{bmatrix} 2 & 1 \\ 1 & 2 \end{bmatrix} \begin{bmatrix} T_5 \\ T_6 \end{bmatrix} \tag{4.57}$$

$$\int (-\eta h a')_h [N]^T T_{w,t} dA = \left(-\frac{\eta h a' L}{12}\right)_h \begin{bmatrix} 2 & 2 & 1 & 1 \\ 1 & 1 & 2 & 2 \end{bmatrix} \begin{bmatrix} T_1 \\ T_2 \\ T_3 \\ T_4 \end{bmatrix} \tag{4.58}$$

$$\int (-\eta h a')_h [N]^T T_{w,t} dA = \left(-\frac{\eta h a' L}{12}\right)_h \begin{bmatrix} 2 & 2 & 1 & 1 \\ 1 & 1 & 2 & 2 \end{bmatrix} \begin{bmatrix} T_7 \\ T_8 \\ T_9 \\ T_{10} \end{bmatrix} \tag{4.59}$$

Similar expressions can be derived for the other terms also for Equations 4.3–4.5. Adding up all the terms obtained, we get algebraic equations corresponding to all nodes, which can be written in a matrix form for an element (Figure 4.1d) of cross flow plate-fin heat exchanger as

$$\begin{bmatrix} C_1 & C_2 & C_3 & C_4 & -2K_2 & -K_2 & 0 & 0 & 0 & 0 & 0 & 0 & 0 & 0 & 0 & 0 \\ C_4 & C_5 & C_2 & C_6 & -K_2 & K_2 & 0 & 0 & 0 & 0 & 0 & 0 & 0 & 0 & 0 & 0 \\ -2K_0 & -K_0 & 4K_0 & -K_0 & 0 & 0 & 0 & 0 & 0 & 0 & 0 & 0 & 0 & 0 & 0 & 0 \\ -K_0 & -2K_0 & -K_0 & 4K_0 & 0 & 0 & 0 & 0 & 0 & 0 & 0 & 0 & 0 & 0 & 0 & 0 \\ -2K_1 & -2K_1 & -K_1 & -K_1 & C_7 & C_8 & -K_1 & -K_1 & -K_1 & -K_1 & 0 & 0 & 0 & 0 & 0 & 0 \\ -K_1 & -K_1 & -2K_1 & -2K_1 & C_9 & C_{10} & -K_1 & -K_1 & -2K_1 & -2K_1 & 0 & 0 & 0 & 0 & 0 & 0 \\ 0 & 0 & 0 & 0 & -2K_2 & -K_2 & C_{11} & C_{12} & C_{13} & C_{14} & -2K_6 & -K_6 & 0 & 0 & 0 & 0 \\ 0 & 0 & 0 & 0 & -K_2 & -2K_2 & C_{14} & C_{15} & C_{12} & C_{16} & K_6 & 2K_6 & 0 & 0 & 0 & 0 \\ 0 & 0 & 0 & 0 & 0 & 0 & -2K_4 & -K_4 & 4K_4 & -K_4 & 0 & 0 & 0 & 0 & 0 & 0 \\ 0 & 0 & 0 & 0 & 0 & 0 & -K_4 & -2K_4 & -K_4 & 4K_4 & 0 & 0 & 0 & 0 & 0 & 0 \\ 0 & 0 & 0 & 0 & 0 & 0 & -2K_5 & -2K_5 & -K_5 & -K_5 & C_{17} & C_{18} & -2K_5 & -2K_5 & -K_5 & -K_5 \\ 0 & 0 & 0 & 0 & 0 & 0 & -K_5 & -K_5 & -2K_5 & -2K_5 & C_{19} & C_{20} & -K_5 & -K_5 & -2K_5 & -2K_5 \\ 0 & 0 & 0 & 0 & 0 & 0 & 0 & 0 & 0 & 0 & -2K_6 & -K_6 & C_{21} & C_{22} & C_{23} & C_{24} \\ 0 & 0 & 0 & 0 & 0 & 0 & 0 & 0 & 0 & 0 & -K_6 & -2K_6 & C_{24} & C_{25} & C_{22} & C_{26} \\ 0 & 0 & 0 & 0 & 0 & 0 & 0 & 0 & 0 & 0 & 0 & 0 & -2K_8 & -K_8 & 4K_8 & -K_8 \\ 0 & 0 & 0 & 0 & 0 & 0 & 0 & 0 & 0 & 0 & 0 & 0 & -K_8 & -2K_8 & -K_8 & 4K_8 \end{bmatrix} \begin{bmatrix} T_1 \\ T_2 \\ T_3 \\ T_4 \\ T_5 \\ T_6 \\ T_7 \\ T_8 \\ T_9 \\ T_{10} \\ T_{11} \\ T_{12} \\ T_{13} \\ T_{14} \\ T_{15} \\ T_{16} \end{bmatrix} = \begin{bmatrix} -Q_t/4 \\ -Q_t/4 \\ -Q_t/4 \\ -Q_t/4 \\ 0 \\ 0 \\ 0 \\ 0 \\ 0 \\ 0 \\ 0 \\ 0 \\ Q_b/4 \\ Q_b/4 \\ Q_b/4 \\ Q_b/4 \end{bmatrix} \tag{4.60}$$

where

$K_0 = (kA_w)_t/6\,L$, $K_1 = (\eta ha)_h/12$, $K_2 = (\eta ha)_h\ /\ 6$, $K_3 = (MC_p)/2$,
$K_4 = (kA_w)_m/6\,L$, $K_5 = (\eta ha)_c/12$, $K_6 = (\eta ha)_c\ /\ 6$, $K_7 = (MC_p)/2$,
$K_8 = (kA_w)_b/6\,L$, $C_1 = 4\,K_0 + 2\,K_1$, $C_2 = 2\,K_1 - K_0$, $C_3 = K_1 - 2\,K_0$, $C_4 = K_1 - K_0$,
$C_5 = 4\,K_0 + K_1$, $C_6 = -2\,K_0 + 2\,K_1$, $C_7 = K_3 + 4\,K_2$, $C_8 = 2\,K_2 - K_3$, $C_9 = 2\,K_2 + K_2$,
$C_{10} = -K_3 + 4\,K_2$, $C_{11} = 4\,K_4 + 2\,K_1 + 2\,K_5$, $C_{12} = -K_4 + 2\,K_1 + 2\,K_5$,
$C_{13} = -2\,K_4 + K_1 + K_5$, $C_{14} = -K_4 + K_1 + K_5$, $C_{15} = 4\,K_4 + K_1 + K_5$,
$C_{16} = -2\,K_4 + 2\,K_1 + 2\,K_5$, $C_{17} = K_7 + 4\,K_6$, $C_{18} = -K_7 + 2\,K_6$, $C_{19} = K_7 + 2\,K_6$,
$C_{20} = -K_7 + 4\,K_6$, $C_{21} = 4\,K_8 + 2\,K_5$, $C_{22} = -K_8 + 2\,K_5$, $C_{23} = -2\,K_8 + K_5$,
$C_{24} = -K_8 + K_5$, $C_{25} = 4\,K_8 + K_5$, and $C_{26} = -2\,K_8 + 2\,K_5$.

The boundary conditions to be satisfied are $(MC_p)T_5 = Q_{h,i}$ and $(MC_p)T_{11} = Q_{c,i}$ for inlet hot and cold fluids and Q_t and Q_b are heat dissipations through the top and bottom plates respectively, which will get cancelled after assembly of all elements, and with the assumptions that top and bottom plates are insulated then Q_t and Q_b become zero. Assembly of the algebraic equations given by Equations 4.1–4.5 in the proper order will give the complete set of algebraic equations for the box element. The nodal equations are developed for all the plate and fluid sub-elements and assembled together to get the element stiffness matrix as shown above. The present element matrix has 16 nodes and therefore it will be a 16 × 16 matrix. The known boundary conditions (inlet temperatures or enthalpies) are incorporated in the above matrix while solving. Details of the assembly procedure and connectivity are available in Segerlind [1]. The element matrices for other pair of stacks are stored in the proper order to obtain the global stiffness matrix.

4.4.2 Counter Flow/Parallel Flow Plate-Fin Heat Exchangers

The discretized model of a counter flow and parallel-flow plate-fin heat exchangers are given in Figures 4.1e and f. The first element given in Figure 4.1f is considered for element matrix formulation. The Galerkin weighted residual integral for Equation (4.19) is written as

$$\int [N]^T \frac{(k_x)\partial^2 T_{w,t}}{L\partial x^2}\partial x - (\vartheta h a')_h \int [N]^T (T_h - T_{w,t})\partial x = 0 \tag{4.61}$$

Using the shape functions (Equations 4.26–4.30), the evaluation of [K] matrix for Equation 4.19 is

$$\int [N]^T k_x A \frac{\partial^2 T_{w,t}}{\partial x^2} dx = \left(\frac{k_x A}{L}\right)\begin{bmatrix} 1 & -1 \\ -1 & 1 \end{bmatrix}\begin{bmatrix} T_1 \\ T_2 \end{bmatrix} \tag{4.62}$$

$$\int \eta h l [N]^T T_h dx = \frac{\eta h l L}{6}\begin{bmatrix} 2 & 1 \\ 1 & 2 \end{bmatrix}\begin{bmatrix} T_3 \\ T_4 \end{bmatrix} \tag{4.63}$$

$$\int (\eta h l)_h [N]^T T_{w,t} dA = \left(\frac{\eta h l L}{6}\right)_h \begin{bmatrix} 2 & 1 \\ 1 & 2 \end{bmatrix}\begin{bmatrix} T_1 \\ T_2 \end{bmatrix} \tag{4.64}$$

Similar expressions can also be derived for the other terms for Equations 4.20–4.23. Assembling all the terms obtained, we get algebraic equations corresponding to the fluid nodes. These sub-elements can be written in a matrix form for an element (Figure 4.1f) of counter flow plate-fin heat exchanger as

$$\begin{bmatrix} J_1 & J_2 & 2H & H & 0 & 0 & 0 & 0 & 0 & 0 \\ J_2 & J_1 & H & 2H & 0 & 0 & 0 & 0 & 0 & 0 \\ 2H & H & J_3 & J_4 & 0 & 0 & 0 & 0 & 0 & 0 \\ HB & 2H & J_5 & J_6 & 0 & 0 & 0 & 0 & 0 & 0 \\ 0 & 0 & 2H & H & J_7 & J_8 & 2H & H & 0 & 0 \\ 0 & 0 & H & 2H & J_8 & J_7 & H & 2H & 0 & 0 \\ 0 & 0 & 0 & 0 & 2H & H & J_3 & J_4 & 2H & H \\ 0 & 0 & 0 & 0 & H & 2H & J_5 & J_6 & H & 2H \\ 0 & 0 & 0 & 0 & 0 & 0 & 2H & H & J_1 & J_2 \\ 0 & 0 & 0 & 0 & 0 & 0 & H & 2H & J_2 & J_1 \end{bmatrix}\begin{bmatrix} T_1 \\ T_2 \\ T_3 \\ T_4 \\ T_5 \\ T_6 \\ T_7 \\ T_8 \\ T_9 \\ T_{10} \end{bmatrix} = \begin{bmatrix} -Q_t \\ -Q_t \\ 0 \\ 0 \\ 0 \\ 0 \\ 0 \\ 0 \\ Q_b \\ Q_b \end{bmatrix} \tag{4.65}$$

where

$K_0 = kA_w/L$, $H = (\eta h a)/6$, $W = MC_p/2$, $J_1 = K_0 - 2H$, $J_2 = -K_0 - H$,
$J_3 = W - 4H$, $J_4 = -W - 2H$, $J_5 = W - 2H$, $J_6 = -W - 4H$, $J_7 = K_0 - 4H$,
and $J_8 = -K_0 - 2H$.

The boundary conditions to be satisfied are $(MC_p)T_3 = Q_{h,i}$ and $(MC_p)T_7 = Q_{c,i}$ for inlet hot and cold fluids. Assembling the algebraic equations given by equations 4.19–4.23 in the proper order will give the complete set of algebraic equations for the box element. The present element matrix has 10 nodes and therefore it will be a 10×10 matrix. The nodal equations are developed for all the plate and fluid sub-elements and assembled together to get the element stiffness matrix as shown above. The known boundary conditions (inlet temperatures or enthalpies) are incorporated in the above matrix while solving. The element matrices for other pair of stacks are stored in the proper order to obtain the global stiffness matrix.

4.4.3 Cross Flow Tube-Fin Heat Exchanger

The model of a cross flow tube-fin heat exchanger is given in Figure 4.1g. The first element given in Figure 4.1 h is considered for element matrix formulation. The Galerkin weighted residual integral for Equation 4.31 is written as

$$\int [N]^T(-\alpha MC_p)_h \frac{\partial T_h}{\partial x}dx + \int [N]^T(\vartheta hP)_h(T_h - T_w)dx = 0 \tag{4.66}$$

Using the shape functions (Equations 4.26–4.30), the evaluation of [K] matrix for Equation 4.31 is as follows:

$$\int [N]^T \frac{(-\alpha MC_p)_h \partial T_h}{\partial x}\partial x = \left(\frac{MC_P}{2}\right)_h \begin{bmatrix} -1 & 1 \\ -1 & 1 \end{bmatrix} \begin{bmatrix} T_1 \\ T_2 \end{bmatrix} \tag{4.67}$$

$$\int \eta hl[N]^T T_h dx = \frac{\eta hlL}{6} \begin{bmatrix} 2 & 1 \\ 1 & 2 \end{bmatrix} \begin{bmatrix} T_3 \\ T_4 \end{bmatrix} \tag{4.68}$$

$$\int (\eta hl)_h [N]^T T_{w,t} dA = \left(\frac{\eta hlL}{6}\right)_h \begin{bmatrix} 2 & 1 \\ 1 & 2 \end{bmatrix} \begin{bmatrix} T_3 \\ T_4 \end{bmatrix} \tag{4.69}$$

Similar expressions can be derived for the other terms also for the Equations (4.31–4.33). Assembling all the terms obtained, thus we get algebraic equations corresponding to the fluid nodes, which can be written in a matrix form for an element (Figure 4.1h) of cross flow tube-fin heat exchanger as follows (before incorporation of boundary conditions):

$$\begin{bmatrix} -W_h+2H_h & W_h+H_h & -2H_h & -H_h & 0 & 0 \\ -W_h+H_h & W_h+2H_h & -H_h & -2H_h & 0 & 0 \\ -2H_h & -H_h & K_0+2H_h+2H_c & K_0+H_h+H_c & -2H_c & -H_c \\ -H_h & 2H_h & -K_0+H_h+H_c & K_0+2H_h+2H_c & -H_c & -2H_c \\ 0 & 0 & -2H_c & -H_c & -W_c+2H_c & W_c+2H_c \\ 0 & 0 & -H_c & -2H_c & -W_c+H_c & W_c+2H_c \end{bmatrix} \begin{bmatrix} T_1 \\ T_2 \\ T_3 \\ T_4 \\ T_5 \\ T_6 \end{bmatrix} = \begin{bmatrix} 0 \\ 0 \\ 0 \\ 0 \\ 0 \\ 0 \end{bmatrix} \tag{4.70}$$

where

$K_0 = kA_w/L$, $H = (\eta hPL)/6$, $W = MC_p/2$

After incorporation of boundary conditions, i.e $(MC_p)T_1 = Q_{h,i}$ and $(MC_p)T_5 = Q_{c,i}$ for inlet hot and cold fluids, the above matrix Equation 4.70 gets changed. Assembly of the algebraic equations given by equations (4.31–4.33) in the proper order will give the complete set of algebraic equations for the box element. The nodal equations are developed for all the plate and fluid sub-elements and assembled together to get the element stiffness matrix as shown above. The element matrices are assembled in the proper order to obtain the global stiffness matrices for cross flow plate-fin, counter flow plate-fin, parallel-flow plate-fin and cross flow tube-fin heat exchangers. These are solved by Gaussian elimination to get the exchanger wall temperature distribution and fluid outlet temperatures. The final sets of simultaneous equations are solved after incorporating the known boundary conditions (inlet temperatures or enthalpies). The outlet temperatures of strip 1 (Figure 4.1b for cross flow plate-fin heat exchanger) will be the inlet temperatures for adjacent strips. Thus by marching in a proper sequence, the temperature distribution in the exchanger is obtained similar to Ravi Kumar et al [5].

4.5 Longitudinal Wall Heat Conduction Effects

4.5.1 General

In above section, the governing equations for cross flow plate-fin, counter flow plate-fin, parallel-flow plate-fin and cross flow tube-fin heat exchangers and the finite element formulations are discussed. In some cases, it has been observed that the longitudinal heat conduction through the wall of an exchanger had a detrimental effect on its thermal performance – Chiou [5]. The rate of longitudinal heat conduction through the exchanger wall depends on the temperature gradient. Therefore, the deterioration of thermal performance of a heat exchanger relates to this temperature gradient. If a temperature gradient is established in the separating walls between fluid flow streams on a heat exchanger, heat transfer by conduction takes place from the hotter to the colder region of the wall, flattens the wall temperature distribution and reduces the performance of the exchanger.

In this section, first the dimensionless groups for two-fluid heat exchanger performance are presented in the presence of longitudinal wall heat conduction. Next the results of finite element models are validated with the available data. Then the study of the influence of longitudinal wall heat conduction on the exchanger performance is extended to cross flow tube-fin heat exchanger, parallel-flow heat exchanger and beyond the available results of cross flow plate-fin heat exchanger, in order to contribute to the literature. Also, FEM analysis has been extended to counter flow heat exchanger including the effects of longitudinal wall heat conduction to demonstrate the capability of the model. Finally, it is indicated that how to take into account this effect for a multi-pass exchanger. In a heat exchanger analysis, the exchanger effectiveness depends upon the number of transfer units NTU and heat capacity rate ratio C^* of the two fluids for a given flow arrangement – Kays and London [2].

$$\varepsilon = \varphi(NTU, C^*, Flow\ arrangement) \tag{4.71}$$

The magnitude of longitudinal heat conduction in the wall depends upon the wall heat conduction capacity and the wall temperature gradient. Let us consider the simple case

(a)

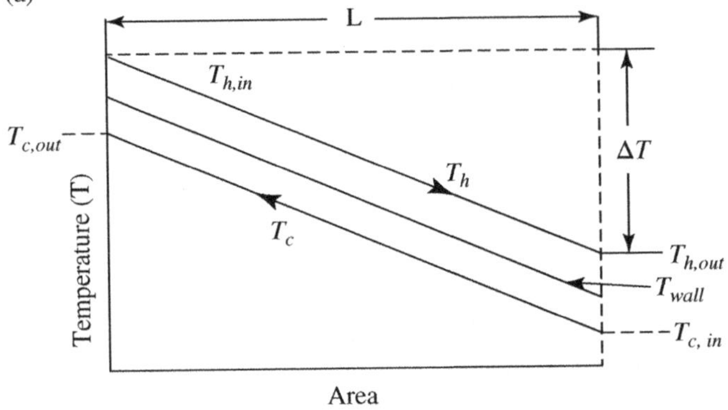

Figure 4.2 (a) Temperature distribution in a counter flow heat exchanger with $C_c = C_h$.

(b)

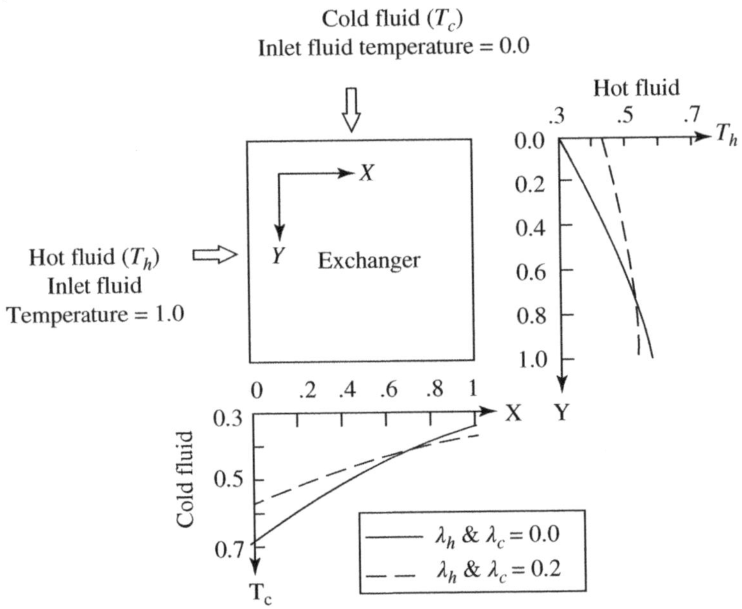

Figure 4.2 (b) Fluid temperature distributions at exits in a cross flow exchanger with zero and finite longitudinal wall heat conduction effect.

of the temperature condition pictured in Figure 4.2a for a direct-transfer type exchanger with $C_{min}/C_{max} = 1$. The temperature difference as shown for hot fluid is of the same magnitude for the cold fluid and also for the wall. The longitudinal conduction rate is

$$Q_w = k_w A \frac{\Delta T}{L} \tag{4.72}$$

The convective heat transfer from the hot fluid to the wall is

$$Q_h = C_h(T_{h,i} - T_{h,0}) \tag{4.73}$$

Similarly, the enthalpy rise of the cold fluid as a result of convection from the wall to the fluid is

$$Q_c = C_c(T_{c,0} - T_{c,i}) \tag{4.74}$$

Thus, the ratios of longitudinal heat conduction in the wall to the convection ratios in the hot and cold fluids are

$$\frac{Q_w}{Q_h} = \frac{k_w A}{LC_h}\frac{\Delta T}{(T_{h,i} - T_{h,o})} \tag{4.75}$$

and

$$\frac{Q_w}{Q_h} = \frac{k_w A}{LC_c}\frac{\Delta T}{(T_{c,o} - T_{c,i})} \tag{4.76}$$

The resulting new dimensionless groups in Equations 4.75 and 4.76 are defined as

$$\lambda_h = \left(\frac{k_w A}{LC}\right)_h \quad \text{and} \quad \lambda_c = \left(\frac{k_w A}{LC}\right)_c \tag{4.77}$$

The wall temperature considering the convective heat transfer between the hot fluid or cold fluid and the wall at any location can be written as

$$Q = (\eta hA)_h(T_h - T_w) = (\eta hA)_c(T_w - T_c) \tag{4.78}$$

or

$$T_w = \frac{T_h + (\eta hA)^* T_c}{1 + (\eta hA)^*} \tag{4.79}$$

whcrc

$$(\eta hA)^* = \frac{(\eta hA)_c}{(\eta hA)_h} \tag{4.80}$$

Hence the wall temperature distribution is between T_h and T_c distributions, and its specific location depends upon the magnitude of $(\eta ha)^*$. Note that sometimes $(\eta ha)^*$ is zero, as in a case of condenser, $T_w = T_h$. Since T_h is approximately constant for a condenser, T_w will also be a constant, indicating no longitudinal temperature gradients in the wall, even though λ will be finite. Thus when $(\eta ha)^*$ is zero or infinity the longitudinal heat conduction is zero. Longitudinal heat conduction effects are maximum for $(\eta ha)^* = 1$. Thus, in the presence of longitudinal heat conduction in the wall, the cross flow plate-fin heat exchanger effectiveness is a function of the following groups:

$$\varepsilon = \varphi(NTU, C^*, \lambda_h, \lambda_c, (\eta hA)^*, Flow\ arrangement) \tag{4.81}$$

Note that for a counter flow or parallel-flow heat exchanger

$$L_h = L_c = L \quad \text{and} \quad A_h = A_c = A \tag{4.82}$$

Equation 4.77 then becomes

$$\lambda_h = \frac{k_w A}{LC_h}; \lambda_c = \frac{k_w A}{LC_c} \tag{4.83}$$

and therefore

$$\frac{\lambda_h}{\lambda_c} = \frac{C_c}{C_h} = C^* \quad \text{for} \quad C_c = C_{min} \tag{4.84}$$

$$= \frac{1}{C^*} \quad \text{for} \quad C_h = C_{min} \tag{4.85}$$

Thus for a counter flow heat exchanger, both λ_h and λ_c are not independent parameters since C^* is already included in Equation 4.81; only one of them is independent. Instead of choosing one of them, alternatively a unique longitudinal conduction parameter λ is defined as follows for a counter flow or parallel-flow heat exchanger.

$$\lambda = \frac{k_w A}{LC_{min}} \tag{4.86}$$

Then, the Equation 4.81 takes the following form for counter flow/parallel-flow heat exchangers

$$\varepsilon = \varphi(NTU, C^*, \lambda, (\eta hA)^*, \ \textit{Flow arrangement}) \tag{4.87}$$

Other parameters that influence longitudinal conduction in direct transfer type heat exchanger are C^*, $(\eta hA)^*$, NTU and flow arrangement. For storage type heat exchangers, an additional parameter is C_r/C_{min}.

The conduction effect factor τ is introduced to illustrate the influence of the two-dimensional longitudinal wall heat conduction on the deterioration of exchanger performance by Chiou [6,7].

$$\tau_L = \frac{\Delta\varepsilon}{\varepsilon} = \frac{\varepsilon_0 - \varepsilon_L}{\varepsilon_0} \tag{4.88}$$

The conduction effect factor directly shows the degree of deterioration of the exchanger heat transfer effectiveness due to the longitudinal wall heat conduction. The larger this factor, the greater the deterioration of the exchanger effectiveness. In this analysis, uniform inlet fluid flow and uniform inlet temperature are assumed.

Let us review typical temperature distribution in the hot fluid and the cold fluid for a cross flow heat exchanger with λ (λ_h or λ_c) = 0.0 and 0.2 respectively, as shown in Figure 4.2b – Kays and London [2]. When λ = 0.2, the exit temperature of the hot fluid is warmer and that of the cold fluid is colder in most regions than those of the respective fluids when λ = 0.0. This results in lower thermal performance of the exchanger when the longitudinal heat conduction in the wall cannot be neglected.

4.5.2 Validation

The accuracy of the solution depends on the number of the elements used. Actually, the number of elements used is determined by a compromise between the accuracy desired and the time required by the computer. Use of a greater number of subdivisions can not only produce results of higher accuracy but also enhance the convergence. For example,

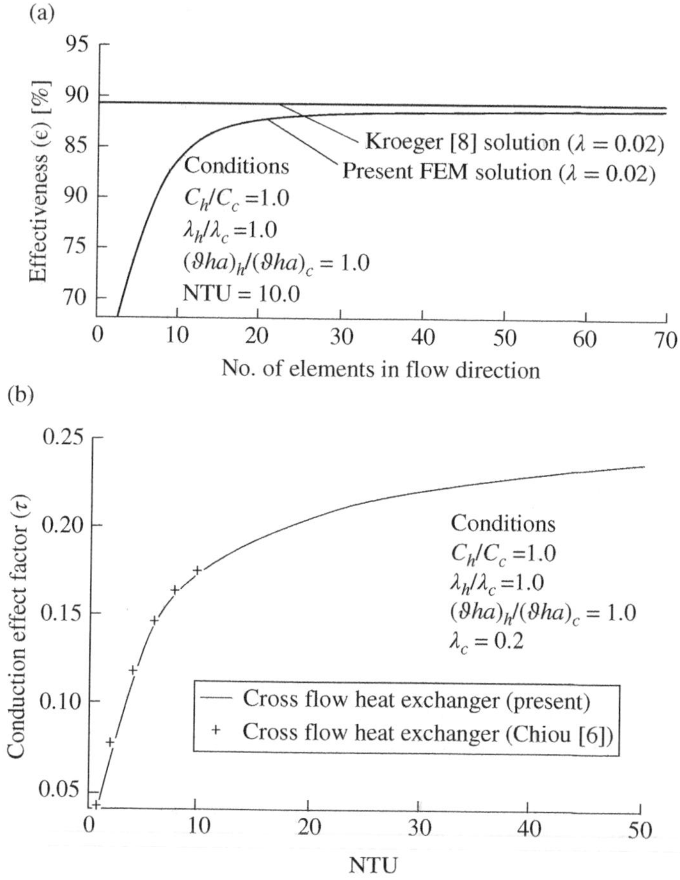

Figure 4.3 Cross flow plate fin heat exchanger: (a) Convergence of solution: counter flow heat exchanger (b) Comparison with available solution: cross flow plate-fin heat exchanger.

as shown in Figure 4.3a, the predicted values approach the available analytical solution by Kroeger [8] for a counter flow heat exchanger. Similarly, the present analysis results have been compared with available data for cross flow plate-fin heat exchanger as shown in Figure 4.3b. The results are in good agreement with the solutions of finite difference method by Chiou [6].

In this analysis, 'air' is used for both hot and cold sides and the unmixed-unmixed type of exchangers is considered. The experimental values of heat transfer coefficients and frictional coefficients provided by Kays and London [2] are used. Analytical solutions without longitudinal wall heat conduction effects are also obtained using the solution procedure given by Kays and London [2].

4.5.3 Cross Flow Plate-Fin Heat Exchanger

In this section, the performance evaluation with effects of longitudinal wall heat conduction for cross flow plate-fin heat exchanger is presented for balanced flow, $C^* = 1$, as well as for unbalanced flows, $C^* \neq 1$. Chiou [6] has provided these effects for $C_x/C_y = 0.25, 0.5$

and 1.0 for cross flow plate-fin heat exchanger. However, Chiou [6] neglected the effects of fluid property variations while investigating the longitudinal wall heat conduction effects. Hence, the exchanger effectiveness of a single pass cross flow plate-fin heat exchanger and its deterioration due to the longitudinal wall heat conduction are determined for a wide range of design/operating parameters using the new method.

In this study, the following ranges of the design and operating parameters of the exchanger are considered:

$C^* = 1.0, 0.8, 0.6, 0.4, 0.2$
$\lambda_h/\lambda_c = 0.5, 1.0, 2.0$
$\lambda_h = 0.0, 0.0125, 0.025, 0.05, 0.1, 0.2$
$(\eta ha)^* = 0.5, 1.0, 2.0$
$NTU = 1$ to 50

The conduction effects are presented as a function of NTU (NTU_{min}) and longitudinal heat conduction parameters (λ) for four magnitudes of $C^* = 1.0, 0.8, 0.6, 0.4$, as shown in Figure 4.4 for balanced flow $C^* = 1.0$ and as shown Figure 4.5 for unbalanced flows $C^* =$ 0.8, 0.6, 0.4. In each case of C^*, there are three magnitudes of $\lambda_h/\lambda_c = 0.5, 1.0, 2.0$. For the case of $C^* = 1.0$, the conduction effect factor increases as the exchanger NTU increases, as shown in Figure 4.4. Also it has been observed that the conduction effect factor, τ increases with increase of conduction parameters (λ's). These results are similar to those observed by another investigator [6]. In the case of unbalanced flow ($C^* = 0.8$) with high λ (say 0.2 or 0.1) values the conduction effect factor (τ) increases with NTU. All other cases of unbalanced flows, the conduction effect factor (τ) generally increases with NTU for lower values of NTU and then it decreases with the increase of NTU as shown in Figure 4.5.

The relations between conduction effect factor (τ) and λ_h/λ_c with NTU are provided in Figure 4.5 (a1–3, b1–3, c1–3) for $(\eta ha)^* = 1.0$. There are some significant variations in performance deteriorations due to longitudinal wall heat conduction with λ_h/λ_c ratio. Hence the present analysis has been carried out by considering the three magnitudes of $\lambda_h/\lambda_c = 0.5, 1.0, 2.0$. Figure 4.5(c1–3) shows that the conduction effect factor is higher for $\lambda_h/\lambda_c = 2.0$. The variations in performance deterioration due to longitudinal wall heat conduction with $(\eta ha)^*$ ratio are also presented in Figure 4.5d. The performance variations with $(\eta ha)^*$ are small enough, can be neglected. These results [9,10] are similar to those observed by other investigators, [6,7,11,14–16]. From the results presented in this section, it is noted that the thermal performance deterioration of cross flow heat exchanger due to longitudinal heat conduction through the wall structure is not always negligible, especially for both the fluids with capacity ratio (C^*) equal to 1.0 and when the longitudinal conduction parameter is large. Information presented in this section, through Figures 4.4 and 4.5 covers the entire range of design and operating conditions of the cross flow plate-fin heat exchanger.

4.5.4 Cross Flow Tube-Fin Heat Exchanger

In cross flow tube-fin heat exchangers, the wall temperature distribution is one-dimensional, and no analysis is available for longitudinal heat conduction effects. Hence, the finite element analysis has been extended to cross flow tube-fin heat exchanger, and

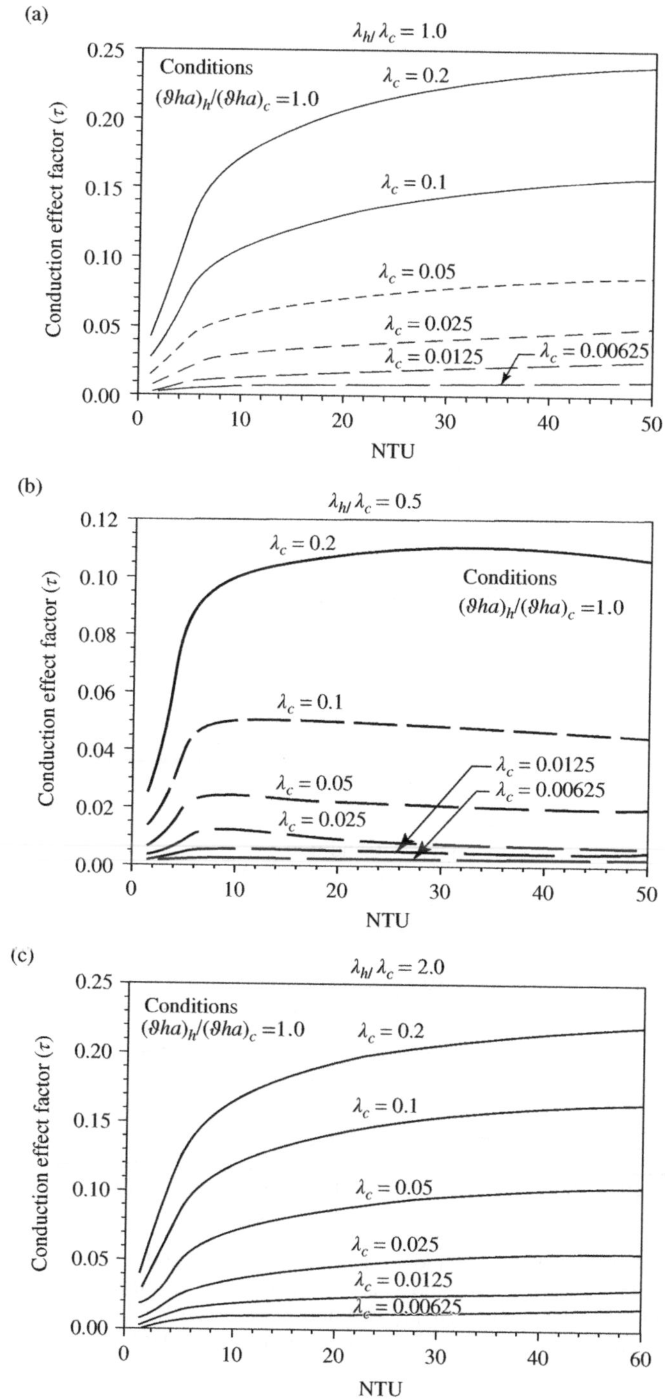

Figure 4.4 LHC effects – cross flow plate fin heat exchanger – $C_{min}/C_{max} = 1.0$.

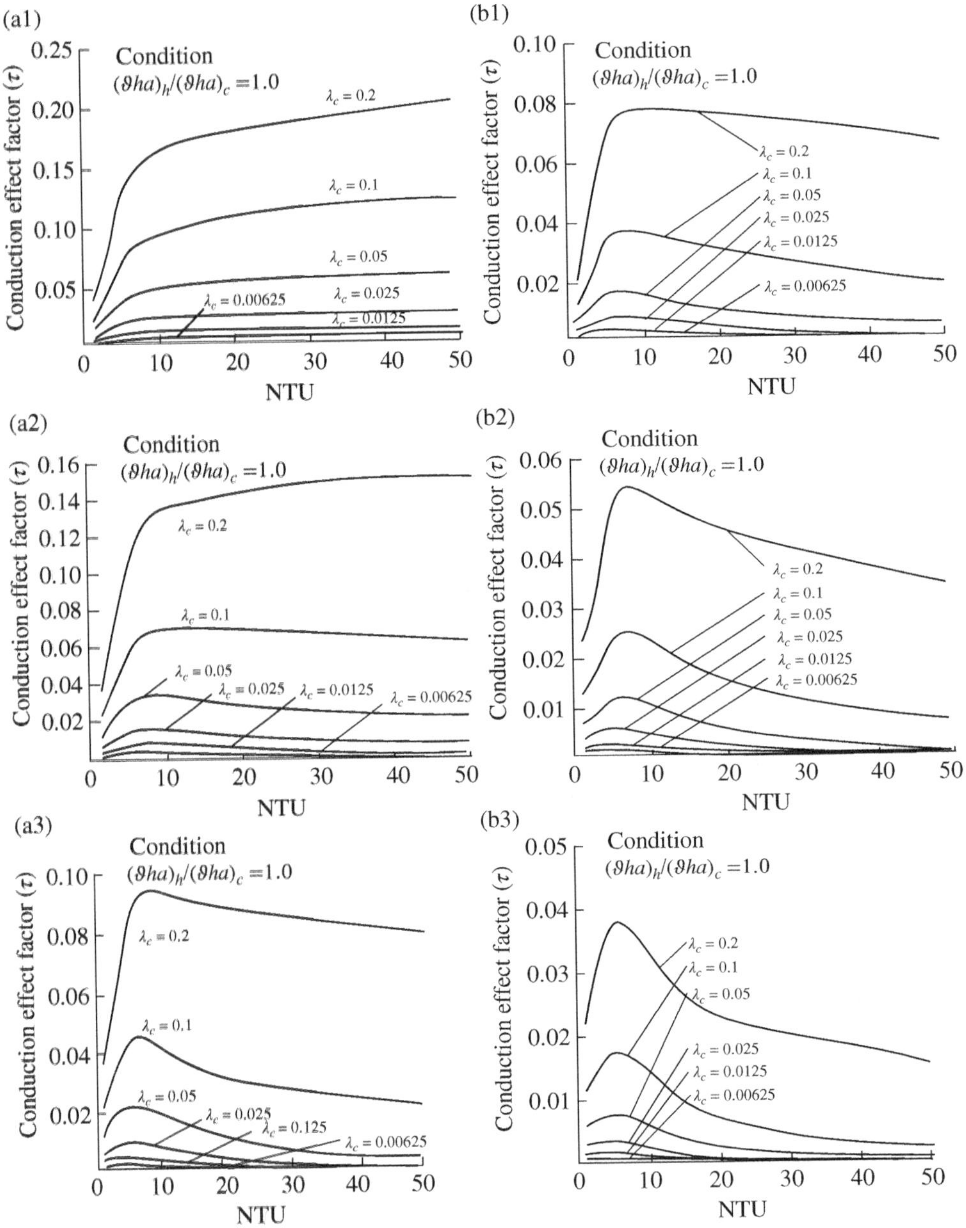

Figure 4.5 LHC effects – cross flow plate fin heat exchanger – (a1) $C_{min}/C_{max} = 0.8$ and $\lambda_h/\lambda_c = 1.0$, (a2) $C_{min}/C_{max} = 0.6$ and $\lambda_h/\lambda_c = 1.0$, (a3) $C_{min}/C_{max} = 0.4$ and $\lambda_h/\lambda_c = 1.0$, (b1) $C_{min}/C_{max} = 0.8$ and $\lambda_h/\lambda_c = 0.5$, (b2) $C_{min}/C_{max} = 0.6$ and $\lambda_h/\lambda_c = 0.5$, (b3) $C_{min}/C_{max} = 0.4$ and $\lambda_h/\lambda_c = 0.5$, (c1) $C_{min}/C_{max} = 0.8$ and $\lambda_h/\lambda_c = 2.0$, (c2) $C_{min}/C_{max} = 0.4$ and $\lambda_h/\lambda_c = 2.0$, (c3) $C_{min}/C_{max} = 0.2$ and $\lambda_h/\lambda_c = 2.0$ (d) Conduction effects with $(\eta hA)_h/(\eta hA)_c$ ratio.

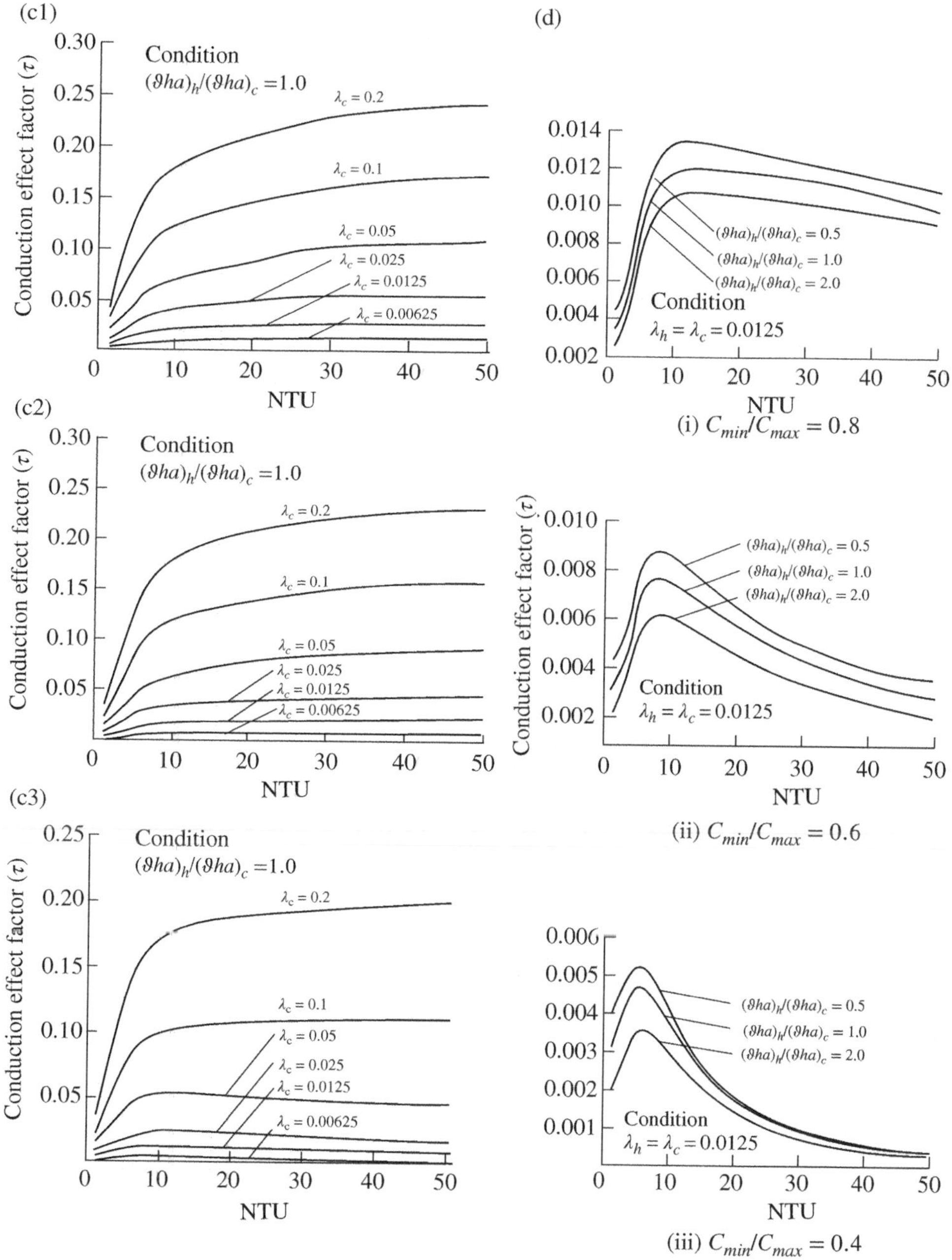

Figure 4.5 (Continued)

the results are presented in Figure 4.6 for both balanced flow, $C^* = 1$, and unbalanced flows $C^* = 0.8, 0.6, 0.4, 0.2$. It is interesting to note that for identical *NTU* and λ, the effect of longitudinal wall heat conduction on the exchanger effectiveness is less for the cross flow tube-fin exchanger than the cross flow plate-fin exchanger. Although this result is surprising, it is expected since the temperature gradient in the tube in a cross flow tube-fin exchanger is one-dimensional compared to that in a cross flow plate-fin heat exchanger.

For the case of $C^* = 1.0$, the conduction effect factor (τ) increases as the exchanger *NTU* increases for high values of λ_h, as shown in Figure 4.6a. For the same case of $C^* = 1.0$, the conduction effect factor (τ) generally increases with *NTU* until *NTU* reaches 8–12, then it remains constant with the increase of *NTU* for low values of λ_h (say $\lambda_h = 0.1$). For the case of $C^* = 0.8$, the conduction effect factor (τ) generally increases with the *NTU* until *NTU* reaches 8–12, then it decreases with increase in *NTU* for all values of λ_h as shown in Figure 4.6b. For the case of $C^* = 0.6$, the conduction effect factor (τ) generally increases with the *NTU* until *NTU* reaches 5, then it decreases with the increase of *NTU* for all values of λ_h as shown in Figure 4.6c. In case of $C^* = 0.4, 0.2$, the conduction effect factor (τ) increases with the increase of *NTU* and diminishes after $NTU = 25$ as shown in Figures 4.6d and e. The relations between $(\eta hA)^*$ and conduction effect factor (τ) with *NTU* are shown in Figure 4.6f. In this case, the variations in performance deteriorations with $(\eta hA)_h/(\eta hA)_c$ ratio are smaller as compared to cross flow plate-fin heat exchanger.

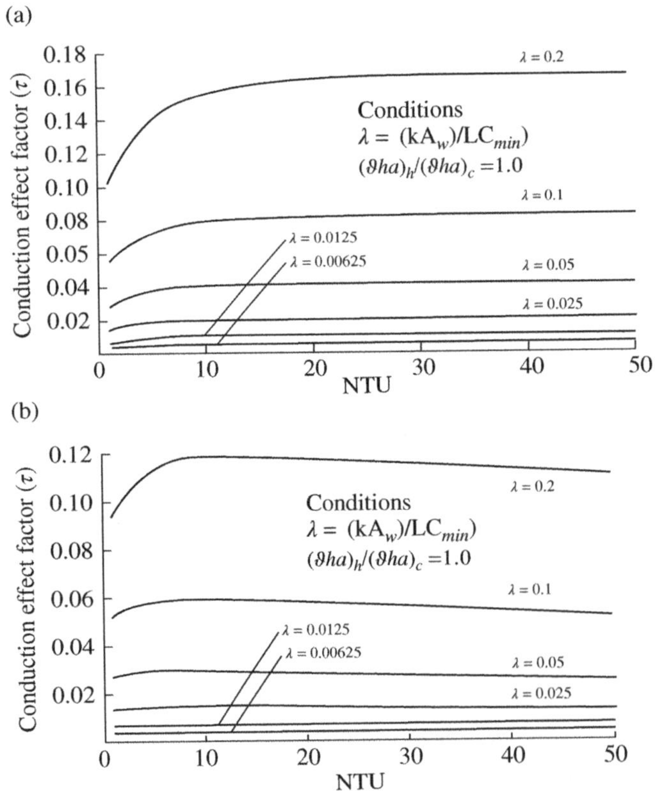

Figure 4.6 LHC effects – cross flow tube fin heat exchanger (a) $C^* = 1.0$ (b) $C^* = 0.8$.

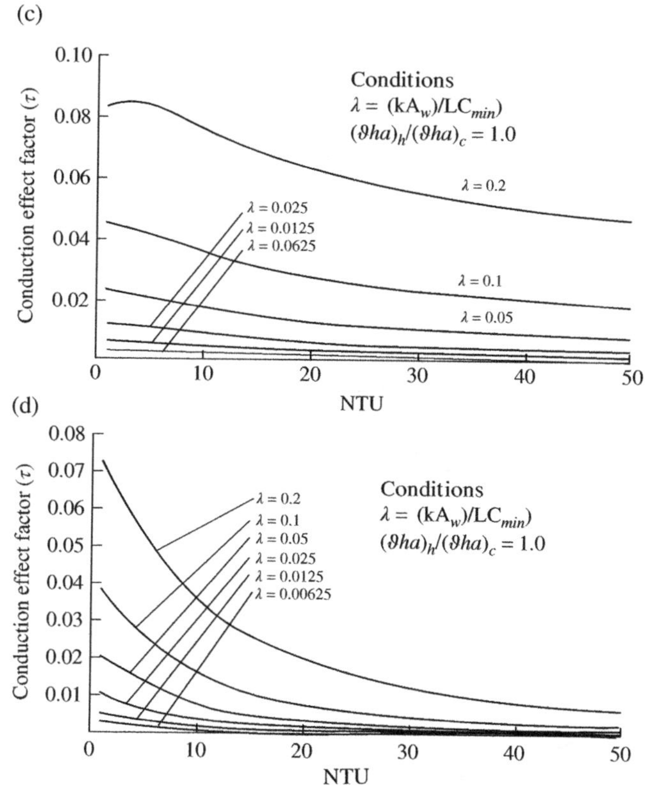

Figure 4.6 (Continued) – cross flow tube fin heat exchanger (c) $C^* = 0.6$ (d) $C^* = 0.4$.

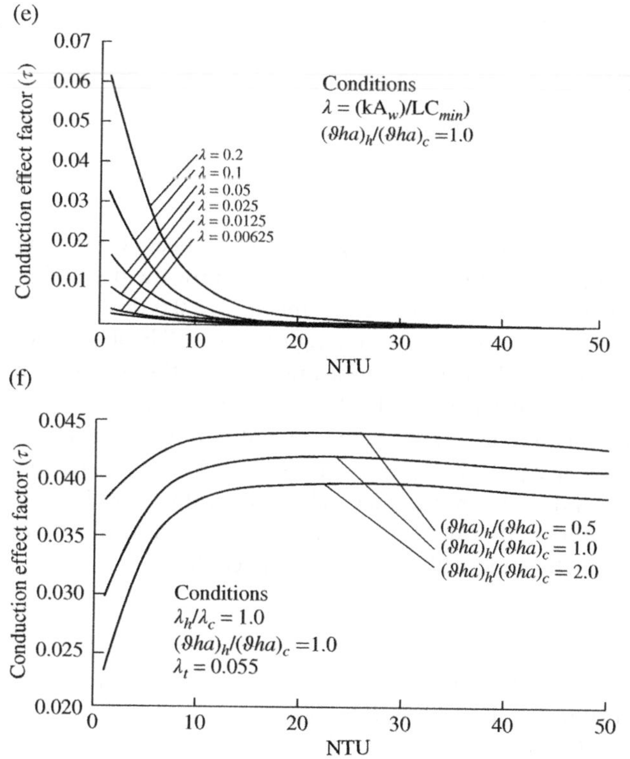

Figure 4.6 (Continued) – cross flow tube fin heat exchanger (e) $C^*= 0.2$ (f) $(\vartheta ha)_h/(\vartheta ha)_c= 0.5, 1.0, 2.0$.

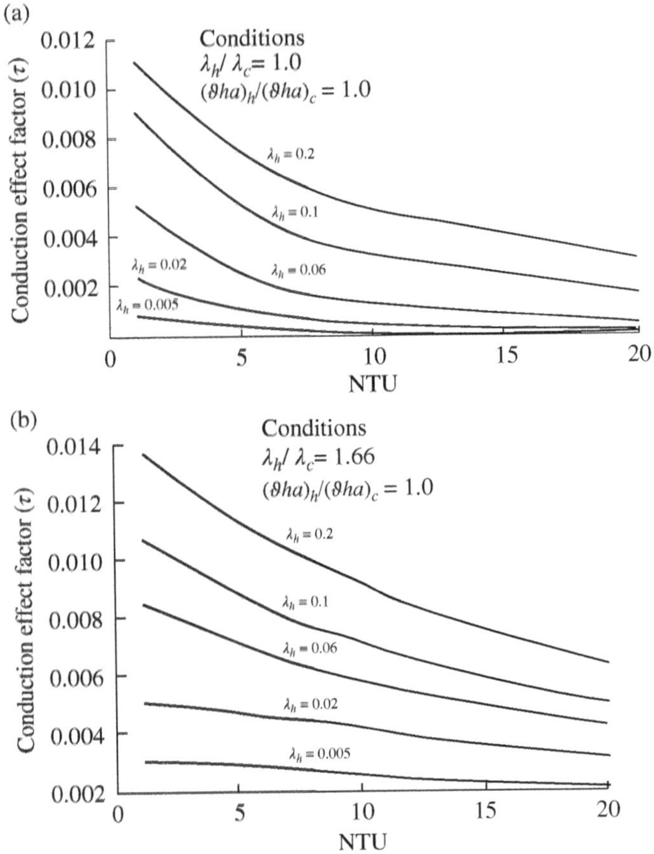

Figure 4.7 LHC effects – parallel flow plate-fin heat exchanger (a) $C^* = 1.0$ (b) $C^* = 0.6$.

4.5.5 Parallel Flow Heat Exchanger

In this section, the performance deteriorations due to longitudinal wall heat conduction effects in parallel-flow heat exchanger are presented in Figure 4.7 to show the prediction capability of the model for both balanced, $C^* = 1$, and unbalanced $C^* = 0.6$ flows. It has been observed that the performance deteriorations due to longitudinal wall heat conduction are negligibly small. In both cases of balanced and unbalanced flows, the conduction effect factor (τ) decreases with increase of *NTU* for all values of λ_h. So far, no study has reported on the longitudinal wall heat conduction problem for parallel-flow heat exchangers of either storage or direct transfer types. However, the present FEM analysis reveals that the performance deteriorations due to longitudinal wall heat conduction of a parallel-flow heat exchanger are much less (1%), when compared with a counter flow and cross flow heat exchangers and may become insignificant as expected.

4.5.6 Counter Flow Heat Exchanger

Closed form analytical solutions are available by Kroeger [8] and Shah [14] for performance variation due to longitudinal wall heat conduction for counter flow heat exchanger for constant heat transfer coefficient cases. However, the present analysis has been extended to counter flow heat exchanger to demonstrate the prediction capability of FEM model. In this analysis, the fluid property variation effects have also been considered. Figure 4.8 presents the relation between the longitudinal conduction effect factor (τ) for the relevant design/operating parameters of the counter flow heat exchanger. As in the cross flow heat exchanger, the most severe performance deterioration is encountered when both fluids have the same heat capacity ratio. In the case of balanced flow, $C^* = 1.0$, the conduction effect factor generally increases with *NTU* until *NTU* reaches 10, then it remains constant with the increase of *NTU* as shown in Figure 4.8a. Whereas, in the case of unbalanced flow, the conduction effect factor generally increases with *NTU* until *NTU* reaches approximately 6–8, then it decreases with the increase of *NTU*, as shown in Figure 4.8b. It has been observed that the behaviour of performance deterioration due to longitudinal heat conduction is similar to that of cross flow heat exchanger, but the magnitudes of the performance deteriorations are less when compared with cross flow heat exchanger. This result is expected since the temperature gradient in the wall in a counter flow exchanger is linear and of lower magnitude compared to that in a cross flow plate-fin heat exchanger.

4.5.7 Relative Comparison of Results

The relative comparison of LHC effects is presented for all four types compact heat exchangers in Figure 4.9 for the same conduction parameter $\lambda = 0.2$. This figure shows that the cross flow plate-fin heat exchanger is having highest performance deterioration and the parallel flow plate-fin heat exchanger is having the lowest performance deterioration, as expected. The performance deteriorations of counter flow plate-fin heat exchanger and cross flow tube-fin heat exchanger lie in between cross flow plate-fin heat exchanger and parallel flow plate-fin heat exchanger as shown in Figure 4.9 for both (a) balanced ($c^* = 1.0$) and (b) unbalanced flows ($c^* = 0.6$).

The relative comparison of longitudinal wall heat conduction effects with C^* is provided in Table 4.1 for two magnitudes of $NTU = 10$ and 50 for reference. For example, the maximum conduction effect factor (τ) values (at $NTU = 10$, $\lambda = 0.2$ and $\lambda_c/\lambda_h = 1.0$) due to longitudinal heat conduction effects are 17.5% for $C^* = 1.0$, 16.9% for $C^* = 0.8$ and 9.3% for $C^* = 0.4$ for cross flow plate-fin heat exchanger. The maximum conduction effect factor (τ) values (at $NTU = 10$, $\lambda = 0.2$ and $\lambda_c/\lambda_h = 1.0$) due to longitudinal heat conduction effects are 3.25% for $C^* = 1.0$, and 2.8% for $C^* = 0.6$ for cross flow tube-fin heat exchanger. The maximum conduction effect factor (τ) values (at $NTU = 10$, $\lambda = 0.2$ and $\lambda_c/\lambda_h = 1.0$) due to longitudinal heat conduction effects are 0.3% for $C^* = 1.0$, and 0.5% for $C^* = 0.6$, which are negligible, for parallel flow heat exchanger.

However, the maximum conduction effect factor (τ) values (at $NTU = 10$, $\lambda = 0.2$ and $\lambda_c/\lambda_h = 1.0$) due to longitudinal heat conduction effects are 3% for $C^* = 1.0$ and 1.5% for $C^* = 0.6$ for counter flow heat exchanger.

4.6 Inlet Flow Non-Uniformity Effects

4.6.1 General

In most heat transfer and pressure drop analysis of heat exchangers, it is presumed that the inlet fluid flow distributions across the exchanger core are uniform. This assumption is generally not realistic under actual operating conditions. For a tube-side flow through a

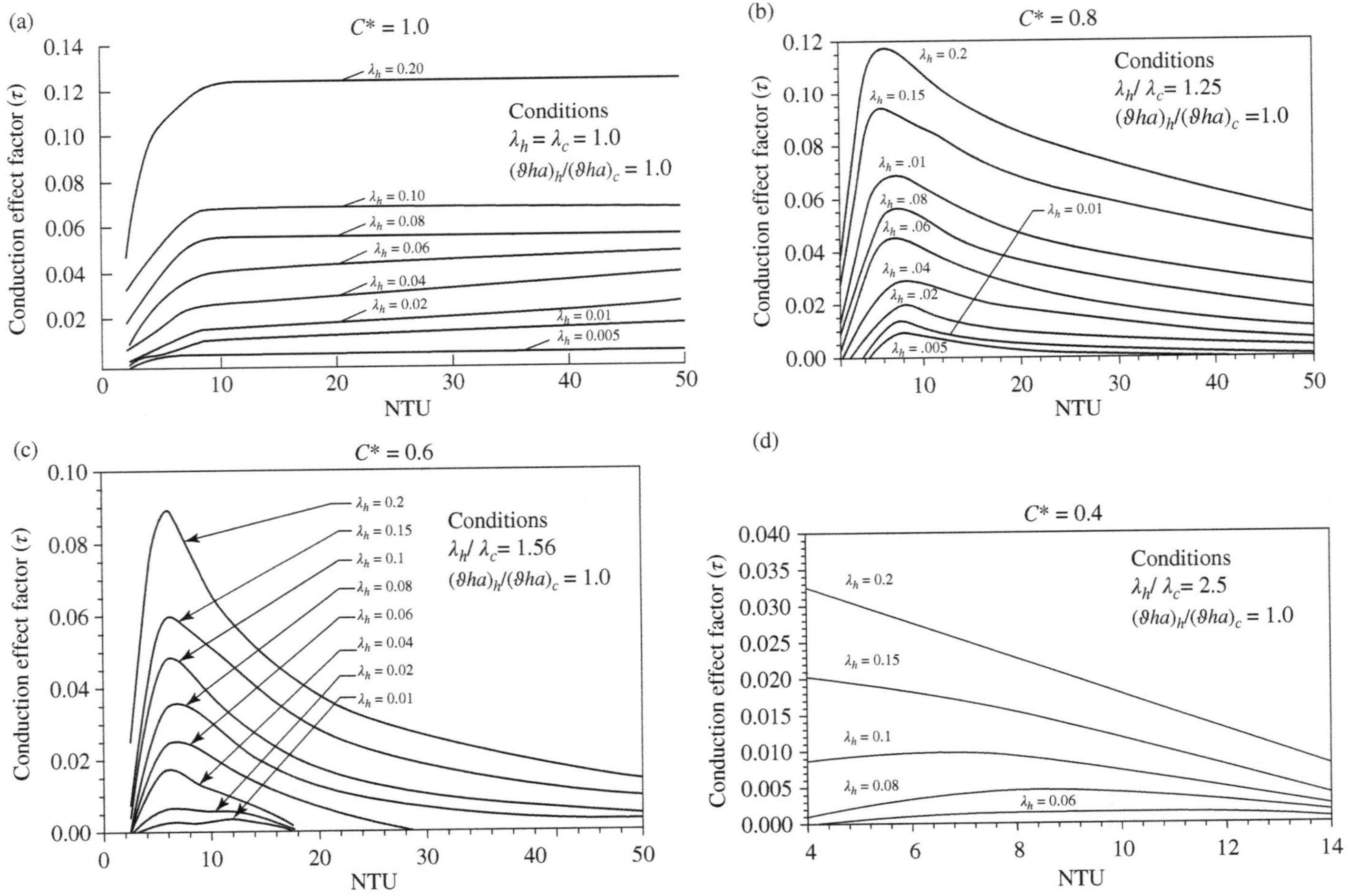

Figure 4.8 LHC effects – Counter flow plate-fin heat exchanger.

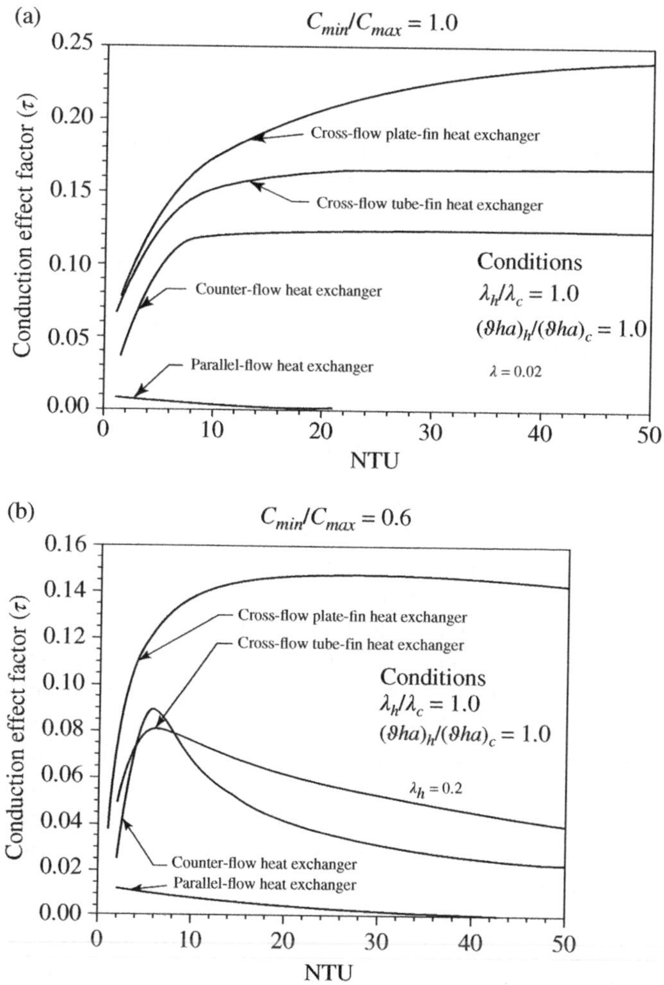

Figure 4.9 LHC effects – Comparison of four types of heat exchangers.

Table 4.1 Summary of results – longitudinal wall heat conduction effects.

		Performance deteriorations (%)			
		NTU = 10		NTU = 50	
Type of exchanger	**C***	$\lambda = 0.05$	$\lambda = 0.20$	$\lambda = 0.05$	$\lambda = 0.20$
Cross flow	1.0	5.00	15.0	9.00	21.0
plate-fin	0.6	3.00	13.5	2.00	15.0
Counter flow	1.0	3.00	12.0	3.50	12.5
plate-fin	0.6	1.50	6.50	0.00	2.00
Cross flow	1.0	3.25	14.5	3.50	15.0
tube-fin	0.6	2.60	11.5	2.50	10.5
Parallel flow	1.0	0.30	0.60	0.00	0.00
plate-fin	0.6	0.50	0.85	0.00	0.00

bundle of tubes, a uniform distribution can mean an equal amount of fluid in each tube. A significant reduction in heat exchanger performance may result when the flow distribution through the core is non-uniform. There are many cases of mal-distribution in heat exchangers, such as:

1) mechanical causes due to the design of headers and inlet ducts influencing the flow distribution
2) self-induced mal-distribution due to the heat transfer process itself, such as the freezing effect in viscous flow coolers or the various types of thermo-acoustic oscillation [17]
3) two-phase flows, which are very difficult to distribute uniformly through the tubes in a tube bundle
4) fouling and/or corrosion affecting the flow distribution.

Two geometrically related mal-distributions are gross and passage-to-passage. In gross mal-distribution, the fluid flow distribution at the core inlet is non-uniform. This occurs because of either poor header design or gross blockage in the core during manufacture. Passage-to-passage mal-distribution occurs within a core in a highly compact exchanger when the small flow passages are not identical as a result of manufacturing tolerances. The objective of the thermal design of a heat exchanger is to determine the most favourable size and configuration of the exchanger core, which meets the demand of the required heat transfer rate within the specified fluid pressure drops, space and cost limitations. Hence, the present analysis has been carried out to determine the effects of gross flow mal-distribution of the cross flow heat exchanger on thermal performance and pressure drop variations for three cases: flow non-uniformity on C_{min} fluid side, on C_{max} fluid side and flow non-uniformity on both fluid sides of the exchanger. In each case, different magnitudes of two-dimensional velocity profiles are considered at the core entrance.

The design of headers and inlet ducts significantly affects the velocity distribution approaching the face of the exchanger core, which is shown in Figure 4.10a. In this type of flow mal-distribution, the variations in flow at the inlet of exchanger core mainly depend on the location of the inlet duct, the ratio of core frontal area to the inlet duct cross-sectional area, the distance of transition duct/header between the core face and the inlet duct and the shape of headers i.e. oblique flow headers or normal flow headers and with/without manifolds. These velocity profiles are either one-dimensional or two-dimensional and either linear or non-linear depending on the flow pattern as discussed. A serious deterioration in heat exchanger performance may result even when the flow distribution through the core is not uniform, particularly when an area increase from the pipe to the core face is 5 to 50 times – Shah [17]. In these cases, the pressure gradient is often higher at the centre than at the edge points of the exchanger core if the inlet duct/pipe is straight. Uniformity of flow distribution over the core is the primary function of the headers.

Based on the concept of Fourier series, a mathematical equation is developed to generate the flow non-uniformity models at exchanger inlet (either hot fluid side or cold fluid side). The heat exchanger core frontal face is a rectangular domain having edges $2a$, $2b$ in the x-y plane as shown in Figure 4.10b and will be a square domain when $a = b$. The lengths with respect to origin 'o' are $x = a$ or $x = -a$ and $y = b$ or $y = -b$.

The cold fluid flows under the influence of a constant pressure gradient in the z direction and its equation is

$$\frac{\partial^2 W_C}{\partial x^2} + \frac{\partial^2 W_C}{\partial y^2} = \frac{1}{\mu}\frac{\partial P_C}{\partial Z} = \text{aconstant} \tag{4.89}$$

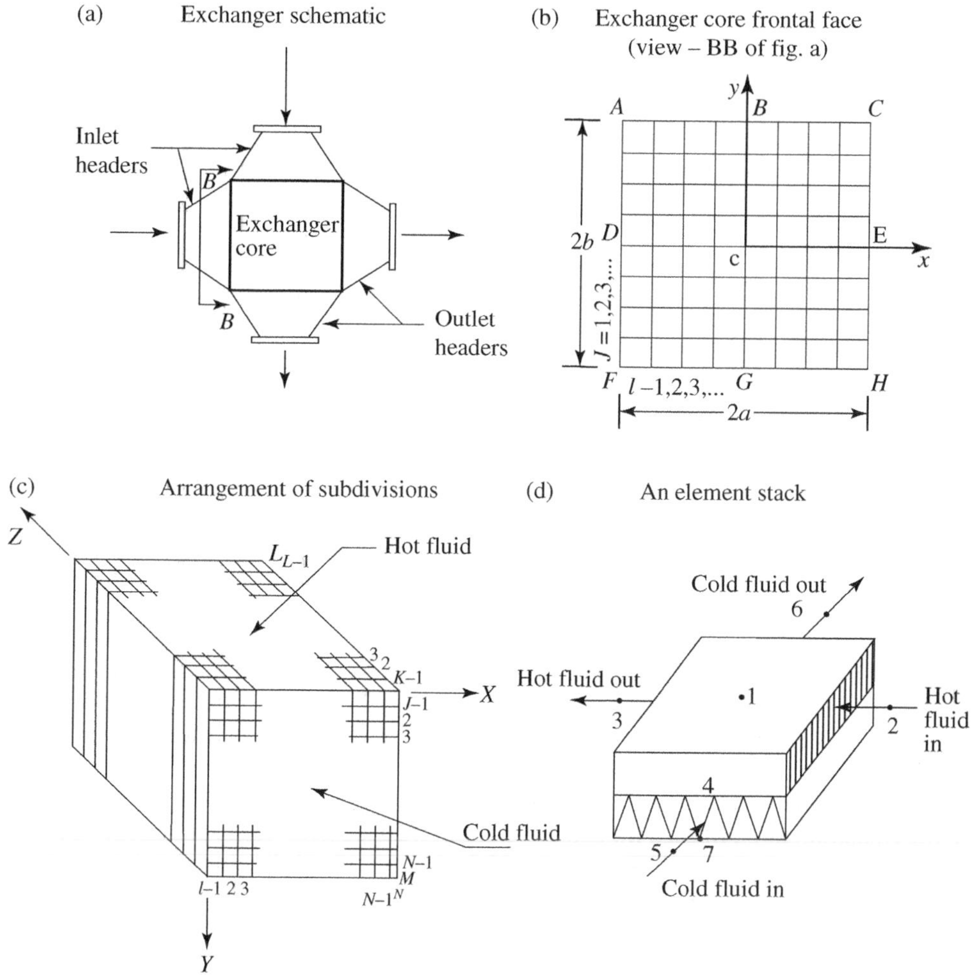

Figure 4.10 Single pass cross flow heat exchanger.

The same equation can be used for other fluid provided the x, y, z coordinates are changed accordingly. Similarly, the hot fluid flows under the influence of a constant pressure gradient in the x direction and its equation is

$$\frac{\partial^2 W_h}{\partial y^2} + \frac{\partial^2 W_h}{\partial z^2} = \frac{1}{\mu}\frac{\partial P_h}{\partial x} = \text{a constant} \tag{4.90}$$

Only one equation can be shown by removing the suffix c or h in Equations 4.89 and 4.90 to avoid repeatability of the equations. Then the equation for the fluid (cold or hot) at the exchanger inlet duct can be represented as:

$$\frac{\partial^2 W}{\partial x^2} + \frac{\partial^2 W}{\partial y^2} = \frac{1}{\mu}\frac{\partial P}{\partial z} = \text{a constant} \tag{4.91}$$

On boundary,

$$\text{at } x = \pm a, \ W(\pm a, y) \tag{4.92}$$

$$\text{at } y = \pm b, \ W(x, \pm b) \tag{4.93}$$

The term $\frac{1}{\mu}\frac{\partial P}{\partial z}$ is a constant that can be regarded as known since the choice of the pressure gradient is in one's control here. So there is only one unknown, $W = W(x,y)$, controlled by Equations 4.91, 4.92 and 4.93 above. The solution to represent the fluid velocity (W) is

$$W(x,y) = \frac{64D}{\pi^4}\sum_{m,n=0}^{\infty}\left\{\frac{(-1)^{m+N+1}}{\left(\frac{2m+1}{a}\right)^2+\left(\frac{2n+1}{b}\right)^2}\frac{\cos\frac{(2m+1)\pi x}{2a}\cos\frac{(2n+1)\pi y}{2b}}{(2m+1)(2n+1)}\right. \tag{4.94}$$

The derivation of above equation is given in Appendix A-1. The arrangement of subdivisions is shown in Figure 4.10c for a single pass cross flow plate-fin heat exchanger. Consider that the fluid (either cold or hot) flowing in the y direction is not uniformly distributed over the exchanger core on the x-z plane. Similarly, the fluid (either hot or cold) flowing in x direction is not uniformly distributed over the exchanger core on the y-z plane as shown in Figure 4.10c.

Equation (4.94) can be used only to generate the possible flow non-uniformity parameters at the entry of exchanger duct. Using the above equations, different types of flow non-uniformity models are generated by considering the different magnitudes of pressure gradients and are shown in Table 4.2. Also, the inlet flow non-uniformity for a typical flow model is shown in Figure 4.11. The velocity at the wall of the inlet duct is zero. The non-zero velocity values in the proposed models are at the points away from the wall of the transition duct. In each model, there are 10 × 10 local flow non-uniformity dimensionless parameters (α's), which correspond to the 10 × 10 subdivisions on the x-z plane perpendicular to the direction of the non-uniform fluid flow. In view of the symmetry of Equation 4.94 with respect to o-x and o-y, only one quarter of the flow non-uniformity parameters (α's) are presented in Table 4.2.

The exchanger thermal performance deterioration factor or flow non-uniformity effect factor (τ_F) is introduced to analyse the two-dimensional flow non-uniformity effects on the deterioration of thermal performance. It is defined as

$$\tau_F = \frac{\Delta\varepsilon}{\varepsilon} = \frac{\varepsilon_0 - \varepsilon_F}{\varepsilon_0} \tag{4.95}$$

The thermal performance deterioration factor (τ_F) directly shows the degree of deterioration of exchanger effectiveness. Analytical solutions without considering the effects

Table 4.2(a) Flow non-uniformity parameters (α's) model – A0.

$I =$	1 ; 10	2 ; 9	3 ; 8	4 ; 7	5 ; 6
$J = 1$; 10	0.100	0.100	0.100	0.100	0.100
2 ; 9	0.100	0.352	0.597	0.819	0.996
3 ; 8	0.100	0.597	1.080	1.523	1.879
4 ; 7	0.100	0.819	1.523	2.177	2.717
5 ; 6	0.100	0.996	1.879	2.717	3.438

Table 4.2(b) Flow non-uniformity parameters (α's) model – A1.

I =	1 ; 10	2 ; 9	3 ; 8	4 ; 7	5 ; 6
J = 1 ; 10	0.500	0.500	0.500	0.500	0.500
2 ; 9	0.500	0.639	0.776	0.899	0.998
3 ; 8	0.500	0.776	1.045	1.291	1.489
4 ; 7	0.500	0.899	1.291	1.655	1.956
5 ; 6	0.500	0.998	1.489	1.956	2.356

Table 4.2(c) Flow non-uniformity parameters (α's) model – A2.

I =	1 ; 10	2 ; 9	3 ; 8	4 ; 7	5 ; 6
J = 1 ; 10	0.600	0.600	0.600	0.600	0.600
2 ; 9	0.600	0.710	0.820	0.919	0.998
3 ; 8	0.600	0.820	1.036	1.234	1.393
4 ; 7	0.600	0.919	1.234	1.526	1.768
5 ; 6	0.600	0.998	1.393	1.768	2.090

Table 4.2(d) Flow non-uniformity parameters (α's) model – A3.

I =	1 ; 10	2 ; 9	3 ; 8	4 ; 7	5 ; 6
J = 1 ; 10	0.700	0.700	0.700	0.700	0.700
2 ; 9	0.700	0.781	0.864	0.939	0.999
3 ; 8	0.700	0.864	1.027	1.177	1.297
4 ; 7	0.700	0.939	1.177	1.398	1.581
5 ; 6	0.700	0.999	1.297	1.581	1.825

Table 4.2(c) Flow non uniformity parameters (α's) model – A4.

I =	1 ; 10	2 ; 9	3 ; 8	4 ; 7	5 ; 6
J = 1 ; 10	0.800	0.800	0.800	0.800	0.800
2 ; 9	0.800	0.855	0.910	0.959	0.999
3 ; 8	0.800	0.910	1.018	1.117	1.197
4 ; 7	0.800	0.959	1.117	1.264	1.385
5 ; 6	0.800	0.999	1.197	1.385	1.546

Table 4.2(f) Flow non-uniformity parameters (α's) model – A5.

I =	1 ; 10	2 ; 9	3 ; 8	4 ; 7	5 ; 6
J = 1 ; 10	0.900	0.900	0.900	0.900	0.900
2 ; 9	0.900	0.923	0.952	0.978	0.999
3 ; 8	0.900	0.952	1.009	1.062	1.104
4 ; 7	0.900	0.978	1.062	1.139	1.203
5 ; 6	0.900	0.999	1.104	1.203	1.289

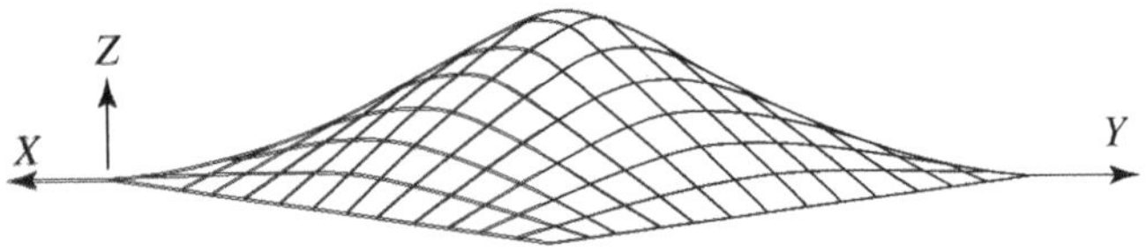

Figure 4.11 Flow non-uniformity model.

of flow non-uniformity are obtained using the solution procedure given by Kays and London [2]. If the temperatures are not known a priori, the iteration is started with assumed outlet temperatures. The new outlet temperatures are calculated and compared with assumed outlet temperatures. The iterations are continued until the convergence is achieved to the fourth digit for all cases. In this analysis, the fluid properties and heat transfer coefficients have been varied from element to element, depending on their local bulk mean temperatures. The heat transfer coefficient is taken from Kays and London [2]. The exchanger is assumed to be ideal with no heat loss to the ambient, and the longitudinal wall heat conduction effects are neglected in this analysis. The temperature distributions on the outlet surfaces of cold and hot fluids are averaged to get the representative outlet mean temperatures for uniform flow cases. In non-uniform flow cases, the temperature distribution in the elemental heat exchangers on the outlet surfaces is obtained using the mass balance:

$$\text{Mean temperature } (\mathrm{T}_{mean}) = \frac{\alpha_1 M_1 T_1 + \alpha_2 M_2 T_2 + \cdots \alpha_n M_n T_n}{\alpha_1 M_1 + \alpha_2 M_2 + \cdots \alpha_n M_n} \tag{4.96}$$

In this study, the ranges of design and operating parameters of the exchanger are

1) $NTU = 1$ to 100
2) C_h/C_c (or C^*) = 0.2, 0.4, 0.6, 0.8 and 1.0
3) $(\eta hA)_h/(\eta hA)_c = 1.$, 0.5 and 2.0

A one-dimensional non-uniformity concept in heat exchangers is simple and straightforward. The one-dimensional inlet velocity distribution can be divided into N step functions, and correspondingly the exchanger is divided into N parallel small exchangers, each representing uniform velocity distribution at the entrance. The analysis of the N exchangers will then provide the overall exchanger effectiveness, which can be compared with the effectiveness if the flows are totally uniform at the exchanger inlet. However, in the present analysis, a two-dimensional non-uniformity concept is considered. When a more complete analysis is attempted, a numerical procedure is feasible.

4.6.2 Validation

The results of the present finite element analysis for flow non-uniformity effects have been compared with available results in Figure 4.12. This comparison of results is made with the available flow non-uniformity models, taken from previous work by Chiou [3]. This comparison is made for FN effect factor for both cases of $C* = 1$ and 0.25 in Figure 4.12a. In this case, the deviation is around 3%. In addition, comparison is made with effectiveness (ε) values, which are more realistic as shown in Figure 4.12b. This accuracy is believed to

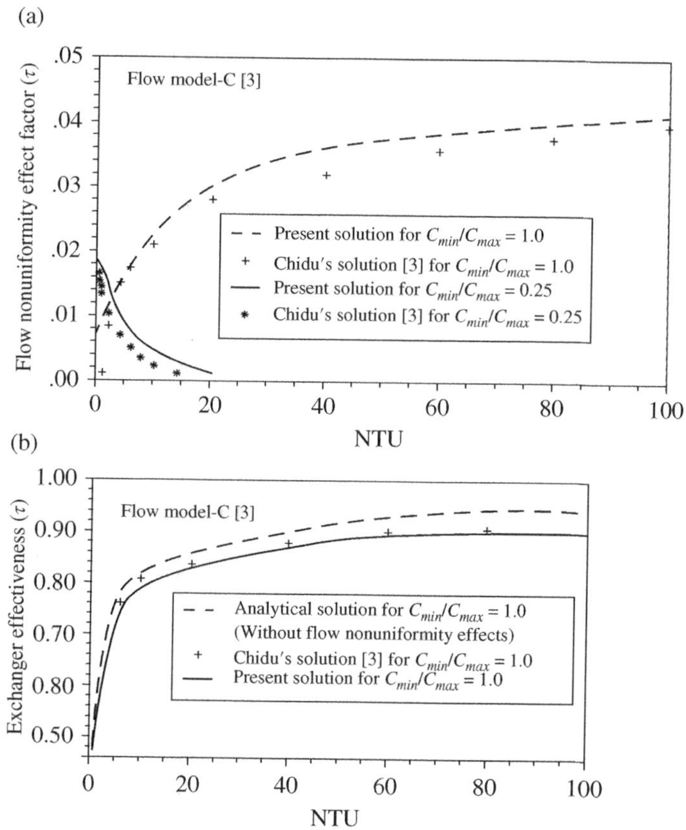

Figure 4.12 Flow non-uniformity effects (comparison of results) – cross flow plate-fin heat exchanger (a) Comparison with flow non-uniformity effect factor (b) Comparison with exchanger effectiveness.

Table 4.3 Flow nonuniformity parameters (α's) – Chiou [3].

I =	1 ; 10	2 ; 9	3 ; 8	4 ; 7	5 ; 6
J = 1 ; 10	0.720	0.720	0.720	0.720	0.720
2 ; 9	0.720	1.000	1.000	1.000	1.000
3 ; 8	0.720	1.000	1.000	1.000	1.000
4 ; 7	0.720	1.000	1.000	1.500	1.500
5 ; 6	0.720	1.000	1.000	1.500	2.180

be sufficient for most engineering applications. The flow model by Chiou [3], which has been used for validation of results, is tabulated in Table 4.3 for ready reference.

4.6.3 Cross Flow Plate-Fin Heat Exchanger

In this section, the performance evaluation with effects of flow non-uniformity on a cross flow heat exchanger is presented, for balanced flow, $C^* = 1$, as well as for unbalanced flows, $C^* \neq 1$. In each case, the influence of flow non-uniformity on thermal performance

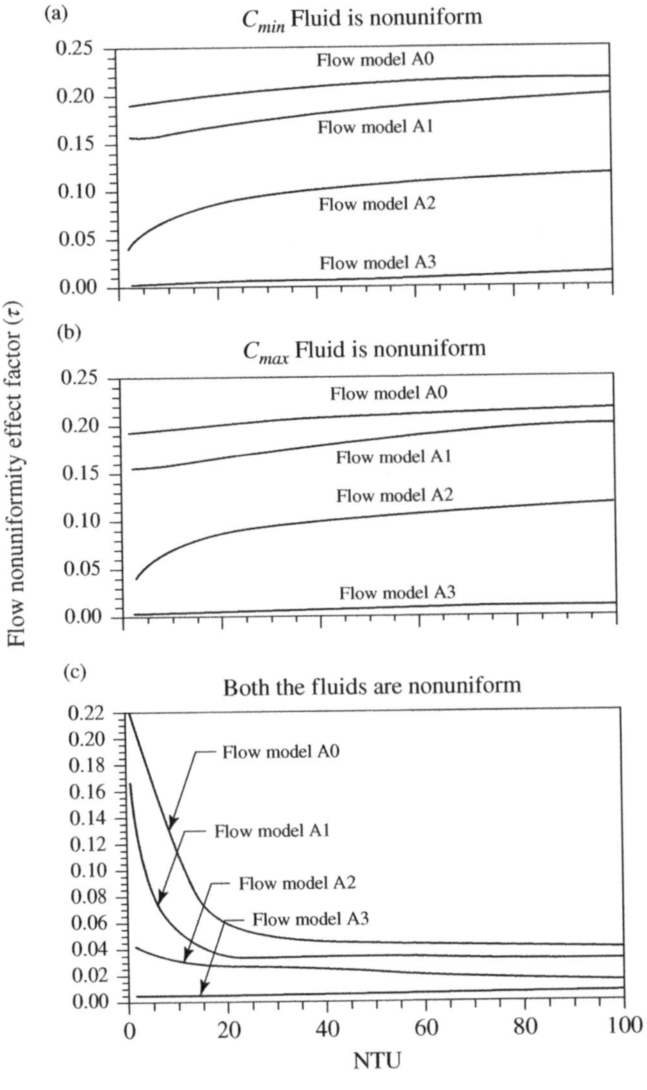

Figure 4.13 Flow non-uniformity effects – $C_{min}/C_{max} = 1.0$ – cross flow plate-fin heat exchanger.

is presented for three cases: flow non-uniformity on minimum fluid capacity (C_{min}) side, flow non-uniformity on maximum fluid capacity (C_{max}) side and flow non-uniformity on both fluid sides. It must be noted that for any flow model, either flow model can be on the hot fluid side or the cold fluid side of the exchanger. The exchanger thermal performance deteriorations are plotted for five magnitudes of $C^* = 1,\ 0.8,\ 0.6,\ 0.4,\ 0.2$ in Figures 4.13–4.17 using the flow non-uniformity models A1–A5.

For $C^* = 1.0$, the τ_F increases as NTU increases when flow non-uniformity is considered on the C_{min} fluid side or C_{max} fluid side. For flow non-uniformity on both fluid sides, the τ_F decreases as the exchanger NTU increases. For unbalanced flows ($C^* \neq 1$), τ_F decreases

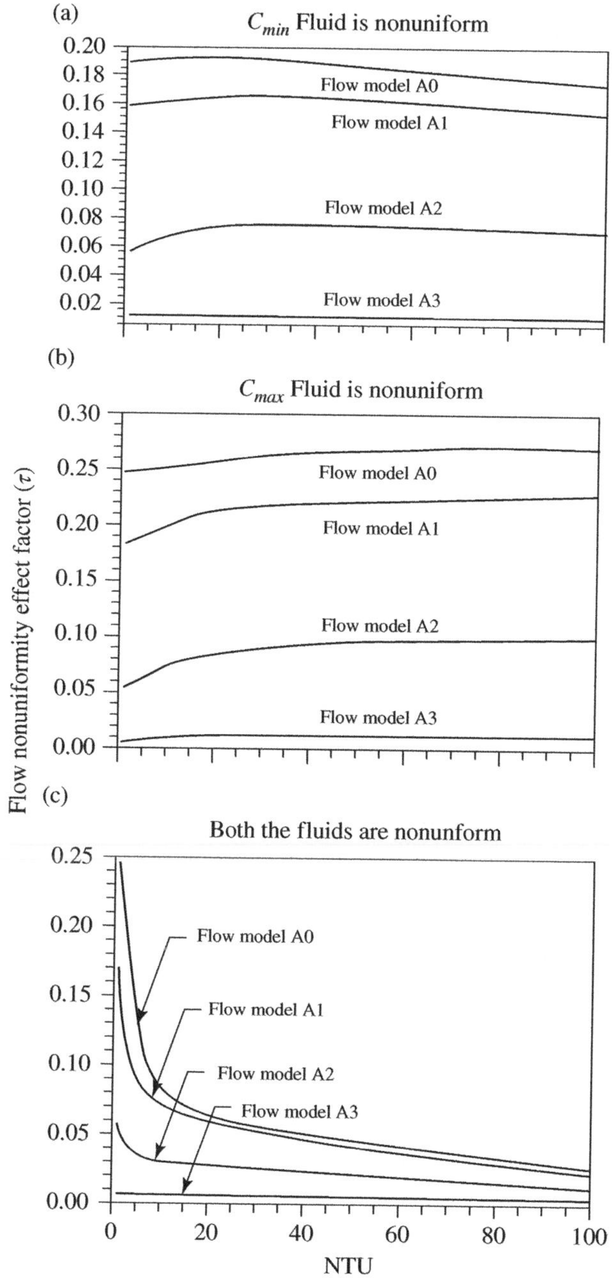

Figure 4.14 Flow non-uniformity effects – $C_{min}/C_{max} = 0.8$ – cross flow plate-fin heat exchanger.

with increase of *NTU*, as shown in Figures 4.14–4.17. It can be seen that the exchanger performance deterioration is quite significant in some applications. For example, the ineffectiveness values (at $NTU = 100$ for flow model A0) are around 19% for $C^* = 1$, 15% for $C^* = 0.8$, 7% for $C^* = 0.6$, 1% for $C^* = 0.4$ and no variation for $C^* = 0.2$. It is noted

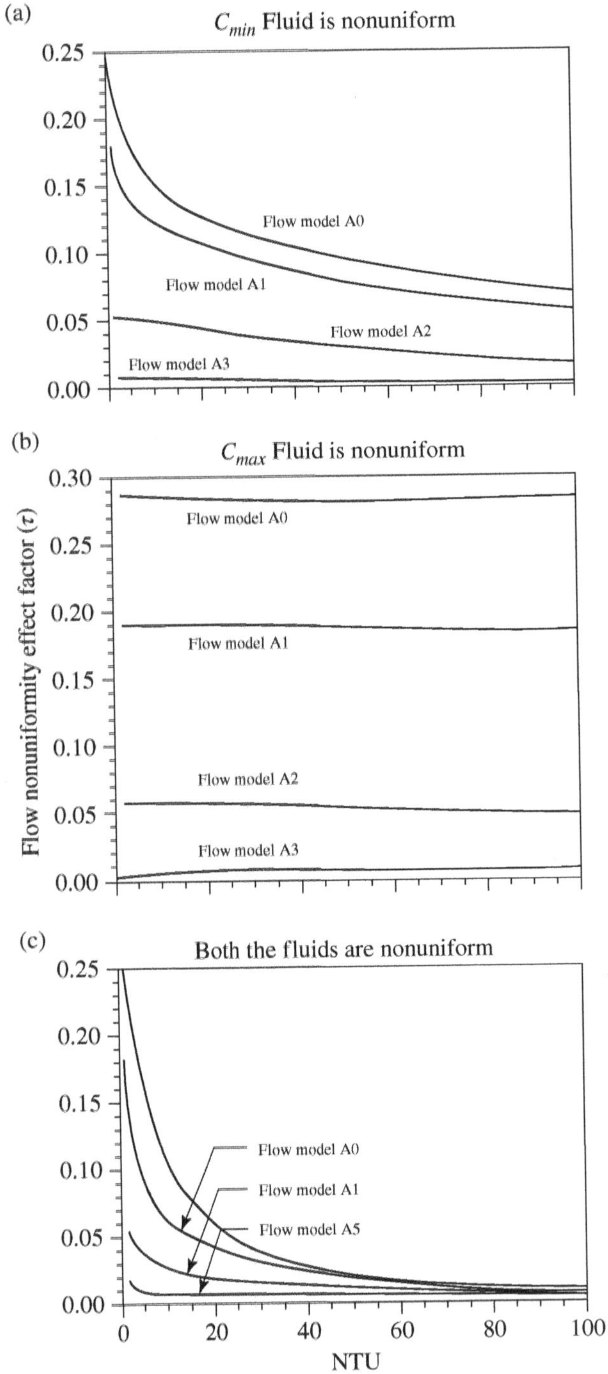

Figure 4.15 Flow non-uniformity effects – $C_{min}/C_{max} = 0.6$ – cross flow plate-fin heat exchanger.

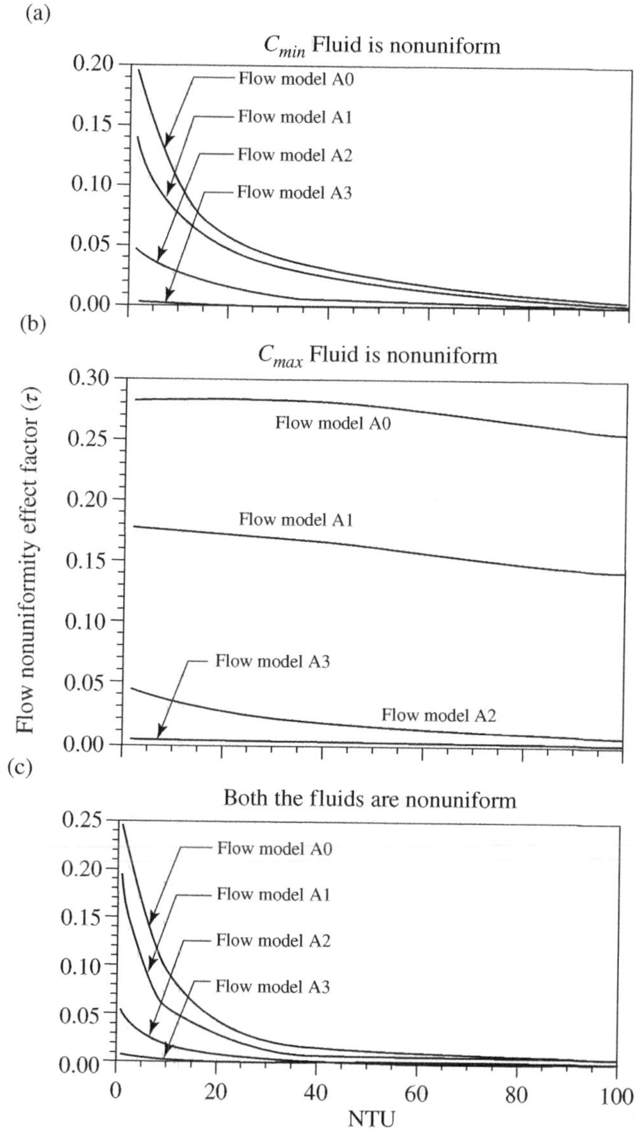

Figure 4.16 Flow non-uniformity effects – $C_{min}/C_{max} = 0.4$ – cross flow plate-fin heat exchanger.

that when the same flow mal-distribution occurs on both fluid sides, the deterioration of the exchanger performance may be lower than those when only one fluid side involves flow mal-distribution. Also, it has been observed that τ_F generally increases as NTU increases in the case of balanced flows, except the case where both fluids are non-uniform, and τ_F decreases with increase of NTU in the case of unbalanced flows as shown in Figure 4.13. It has been observed that τ_F decreases as NTU increases when the flow non-uniformity is considered on both the fluids, as shown in Figure 4.13c. In the case of balanced flows, the C_{min} is equal to C_{max} and hence the performance

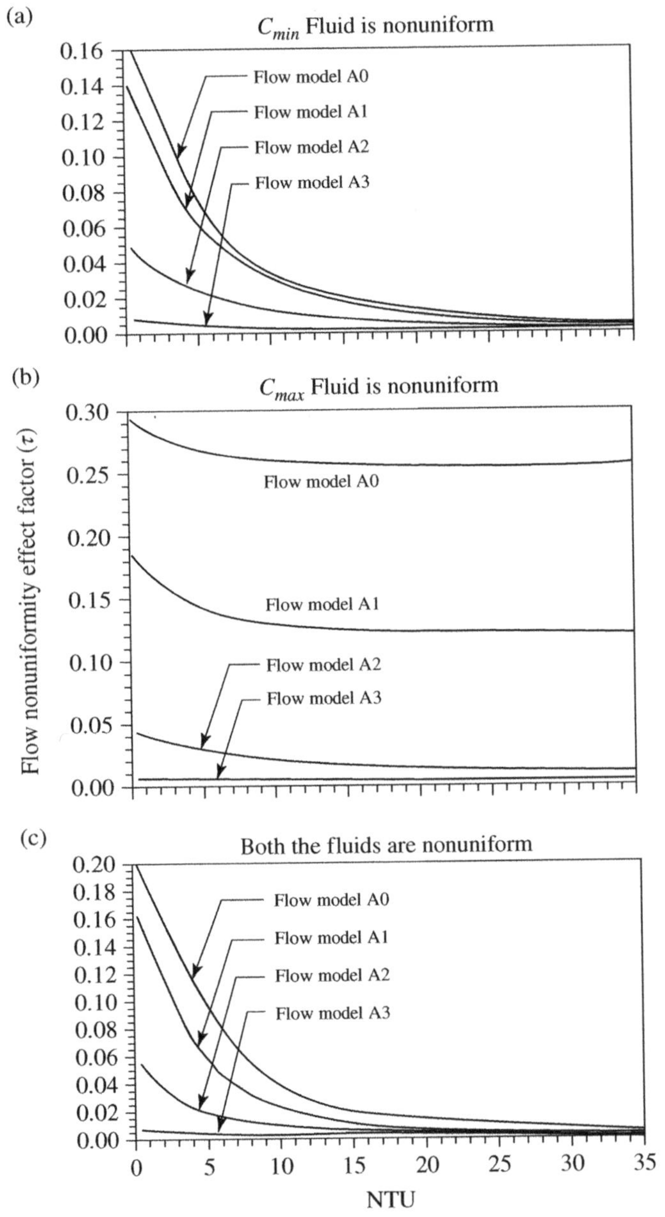

Figure 4.17 Flow non-uniformity effects – $C_{min}/C_{max} = 0.2$ – cross flow plate-fin heat exchanger.

deterioration values are equal in both cases, as expected, as shown in Figure 4.13. Information presented in these figures is not restricted to models considered in this analysis, but the results can be interpolated for other similar flow models. The deterioration of thermal performance of single pass cross flow heat exchangers due to flow non-uniformity presented [18–21] in Figures 4.13–4.17 is generally similar to those reported in previous investigations by Chiou [3,22–24]. However, the type of the fluid maldistribution considered in this investigation is basically different from those reported previously; direct comparison of all these results is thus not possible.

4.6.4 Cross Flow Tube-Fin Heat Exchanger

In this section, the performance evaluation with effects of flow non-uniformity in cross flow tube-fin heat exchanger is presented for balanced flow, $C^* = 1$. The convective heat transfer coefficient (h) is taken from the experimental values of Kays and London [2]. The exchanger thermal performance deteriorations are plotted as a function of NTU ($NTU_{overall}$) for three magnitudes of $C^* = 1.0, 0.6, 0.2$ in Figures 4.18–4.20 using the flow models A0, A1 and A5. In each graph, there are three magnitudes of pressure gradients (say max, mini and medium variations) at the inlet of the centre of exchanger core. In each case, the influence of flow non-uniformity on thermal performance is presented for the following conditions: flow non-uniformity on maximum fluid capacity (C_{max}) side, flow non-uniformity on minimum fluid capacity (C_{min}) side and flow non-uniformity on both fluid sides.

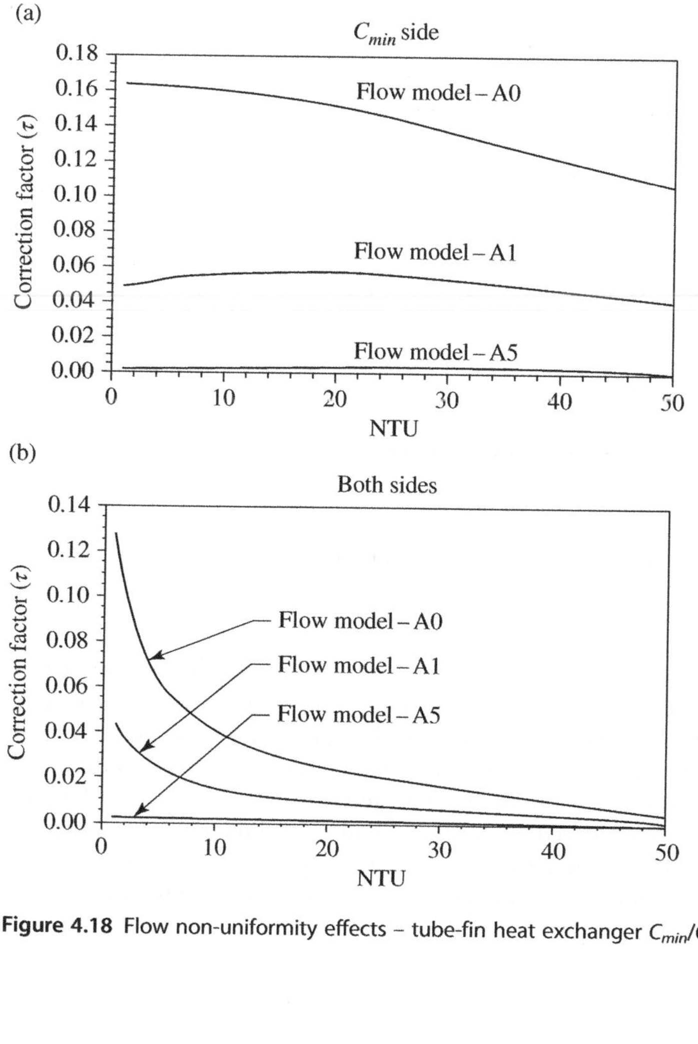

Figure 4.18 Flow non-uniformity effects – tube-fin heat exchanger $C_{min}/C_{max} = 1.0$.

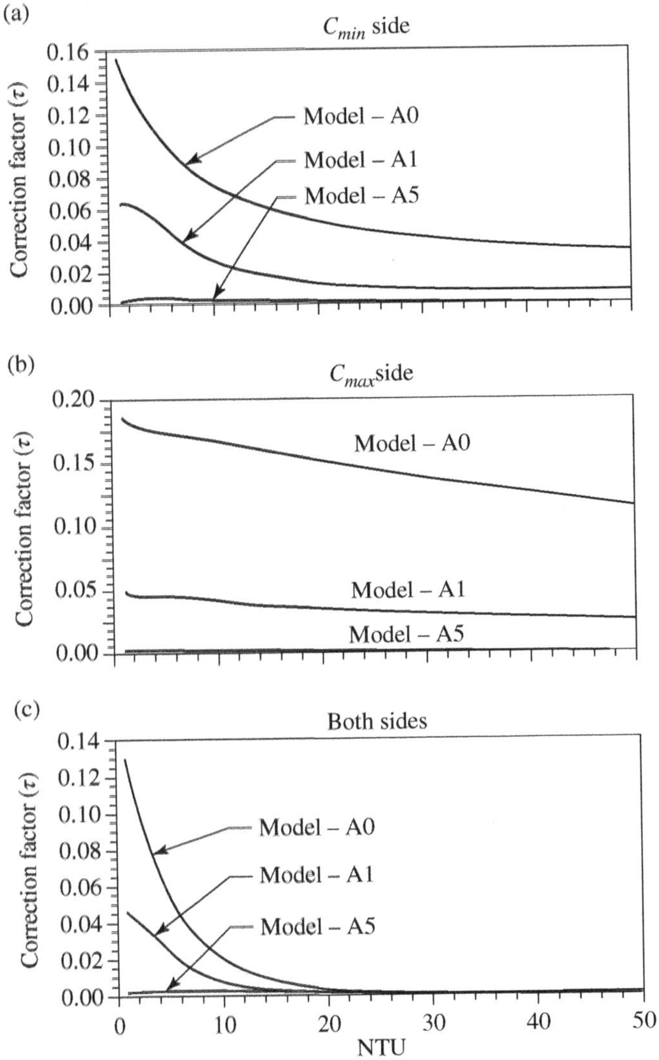

Figure 4.19 Flow non-uniformity effects – tube-fin heat exchanger $C_{min}/C_{max} = 0.6$.

In the case of $C^* = 1.0$, the τ_F decreases with increase in NTU when flow non-uniformity is considered either on the C_{min} fluid side or on both fluid sides. However, in the case of balanced flows, $C^* = 1.0$, the variations in thermal performance with variation of NTU is very much less when the flow non-uniformity is considered on the C_{min} fluid side. It has been observed that the performance due to flow non-uniformity reduces when the flow non-uniformity models are considered on both hot and cold fluid sides, as shown in Figure 4.18b. In the case of unbalanced flows, $C^* \neq 1$, the performance variations are similar to those of balanced flows, $C^* = 1.0$, as shown in Figures 4.19 and 4.20. It can

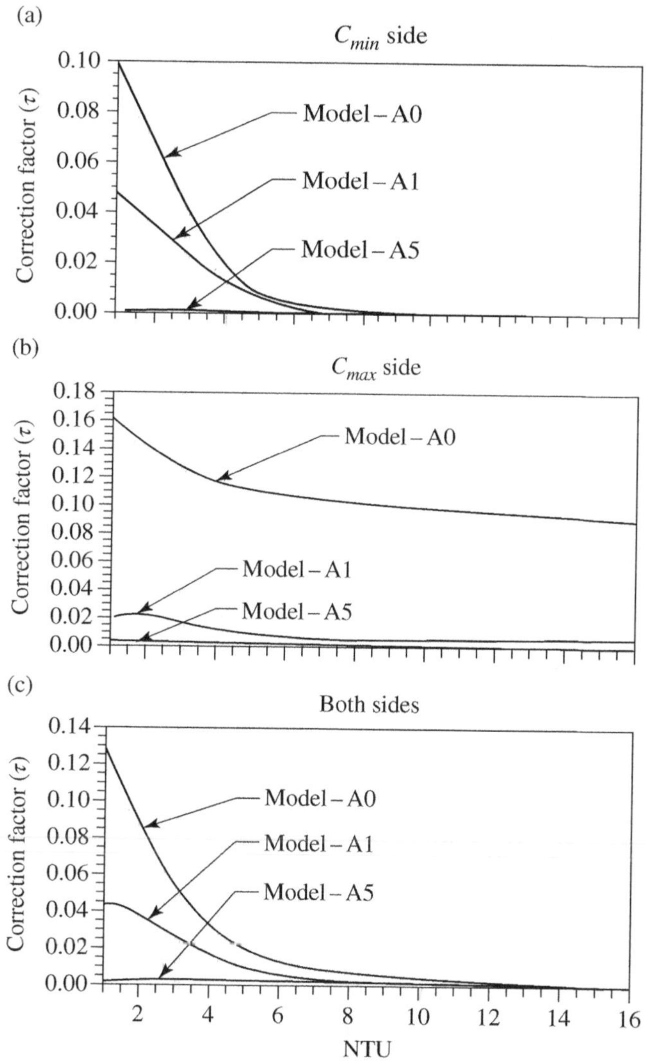

Figure 4.20 Flow non-uniformity effects – tube-fin heat exchanger $C_{min}/C_{max} = 0.2$.

be seen that the exchanger performance deterioration is quite significant in some typical applications especially when *NTU* is lower (<10). It is noted that when the same flow mal-distribution occurs on both fluid sides, the deterioration of the exchanger performance may be lower than those when only one fluid side involves flow mal-distribution, as shown in Figure 4.21. Also, the relative comparison of results of flow non-uniformity has been shown in Figure 4.22 for balanced ($C^* = 1$) and unbalanced flows ($C^* = 0.6, 0.2$) of tube-fin heat exchanger. It has been observed that the performance deteriorations are higher for balanced flows when compared to the unbalanced flows.

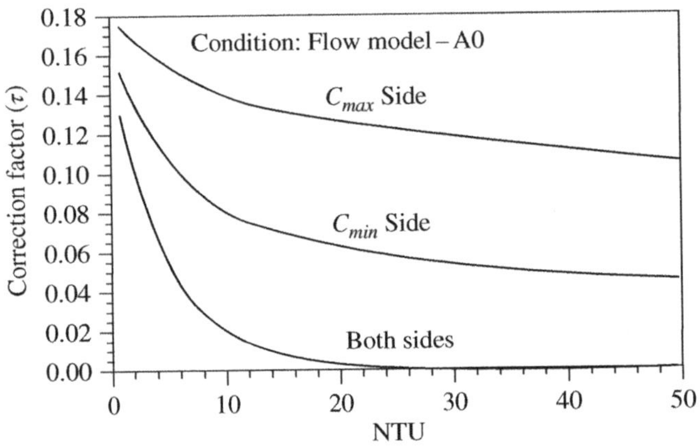

Figure 4.21 Flow non-uniformity effects – tube-fin heat exchanger $C* = 0.6$ – comparison with fluid capacity rates.

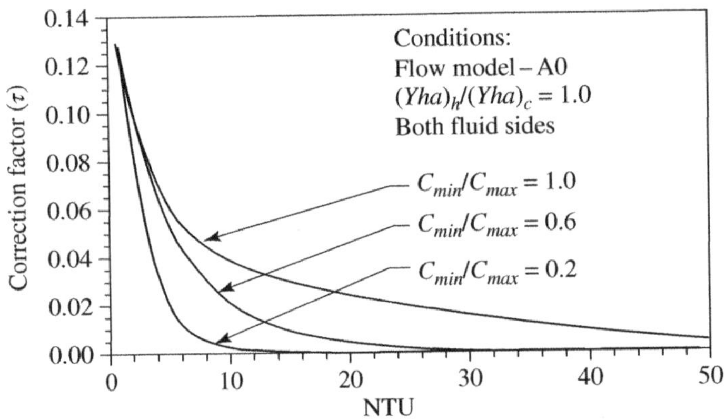

Figure 4.22 Flow non-uniformity effects – tube-fin heat exchanger – comparison with $C*$.

4.6.5 Pressure Drop Variations – Flow Non-Uniformity

The ratio of heat transferred to the pumping power required is an important factor in economizing an exchanger. Pumping power depends upon the pressure drop required to pump the fluid at the required velocity. Hence, it is necessary to estimate the pressure drop along with thermal performance. Pressure drops in the fluids are evaluated using the following formula from Kays and London [2]:

$$\text{Pressure drop}(\Delta P) = \frac{G^2 V_1}{2g}\left[(1+\sigma^2)\left(\frac{V_2}{V_1}-1\right)+f\frac{SV_m}{S_f V_1}\right] \tag{4.97}$$

As given by Kays and London [2], the friction factors are evaluated using Reynolds number vs friction factor graphs. The entrance and exit losses are not accounted for while

Table 4.4 Exchanger core outlet pressure variations with model – A0.

I	1	2	3	4	5	6	7	8	9	10
$J=1$	99.97	99.97	99.97	99.97	99.97	99.97	99.97	99.97	99.97	99.97
2	99.97	99.33	97.98	96.23	94.60	94.29	95.56	97.39	99.06	99.97
3	99.97	98.05	93.16	85.96	78.63	77.18	82.98	90.69	97.06	99.97
4	99.97	96.52	86.76	70.79	51.76	47.84	64.25	81.86	94.69	99.97
5	99.97	95.29	81.17	55.36	19.20	19.00	43.89	74.15	92.78	99.97
6	99.97	95.29	81.17	55.36	19.20	19.00	43.89	74.15	92.78	99.97
7	99.97	96.52	86.76	70.79	51.76	47.84	64.25	81.86	94.69	99.97
8	99.97	98.05	93.16	85.96	78.63	77.18	82.98	90.69	97.06	99.97
9	99.97	99.33	97.98	96.23	94.60	94.29	95.56	97.39	99.06	99.97
10	99.97	99.97	99.97	99.97	99.97	99.97	99.97	99.97	99.97	99.97

Conditions:

a) Inlet pressure = 100 kPa
b) Outlet pressure (if flow is uniform) = 95.17 kPa.
c) $NTU = 8.0$, $C^* = 1.0$ and $(\eta ha)^* = 1.0$

Table 4.5 Exchanger core outlet pressure variations with model – A1.

I	1	2	3	4	5	6	7	8	9	10
$J=1$	99.00	98.90	98.90	98.90	98.85	98.81	98.80	98.75	98.70	98.70
2	99.00	98.07	96.88	95.63	94.56	94.29	94.95	96.05	97.35	98.54
3	99.00	97.03	94.02	90.46	87.26	86.54	88.77	92.20	95.75	98.47
4	99.00	96.05	90.99	84.57	78.55	77.29	81.75	88.15	94.26	98.43
5	99.00	95.33	88.64	79.78	71.10	69.37	76.02	85.01	93.17	98.42
6	99.00	95.33	88.64	79.78	71.11	69.37	76.02	85.01	93.17	98.42
7	99.00	96.05	90.99	84.57	78.55	77.29	81.75	88.15	94.26	98.43
8	99.00	97.03	94.02	90.46	87.26	86.54	88.77	92.20	95.75	98.47
9	99.00	98.07	96.88	95.63	94.56	94.29	94.95	96.05	97.35	98.54
10	99.00	98.90	98.90	98.90	98.85	98.81	98.80	98.75	98.70	98.70

Conditions:

a) Inlet pressure = 100 kPa
b) Outlet pressure (if flow is uniform) = 95.17 kPa.
c) $NTU = 8.0$, $C^* = 1.0$ and $(\eta ha)^* = 1.0$

investigating the pressure drops in this analysis. The elemental exchanger core outlet pressures are evaluated for models A0, A1, A_5 and tabulated in Tables 4.7–4.9. It can be seen from these tables that there is a significant variation in pressure drops due to flow non-uniformity, as there are considerable variations in velocities. For example, the maximum pressure drop values are: 17 times (for model A0), 6 times (for model

Table 4.6 Exchanger core outlet pressure variations with model – A5.

I	1	2	3	4	5	6	7	8	9	10
$J = 1$	96.61	96.38	96.16	96.00	96.00	95.60	95.43	95.29	95.18	95.06
2	96.61	96.00	95.41	94.89	94.45	94.21	94.19	94.33	94.61	95.00
3	96.61	95.69	94.73	93.86	93.14	92.83	92.99	93.44	94.13	94.96
4	96.61	95.45	94.19	92.99	92.02	91.65	91.98	92.73	93.77	94.93
5	96.61	95.29	93.83	92.41	91.26	90.86	91.30	92.25	93.53	94.91
6	96.61	95.29	93.83	92.41	91.26	90.86	91.30	92.25	93.53	94.91
7	96.61	95.45	94.19	92.99	92.02	91.65	91.98	92.73	93.77	94.93
8	96.61	95.69	94.73	93.86	93.14	92.83	92.99	93.44	94.13	94.96
9	96.61	96.00	95.41	94.89	94.45	94.21	94.19	94.33	94.61	95.00
10	96.61	96.38	96.16	96.00	96.00	95.60	95.43	95.29	95.18	95.06

Conditions:

a) Inlet pressure = 100 kPa
b) Outlet pressure (if flow is uniform) = 95.17 kPa.
c) $NTU = 8.0$, $C^* = 1.0$ and $(\eta ha)^* = 1.0$

Table 4.7 Temperature non-uniformity parameters (β's) – Chiou [26].

$I =$	1	2	3	4	5	6	7	8	9	10
$J = 1$	0.918	0.934	0.960	0.976	0.985	0.993	1.001	1.010	1.035	1.043
2	0.918	0.943	0.968	0.985	0.993	1.001	1.010	1.026	1.043	1.051
3	0.918	0.943	0.968	0.985	0.993	1.001	1.018	1.026	1.043	1.043
4	0.918	0.943	0.968	0.985	0.993	1.001	1.018	1.026	1.051	1.068
5	0.918	0.943	0.968	0.985	0.993	1.001	1.018	1.026	1.051	1.076
6	0.918	0.918	0.918	0.951	0.985	0.993	1.001	1.010	1.051	1.076
7	0.918	0.918	0.918	0.951	0.985	0.993	1.001	1.010	1.051	1.085
8	0.918	0.918	0.918	0.985	0.993	1.010	1.018	1.035	1.068	1.102
9	0.918	0.918	0.918	1.001	1.010	1.018	1.035	1.043	1.076	1.102
10	0.918	0.960	1.001	1.043	1.068	1.076	1.102	1.119	1.136	1.226

A1) and 2 times (for model A5) higher as compared to the uniform flow cases. The pressure drop for uniform flow case is 4.8 kPa with the inlet pressure of 100 kPa. The pumping power required is much higher for the flow non-uniformity case than the uniform flow case.

Fleming [25] recommended some important considerations for the header and the manifold design to get minimum and uniform pressure drop in the exchanger core. Fleming [25] has claimed that no analysis is available for an increase in the pressure drop due to cross flow mal-distribution. Since such non-uniform flow is associated with poor header

Table 4.8 Effects of longitudinal wall heat conduction and flow non-uniformity cross flow plate-fin heat exchanger.

		Performance variations (%)													
		NTU = 10.0							NTU = 50.0						
		LHC				LHC + FN			LHC				LHC + FN		
Models	C^*	$\lambda = 0.05$	$\lambda = 0.1$	$\lambda = 0.2$	FN	$\lambda = 0.05$	$\lambda = 0.1$	$\lambda = 0.2$	$\lambda = 0.05$	$\lambda = 0.1$	$\lambda = 0.2$	FN	$\lambda = 0.05$	$\lambda = 0.1$	$\lambda = 0.2$
	1.0	5.50	10.5	17.5	7.00	10.5	15.0	23.0	6.00	14.0	22.5	11.5	8.00	9.00	18.0
A1	0.6	3.40	7.00	13.5	5.00	6.00	12.0	22.0	2.75	6.75	15.3	3.50	4.00	6.00	10.5
	0.2	0.95	2.00	5.00	2.00	3.00	7.00	14.0	0.05	0.55	2.85	0.03	1.00	1.75	2.25
	1.0	5.50	10.5	17.5	0.50	8.90	14.0	21.0	6.00	14.0	22.5	1.20	6.00	10.0	17.5
A5	0.6	3.40	7.00	13.5	1.00	5.00	10.0	19.5	2.75	6.75	15.3	0.50	3.00	5.00	9.50
	0.2	0.95	2.00	5.00	0.50	2.00	5.00	11.0	0.05	0.55	2.85	0.03	0.70	1.50	2.00

Conditions: C_{min}/C_{max} (C^*) = 1.0, $(\eta ha)^*$ = 1.0, λ_c/λ_h = 1.0 and flow non-uniformity on C_{max} fluid side

Table 4.9 Effects of longitudinal wall heat conduction and flow non-uniformity cross flow tube-fin heat exchanger.

		Performance variations (%)													
		NTU = 10.0							NTU = 50.0						
		LHC				LHC + FN			LHC				LHC + FN		
Models	C^*	$\lambda = 0.05$	$\lambda = 0.1$	$\lambda = 0.2$	FN	$\lambda = 0.05$	$\lambda = 0.1$	$\lambda = 0.2$	$\lambda = 0.05$	$\lambda = 0.1$	$\lambda = 0.2$	FN	$\lambda = 0.05$	$\lambda = 0.1$	$\lambda = 0.2$
	1.0	3.75	7.90	17.0	5.95	9 50	12.5	19.5	4.00	8.00	17.5	4.95	7.50	10.0	15.5
A1	0.6	1.85	3.50	7.50	2.25	3 80	5.50	8.35	1.50	3.00	5.00	1.80	2.00	3.35	6.00
	0.2	0.60	0.85	1.00	1.20	0 03	0.05	0.07	0.0	0.0	0.0	0.0	0.0	0.0	0.0
	1.0	3.75	7.90	17.0	0.15	4 50	8.25	15.5	4.00	8.00	17.5	0.05	2.75	5.75	13.2
A5	0.6	1.85	3.50	7.50	0.10	1 60	2.60	6.00	1.50	3.00	5.00	0.0	0.50	1.00	3.00
	0.2	0.60	0.85	1.00	0.05	0.02	0.04	0.06	0.0	0.0	0.0	0.0	0.00	0.00	0.00

Conditions: C^* = 1.0, $(\eta ha)^*$ = 1.0, and flow non-uniformity on C_{min} fluid side design or gross core blockage, the static pressure distribution at the core inlet and outlet faces may not be uniform. Hence, no simple modelling for the pressure drop evaluation is possible. As a conservative approach, he has recommended that the core pressure drop be evaluated based on the highest velocity component of the flow mal-distribution.

4.7 Inlet Temperature Non-Uniformity Effects

4.7.1 General

For various reasons, the inlet temperatures of the fluid flowing through a great number of heat exchangers are usually not uniformly distributed over the exchanger cores under actual operating conditions. However, in the thermal analysis of heat exchangers, it is normally assumed that the inlet temperature of the fluid flowing into the exchanger is uniformly distributed over the exchanger core. Hence, the FEM analysis has been extended to inlet temperature non-uniformity effects in cross flow plate-fin and tube-fin exchangers using the governing equations and FEM formulations provided earlier. This analysis is carried out with uniform flow through the exchanger core and neglecting the longitudinal wall heat conduction effects in the fluid flow direction.

In Section 4.6, a mathematical equation to generate the flow non-uniformity models was presented. This has been modified to get the temperature distribution at the inlet of the exchanger core:

$$\frac{\partial^2\theta}{\partial x^2}+\frac{\partial^2\theta}{\partial y^2}=\text{Constant} \tag{4.98}$$

where

$$\theta=\frac{T_{nonuniform}}{T_{uniform}} \tag{4.99}$$

The solution to represent the temperature distribution (θ) is

$$\theta(x,y)=\frac{64D}{\prod^4}\sum_{m,n=0}^{\infty}\frac{(-1)^{m+n+1}}{\left(\frac{2m+1}{a}\right)^2\left(\frac{2n+1}{b}\right)^2(2m+1)(2n+1)}\frac{\cos(2m+1)\prod x}{2a}\frac{\cos(2n+1)\prod y}{2b} \tag{4.100}$$

The above equation can be used to generate the temperature non-uniformity models which are the same as flow non-uniformity models, given in Table 4.2. These tables contain non-dimensional parameters (defined in Equations 4.43 & 4.44) which can be used for inlet flow as well as temperature distribution at the exchanger core, if the heating point is at the centre of the inlet duct.

The exchanger thermal performance correlation factor (τ_T) is introduced, as given earlier, to illustrate the influence of two-dimensional temperature non-uniformity on thermal performance. It is defined as

$$\tau_T=\frac{\varepsilon_0-\varepsilon_T}{\varepsilon_0} \tag{4.101}$$

The thermal performance correction factor (τ_T) directly shows the degree of deterioration or augmentation of the exchanger effectiveness. This factor can be either positive or negative as presented by Chiou [26]. If the correction factor (τ_T) is positive, it indicates that the effectiveness of that exchanger with non-uniform inlet temperature/flow distribution is less than that of the exchanger with the arithmetic average inlet temperature/flow as its fictitious uniform inlet temperature. Thus, if the correction factor (τ_T) is positive, the non-uniformity of the inlet fluid flow/temperature has a deterioration effect on the thermal performance of the exchanger. On the other hand, if the correction factor (τ_T) is negative, the non-uniformity of the inlet fluid flow/temperature has an augmentation effect on the exchanger thermal performance. Analytical solutions without considering the effects of temperature non-uniformity are obtained using the solution procedure given by Kays and London [2]. In this analysis, the variations of fluid properties have been taken into consideration. Since the temperature distribution through the exchanger core is not uniform, the effects of variable heat transfer coefficients have been taken in this analysis.

4.7.2 Validation

The present model has been validated with the available numerical solutions provided by Chiou [26]. The present FEM analysis has been verified with Chiou's temperature non-uniformity model (Chiou [26]). The same temperature non-uniformity model has been tabulated in Table 4.7 for easy reference. Figure 4.23 gives the comparison of present FEM results with available results for a cross flow plate-fin heat exchanger. The agreement is found to be good. Chiou [26] has extended his numerical procedures to an engine radiator (Gaddis et al. [27]). The FEM model introduced for single pass cross flow plate-fin and tube-fin heat exchangers predicts thermal performance variations due to temperature non-uniformity, which are in good agreement with the available numerical solutions.

4.7.3 Cross Flow Plate-Fin Heat Exchanger

In this section, the performance evaluation with effects of inlet temperature non-uniformity in cross flow plate-fin heat exchanger is presented, for balanced flow, $C^* = 1$ as well as unbalanced flows $C^* \neq 1$ The exchanger thermal performance variations

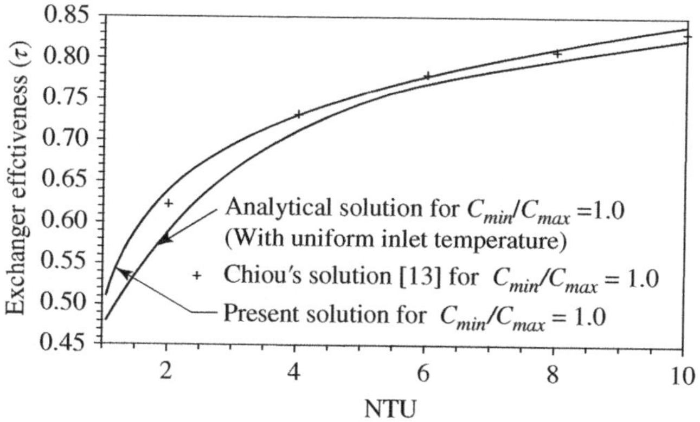

Figure 4.23 Temperature non-uniformity effects – comparison of results.

are plotted as a function of *NTU* ($NTU_{overall}$) and correction factor(τ_T) for five magnitudes of $C^* = 1.0, 0.8, 0.6, 0.4, 0.2$ using the models A1, A2, A3, A4, A5 for the following cases:

1) temperature non-uniformity on hot fluid side
2) temperature non-uniformity on cold fluid side
3) temperature non-uniformity on both fluid side

The temperature non-uniformity effects are compared with Chiou [26] results and found to be matched well, as shown in Figure 4.23. It can be seen that the exchanger thermal performance deterioration due to temperature non-uniformity is quite significant in some typical applications. It is also observed that there is some augmentation in performance, as shown in Figures 4.24–4.28. In the case of temperature non-uniformity on the hot fluid side for $C^* = 1, 0.8, 0.6, 0.4, 0.2$ the correction factor (τ_T) decreases with *NTU* until *NTU* reaches 40, when it remains constant with further increase of *NTU*. However, it has been observed that the correction factors become negative (i.e. enhancement in performance) beyond $NTU = 10$ for both balanced and unbalanced cases.

In the case of temperature non-uniformity on the cold fluid side for $C^* = 1.0, 0.8, 0.6, 0.4$, the correction factor decreases as *NTU* increases. However, in some cases, there is no significant variation in performance between the models, as shown in Figures 4.24–4.26 and 4.28c. In the case of temperature non-uniformity on the hot fluid side for $C^* = 0.4, 0.2$, the correction factor decreases with increase of *NTU* and it becomes insignificant after $NTU = 8$. In the case of temperature non-uniformity on the cold fluid side for $C^* = 0.4, 0.2$, the correction factors are negative (i.e. enhancement in performance), as shown in Figures 4.27b and 4.28b. In these cases, the performance enhancement values are higher for lower *NTU*s. Similarly, the performance variations due to temperature non-uniformity have been studied by taking the temperature non-uniformity on both cold and hot fluid sides, as shown in Figures 4.24b, 4.25c, 4.26c, 4.27c and 4.28c.

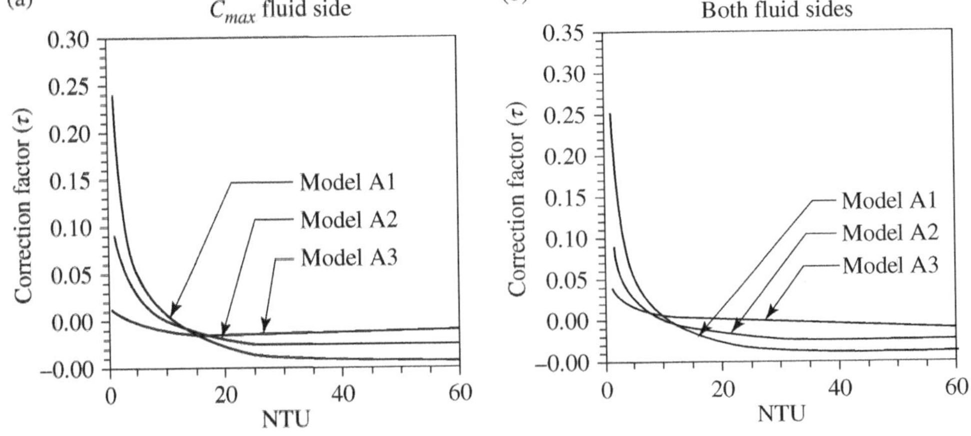

Figure 4.24 Temperature non-uniformity effects – plate-fin heat exchanger $C_{min}/C_{max} = 1$.

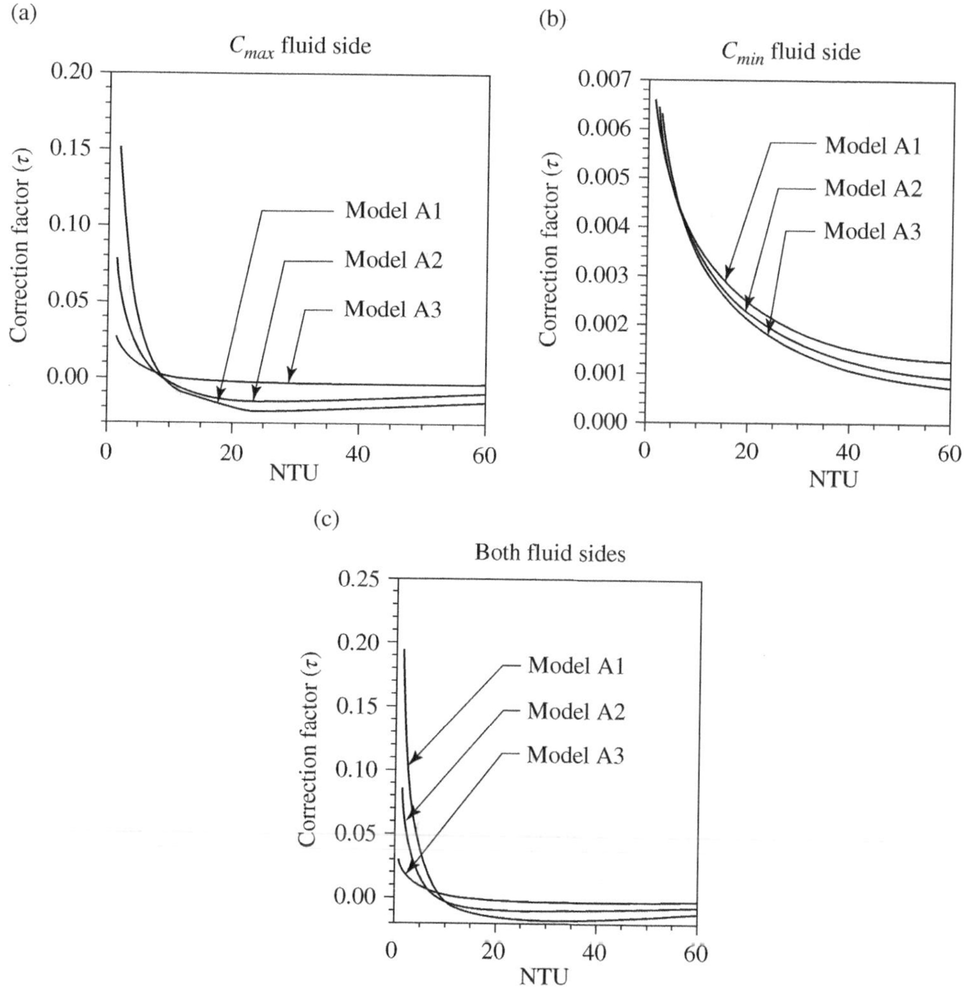

Figure 4.25 Temperature non-uniformity effects – plate-fin heat exchanger $C_{min}/C_{max} = 0.8$.

However, there is no significant variation in performance between the models. The performance variations in thermal performance variations of a cross flow plate-fin heat exchanger due to inlet temperature non-uniformity are presented in Figures 4.24–4.28. These results (Ranganayakulu et al. [21,28]) are generally similar to those reported in previous investigations (Chiou et al. [26], Gaddis et al. [27]). It should be noted that Yamasheta et al. [29] studied the performance of cross flow heat exchangers with variable physical properties, but it has been found in our analysis that the effects of variable physical properties are negligible for air media. Further Ravikumar et al. [30] analyzed compact heat exchangers using FEM without considering the above LHC, FN and TN effects. However, the type of the temperature non-uniformities considered in this investigation are basically different from those reported previously; direct comparison of all these results is thus not possible.

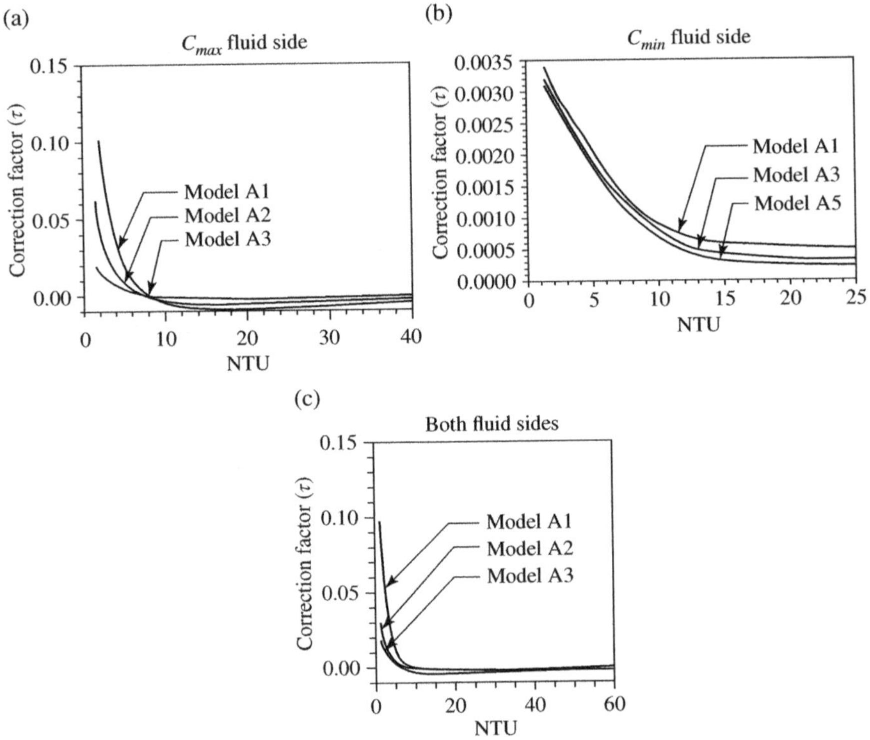

Figure 4.26 Temperature non-uniformity effects – plate-fin heat exchanger $C_{min}/C_{max} = 0.6$.

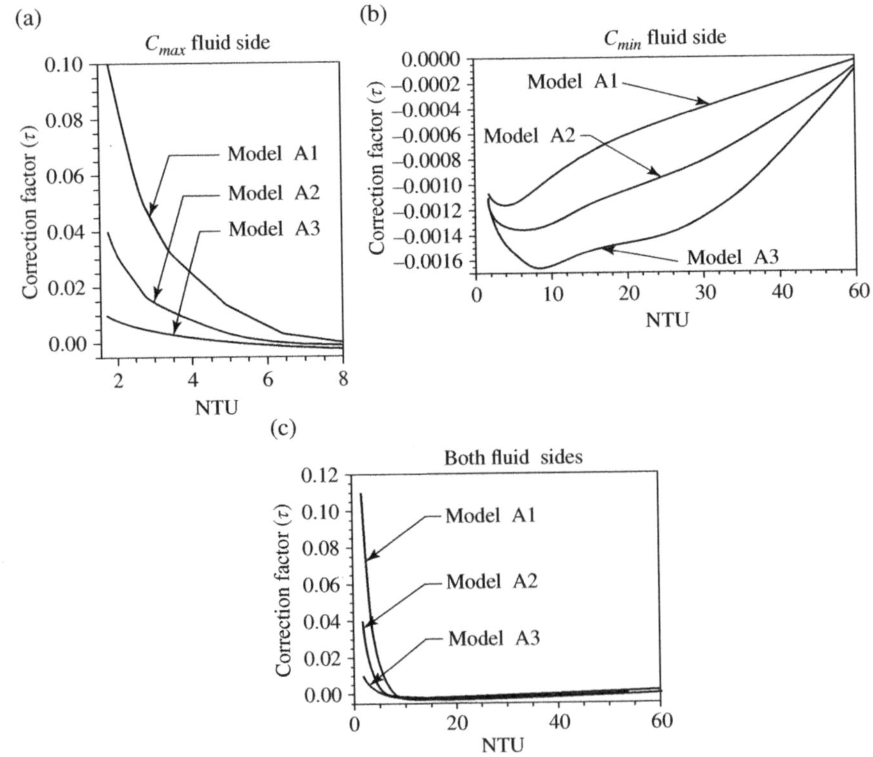

Figure 4.27 Temperature non-uniformity effects – plate-fin heat exchanger $C_{min}/C_{max} = 0.4$.

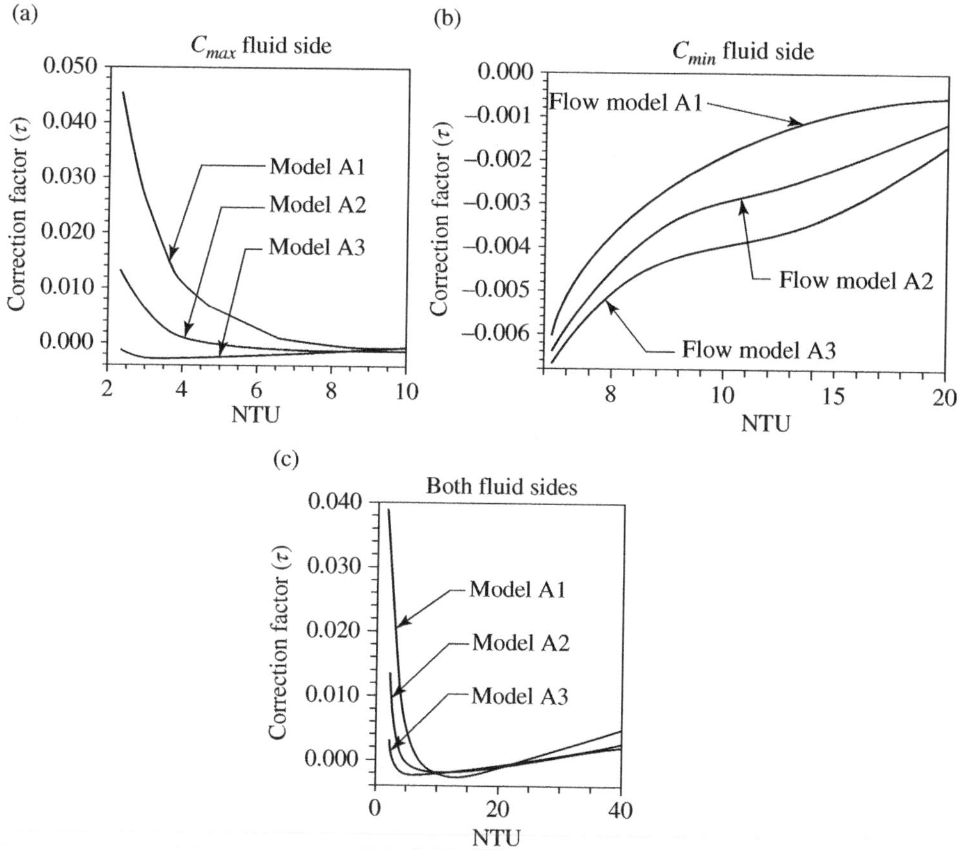

Figure 4.28 Temperature non-uniformity effects – plate-fin heat exchanger $C_{min}/C_{max} = 0.2$.

4.7.4 Cross Flow Tube-Fin Heat Exchanger

In this section, the performance evaluation with effects of temperature non-uniformity in a cross flow tube-fin heat exchanger is presented for balanced flow, $C^* = 1$, as well as for unbalanced flows $C^* \neq 1$. The exchanger thermal performance variations are plotted as a function of *NTU* ($NTU_{overall}$) and correction factor (τ_T) for three magnitudes of $C^* = 1.0, 0.6, 0.2$ using the temperature models A0, A1 and A5 tabulated in Table 4.2.

Figures 4.29–4.33 show the performance variations due to temperature non-uniformity for cross flow tube-fin heat exchangers. In each figure, there are three cases: temperature non-uniformity on hot fluid side, temperature non-uniformity on cold fluid side and temperature non-uniformity on both fluid sides. It can be seen from these figures that the performance deterioration decreases with increase of *NTU* for all cases of balanced ($C^* = 1$) and unbalanced ($C^* = 0.6$) flows. In the case of temperature non-uniformity for unbalanced flow of $C^* = 0.2$, the correction factor

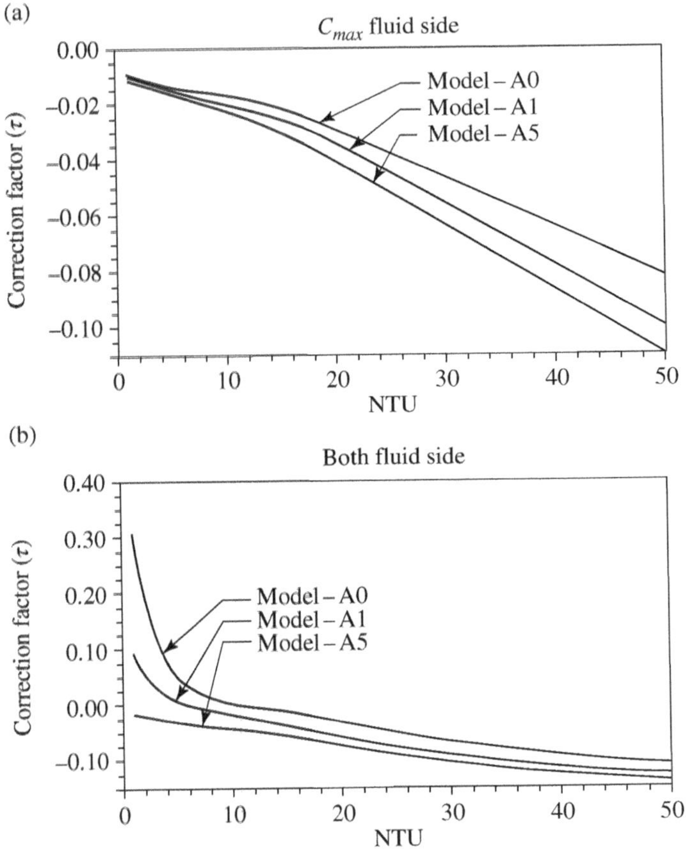

Figure 4.29 Temperature non-uniformity effects – tube-fin heat exchanger $C_{min}/C_{max} = 1.0$.

decreases with increase of *NTU* until *NTU* equals 5 and then increases with increase of *NTU*, as shown in Figure 4.31. It is also noted that there is significant augmentation (up to 10%) in thermal performance beyond $NTU = 5$. However, there is no significant variation in performance with temperature non-uniformity models for all cases of balanced and unbalanced flows. The relative comparison of temperature non-uniformity effects with C^* has been shown in Figure 4.32. This figure shows that the performance variations are higher for balanced flow, $C^* = 1.0$, as compared to those of unbalanced flows, $C^* = 0.6$, 0.4, in the regions of lower *NTU*. However, in the regions of higher *NTU*, the performance variations are lower for balanced flows when compared with unbalanced flows, as shown in Figure 4.32. Also, the performance variations with fluid capacity rates have been shown in Figure 4.33. It has been observed that the performance variations are lower when the temperature non-uniformity is considered on the C_{max} fluid side, as shown in Figure 4.33. Figure 4.34 also presents the relative comparison of temperature non-uniformity effects for $C_{min}/C_{max} = 0.6$ for a tube-fin heat exchanger.

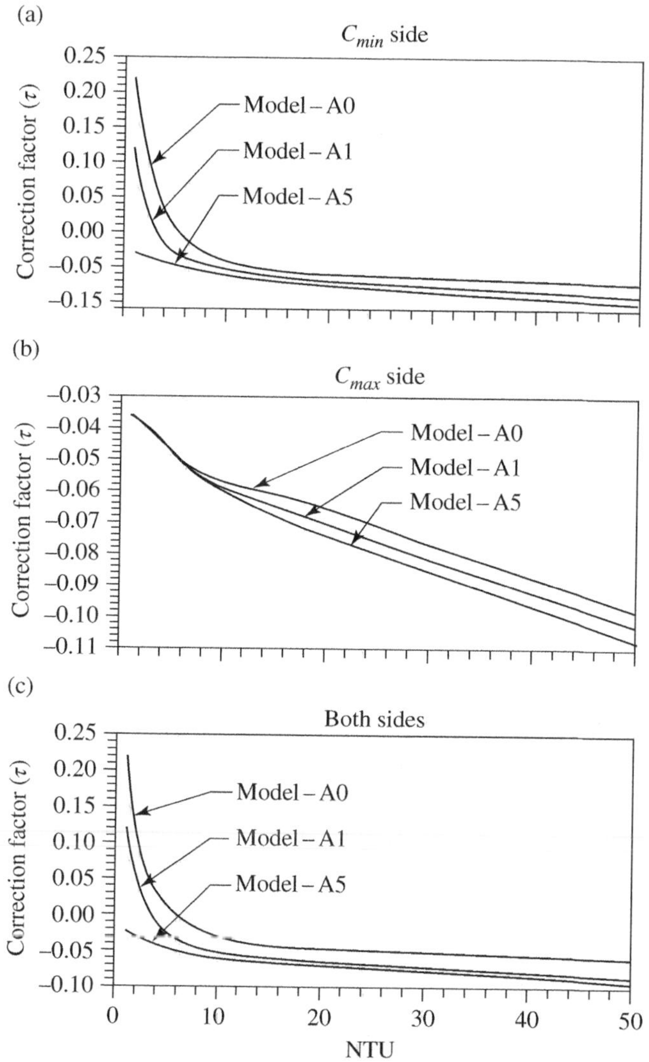

Figure 4.30 Temperature non-uniformity effects – tube-fin heat exchanger $C_{min}/C_{max} = 0.6$.

4.8 Combined Effects of Longitudinal Heat Conduction, Inlet Flow Non-Uniformity and Temperature Non-Uniformity

4.8.1 General

In earlier sections, the individual effects of longitudinal wall heat conduction, inlet fluid flow non-uniformity and inlet temperature non-uniformity in cross flow plate-fin and tube-fin heat exchangers have been studied in detail. In actual practice, heat exchangers may be subjected to longitudinal wall heat conduction, inlet flow non-uniformity and

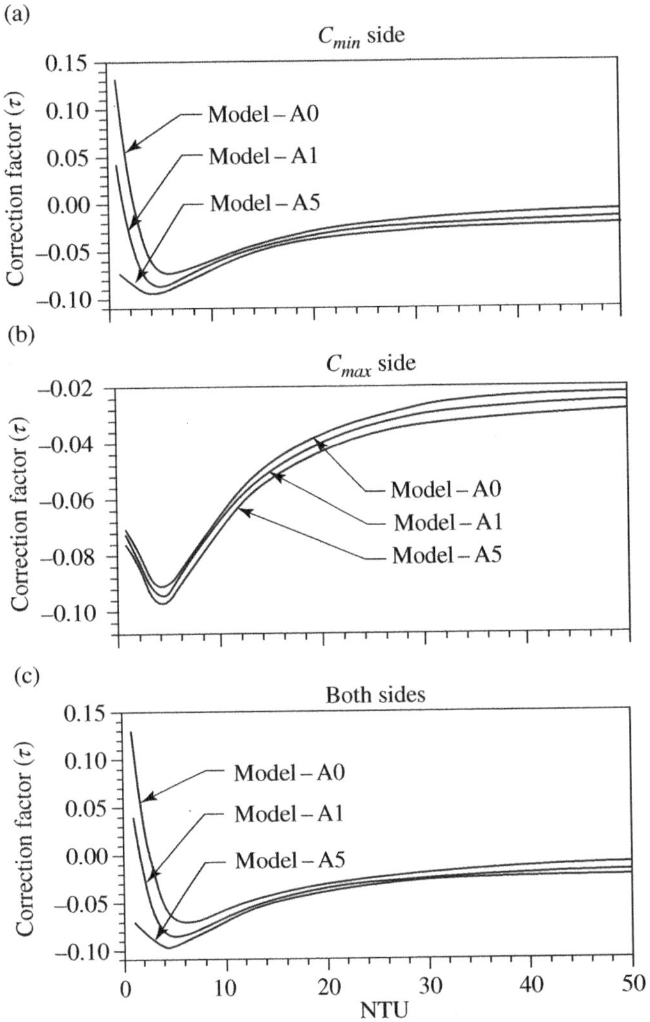

Figure 4.31 Temperature non-uniformity effects – tube-fin heat exchanger $C_{min}/C_{max} = 0.2$.

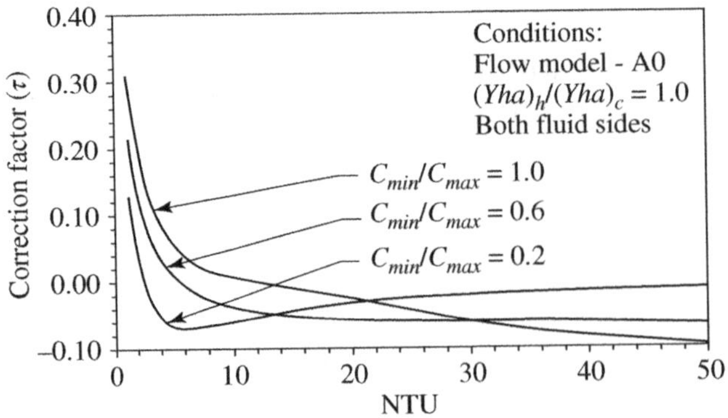

Figure 4.32 Temperature non-uniformity effects – tube-fin heat exchanger comparison with C^*.

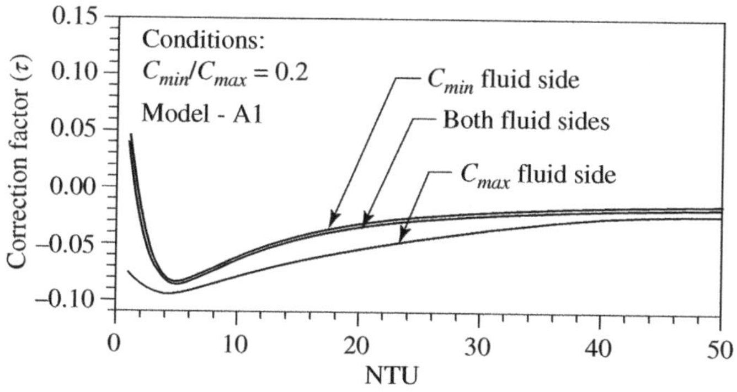

Figure 4.33 Temperature non-uniformity effects –tube-fin heat exchanger – comparison with fluid capacity rates.

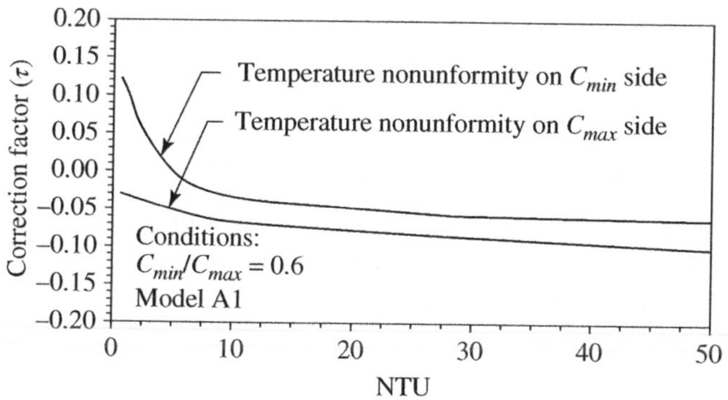

Figure 4.34 – Temperature non-uniformity effects – relative comparison for tube-fin heat exchanger.

inlet temperature non-uniformity together. Hence, the present FEM analysis has been extended to the combined effects of longitudinal heat conduction, inlet flow non-uniformity and temperature non-uniformity.

4.8.2 Validation

Literature on the investigation of combined effects of longitudinal wall heat conduction and flow non-uniformity for direct transfer type cross flow heat exchangers is very limited by Chiou et al. [6]. The present FEM analysis has been compared with available results by Chiou et al. [6] of combined effects of longitudinal wall heat conduction and flow non-uniformity for cross flow plate-fin heat exchanger, as shown in Figure 4.35. This comparison is found to be good. The accuracy is believed to be sufficient for most of the engineering applications.

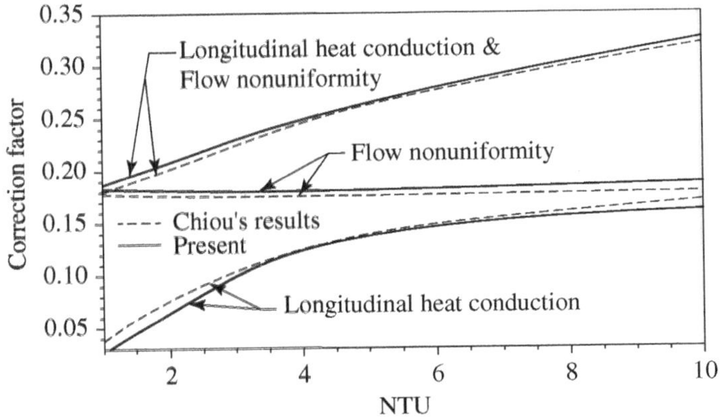

Figure 4.35 – Longitudinal heat conduction and inlet flow non-uniformity effects – plate-fin heat exchanger – comparison of results.

However, there is no information available till now for the following combinations:

1) combined effects of longitudinal heat conduction and temperature non-uniformity
2) combined effects of longitudinal heat conduction, flow non-uniformity and temperature non-uniformity
3) combined effects of flow non-uniformity and temperature non-uniformity

Hence, direct comparison is not possible for the above combinations of the combined effects. However, the present FEM results are validated with individual effects of longitudinal heat conduction, flow non-uniformity and temperature non-uniformity for cross flow plate-fin heat exchanger.

4.8.3 Combined Effects of Longitudinal Wall Heat Conduction and Inlet Flow Non-Uniformity

In this section, the performance evaluation with the combined effects of longitudinal heat conduction and flow non-uniformity is presented for cross flow plate-fin and tube-fin heat exchangers. Another exchanger thermal performance deterioration factor, $\tau_{L,F}$, is introduced to illustrate the influence of the combined effects of longitudinal heat conduction and flow non-uniformity.

It is defined as

$$\tau_{L,F} = \frac{\varepsilon_0 - \varepsilon_{L,F}}{\varepsilon_0} \tag{4.102}$$

4.8.3.1 Cross Flow Plate-Fin Heat Exchanger – Combined Effects (LHC, FN)

The same fluid flow non-uniformity models generated earlier are used for the combinations of longitudinal heat conduction and flow non-uniformity in cross flow plate-fin heat exchangers in this analysis. The performance evaluation with the effects of longitudinal heat conduction and flow non-uniformity is presented for balanced flow, as well as unbalanced flows. The exchanger thermal performance deteriorations are plotted as a function of NTU ($NTU_{overall}$) for three magnitudes of $C_{min}/C_{max} = 1.0, 0.6, 0.2$ for the following cases:

1) combined effects of longitudinal heat conduction and flow non-uniformity on C_{min} fluid side
2) combined effects of longitudinal heat conduction and flow non-uniformity on C_{max} fluid side
3) combined effects of longitudinal heat conduction and flow non-uniformity on both C_{max} and C_{min} fluid sides

Figures 4.36–4.38 show the combined effects of longitudinal heat conduction and flow non-uniformity in cross flow plate-fin heat exchanger. In each case, there are three figures (a, b, c) considering $\lambda = 0.2, 0.1, 0.05$. It can be seen from these figures that the performance deteriorations ($\tau_{L,F}$) increase with increase of λ for both balanced and unbalanced flows of the exchanger. In the case of $C^* = 1$, the performance deterioration increases with increase of NTU until NTU reaches 8 and then decreases with increase of NTU. However, the performance deterioration decreases with increase of NTU for all cases of unbalanced flows. It has been observed that the effects of longitudinal heat

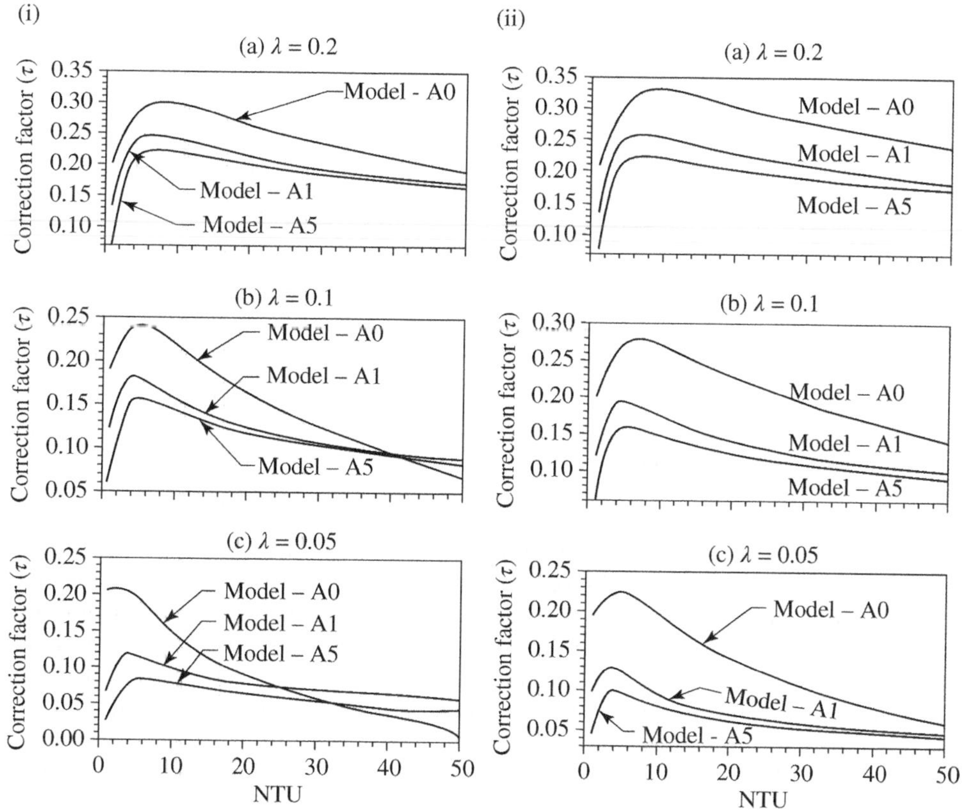

Figure 4.36 Combined effects of LHC and Flow non-uniformity – plate-fin heat exchanger: (i) C_{max} fluid side (ii) both fluid side: $C_{min}/C_{max} = 1.0$.

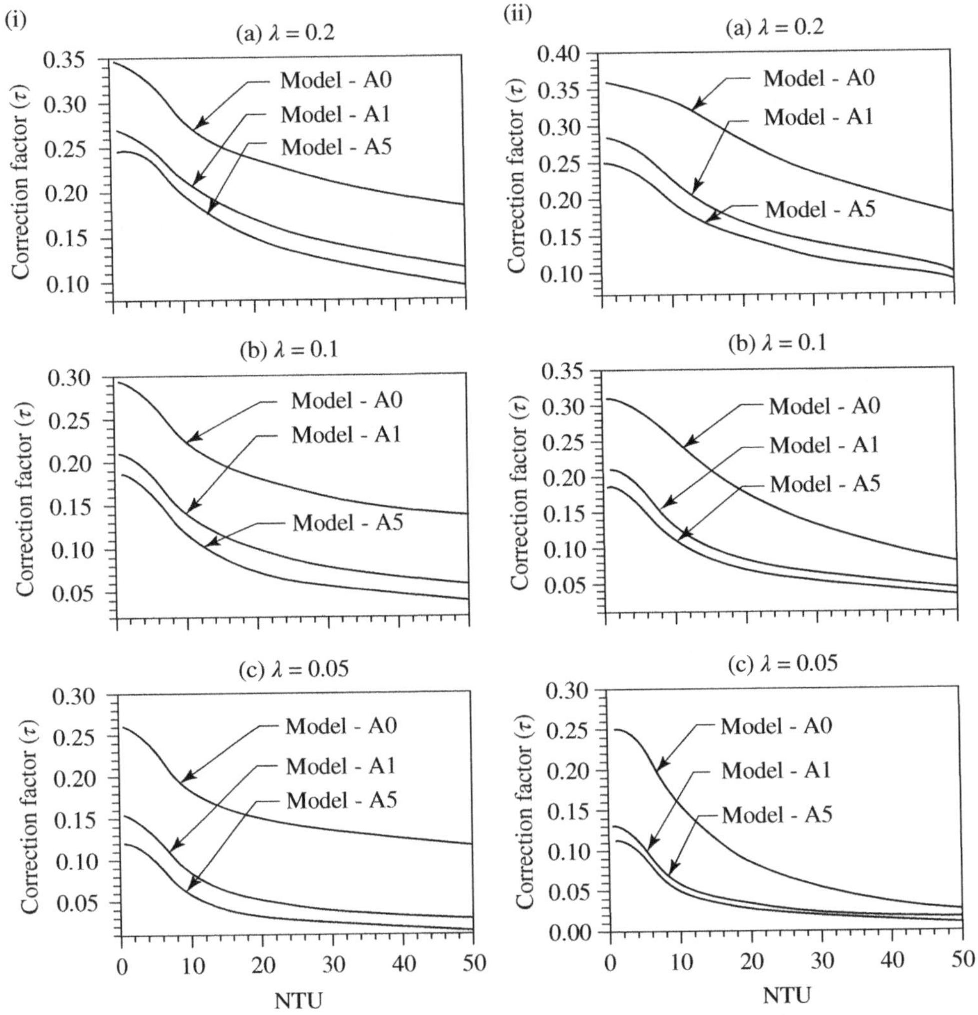

Figure 4.37 Combined effects of LHC and flow non-uniformity –plate-fin heat exchanger: (i) C_{max} fluid side (ii) C_{min} fluid side (iii) both fluid side: $C_{min}/C_{max} = 0.6$.

conduction and flow non-uniformity on the deterioration of exchanger performance tend to augment each other in the regions of lower *NTU*, but tend to eliminate each other in the regions of higher *NTU*. For example, the performance deteriorations due to individual effects of longitudinal heat conduction ($\lambda = 0.05$) and flow non-uniformity at $NTU = 10$ for $C^* = 1.0$ for model A1 are 5% and 7% respectively, whereas the performance deterioration for the combined effects of longitudinal heat conduction and flow non-uniformity is around 10%. Similarly, the performance deterioration due to individual effects of longitudinal heat conduction and flow non-uniformity at $NTU = 50$ for model A1 are 9% and 10% respectively, whereas the performance deterioration due to the combined effects of longitudinal wall heat conduction and flow non-uniformity is 7%.

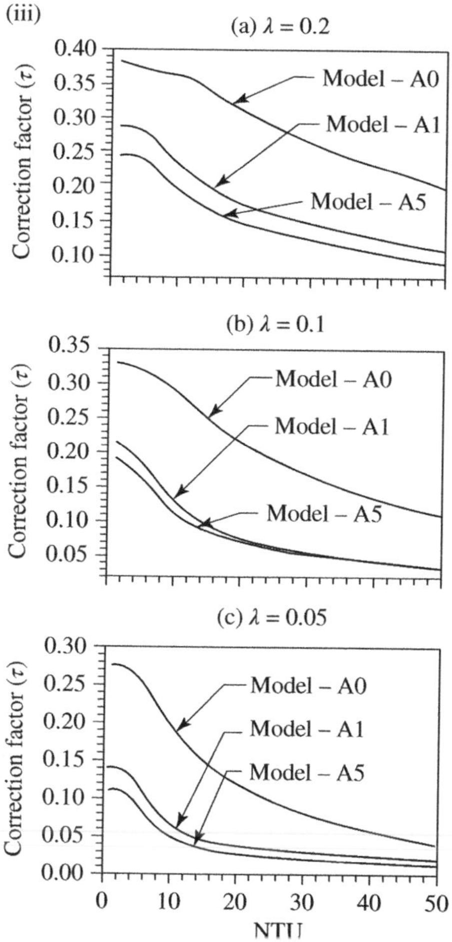

Figure 4.37 (Continued)

The relative comparison of results is tabulated in Table 4.8. In this table, the combined effects of longitudinal wall heat conduction and flow non-uniformity are compared with the individual effects of longitudinal wall heat conduction and flow non-uniformity for cross flow plate-fin heat exchanger. In this table, the performance variations (in %) is compared for two typical cases of $NTU = 10$ and 50 using the flow model A1 and flow model A5. It can be seen from this table that the performance deterioration increases further when the longitudinal wall heat conduction and flow non-uniformity are considered together. In the case of the combined effects of longitudinal wall heat conduction and flow non-uniformity, the performance deterioration increases by 1% to 9% (max) when compared to its individual effects of longitudinal wall heat conduction and flow non-uniformity, as shown in Table 4.8.

The relation between the ratio of C^* and correction factor ($\tau_{L,F}$) is shown in Figure 4.39 for two cases: (a) with λ variation and (b) with C^*. Figure 4.39a shows that correction

(i)

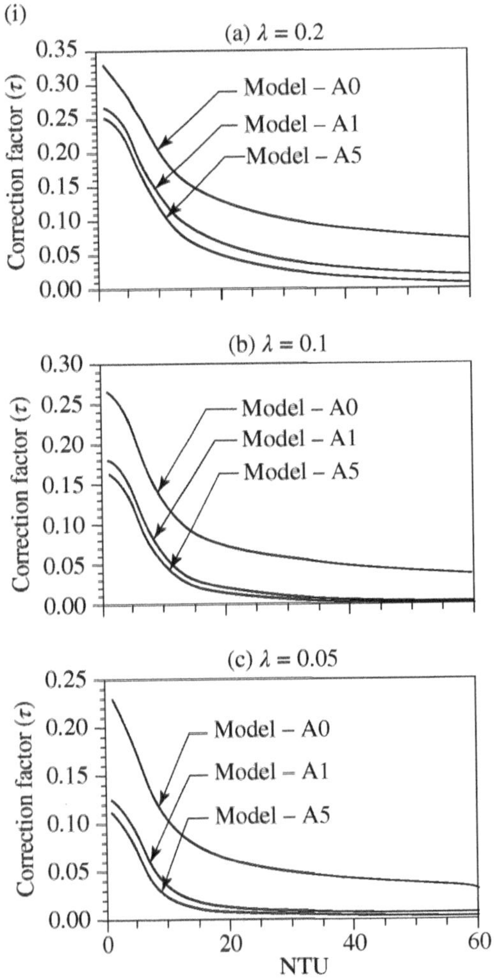

Figure 4.38 Combined effects of LHC and Flow non-uniformity – plate-fin heat exchanger: (i) C_{max} fluid side (ii) C_{min} fluid side (iii) both fluid side: $C_{min}/C_{max} = 0.2$.

factor is higher when λ is higher. For example, the correction factors at $NTU = 10$, are 23.5% for $\lambda = 0.2$, 15% for $\lambda = 0.1$ and 8% for $\lambda = 0.05$ for flow model A5. Figure 4.39b shows that the performance deteriorations are higher for balanced flows as compared to that of unbalanced flows. For example, the performance deteriorations (at $NTU = 10$ and $\lambda = 0.1$) are around 15.5% for $C^* = 1.0$, 13% for $C^* = 0.6$ and 8.5% for $C^* = 0.2$ for flow model A1 and when the flow non-uniformity is considered on C_{max} fluid side. Information presented in these figures is not restricted to the models considered in this analysis, but the results can be interpolated for other similar flow models. For example, the results can be interpolated for any intermediate flow mal-distribution model, between the curves, as the upper curve is showing higher performance deviations and lower curve is showing the minimum performance deviations. The deterioration of

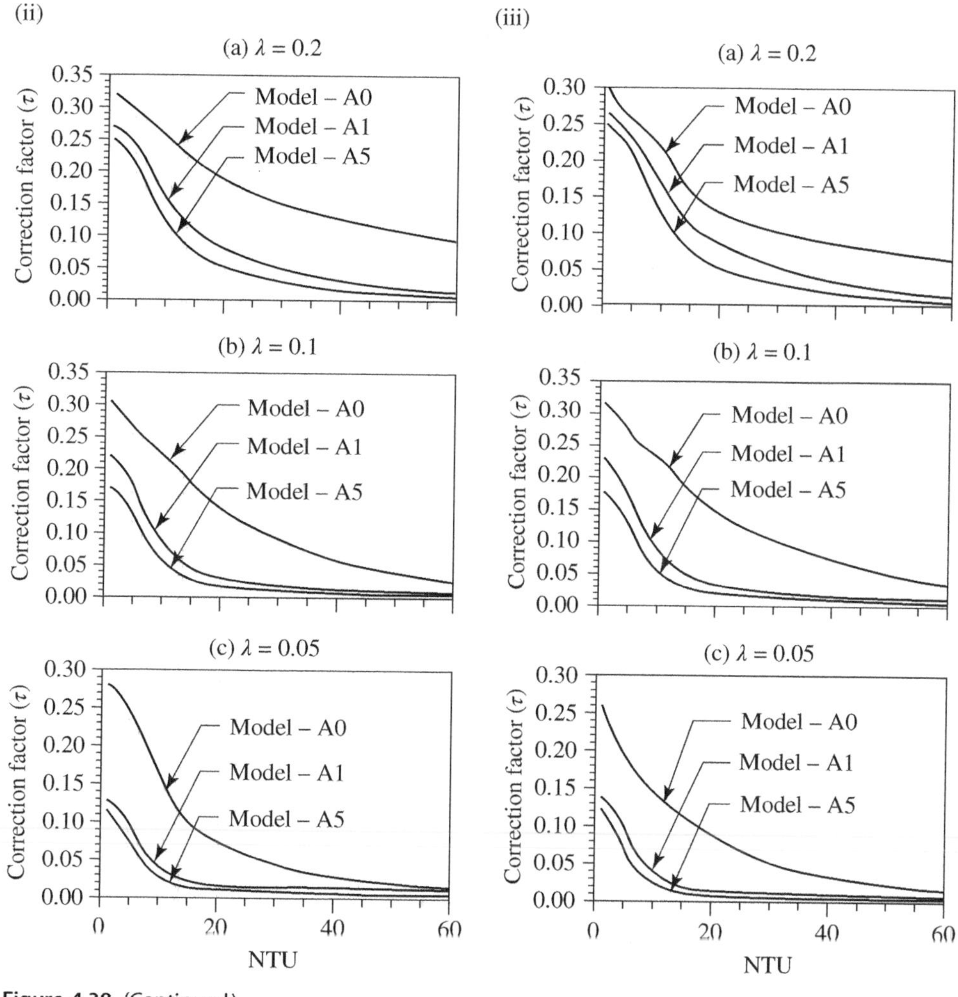

Figure 4.38 (Continued)

thermal performance of single pass cross flow plate-fin heat exchanger due to the combined effects of longitudinal wall heat conduction and flow non-uniformity presented in this section (Ranganayakulu [31–34]) is generally similar to those reported in previous investigations (Chiou [6]). However, the type of the fluid mal-distributions considered in this investigation are basically different from those reported previously; direct comparison of all these results is thus not possible.

4.8.3.2 Cross Flow Tube-Fin Heat Exchanger – Combined Effects (LHC, FN)

In this section, the performance evaluation with the effects of longitudinal wall heat conduction and flow non-uniformity in cross flow tube-fin compact heat exchanger is presented for balanced flows and unbalanced flows. The exchanger thermal performance

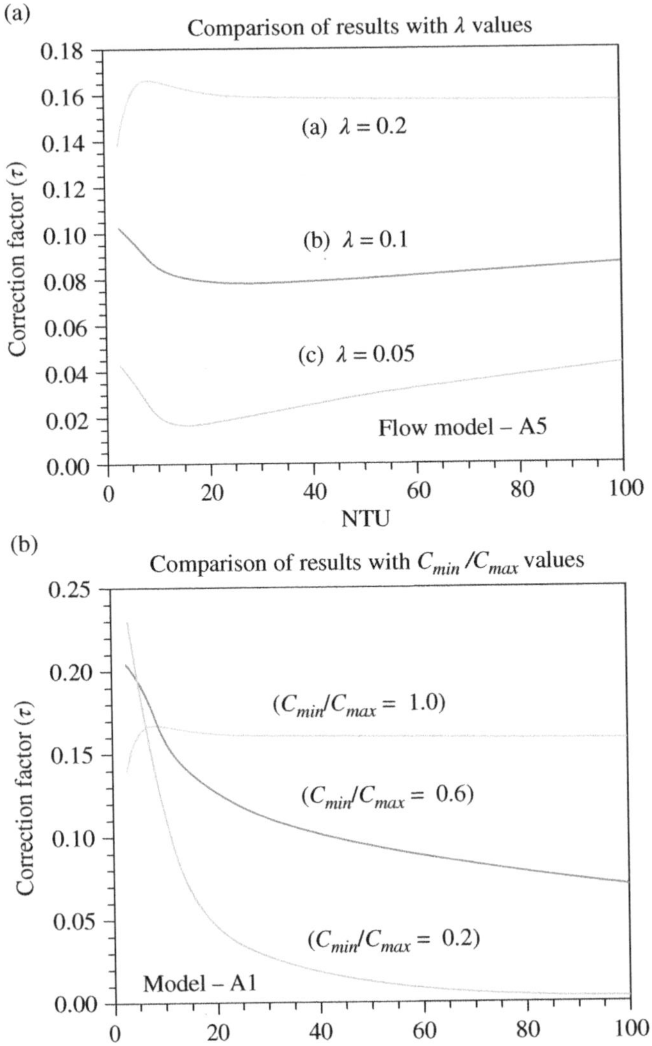

Figure 4.39 Comparison of combined effects of LHC and flow non-uniformity – cross flow plate-fin heat exchanger – both fluid sides.

deteriorations are plotted as a function of NTU ($NTU_{overall}$) for three magnitudes of $C^* =$ 1.0, 0.6, 0.2 using the same models (A0, A1, A5) for the following cases:

1) combined effects of longitudinal heat conduction and flow non-uniformity on C_{min} fluid side
2) combined effects of longitudinal heat conduction and flow non-uniformity on C_{max} fluid side
3) combined effects of longitudinal heat conduction and flow non-uniformity on both hot and cold fluid side

Figures 4.40–4.44 show the combined effects of longitudinal heat conduction and flow non-uniformity in cross flow tube-fin heat exchanger. In each case, there

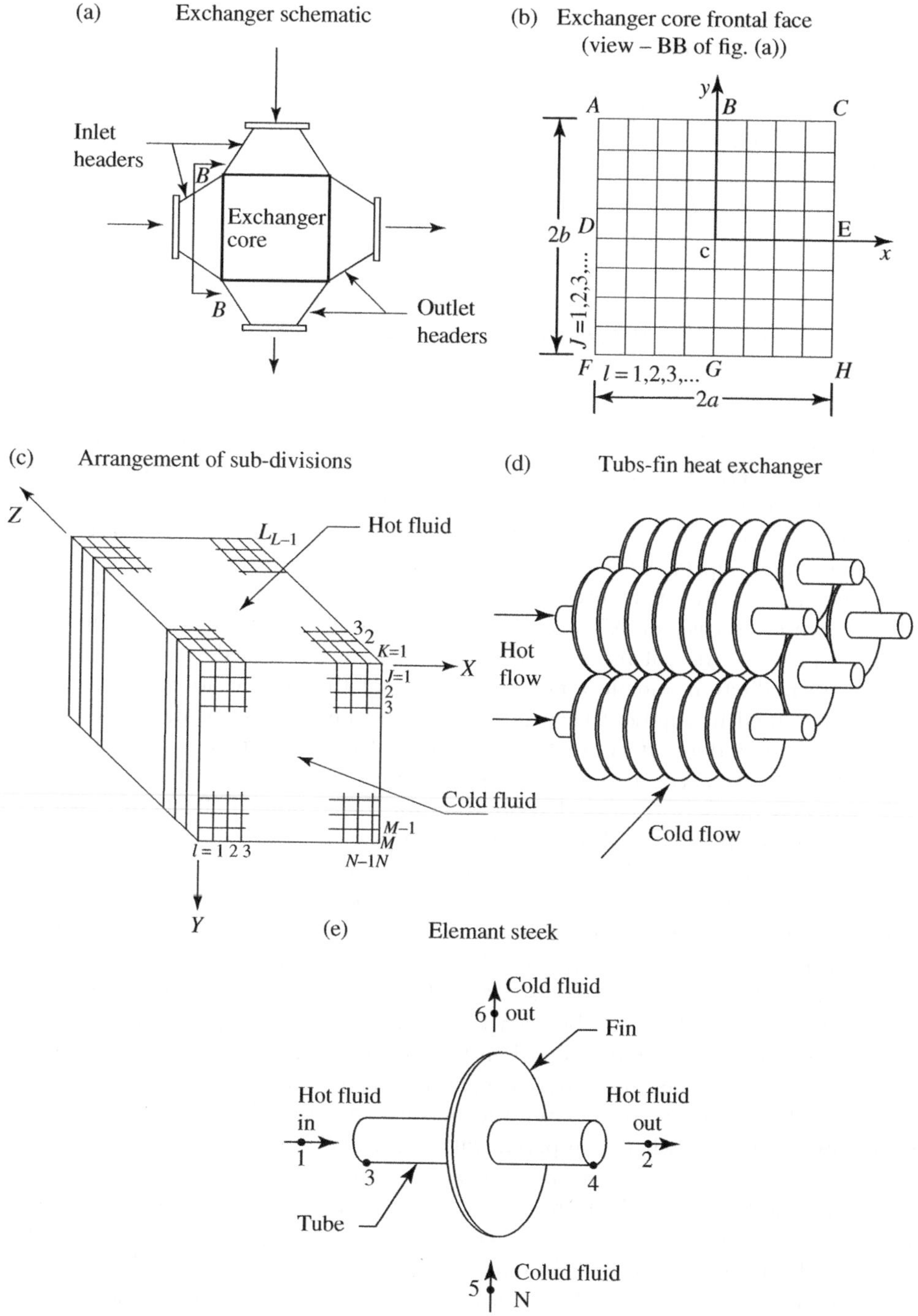

Figure 4.40 Single pass cross flow tube-fin heat exchanger.

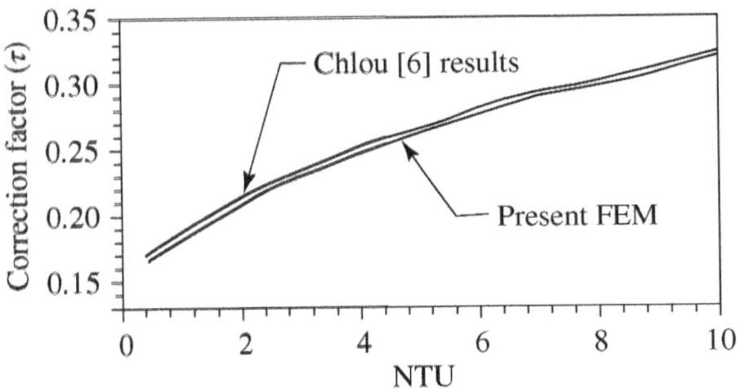

Figure 4.41 Combined effects of LHC and flow non-uniformity – tube-fin heat exchanger – comparison of results.

are three figures (a, b, c) considering λ = 0.2, 0.1, 0.05. It can be seen from these figures that the performance deteriorations ($\tau_{L,F}$) increase with increase of λ for both balanced and unbalanced flows of the exchanger as in the case of plate-fin heat exchanger. In the case of $C^* = 1$, the performance deterioration increases with increase of NTU (up to $NTU = 10$) and then decreases with increase of NTU after $NTU = 10$, as shown in Figure 4.42. However, the performance deteriorations decrease with increase of NTU for all cases of unbalanced flows, as shown in Figures 4.43–4.44. Also, it has been observed that the combined effects of longitudinal heat conduction and flow non-uniformity on the deterioration of exchanger performance tend to augment each other. For example, the performance deteriorations due to individual effects of longitudinal heat conduction (λ = 0.05) and flow non-uniformity at $NTU = 10$ for $C^* = 1.0$ for model A1 are 3.25% and 5.7% respectively, whereas the performance deterioration for the combined effects of longitudinal heat conduction and flow non-uniformity is around 18.5%. The relative comparison of results is tabulated in Table 4.9. In this table, the combined effects of longitudinal wall heat conduction and flow non-uniformity are compared with individual effects of longitudinal heat conduction and flow non-uniformity for cross flow tube-fin heat exchanger. In the table the performance variations (in %) have been compared for two typical cases of $NTU = 10$ and 50 using the flow model A1 and flow model A5. It can be seen from the table that the performance deterioration generally increases further when the longitudinal wall heat conduction and flow non-uniformity are considered together as in the case of plate-fin heat exchanger. However, in some cases the performance deteriorations decreases for the combined effects when compared with individual effects of longitudinal wall heat conduction and flow non-uniformity. In the case of the combined effects of longitudinal wall heat conduction and flow non-uniformity, the performance deteriorations increases by 1–5% (max) as compared to its individual effects of longitudinal wall heat conduction and flow non-uniformity, as shown in Table 4.9.

In general, it has been observed that the performance deteriorations are higher for balanced flows when compared to unbalanced flows. For example, the performance deteriorations (at $NTU = 10$ and λ = 0.05) are around 9% for $C^* = 1.0$, 6.5% for $C^* = 0.6$ and 1.2% for $C^* = 0.2$ for flow model A1 and when the flow non-uniformity is considered on C_{max} fluid side. Also, it has been observed that the correction factor ($\tau_{L,F}$) increases with the increase of longitudinal heat conduction parameter for all cases. For example, the

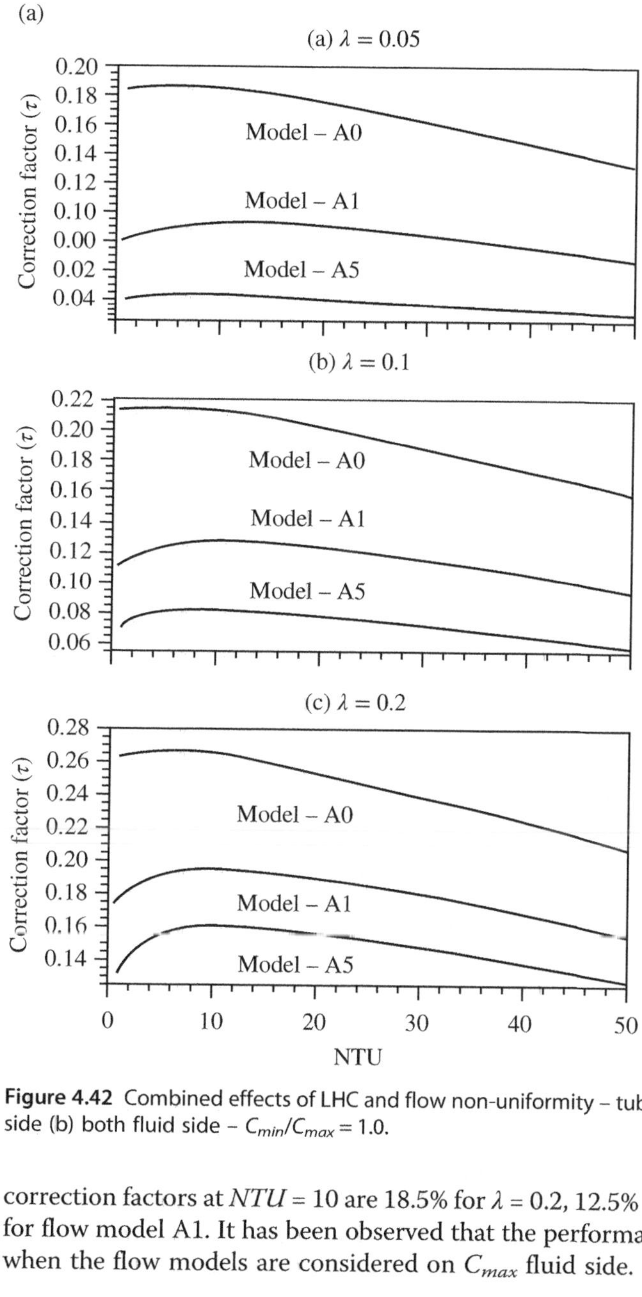

Figure 4.42 Combined effects of LHC and flow non-uniformity – tube-fin heat exchanger (a) C_{max} fluid side (b) both fluid side – $C_{min}/C_{max} = 1.0$.

correction factors at $NTU = 10$ are 18.5% for $\lambda = 0.2$, 12.5% for $\lambda = 0.1$ and 9% for $\lambda = 0.05$ for flow model A1. It has been observed that the performance deteriorations are higher when the flow models are considered on C_{max} fluid side.

4.8.4 Combined Effects of Longitudinal Wall Heat Conduction, Inlet Flow Non-Uniformity and Temperature Non-Uniformity

In the previous sections (4.5, 4.6 & 4.7) the individual effects of longitudinal heat conduction, flow non-uniformity are studied. In this section, the performance evaluation with the combined effects of longitudinal heat conduction, flow non-uniformity and

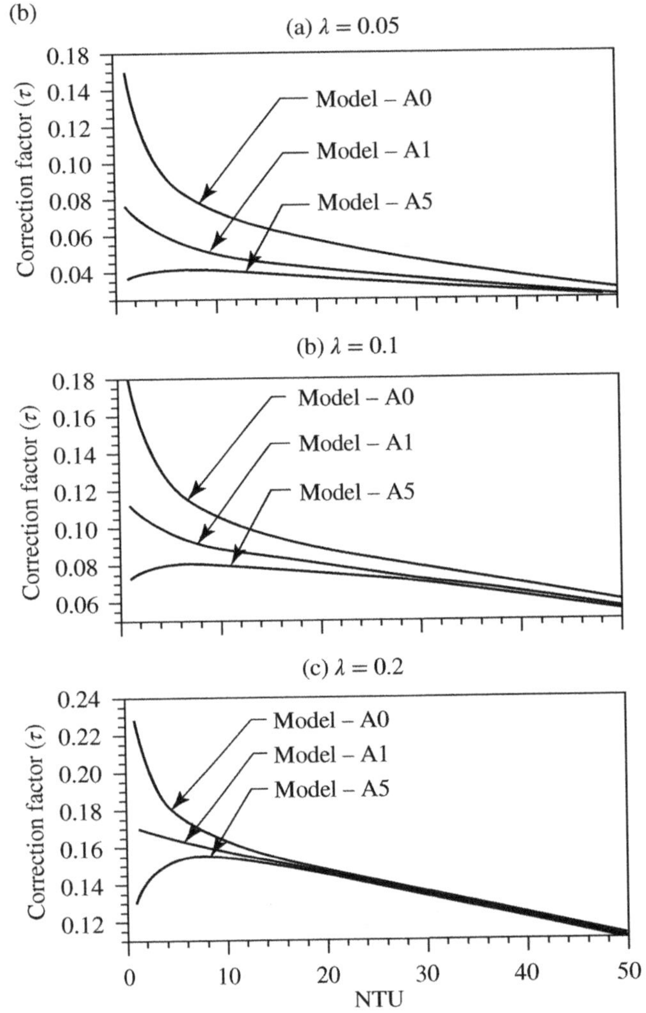

Figure 4.42 (Continued)

temperature non-uniformity is presented for cross flow plate-fin and tube-fin heat exchangers. Another exchanger thermal performance correction factor $\tau_{L,F,T}$ is introduced to illustrate the influence of the combined effects of longitudinal heat conduction for cross flow plate-fin and tube-fin heat exchangers. It is defined as

$$\tau_{L,F,T} = \frac{\varepsilon_0 - \varepsilon_{L,F,T}}{\varepsilon_0} \tag{4.103}$$

Here, the correction factor can be positive or negative. The present section is devoted to the study of the combined effects of longitudinal heat conduction, flow non-uniformity and temperature non-uniformity.

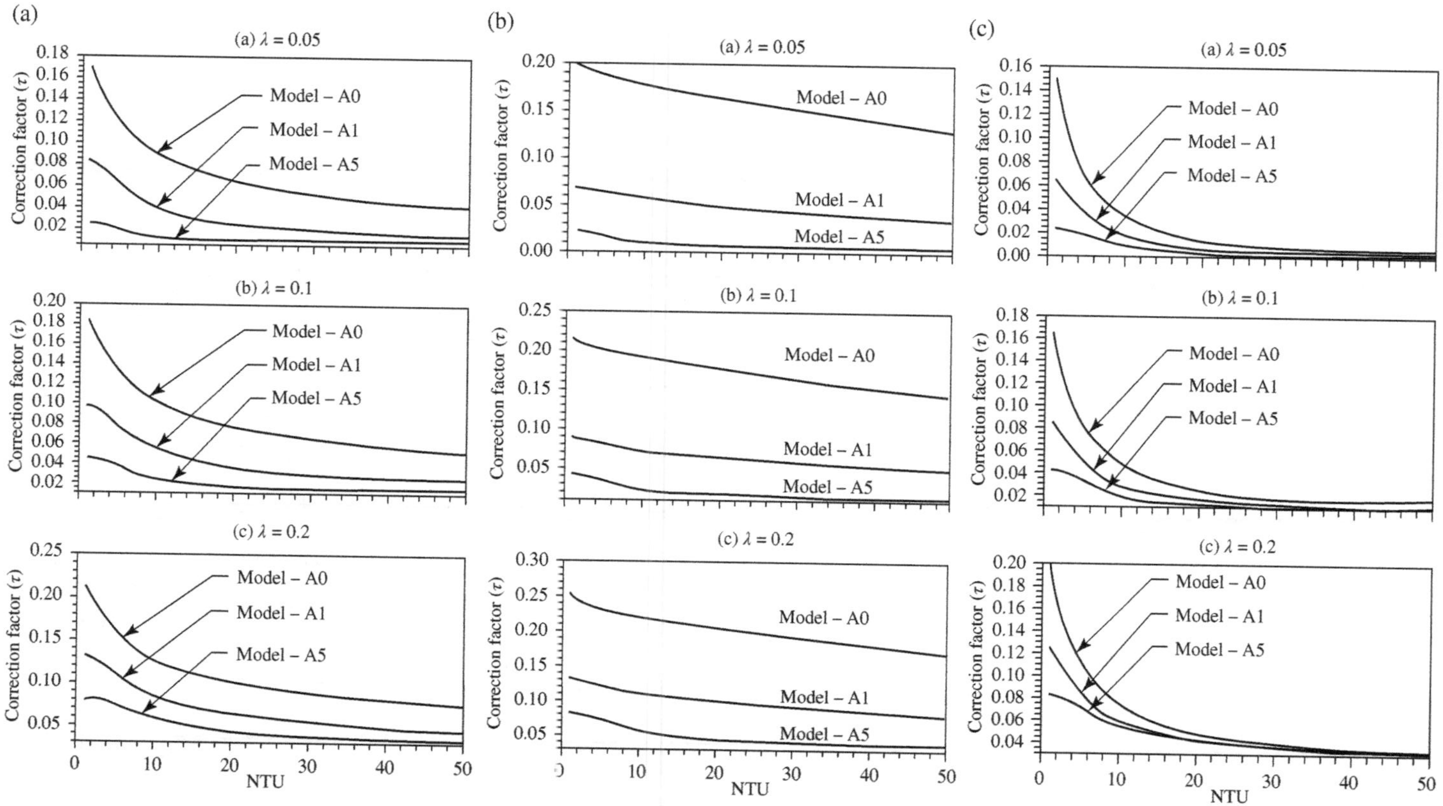

Figure 4.43 Combined effects of LHC and flow non-uniformity – Tube-fin heat exchanger (a) C_{max} fluid side (b) C_{min} fluid side (c) both fluid side – $C_{min}/C_{max} = 0.6$

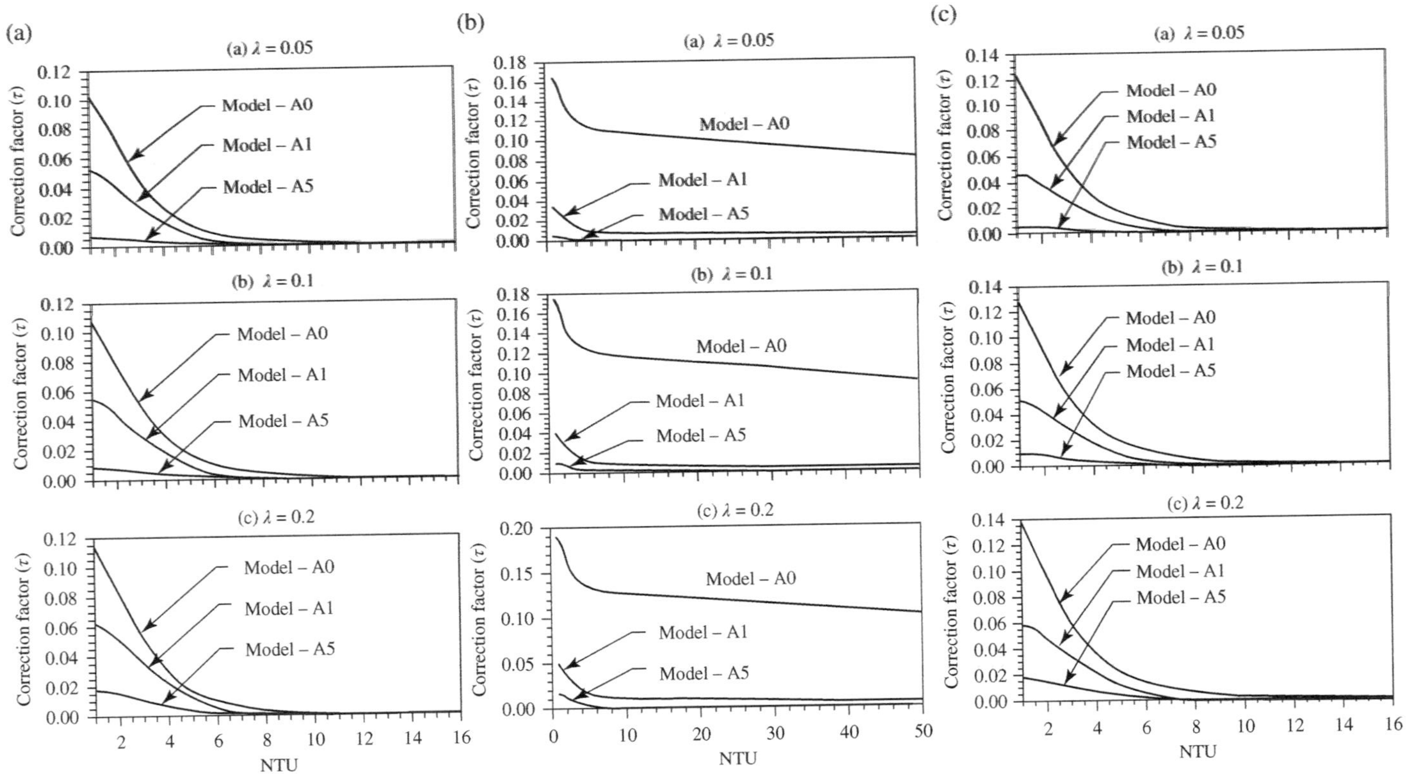

Figure 4.44 Combined effects of LHC and flow non-uniformity – Tube-fin heat exchanger (a) C_{max} fluid side (b) C_{min} fluid side (c) both fluid side – $C_{min}/C_{max} = 0.2$.

4.8.4.1 Cross Flow Plate-Fin Heat Exchanger – Combined Effects (LHC, FN, TN)

The same temperature/flow non-uniformity models generated earlier (Section 4.7) are used for the combined effects of longitudinal heat conduction, flow non-uniformity and temperature non-uniformity in cross flow plate-fin heat exchanger in this analysis. The performance evaluation with the effects of longitudinal heat conduction, flow non-uniformity and temperature non-uniformity is presented for both balanced flows and unbalanced flows. The exchanger thermal performance deteriorations are plotted as a function of NTU ($NTU_{overall}$) for three magnitudes of $C_{min}/C_{max} = 1.0$, 0.6, 0.2 for the following cases:

1) combined effects of longitudinal heat conduction, flow non-uniformity and temperature non-uniformity on C_{min} fluid side
2) combined effects of longitudinal heat conduction, flow non-uniformity and temperature non-uniformity on C_{max} fluid side
3) combined effects of longitudinal heat conduction, flow non-uniformity and temperature non-uniformity on both hot and cold fluid sides

Figures 4.45–4.47 shows the combined effects of longitudinal heat conduction, flow non-uniformity and temperature non-uniformity in cross flow plate-fin heat exchanger. In

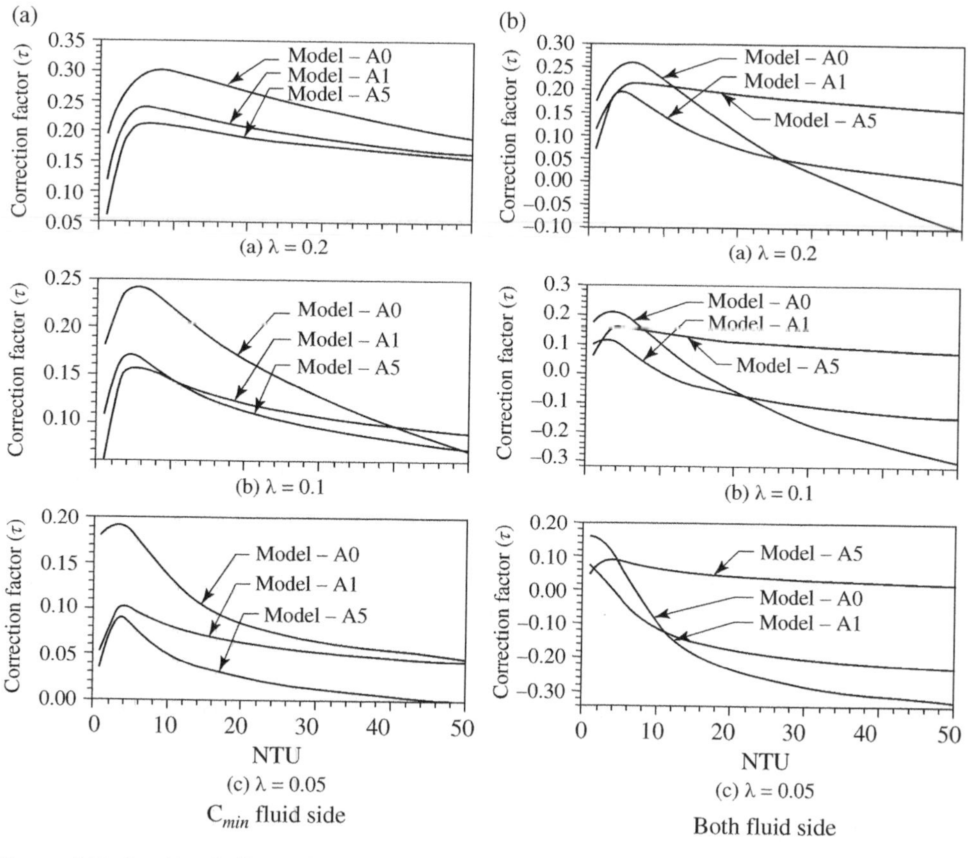

Figure 4.45 Combined effects of LHC, FN and TN – plate-fin heat exchanger – ($C_{min}/C_{max} = 1.0$).

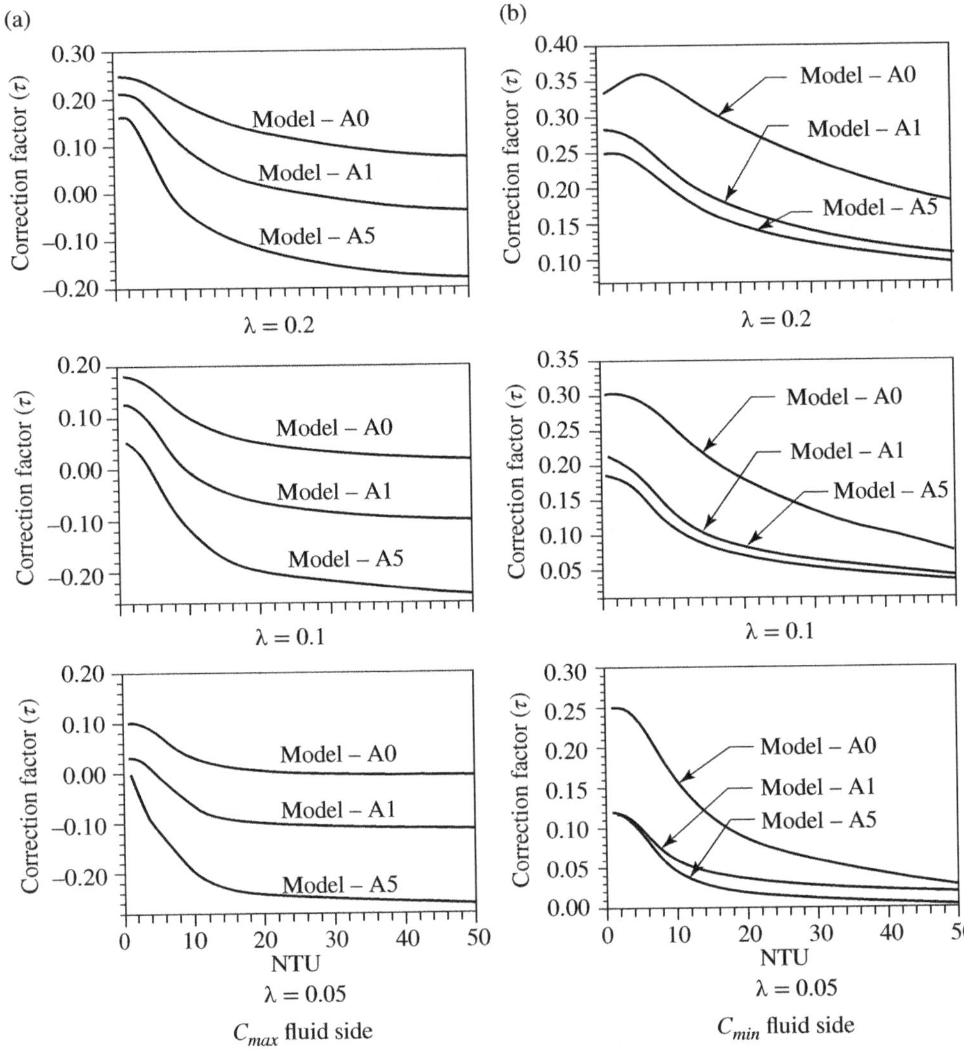

Figure 4.46 Combined effects of LHC, FN and TN – plate-fin heat exchanger – (C_{min}/C_{max} = 0.6).

each case, there are three figures (a, b, c) considering λ = 0.05, 0.1, 0.2. It has been observed that the performance variations increase with increase of λ for both balanced and unbalanced flows of plate-fin heat exchangers. In the case of balanced flows of plate-fin heat exchanger, the performance deterioration increases with increase in *NTU* (up to *NTU* = 5) and then decreases with increase of *NTU* for balanced flows, as shown in Figure 4.45. However, the performance variations decreases with increase of *NTU* for the cases of unbalanced flow, C^* = 0.6, as shown in Figures 4.46–4.47. In the case of unbalanced flow, C^* = 0.6, the correction factor decreases with increase of *NTU* until *NTU* reaches 10 and then increases with increase of *NTU*. It is also noted that there is significant augmentation in thermal performance (up to 20%), when the λ is low in

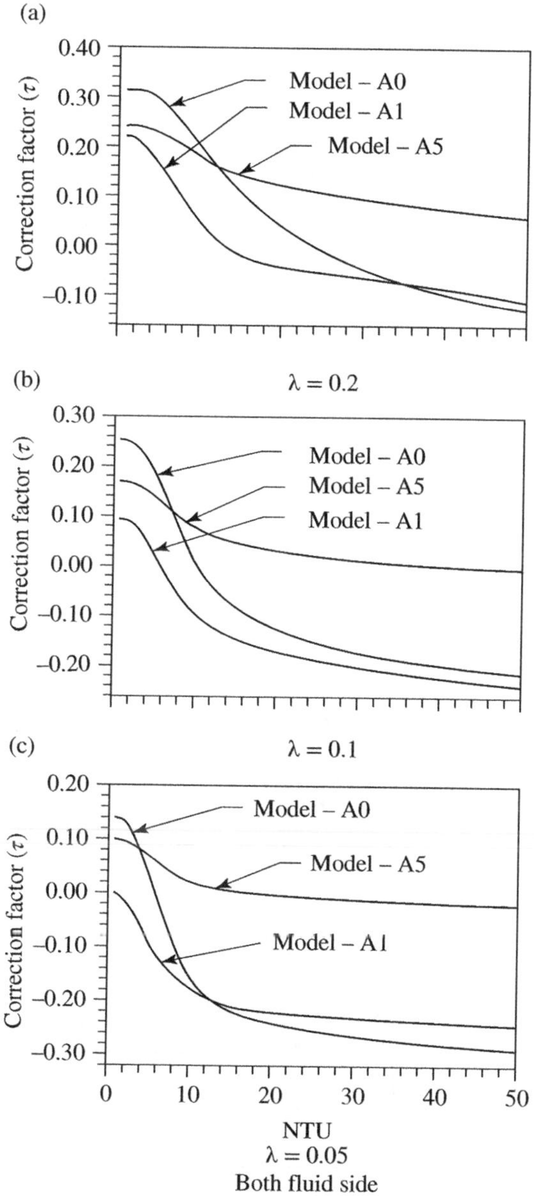

Figure 4.46 (Continued)

some cases. This observation indicates that the combined effects of longitudinal wall heat conduction, flow non-uniformity and temperature non-uniformity on the deterioration of exchanger performance tend to eliminate each other in the region of large *NTU*, but tend to augment each other in the region of small *NTU*. It has been observed that the performance variations are higher when the flow/temperature models are considered on the C_{min} side. Also, it has been observed that the performance deteriorations are more

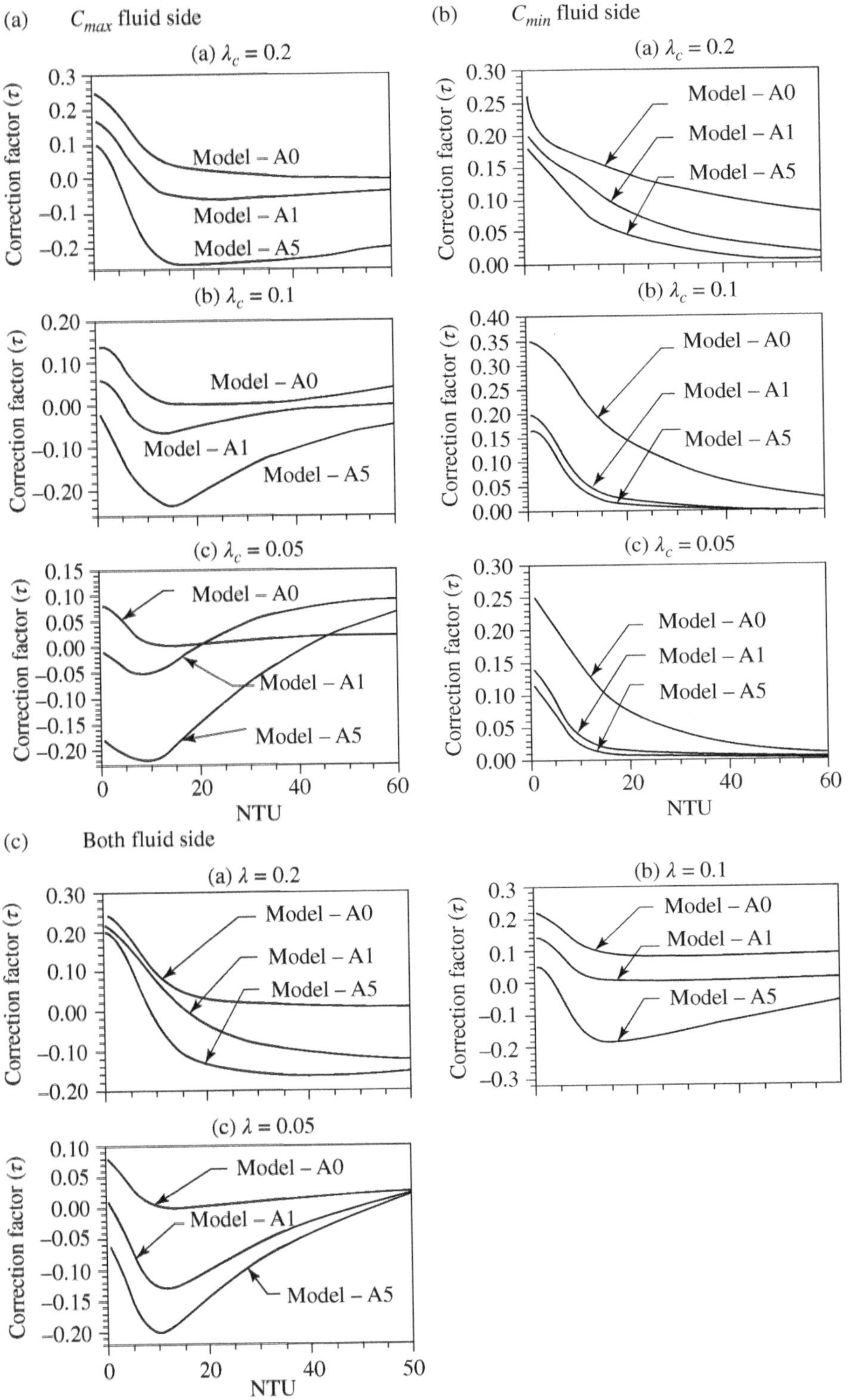

Figure 4.47 Combined effects of LHC, FN and TN – plate-fin heat exchanger – ($C_{min}/C_{max} = 0.2$).

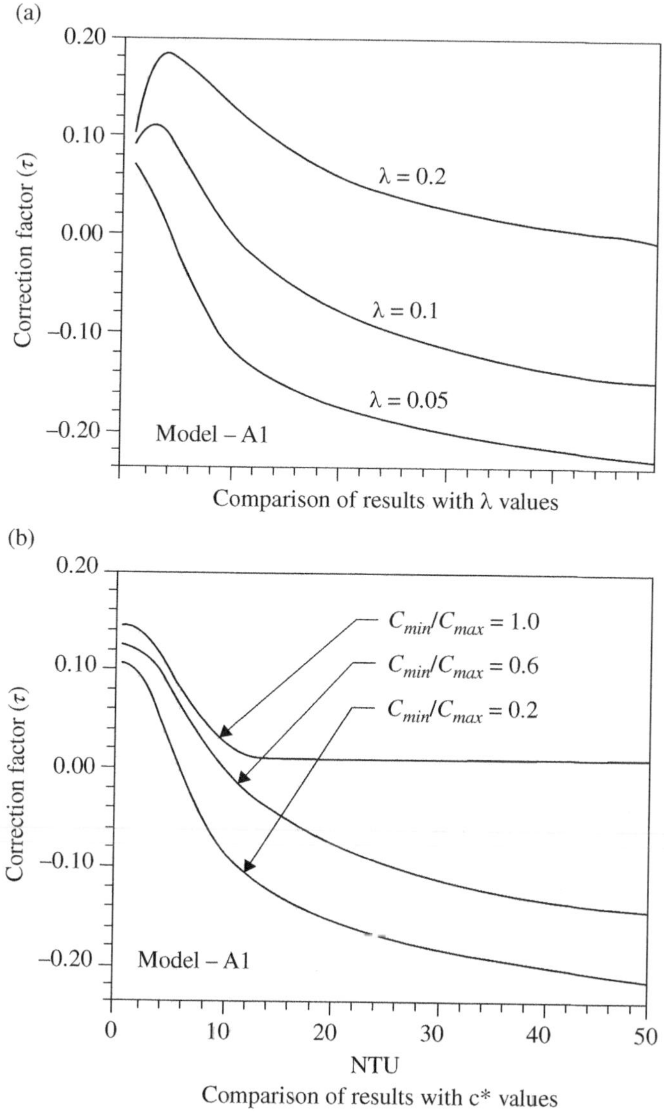

Figure 4.48 Combined effects of LHC, FN and TN – plate-fin heat exchanger – both fluid sides.

for balanced flows as compared to that of unbalanced flows. The relative comparison between the ratio of C_{min}/C_{max} and for three λ values versus correction factor ($\tau_{L,F,T}$) are shown in Figure 4.47 for the combined effects of longitudinal heat conduction, flow non-uniformity and temperature non-uniformity for cross flow plate-fin heat exchanger.

The relative comparison of results is tabulated in Table 4.10. In this table, the combined effects of longitudinal wall heat conduction, flow non-uniformity and temperature non-uniformity are compared with individual effects of longitudinal heat conduction, flow non-uniformity and temperature non-uniformity for cross flow plate-fin heat exchanger. For example, the performance variations due to individual effects of

Table 4.10 Effects of longitudinal wall heat conduction, flow and temperature non-uniformity cross flow plate-fin heat exchanger.

Models	C^*	Performance variations (%)																
		NTU = 10.0								NTU = 50								
		LHC					LHC + FN + TN			LHC					LHC + FN + TN			
		$\lambda = 0.05$	$\lambda = 0.1$	$\lambda = 0.2$	FN	TN	$\lambda = 0.05$	$\lambda = 0.1$	$\lambda = 0.2$	$\lambda = 0.05$	$\lambda = 0.1$	$\lambda = 0.2$	FN	TN	$\lambda = 0.05$	$\lambda = 0.1$	$\lambda = 0.2$
	1.0	5.50	10.5	17.5	7.00	0.55	8.00	14.2	22.0	6.00	14.0	22.5	4.95	11.5	5.00	10.0	17.5
A1	0.6	3.40	7.00	13.5	5.00	0.10	5.70	12.5	21.5	2.75	6.75	15.3	1.80	3.50	3.00	5.00	11.0
	0.2	0.95	2.00	5.00	2.00	−0.2	3.50	7.50	14.0	0.05	0.55	2.85	0.0	0.03	0.75	1.00	2.50
	1.0	5.50	10.5	17.5	0.50	0.50	5.00	14.0	20.5	6.00	14.0	22.5	0.05	1.20	0.55	6.00	17.0
A5	0.6	3.40	7.00	13.5	1.00	0.07	4.00	10.7	19.5	2.75	6.75	15.3	0.0	0.50	0.45	4.00	10.0
	0.2	0.95	2.00	5.00	0.50	0.40	2.50	5.50	11.8	0.05	0.55	2.85	0.0	0.03	0.25	0.50	1.00

Conditions: $C^* = 1.0$, $(\eta ha)^* = 1.0$, $\lambda_c/\lambda_h = 1.0$ and flow/temperature non-uniformity on C_{min} fluid side longitudinal wall heat conduction ($\lambda = 0.05$), flow non-uniformity and temperature non-uniformity at $NTU = 10$ for $C^* = 1.0$ for model A1 are 5.5%, 7% and 0.55% respectively, whereas the performance variation for the combined effects of longitudinal wall heat conduction, flow non-uniformity and temperature non-uniformity is around 8%. It can be seen from the table that the performance variations generally increases when the longitudinal wall heat conduction, temperature non-uniformity and flow non-uniformity are considered together. In the case of the combined effects of longitudinal wall heat conduction, temperature non-uniformity and flow non-uniformity, the performance variations increase by 1–4.5% (max) as compared to its individual effects of longitudinal wall heat conduction, temperature non-uniformity and flow non-uniformity, as shown in Table 4.10.

4.8.4.2 Cross Flow Tube-Fin Heat Exchanger - Combined Effects (LHC, FN, TN)

The same temperature/flow non-uniformity models generated earlier are used for the combined effects of longitudinal heat conduction, flow non-uniformity and temperature non-uniformity in cross flow tube-fin heat exchanger in this analysis. The performance evaluation with the effects of longitudinal heat conduction, flow non-uniformity and temperature non-uniformity is presented for both balanced flows and unbalanced flows. The exchanger thermal performance deteriorations are plotted as a function of NTU ($NTU_{overall}$) for three magnitudes of $C_{min}/C_{max} = 1.0$, 0.6, 0.2 for the following cases:

1) combined effects of longitudinal heat conduction, flow non-uniformity and temperature non-uniformity on C_{min} fluid side
2) combined effects of longitudinal heat conduction, flow non-uniformity and temperature non-uniformity on C_{max} fluid side
3) combined effects of longitudinal heat conduction, flow non-uniformity and temperature non-uniformity on both hot and cold fluid sides

Figues 4.49–4.56 show the combined effects of longitudinal heat conduction, flow non-uniformity and temperature non-uniformity in cross flow tube-fin heat exchanger. In each case, there are three figures (a, b, c) considering λ = 0.05, 0.1, 0.2. It has been observed that the performance variations increase with increase of λ for both balanced and unbalanced flows of tube-fin heat exchangers as in the case of plate-fin heat exchanger. The correction factor ($\tau_{L,F,T}$) decreases with increase of NTU for all cases of balanced and unbalanced flows, as shown in Figures 4.49–4.56. It is noted that there is significant augmentation in thermal performance (up to 10%) when λ is low, as shown in Figure 4.51–4.52. This augmentation (Ranganayakulu et al. [35,36]) in performance is due to temperature non-uniformity as observed by earlier investigators (Chiou [26] and Gaddis [27]). This observation indicates that the effects of longitudinal heat conduction, flow non-uniformity and temperature non-uniformity on the deterioration of exchanger performance tend to eliminate each other in the region of large NTU, but tend to augment each other in the region of small NTU. It has been observed that the performance variations are higher when the flow/temperature models are considered on C_{min} side. Also, it has been observed that the performance deteriorations are more for balanced flows as compared to that of unbalanced flows.

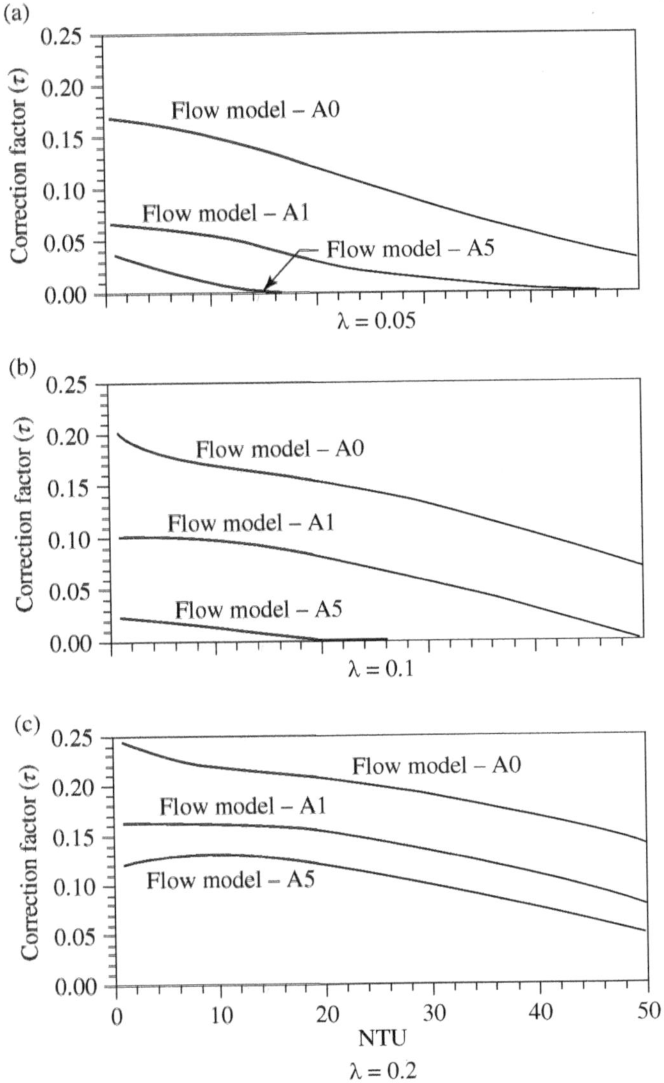

Figure 4.49 Combined effects of LHC, FN and TN tube-fin heat exchanger – C_{min} fluid side (C_{min}/C_{max} = 1.0).

The relative comparison of results is tabulated in Table 4.11. In this table, the combined effects of longitudinal wall heat conduction, flow non-uniformity and temperature non-uniformity are compared with individual effects of longitudinal heat conduction, flow non-uniformity and temperature non-uniformity for cross flow tube-fin heat exchanger. The performance variations due to individual effects of longitudinal heat conduction (at λ = 0.05), flow non-uniformity and temperature non-uniformity at $NTU = 10$ for $C^* = 1.0$ for model A1 are 3.75%, 5.95% and −2% respectively, whereas the performance variation for the combined effects of longitudinal heat conduction, flow non-uniformity and temperature non-uniformity is around 6%.

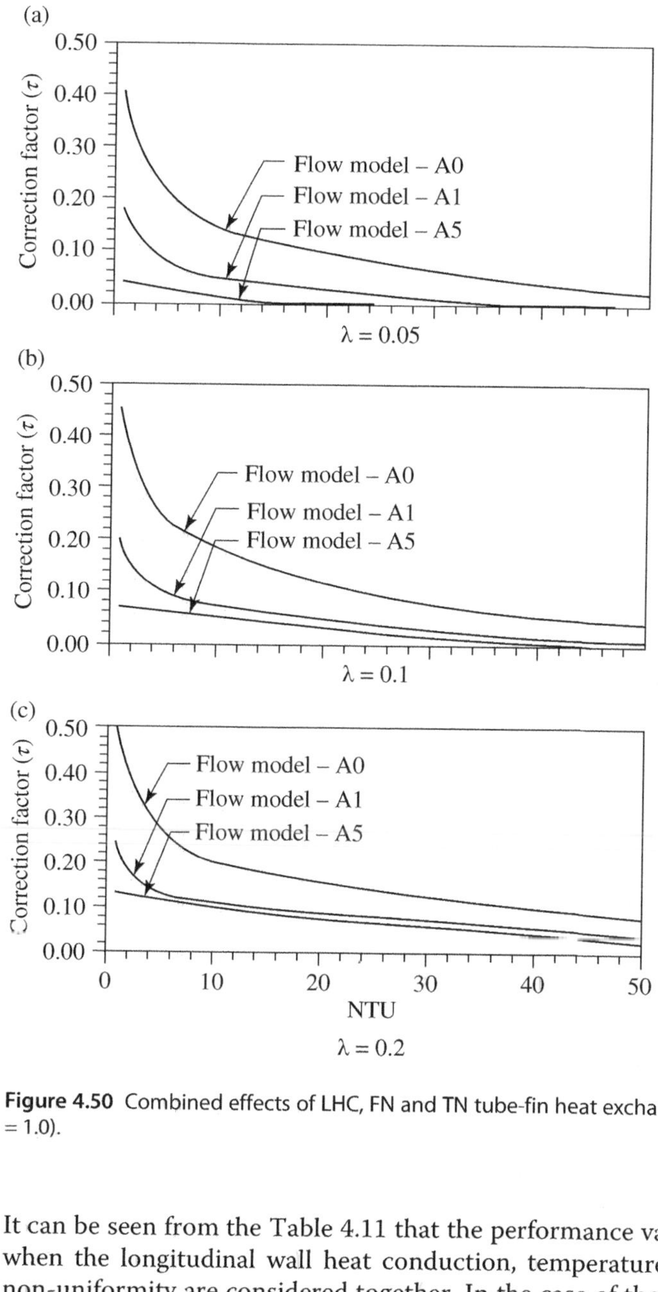

Figure 4.50 Combined effects of LHC, FN and TN tube-fin heat exchanger – both fluid side (C_{min}/C_{max} = 1.0).

It can be seen from the Table 4.11 that the performance variations generally decrease when the longitudinal wall heat conduction, temperature non-uniformity and flow non-uniformity are considered together. In the case of the combined effects of longitudinal wall heat conduction, temperature non-uniformity and flow non-uniformity, the performance variations decreases by 1–14.5% (max), when compared with its individual effects of longitudinal wall heat conduction, temperature non-uniformity and flow non-uniformity, as shown in Table 4.11. In addition, comparison of individual effects with combined effects of LHC, FN and TN for tube-fin heat exchanger is also presented in Figure 4.57.

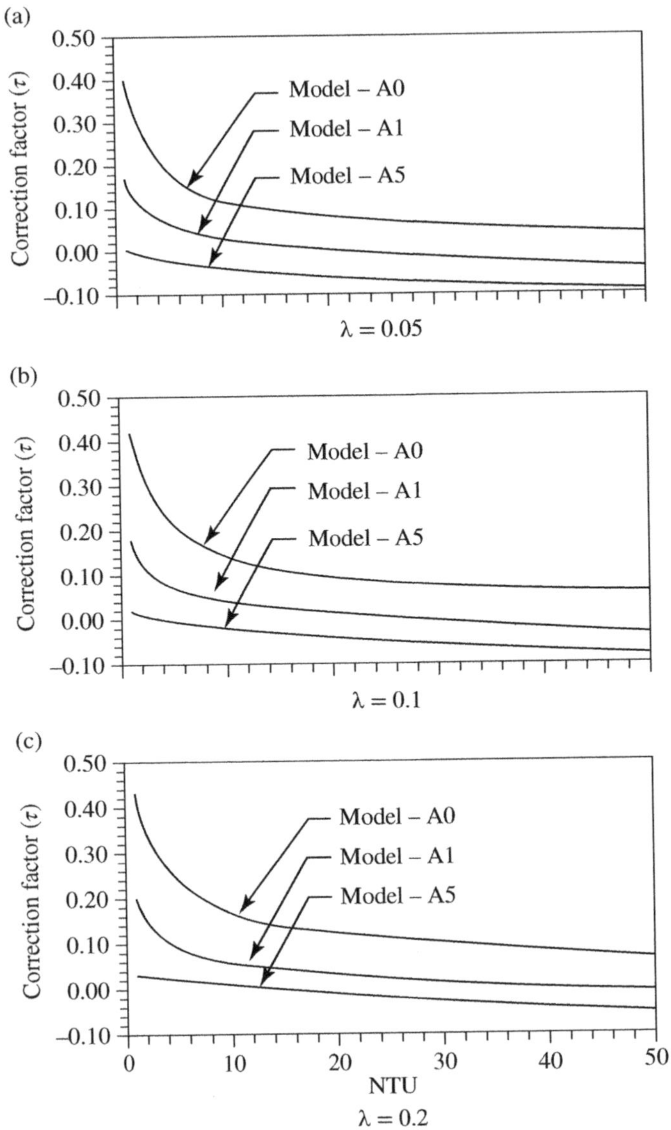

Figure 4.51 Combined effects of LHC, FN and TN – tube-fin heat exchanger – C_{min} fluid side – ($C_{min}/C_{max} = 0.6$).

4.8.5 Combined Effects of Inlet Flow Non-Uniformity and Temperature Non-Uniformity

In some cases, even though the longitudinal heat conduction effects may be negligible (if λ is very low), the exchanger may be subjected to both flow and temperature non-uniformity. Hence, there is a need to analyse the exchanger performance with the combined effects of temperature non-uniformity and flow non-uniformity. Earlier,

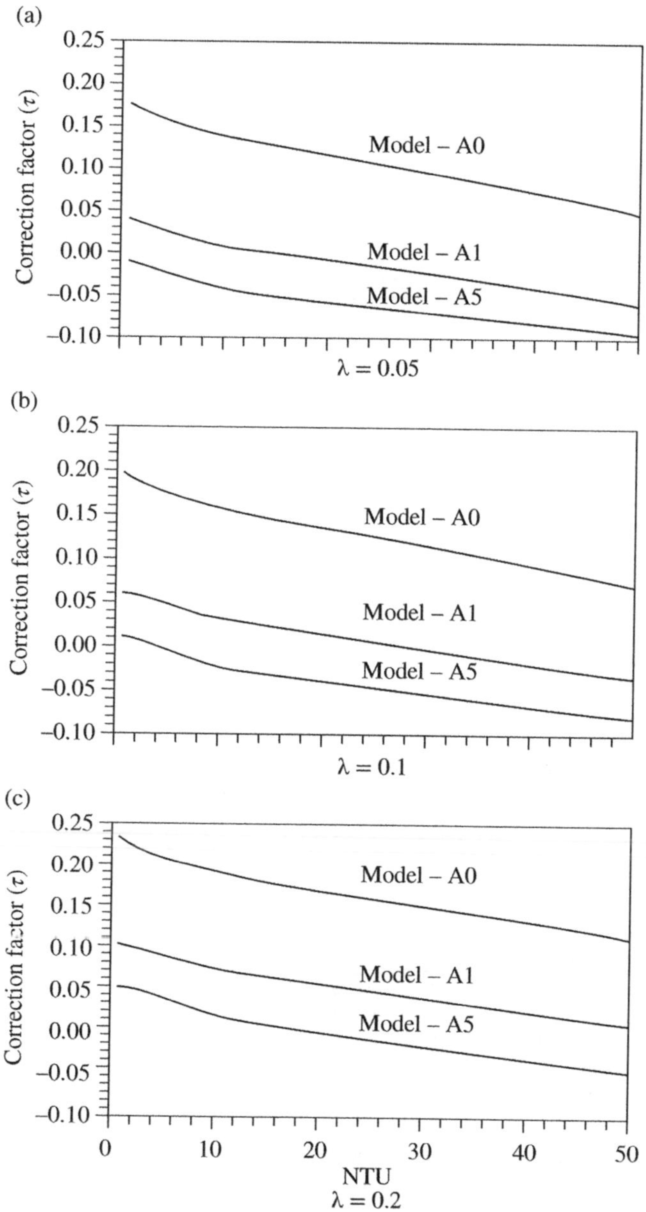

Figure 4.52 Combined effects of LHC, FN and TN tube-fin heat exchanger – C_{max} fluid side (C_{min}/C_{max} = 0.6).

the individual effects of flow non-uniformity and temperature non-uniformity in compact heat exchangers were studied. In this section, the performance evaluation with the combined effects of flow non-uniformity and temperature non-uniformity is presented by neglecting the effects of longitudinal wall heat conduction for cross flow plate-fin and tube-fin heat exchangers. Another exchanger thermal performance correction factor

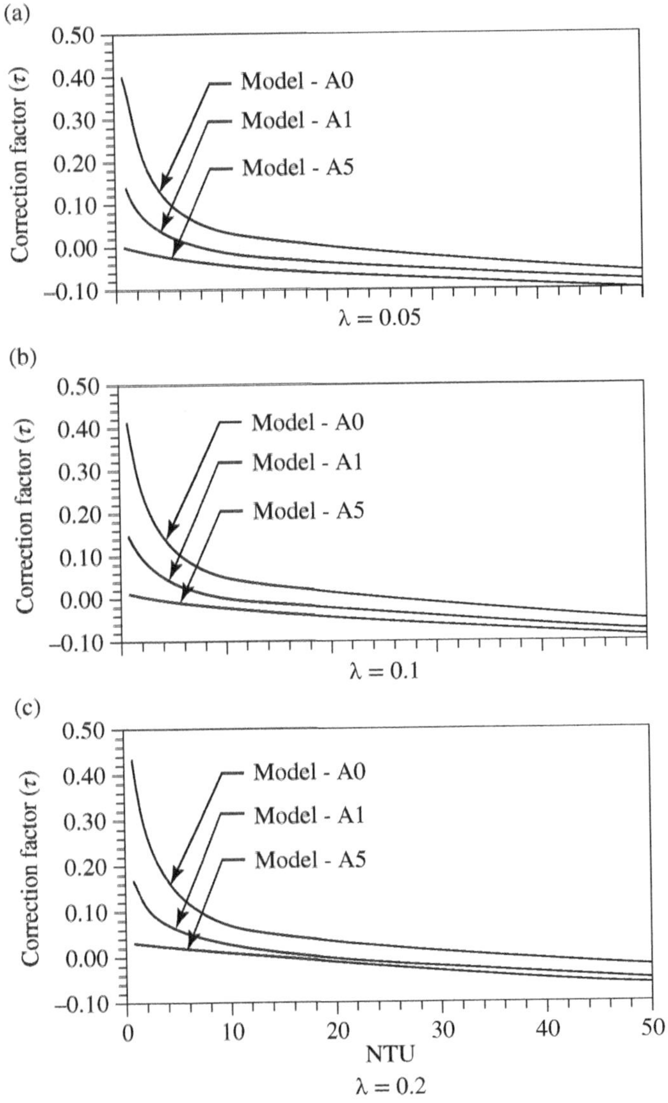

Figure 4.53 Combined effects of LHC, FN and TN tube-fin heat exchanger – both fluid side ($C_{min}/C_{max} = 0.6$).

($\tau_{F,T}$) is introduced to illustrate the influence of combined effects of flow non-uniformity and temperature non-uniformity for cross flow plate-fin and tube-fin heat exchangers. It is defined as

$$\tau_{F,T} = \frac{\varepsilon_0 - \varepsilon_{F,T}}{\varepsilon_0} \qquad (4.104)$$

The present section is devoted to the study of combined effects of flow non-uniformity and temperature non-uniformity in the following heat exchangers.

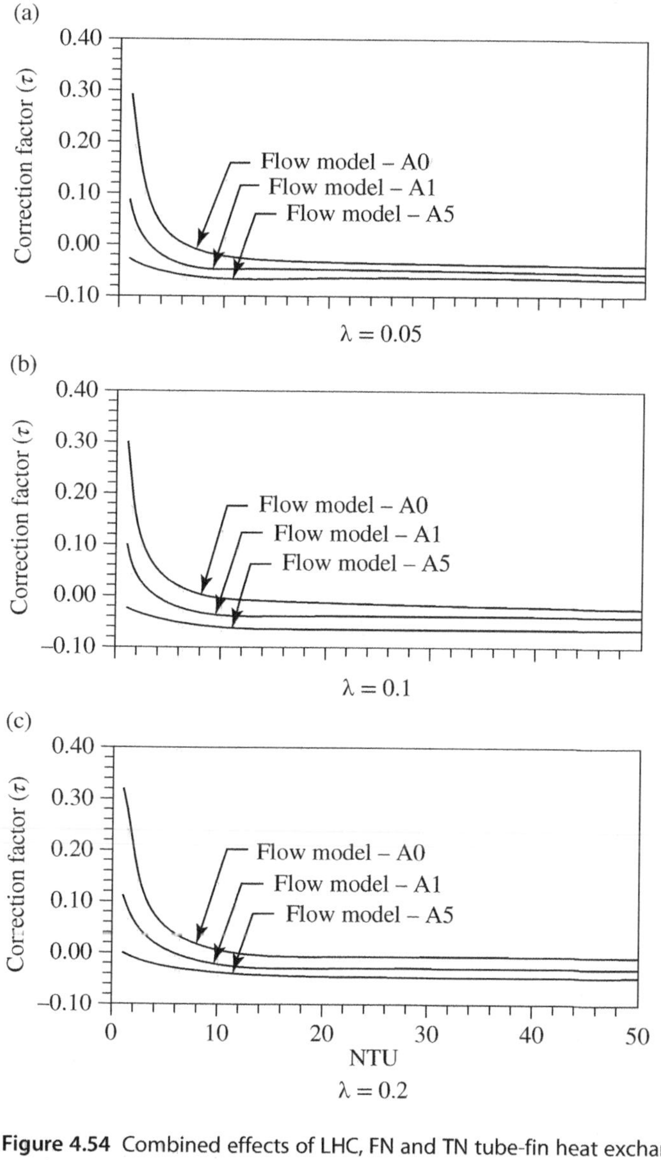

Figure 4.54 Combined effects of LHC, FN and TN tube-fin heat exchanger – C_{min} fluid side (C_{min}/C_{max} = 0.2).

4.8.5.1 Cross Flow Plate-Fin Heat Exchanger

The flow/temperature non-uniformity models considered earlier for individual effects of flow non-uniformity and temperature non-uniformity are solved again for the combined effects of flow non-uniformity and temperature non-uniformity in cross flow plate-fin heat exchanger. In this section, the performance evaluation with the effects of flow and temperature non-uniformity in cross flow plate-fin heat exchanger is presented, for balanced flows, $C^* = 1$, as well as for unbalanced flows, $C^* \neq 1$. The exchanger thermal

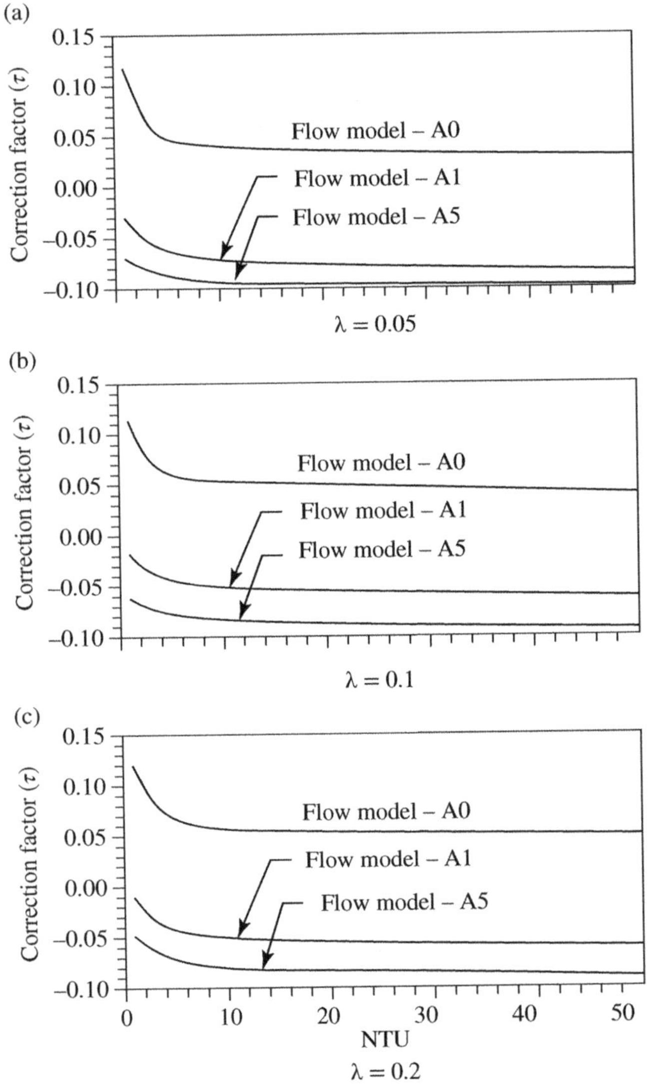

Figure 4.55 Combined effects of LHC, FN and TN tube-fin heat exchanger – C_{max} fluid side (C_{min}/C_{max} = 0.2).

performance deteriorations are plotted as a function of *NTU* ($NTU_{overall}$) for five magnitudes of C^* = 1.0, 0.8, 0.6, 0.4, 0.2 for the following cases:

1) both flow and temperature non-uniformity on C_{min} fluid side
2) both flow and temperature non-uniformity on C_{max} fluid side
3) both flow and temperature non-uniformity on both C_{min} and C_{max} fluid side
4) flow non-uniformity on C_{min} fluid side and temperature non-uniformity on C_{max} fluid side

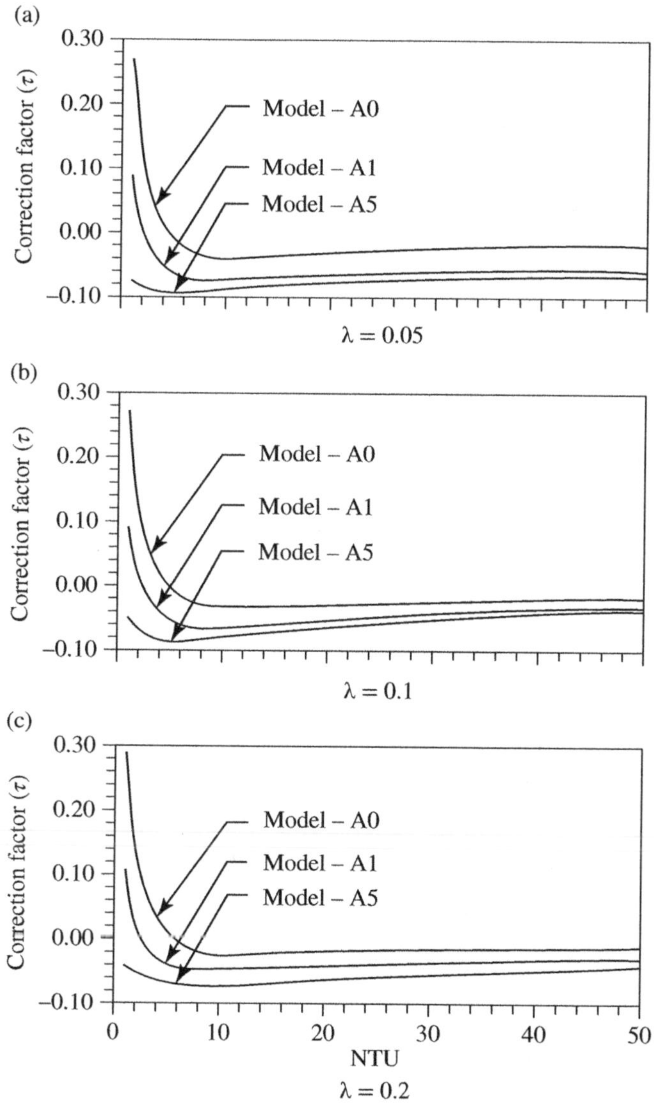

Figure 4.56 Combined effects of LHC, FN and TN tube-fin heat exchanger – both fluid side (C_{min}/C_{max} = 0.2).

5) flow non-uniformity on C_{max} fluid side and temperature non-uniformity on C_{min} fluid side

In the case of the individual effect of flow non-uniformity, some significant deterioration in performance is observed, as discussed earlier. In the case of individual effect of temperature non-uniformity, some significant deteriorations in performance in the regions of smaller *NTU*s and some augmentation in performance in the regions of higher *NTU*s are observed. However, in the case of combined effects of flow non-uniformity and

Table 4.11 Effects of longitudinal wall heat conduction, flow and temperature non-uniformity cross flow tube-fin heat exchanger.

Models	C^*	Performance variations (%)															
		NTU = 10.0								NTU = 50							
		LHC					LHC + FN + TN			LHC					LHC + FN + TN		
		$\lambda = 0.05$	$\lambda = 0.1$	$\lambda = 0.2$	FN	TN	$\lambda = 0.05$	$\lambda = 0.1$	$\lambda = 0.2$	$\lambda = 0.05$	$\lambda = 0.1$	$\lambda = 0.2$	FN	TN	$\lambda = 0.05$	$\lambda = 0.1$	$\lambda = 0.2$
A1	1.0	3.75	7.90	17.0	5.95	−2.0	6.00	9.90	14.0	4.00	17.5	5.00	4.95	−10.	0.0	0.50	8.00
	0.6	1.85	3.50	7.50	2.25	−3.5	4.00	4.50	8.00	1.50	5.00	−4.0	1.80	−8.5	−4.0	−3.00	0.50
	0.2	0.60	0.85	1.00	1.20	−7.0	−5.0	−4.5	−4.0	0.00	0.00	−2.0	0.00	−3.0	−2.0	−1.75	−1.50
A5	1.0	3.75	7.90	17.0	0.15	−2.2	1.00	1.50	2.50	4.00	17.5	4.50	0.05	−11.	−2.0	−1.00	5.00
	0.6	1.85	3.50	7.50	0.10	−5.0	−2.5	0.00	1.00	1.50	5.00	−5.0	0.00	−9.5	−8.0	−7.00	−5.00
	0.2	0.60	0.85	1.00	0.05	−7.5	−3.5	−3.0	−2.5	0.00	0.00	−2.5	0.00	−3.5	−2.2	−2.00	−1.75

Conditions: $C^* = 1.0$, $(\eta hA)^* = 1.0$, and flow/temperature non-uniformity on the C_{min} fluid side

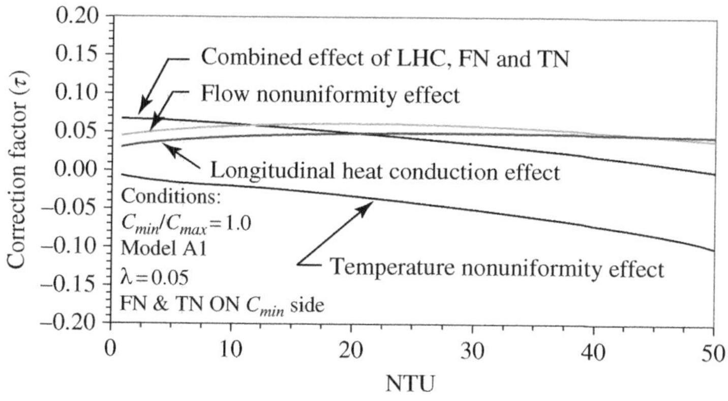

Figure 4.57 Comparison of individual effects with combined effects of LHC, FN and TN for tube-fin heat exchanger.

temperature non-uniformity, the performance deteriorations increase further, as shown in Figures 4.58–4.63. It has been observed that the correction factor ($\tau_{F,T}$) decreases with increase of *NTU* for all cases of balanced and unbalanced flows of heat exchanger, when flow and temperature non-uniformity are considered on C_{min} fluid side or on both fluid sides. However, the correction factor ($\tau_{F,T}$) increases with increase of *NTU*, when the flow/temperature non-uniformity is considered on the C_{max} fluid side. Figure 4.64 presents relative comparison of FN and TN results for cross flow plate-fin heat exchanger.

The various combinations of flow and temperature non-uniformities have been considered for relative comparison, as shown in Figure 4.64 for the unbalanced flow of $C^* = 0.6$. It has been observed that there is some significant difference in performance variations. The performance deteriorations are higher when the flow non-uniformity on C_{max} fluid side and temperature non-uniformity on C_{min} fluid side are considered. However, it has been observed that there is some augmentation in performance, when the flow non-uniformity on C_{min} fluid side and temperature non-uniformity on C_{max} fluid side are considered, as shown in Figure 4.64.

4.8.5.2 Cross Flow Tube-Fin Heat Exchanger

The flow/temperature non-uniformity models considered earlier for individual effects of flow non-uniformity and temperature non-uniformity are solved again for the combined effects of flow non-uniformity and temperature non-uniformity in cross flow tube-fin heat exchanger. In this section, the performance evaluation with the effects of flow and temperature non-uniformity in cross flow tube-fin heat exchanger is presented, for balanced flows, $C^* = 1$, as well as for unbalanced flows, $C^* \neq 1$. The exchanger thermal performance deteriorations are plotted as a function of *NTU* ($NTU_{overall}$) for three magnitudes of $C^* = 1.0, 0.6, 0.2$ for the following cases:

1) both flow and temperature non-uniformity on C_{min} fluid side
2) both flow and temperature non-uniformity on C_{max} fluid side
3) both flow and temperature non-uniformity on both C_{min} and C_{max} fluid sides

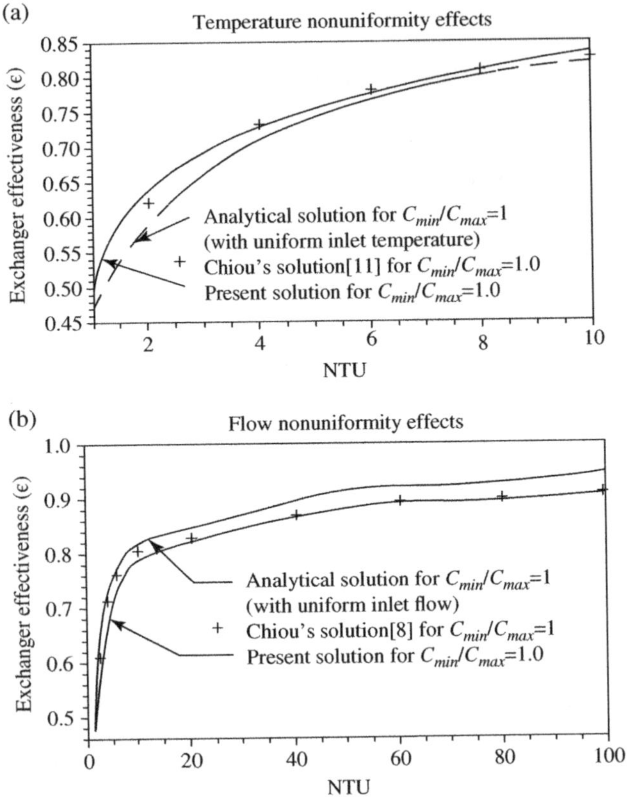

Figure 4.58 Comparison results of temperature non-uniformity effects and flow non-uniformity effects.

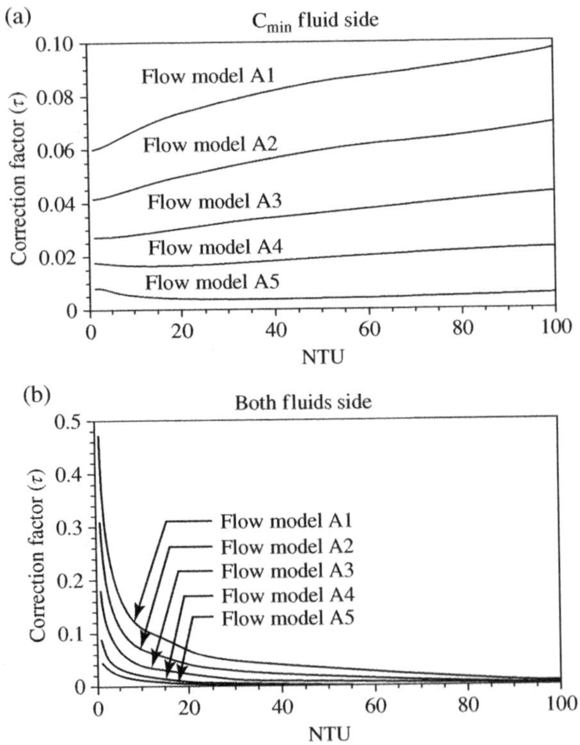

Figure 4.59 Flow and Temperature non-uniformity effects - Cross flow plate-fin heat exchanger $C_{min}/C_{max} = 1$.

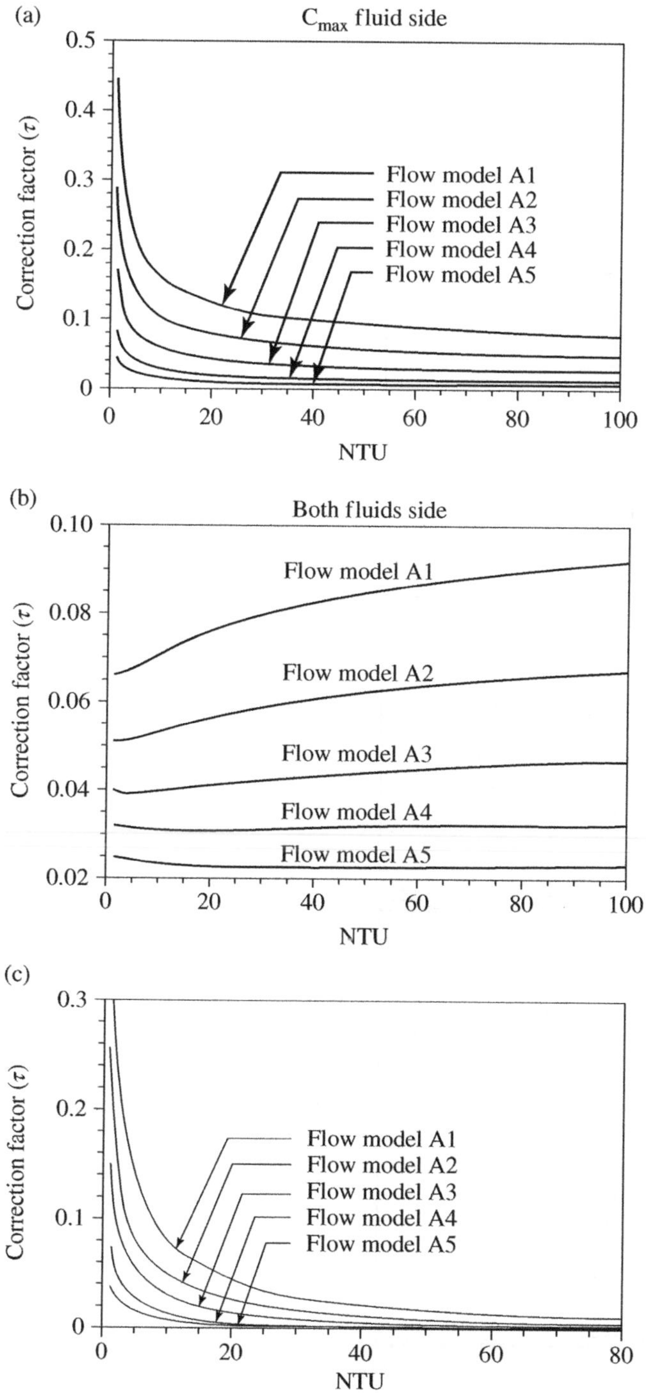

Figure 4.60 Flow and temperature non-uniformity effects – cross flow plate-fin heat exchanger $C_{min}/C_{max} = 0.8$.

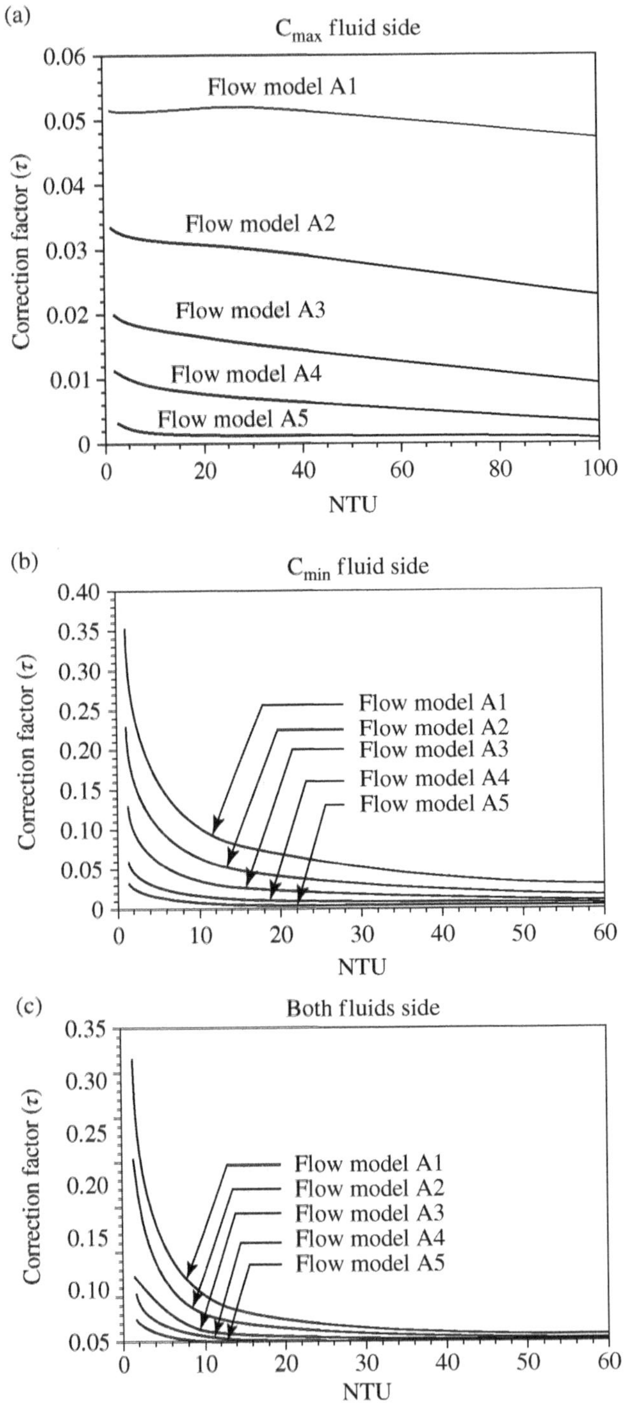

Figure 4.61 Flow and temperature non-uniformity effects – cross flow plate-fin heat exchanger $C_{min}/C_{max} = 0.6$.

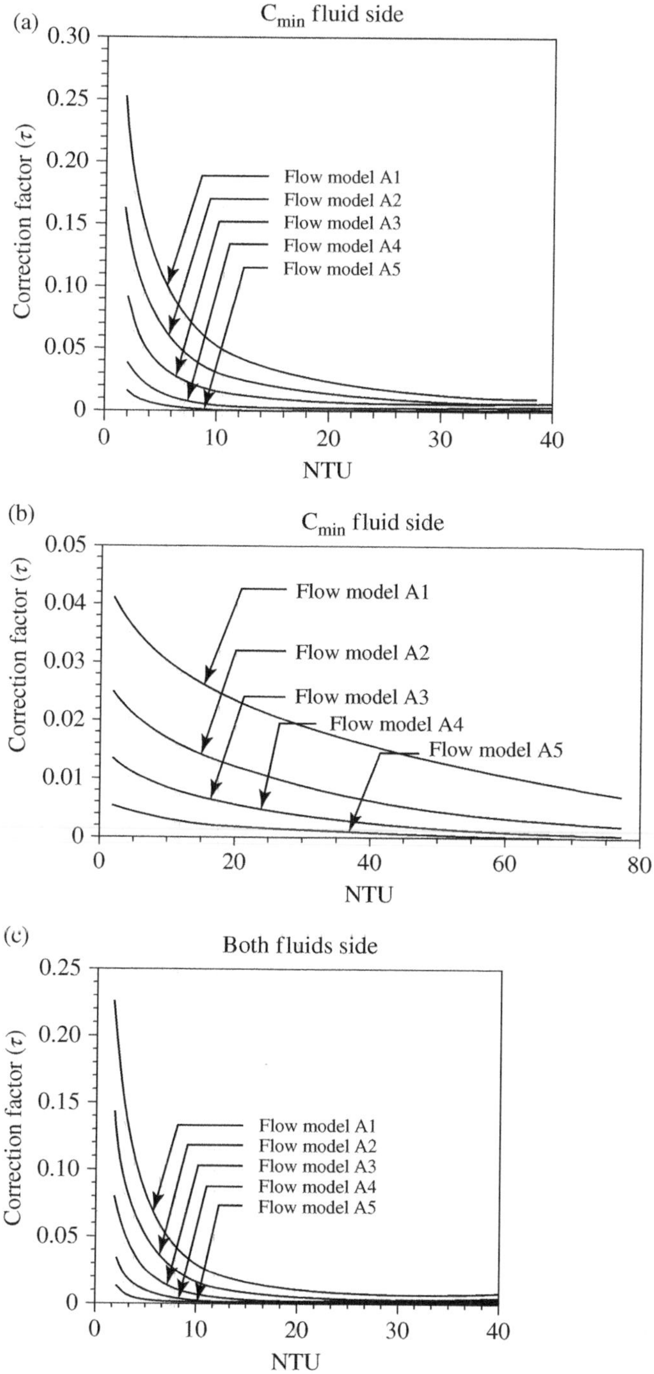

Figure 4.62 Flow and temperature non-uniformity effects – cross flow plate-fin heat exchanger $C_{min}/C_{max} = 0.4$.

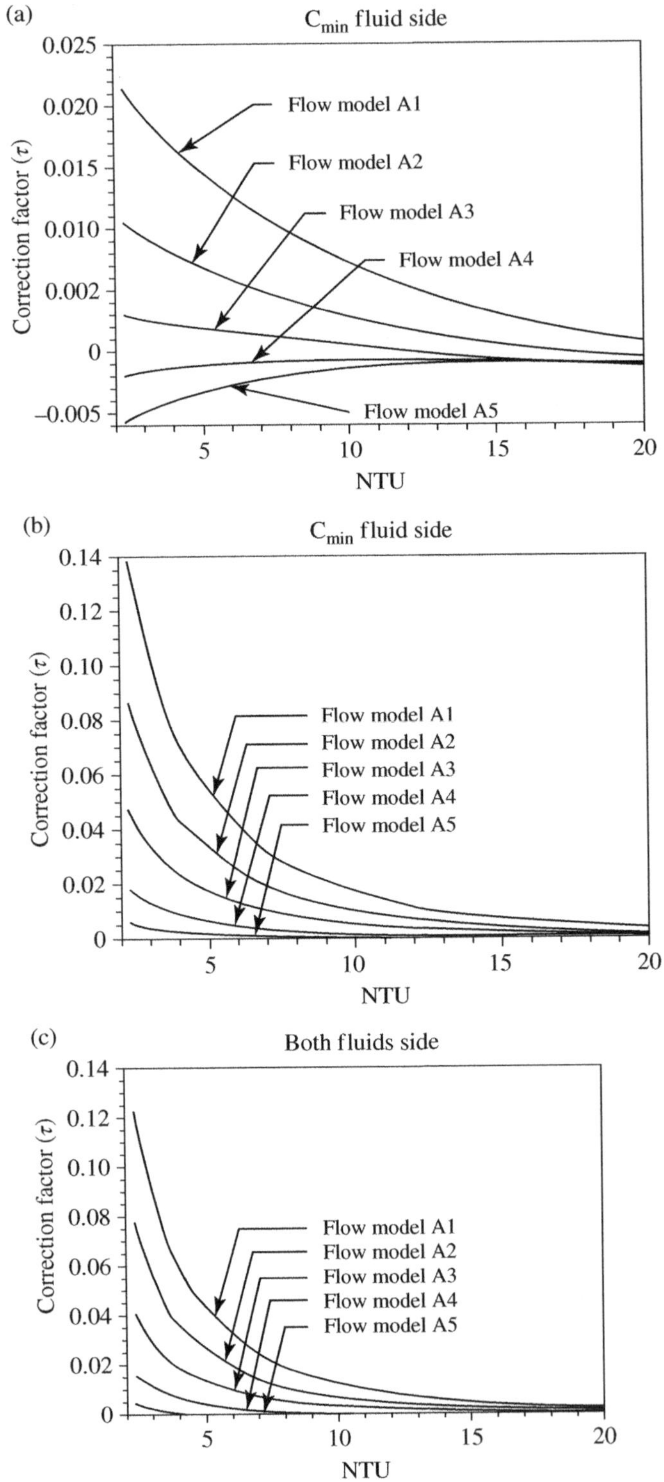

Figure 4.63 Flow and temperature non-uniformity effects – cross flow plate-fin heat exchanger $C_{min}/C_{max} = 0.2$.

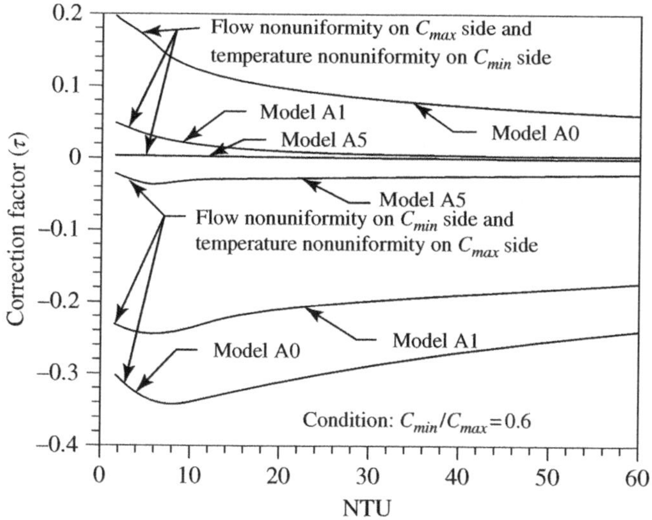

Figure 4.64 Relative comparison of FN and TN results – cross flow plate-fin heat exchanger.

In the case of individual effect of flow non-uniformity, it has been observed that there is some significant deterioration in performance, as discussed earlier. In the case of individual effect of temperature non-uniformity, some significant deterioration in performance in the regions of smaller *NTUs* and some augmentation in performance in the regions of higher *NTUs* are observed. However, in the case of combined effects of flow non-uniformity and temperature non-uniformity, the performance deteriorations increase further, as shown in Figures 4.65–4.67. Also, it has been observed that the correction factor ($\tau_{F,T}$) decreases with increase of *NTU* for all cases of $C^* = 1.0$ and $C^* = 0.6$. However, the correction factor ($\tau_{F,T}$) decreases with increase of *NTU* until *NTU* reaches 8 and then increases with increase of *NTU*. Also, it has been observed that the performance deteriorations are higher for balanced flows as compared to that of unbalanced flows.

4.9 FEM Analysis of Micro Compact Heat Exchangers

In recent years, with the rapidly emerging of micro electromechanical systems (MEMS), many micro-fabrication technologies have been developed and are being introduced in the field of heat transfer engineering to realize micro-channel devices. Compared with conventional heat exchangers, the main advantage of micro compact heat exchangers (MHE) is their extremely high heat transfer area per unit volume. However, the well-known benefits of small channels with respect to heat transfer must be weighed against the cost of the steep pressure gradient associated with flow through micro-channels. Thus both the heat transfer and the flow resistance should be considered in the design of MHEs. Generally micro heat exchanger can be applied in many important fields: micro-electronics, aviation and aerospace, medical treatment, biological engineering,

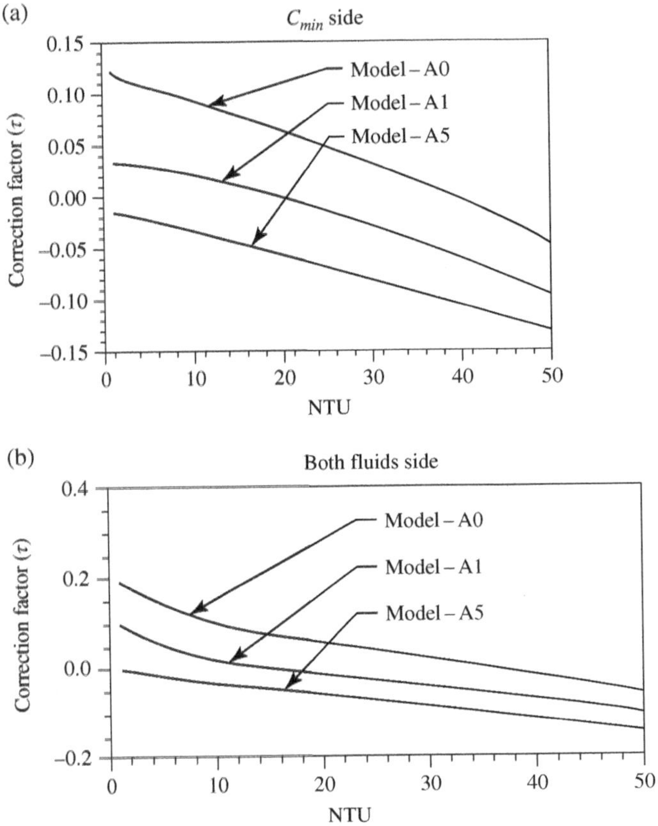

Figure 4.65 Combined effects of flow and temperature non-uniformity – tube-fin heat exchanger $C_{min}/C_{max} = 1.0$.

materials sciences, cooling of high temperature superconductors, thermal control of film deposition and cooling of powerful laser mirrors. Compared with conventional heat-exchangers, the main advantage of micro heat exchangers is their extremely high heat transfer area per unit volume.

A basic schematic of the cross flow MHE is shown in Figure 4.68. This MHE is designed based upon the performance criteria of a car radiator, where its main function is for the dissipation of heat to air to prevent the engine from overheating. Figure 4.69 shows dimensions that specify the internal geometry of the heat exchanger while details for other parameters associated with the basic model of the MHE are given in Table 4.12. The general structure of a micro-channel is shown in Figure 4.68.

The first step for the numerical solution of any differential equation is to discretize the domain over which the solution is sought. The MHE is divided into a number of elements along the z direction, i.e. length of air channel. The actual unit cell for analysis is very complex. In order to reduce modelling and computational effort, a simplified element has been made. Figure 4.70 shows the simplified model along with the nodes and the heat transfer mode. The walls are discretized by four noded bilinear elements. For each

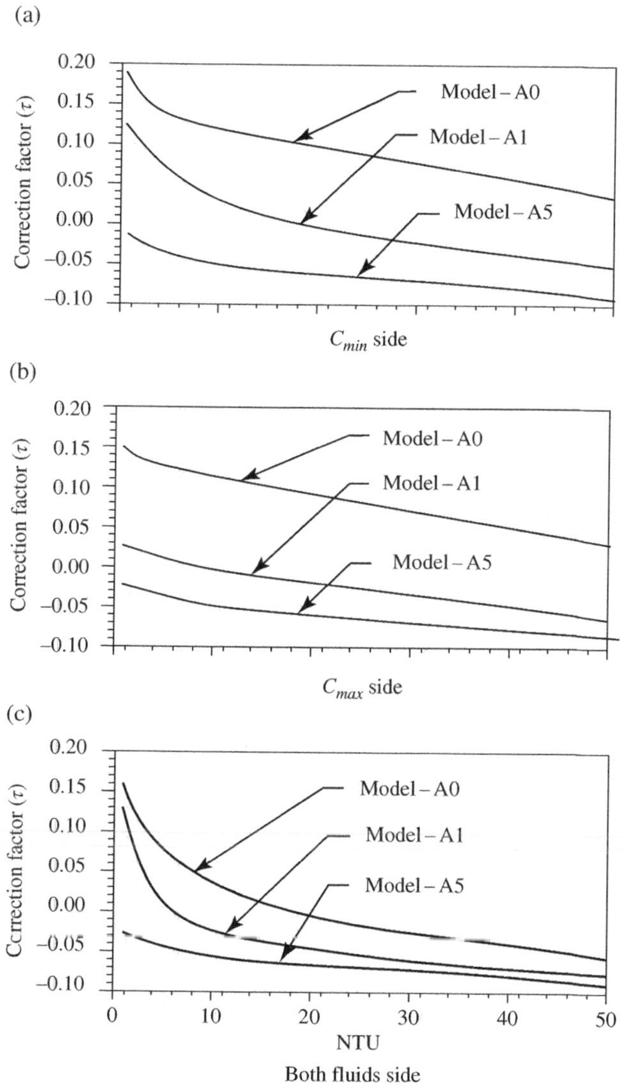

Figure 4.66 Combined effects of flow and temperature non-uniformity – tube-fin heat exchanger $C_{min}/C_{max} = 0.6$.

wall, heat transfer by conduction is only considered in two directions: x and z axis. However, heat transfer by conduction in the direction of that y axis will also be considered for the two vertical walls at both the left- and the right-hand sides. The walls made up of nodes 1, 2, 6, 5 and 4, 3, 7, 8 are treated as fins for the transfer of heat. Air and water streams are discretized by two noded linear elements.

Both the walls made up of nodes 9–12 and 13–16 receive heat from the hot water flowing in the water channels by means of convection. The heat will then be transferred to both the walls made up of nodes 1–4 and 5–8 by conduction in the y direction. These two new walls will dissipate heat to the air streams through convection and the fin effect.

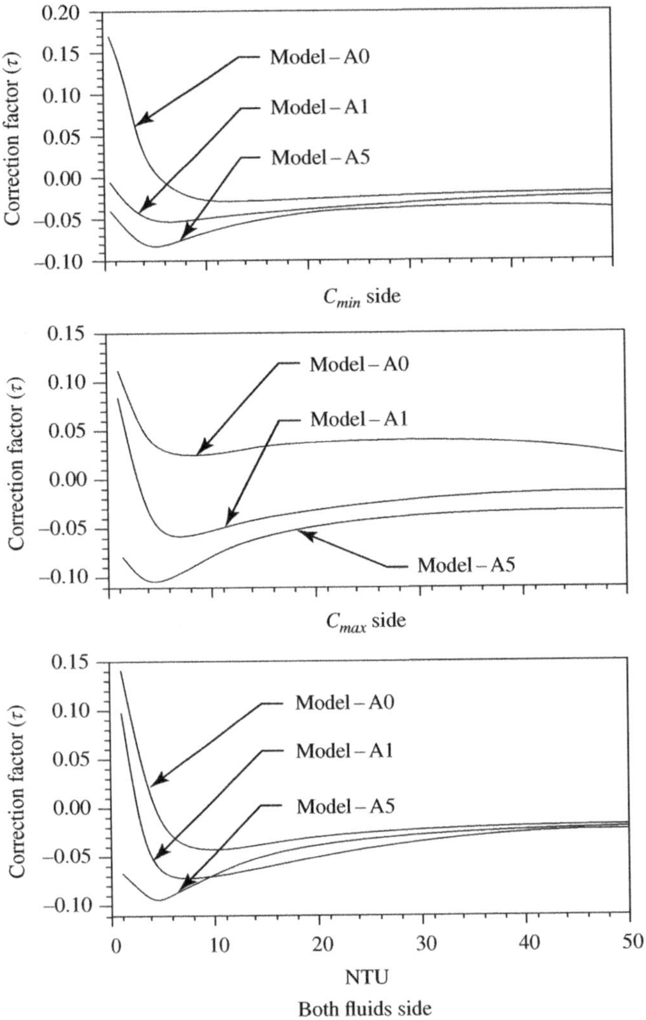

Figure 4.67 Combined effects of flow and temperature non-uniformity – tube-fin heat exchanger $C_{min}/C_{max} = 0.2$.

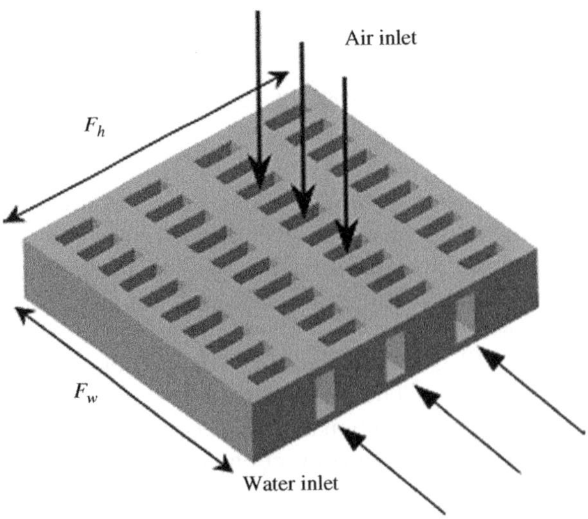

Figure 4.68 General structure of a micro compact heat exchanger.

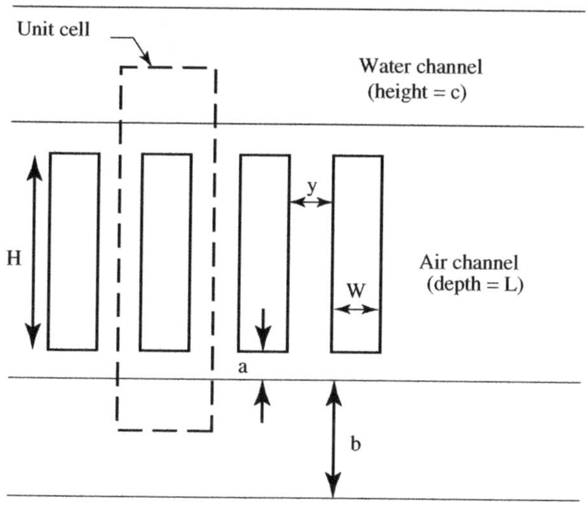

Figure 4.69 Internal geometry for micro-compact heat exchanger.

Table 4.12 Parameters associated with the basic model of micro compact heat exchanger.

Parameters	Value
K (W/mK)	0.20
w (µm)	200
H (µm)	750
Y (µm)	200
L (µm)	1443
A (µm)	125
B (µm)	500
c (µm)	1200
V_a (m/s)	4–12

4.9.1 Governing Equations and Finite Element Formulation

Considering the control volume in the left wall of Figure 4.70 and the energy balance including the conduction and convection heat transfer yields the governing equation as per (Lee et al. [16])

$$Ka_1\frac{\partial^2 T}{\partial x^2}+Ka_1\frac{\partial^2 T}{\partial z^2}-h_a\left(\frac{w}{w_c}\right)\left[T_{1,2,3,4}-T_{17,18}\right]-\frac{K}{a}\left[T_{1,2,3,4}-T_{9,10,11,12}\right]-\eta_f\left(\frac{h_a}{w_c}\right)\left(\frac{H}{2}\right)\left[T_{1,2}-T_{17,18}\right]-\eta_f\left(\frac{h_a}{w_c}\right)\left(\frac{H}{2}\right)\left[T_{3,4}-T_{17,18}\right]=0 \quad (4.105)$$

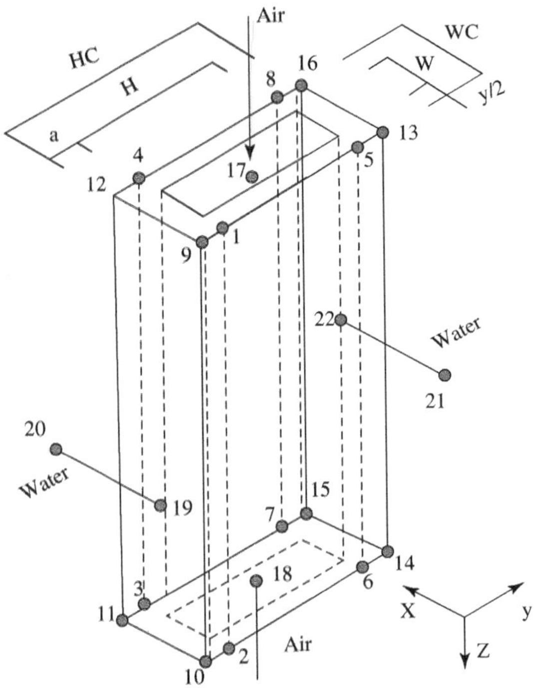

Figure 4.70 Simplified unit cell with nodes for micro compact heat exchanger.

For the outer left wall (nodes 9–12), the heat transferred from the water and the heat loss to the inner left wall can be given by:

$$Ka_1\frac{\partial^2 T}{\partial x^2}+Ka_1\frac{\partial^2 T}{\partial z^2}-h_w\left[T_{9,10,11,12}-T_{19,20}\right]+\frac{K}{a}\left[T_{1,2,3,4}-T_{9,10,11,12}\right]=0 \qquad (4.106)$$

Similarly, the governing equations for the right-hand side inner (nodes 5, 6, 7, 8) and outer (nodes 13–16) wall can be written as:

$$Ka_1\frac{\partial^2 T}{\partial x^2}+Ka_1\frac{\partial^2 T}{\partial z^2}-h_a\left(\frac{w}{w_c}\right)\left[T_{5,6,7,8}-T_{17,18}\right]-\frac{K}{a}\left[T_{5,6,7,8}-T_{13,14,15,16}\right]-\eta_f\left(\frac{h_a}{w_c}\right)\left(\frac{H}{2}\right)\left[T_{5,6}-T_{17,18}\right]-\eta_f\left(\frac{h_a}{w_c}\right)\left(\frac{H}{2}\right)\left[T_{7,8}-T_{17,18}\right]=0 \qquad (4.107)$$

$$Ka_1\frac{\partial^2 T}{\partial x^2}+Ka_1\frac{\partial^2 T}{\partial z^2}-h_w\left[T_{13,14,15,16}-T_{21,22}\right]+\frac{K}{a}\left[T_{5,6,7,8}-T_{13,14,15,16}\right]=0 \qquad (4.108)$$

The heat transferred from the left-hand side water element to the outer left wall yields the governing equation of water and can be expressed as

$$-M_wC_{pw}\left[\frac{T_{19}-T_{20}}{w_c}\right]-h_wL\left[T_{9,10,11,12}-T_{19,20}\right]=0 \qquad (4.109)$$

Similarly for the water element on the right-hand side, the governing equation can be expressed as

$$-M_w C_{pw}\left[\frac{T_{21}-T_{22}}{w_c}\right]-h_w L\left[T_{13,14,15,16}-T_{21,22}\right]=0 \quad (4.110)$$

The heat transferred from the inner wall and fins to the air stream element for both sides can be computed by

$$-M_a C_{pa}\left[\frac{T_{17}-T_{18}}{L}\right]-h_a W\left[T_{1,2,3,4}-T_{17,18}\right]-h_a W\left[T_{1,2,3,4}-T_{17,18}\right]$$
$$-\eta_f h_a\left(\frac{H}{2}\right)\left[T_{1,2}-T_{17,18}\right]-\eta_f h_a\left(\frac{H}{2}\right)\left[T_{3,4}-T_{17,18}\right]-\eta_f h_a\left(\frac{H}{2}\right)\left[T_{5,6}-T_{17,18}\right]$$
$$-\eta_f h_a\left(\frac{H}{2}\right)\left[T_{7,8}-T_{17,18}\right]=0$$

(4.111)

The finite element formulation from equations (4.105) to (4.111) was carried out using the Galerkin weighted residual integrals.

The following section shows equations and calculations used to obtain the unknown values for the finite element formulation.

1) Wall thickness

$$a_1=\frac{a}{2} \quad (4.112)$$

2) Average temperatures are used

$$T_{i,j}=\frac{1}{2}\left(T_i+T_j\right) \quad (4.113)$$

$$T_{p,q,r,s}=\frac{1}{4}\left(T_p+T_q+T_r+T_s\right) \quad (4.114)$$

3) The wall temperature is assumed to vary as

$$T_{wall}=N_i T_i+N_j T_j+N_k T_k+N_m T_m \quad (4.115)$$

where the shape functions N for a general quadrilateral element (the rectangular element being a particular case) are given by the following expressions (Segerlind [1]).

$$N_i=\left(1-\frac{z}{L}\right)\left(1-\frac{x}{w_c}\right) \quad (4.116)$$

$$N_j=\left(\frac{z}{L}\right)\left(1-\frac{x}{w_c}\right) \quad (4.117)$$

$$N_k=\frac{zx}{L\times w_c} \quad (4.118)$$

$$N_m=\left(\frac{x}{w_c}\right)\left(1-\frac{z}{L}\right) \quad (4.119)$$

4) The fluids temperature are assumed to vary as

$$T_{fluid} = N_p T_p + N_q T_q \tag{4.120}$$

where the shape function N can be expressed as (Segerlind [1])

For air | For Water

$$N_p = 1 - \frac{z}{L} \qquad N_p = 1 - \frac{x}{w_c} \tag{4.121}$$

$$N_q = \frac{z}{L} \qquad N_q = \frac{x}{w_c} \tag{4.122}$$

5) Estimation of the total number of air channels N can be obtained by dividing the total frontal area of the heat exchanger by the unit cell area as given by

$$N = \frac{(F_w \times F_h)}{(b + 2a + H)(y + w)} \tag{4.123}$$

Effectiveness of the MHE can be calculated using Equation (4.124). Different fluid flow ratio as given by Equation (4.125) is simulated to study their effects on the effectiveness of the MHE.

$$\begin{aligned} Effectiveness &= \frac{Actual\ heat\ transfered}{Max\ heat\ transfer} \times 100 \\ Effectiveness &= \frac{T_{air,outlet} - T_{air,inlet}}{T_{water,inlet} - T_{air,inlet}} \times 100 \end{aligned} \tag{4.124}$$

$$Capacity\ ratio = \frac{C_{min}}{C_{max}} = \frac{m_{air} \times C_{p,air}}{m_{water} \times C_{p,water}} \tag{4.125}$$

In order to present a dimensionless study, the number of transfer units NTU is introduced and can be computed using

$$NTU = \frac{UA}{C_{min}} = \frac{Q}{\Delta T_{max} \times C_{min}} = \frac{Q}{(T_{water,inlet} - T_{air,inlet}) \times C_{min}} \tag{4.126}$$

The objective of the optimization is to maximize the effectiveness of the micro compact heat exchanger. GA will generate the initial population for parameters L, c, w, H, y, a, b and capacity ratio, and this data is then fed into the trained ANN algorithm as input variables to predict the performance of the MHE, namely the effectiveness. The predicted effectiveness of the MHE will be evaluated by GA and is given a fitness score which is then measured, recorded, ranked and compared with previous iteration results. The top grade chromosomes (highest value of fitness score) will be selected and allowed to reproduce with other individuals in the population, where they will undergo the crossover and mutation processes. The population of the newly recombined and mutated chromosomes becomes the new input parameters for the trained ANN, which subsequently predicts the effectiveness for a new iteration in GA. Figure 4.71 shows predictions using GA to find the maximum effectiveness for a variation in w. The maximum effectiveness obtained is 24.776% for initial design.

Table 4.13 shows the GA optimized design parameters for the MHE where the maximum effectiveness obtained is 59.191% as opposed to the maximum effectiveness of 24.776% obtained from initial design.

The performance of the MHE has been determined using FEM for a limited number of parameters (Lee et al. [16]). The present FEM study results have been compared with neuro-genetic optimization methodology and the results matched very well. GA improves the performance of the MHE by proposing a new set of design parameters as indicated in Lee et al. [16]. The maximum effectiveness obtained through GA is 59.191%, as opposed to the effectiveness of 24.776% obtained from FEM simulation, thus yielding an improvement of almost 35%. The above analysis is similar to various types of heat exchanger analysis presented in Dakshina Murthy et al. [37] Ravi Kumar et al. [38,39], Seetharamu [40], Quadir et al. [41], Jeevan et al. [42] and Pradeep Hegde et al. [43].

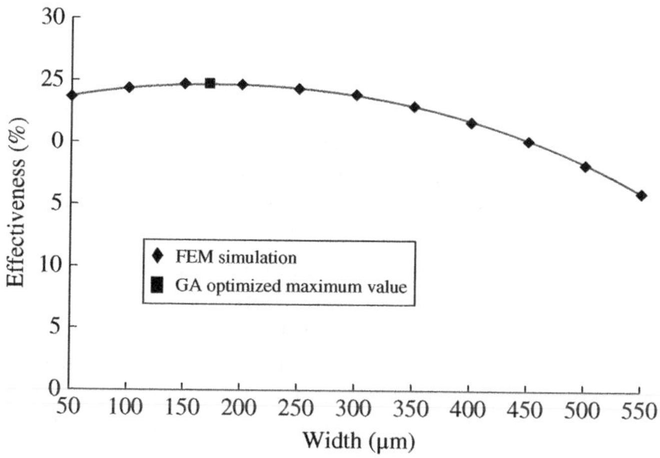

Figure 4.71 Maximum effectiveness obtained using GA for a variation in *w* for initial design.

Table 4.13 Optimized design parameters of MHE obtained using GA.

Design parameters	Value
w (μm)	300
H (μm)	997
y (μm)	410
L (μm)	1992
a (μm)	73
b (μm)	139
c (μm)	1185
capacity ratio	0.22
effectiveness (%)	59.19

4.10 Influence of Heat Conduction from Horizontal Tube in Pool Boiling

4.10.1 General

In pool boiling, the electrically heated tube releases the energy non-uniformly to the liquid, caused by different surface roughness and flowing liquid. The heat transfer coefficient therefore varies with axial and azimuthal position on the tube. Hence a finite element analysis has been carried out on a horizontal 25.4 mm copper tube for evaporation in pool boiling for all possible cases of analysis of conduction heat transfer in the wall (one-dimensional, two-dimensional and three-dimensional). A test tube is made with different surface structures, tested and analysed for heat conduction effects.

The accurate prediction of the thermal performance of an evaporator in the design stage is highly desirable for most applications. Many experimental investigations are available on nucleate boiling heat transfer in the literature pertaining to the influence of the tube diameter, tube material and its roughness, saturation pressure, heat flux and fluid properties. Investigations of the heat conduction effect in the wall of evaporator tubes are very limited. Two-dimensional heat conduction exists in the wall because of the favourable heat transfer conditions on the lower parts of the tube surface causing heat to flow partially in the azimuthal direction through the tube wall and increasing the radial heat flux on the surface. The enhancement below the flank of the tube results from the prolonged contact of the bubbles with the thin superheated liquid layer in the immediate vicinity of the tube wall – either at the nucleation site itself (bottom) or when sliding upwards – due to the component of the gravitational force directed towards the tube surface. There will be a considerable variation in local wall superheats ΔT because of the two-dimensional heat conduction in azimuthal and radial directions of the tube, which are measured and reported by Gorenflo [44], Hübner et al. [45] and Luke et al. [46].

In addition, there will be a temperature gradient in the longitudinal direction of the tube wall if sections with different surface treatment exist on the tube surface. Hence there is a need to develop a model that can calculate the temperature distributions of the evaporator tube in all three directions: radial, azimuthal and axial. The longitudinal heat conduction effects have been well characterized and their effects have been calculated using finite element method for plate-fin and tube-fin heat exchangers (Ranganayakulu et al. [9]). However, there is no information available on the prediction of three-dimensional heat conduction of evaporator tubes in pool boiling. Compared with the many experimental investigations on nucleate boiling heat transfer in the literature, the numerical papers devoted to heat conduction effects are limited. According to Dhir [47], who summarizes the state of the art, four different mechanisms are decisive for the heat transfer. These are: time-dependent heat conduction, forced convection close to growing bubble, evaporation at the liquid-vapour interface and natural convection on the heated wall in the liquid pool area. The contribution of each of these mechanisms to the overall heat transfer depends on the input heat flux, the physical properties of the fluid and the geometry of the heat source. In that analysis, a two-dimensional heat conduction of a heated plate is considered.

However, the temperature distributions in all three directions of the heated wall of an evaporator tube depend on the input heat flux, reduced pressures $p^* = p_s/p_c$ (p_c = critical pressure) and geometry of the surface structure as is the case for a copper tube with 25.4 mm outside diameter under investigation. In a continuous process of experiments have been conducted for different tube diameters (Kaupmann et al. [48]) with different surface finishes, as presented by Luke et al. [49] for different working fluids and similar works by Dieter Gorenflo et al. [50] and Dieter Gorenflo et al. [51]. The various parts of the test copper tube with heaters and thermocouples are soft soldered throughout the tube length in a glove box with reducing atmosphere in order to get a homogeneous connection of all parts, as shown in Figure 4.72. The details of copper tube are shown in Figure 4.72a. This tube is made up of three different surface structures: primary surface (sand blasted), secondary surface and main surface (sand blasted and grooved). There are 12 thermocouples in the secondary section arranged equidistant in azimuthal direction, 24 thermocouples in the main section arranged equidistant in azimuthal direction and 36 thermocouples at the primary section arranged equidistant in azimuthal direction 12 left and 24 right of the centre, as shown in sectional view of Figure 4.72b.

These test results are tabulated in Table 4.14 for two typical reduced pressures of p^* = 0.05, 0.1 for easy reference. In this table, the boiling heat transfer coefficient (α) values are obtained by taking average values of thermocouples for corresponding heat flux (q) input values, which are assumed to be uniform on the whole surface not considering axial and azimuthal heat conduction.

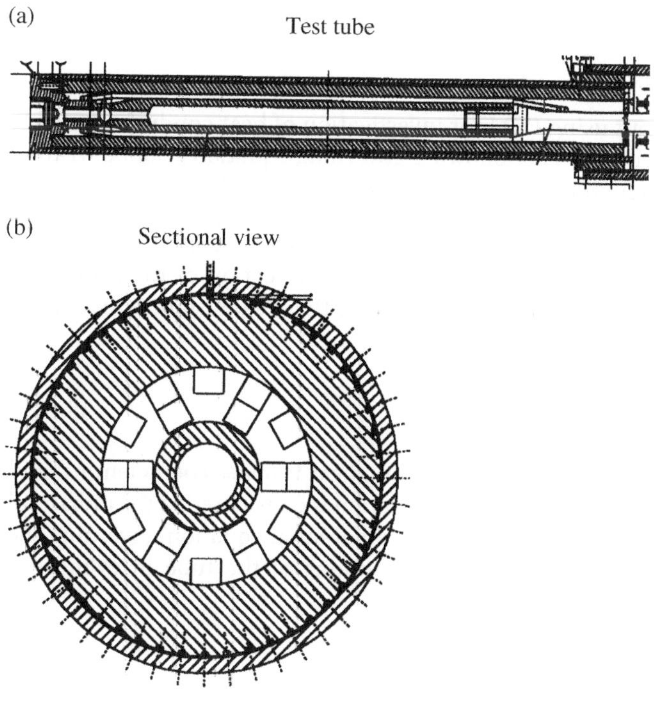

Figure 4.72 Details of copper tube, 25.4 mm OD (Ranganayakulu et al. [57, 58])

Table 4.14 Experimental values of the heat transfer coefficient $\alpha_1, \alpha_2, \alpha_3$ for propane. (Danger et al. [52])

p^*	T_∞ (K)	q (kW/m^2)	Primary section α_1 (W/m^2 K)	Secondary section α_2 (W/m^2 K)	Main section α_3 (W/m^2 K)
0.05	263.30	20.0	2439.5	2852.6	2963.1
		10.0	1403.4	1607.4	1714.4
		5.0	724.8	843.6	912.7
		1.5	341.75	405.3	428.9
		0.5	270.93	242.4	242.9
		0.1	196.97	176.9	176.2
0.10	282.54	70.0	11211.3	12671.0	12493.6
		50.0	8100.9	9143.2	9035.8
		20.0	3672.1	4111.2	4215.9
		10.0	2089.6	2350.7	2488.4
		5.0	1170.1	1315.5	1439.1
		1.5	432.7	523.6	576.3
		0.5	259.6	297.4	309.5
		0.1	199.5	175.2	176.0

4.10.2 Governing Equations

An analysis has been carried out using FEM for investigation of heat conduction effects on a horizontal tube in pool boiling. The following governing equations can be written, based on the type of heat conduction in the horizontal evaporator tube as:

1) In the case of long thin tubes with variable boiling heat transfer coefficient (α), one-dimensional heat conduction is considered in the axial direction of the horizontal evaporator tube. The differential equation governing the steady-state temperature distribution along a horizontal tube is

$$\frac{\partial}{\partial z}\left(k\frac{\partial T}{\partial z}\right) = 0 \tag{4.127}$$

2) In the case of long thick tubes with variable boiling heat transfer coefficient (α) and with axial symmetry, two-dimensional heat conduction is considered in the radial and axial directions (r, z) of the horizontal tube. The differential equation governing the steady-state two-dimensional temperature distributions of the tube in cylindrical coordinates is

$$\frac{1}{r}\frac{\partial}{\partial r}\left(kr\frac{\partial T}{\partial r}\right) + \frac{\partial}{\partial z}\left(k\frac{\partial T}{\partial z}\right) = 0 \tag{4.128}$$

3) In the case of thick tubes with uniform boiling heat transfer coefficient (α) in the axial direction, two-dimensional heat conduction is considered in radial and

azimuthal directions (r, φ) of the horizontal tube because of the favourable heat transfer conditions on the lower part of the tube surface. The differential equation governing the steady state two-dimensional temperature distribution of the tube in x, y coordinates is

$$\frac{\partial}{\partial x}\left(k\frac{\partial T}{\partial x}\right)+\frac{\partial}{\partial y}\left(k\frac{\partial T}{\partial y}\right)=0 \tag{4.129}$$

4) In the case of long thick tubes with variable boiling heat transfer coefficient (α), a three-dimensional heat conduction exists in radial, azimuthal and axial direction (r, ϕ, z) of the horizontal tube. The differential equation governing the steady-state three-dimensional temperature distribution of the tube without heat sources or sinks and with constant thermal conductivity in Cartesian coordinates (x, y, z) is

$$\left(\frac{\partial}{\partial x}\left(k\frac{\partial T}{\partial x}\right)+\frac{\partial}{\partial y}\left(k\frac{\partial T}{\partial y}\right)+\frac{\partial}{\partial z}\left(k\frac{\partial T}{\partial z}\right)\right)=0 \tag{4.130}$$

The above equations incorporate the following boundary and initial conditions.

Since the above differential equations are second-order, at least two boundary conditions need to be specified. The possible boundary conditions are

a)

$$T(x,y,z)=T_0 \qquad \text{on surface} S_1 \tag{4.131}$$

(temperature specified on surface 1)

b)

$$k\left(\frac{\partial T}{\partial x}\right)l_x+k\left(\frac{\partial T}{\partial y}\right)l_y+k\left(\frac{\partial T}{\partial z}\right)l_z-\alpha\left(T-T_\infty\right)=0 \text{ on surface} S_2 \tag{4.132}$$

(convective/boiling boundary condition on surface 2)

c)

$$k\left(\frac{\partial T}{\partial x}\right)l_x+k\left(\frac{\partial T}{\partial y}\right)l_y+k\left(\frac{\partial T}{\partial z}\right)l_z+q=0 \text{ on surface} S_3 \tag{4.133}$$

(input heat flux specified on surface 3)

l_x, l_y, l_z = direction cosines of the outward drawn normal (n) to the boundary.

4.10.3 Finite Element Analysis

The finite element equations for the heat conduction problems can be derived either by using a variational approach or by using a Galerkin approach (Rao [53], Comini [54], Huebner et al. [55]). If a variational approach is used for deriving the system of equations, there is need to specify only the essential boundary conditions and the natural boundary conditions will be implicitly satisfied in the solution procedure. In this analysis a uniform input heat flux has been assumed at the radius of r = 7.05 mm neglecting variations in the heaters, which are placed equidistant in concentric inner circle, as shown in Figure 4.72b. The heat transfer through the ends of the heated tube is neglected. The heat transfer coefficient is set proportional to the heat flux at the surface to the exponent 'n' ($\alpha \approx q^n$), with $n = 0.843$ for the primary section and $n = 0.814$ for the main and secondary sections. The test tube has been modelled in the following cases depending upon the type of the problem.

4.10.3.1 One-Dimensional Case

The schematic diagram of a 25.4 mm diameter copper tube is shown in Figure 4.73a along with details of lengths of primary section, secondary section and main section for one-dimensional heat conduction analysis using Equation 4.127. In this figure only the heated length of the tube has been taken and divided into several finite elements for one-dimensional analysis, as shown in Figure 4.73(b).

Idealize the tube into several finite elements (numbered as 1, 2, 3, ...n), as shown in Figure 4.73b by assuming a linear temperature variation inside any element 'e' as

$$T^{(e)}(x) = a_1 + a_2\,x \tag{4.134}$$

The constants a_1 and a_2 can be expressed in terms of the unknown nodal temperature $T^{(e)}$.

4.10.3.2 Two-Dimensional Case (Axial and Radial)

Here the 2D case of Equation 4.130 is considered, taking radial and axial (r, z) coordinates as the axi-symmetric case. Replace the tube body of revolution by an assembly of ring elements, as shown in Figure 4.74a for a two-dimensional heat conduction analysis. The cross-section (A-A) of the ring is a triangle having nodes i, j, k, as shown in Figure 4.74b.

Idealize the tube into several finite elements, as shown in Figure 4.74c. Assume a linear temperature variation inside any element 'e' as

$$T^{(e)}(r,z) = a_1 + a_2\,r + a_3 z \tag{4.135}$$

with $a_i = a_i(\{T_m, r_m, z_m\})$ for $m = i, j, k$.

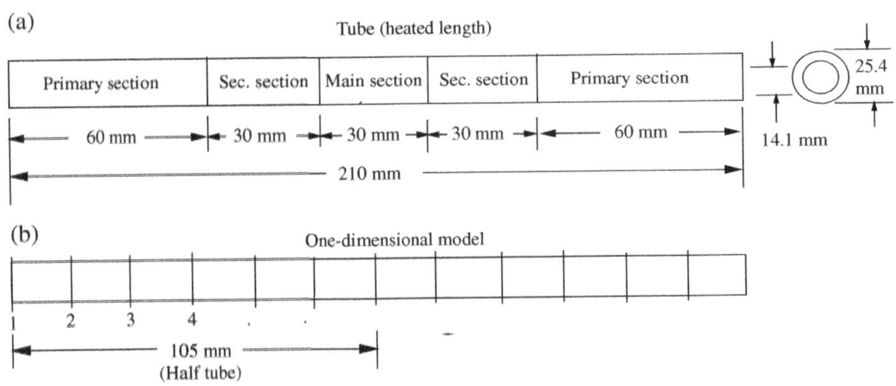

Figure 4.73 Details of section lengths and one-dimensional model of the tube.

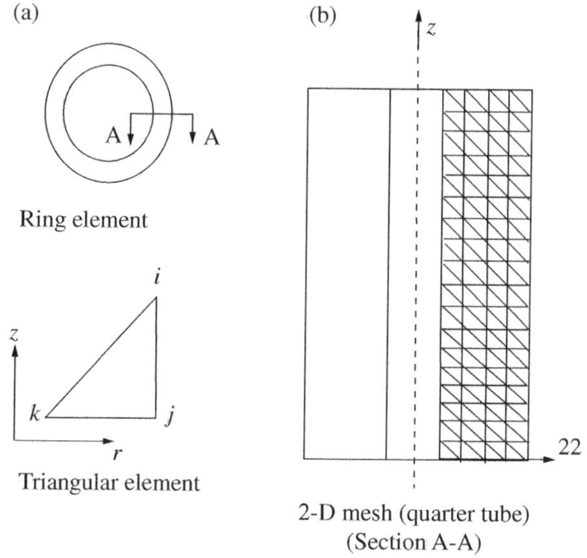

Figure 4.74 Two-dimensional mesh (axial and radial direction).

4.10.3.3 Two-Dimensional Case (Azimuthal and Radial)

In the case of two-dimensional heat conduction in azimuthal and radial directions, Equation 4.131 can be used for steady state analysis. Here the radial and azimuthal (r, ϕ) directions can be formulated in Cartesian coordinates (x, y) as well.

Idealize the solution region with triangular elements, as shown in Figure 4.75a, by assuming a linear temperature variation inside element '*e*' similar to 2D (axial and radial) case as

$$T^{(e)}(x,y) = a_1 + a_2 x + a_3 y \tag{4.136}$$

with $a_i = a_i(\{T_m, x_m, y_m\})$ for $m = i, j, k$.

Here the radial and azimuthal (r, φ) directions using triangular elements, as shown in Figure 4.75b.

4.10.3.4 Three-Dimensional Case

Here the 3D heat conduction case of Equation 4.130 is used taking radial, azimuthal and axial (r, ϕ, z) in Cartesian coordinates (x, y, z) as well. Replace the tube body of revolution by an assembly of tetrahedron 3D elements, as shown in Figure 4.76a for three-dimensional heat conduction analysis. The tube is divided into a number of trapezoidal boxes for clarity of mesh, as shown in Figure 4.76b. Each trapezoidal box consists of five tetrahedral elements each, as shown in Figure 4.76c. Assume a linear temperature variation inside any element '*e*' so that the temperature $\vec{T}$ can beexpressed as

$$T^{(e)} = [N]\ \vec{T} = [N_i, N_j, N_k, N_l] \cdot [T_i, T_j, T_k, T_l]^T \tag{4.137}$$

where $N_i = L_1, N_j = L_2, N_k = L_3, N_l = L_4$ are the natural coordinates L_1, L_2, L_3, L_4, related to the global Cartesian coordinates of the nodes i, j, k, l.

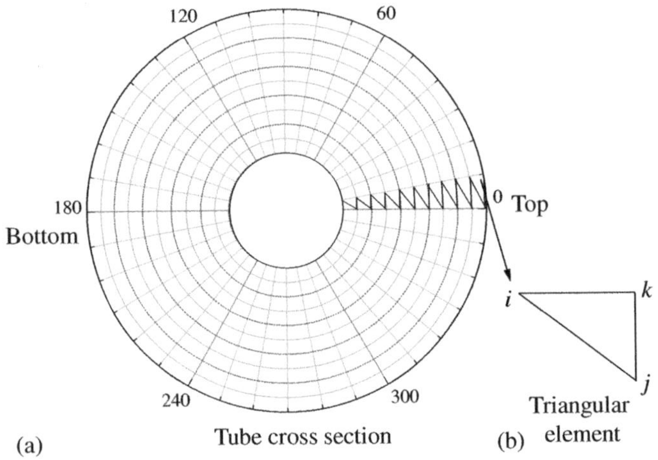

Figure 4.75 Two-dimensional model (radial and azimuthal directions).

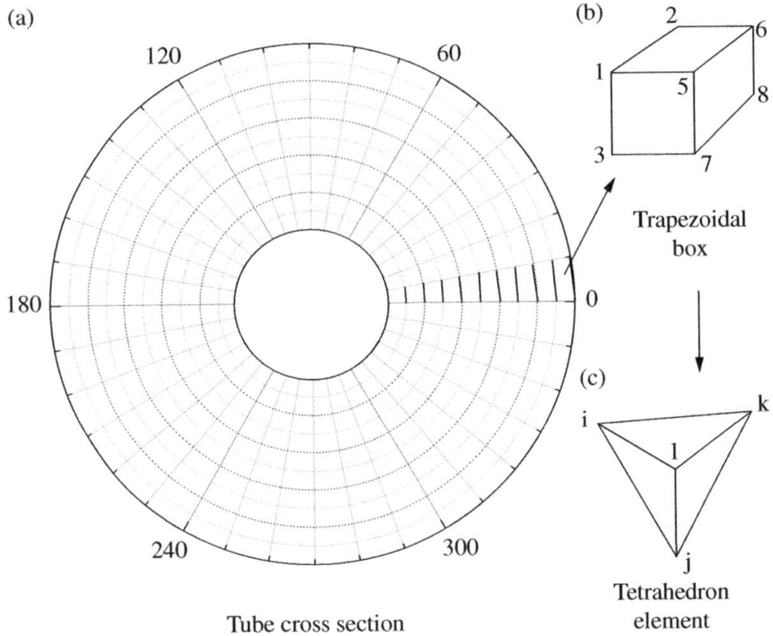

Figure 4.76 Three-dimensional model.

4.10.4 Results

After studying the measured data of the test tube, it has been observed that there is significant variation of superheat (ΔT) values in the axial direction in addition to the azimuthal variation. As observed earlier by Gorenflo [44] and Hübner et al. [45], the heat transfer rate is higher near the bottom part of the tube than at the top part of the tube. The heat transfer variation in azimuthal and axial directions is significant in moderate heat fluxes, especially at reduced pressures $p^* = 0.05$ and 0.1. Hence these two cases have been analysed for heat conduction effects as sample cases. Using the test data of these two cases (Table 4.14), a diagram is plotted taking the heat transfer coefficients (α) of the main section against the heat flux (q) in Figure 4.77a for easy reference. However, it can be seen that the secondary section α is about 3–8% less than the main section, and the primary section α is about 10–20% less than the main section for high q cases, as indicated in Figure 4.77b.

The accuracy of the FEM solution depends on the number of the elements used. Actually, this number is determined by a compromise between the accuracy desired and the time required by the computer. Use of a greater number of subdivisions can produce results of a higher accuracy. Here it is found that a minimum of 70,000 elements (16,665 nodes) are required in 3D analysis (for quarter tube) for the desired accuracy. The approximate CPU time for one case is 13 seconds for 1D, 4 minutes for 2D and as high as 22 hours for 3D analysis with reasonable accuracy using a Sun Sparc Ultra 5 (270 MHz) computer. Moreover in FEM, the accuracy not only depends on the number but also on the type of elements used, the boundary conditions and the type of problem such as 1D, 2D or 3D. In this analysis the measured superheat (ΔT) values have been used as known boundary conditions, as shown in Figure 4.78. In this figure, the ΔT variation versus tube length has been shown along with experimental values (given at $z = 0$ mm, 75 mm and 105 mm) as boundary conditions for $p^* = 0.1$ and 0.05 as validation of finite element analysis. In this figure, the temperature distribution is as expected and the accuracy found to be good.

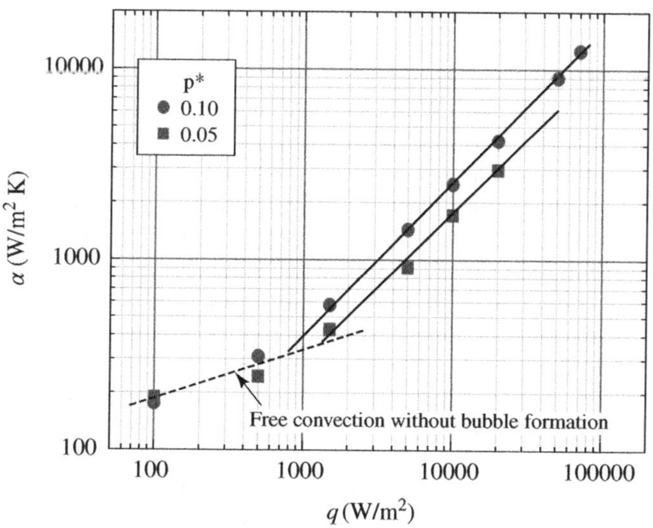

Figure 4.77 (a) Heat transfer coefficient (α) of main section versus input heat flux (q) using experimental data.

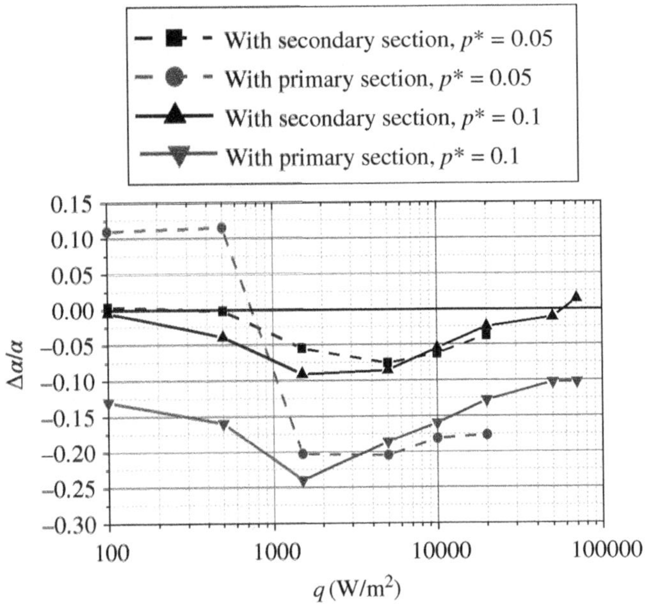

Figure 4.77 (b) Variation of α in primary and secondary sections with respect to the main section.

4.10.4.1 One-Dimensional Heat Conduction Case

The results of the 1D case are plotted in Figures 4.79 and 4.80 for reduced pressures, p^* = 0.1 and 0.05. In this analysis, the averaged experimental α values have been given as initial secondary boundary conditions. In these cases, the experimental thermocouple data in the azimuthal direction has been averaged and used as convective/boiling boundary conditions since it is a one-dimensional heat conduction case. The radial q has been calculated taking the axial q into account in the axial direction, then the new α has been calculated for every element and iterated until the variation in α minimizes. In this analysis, only half of the tube has been taken since it is symmetric in the axial direction. It has been observed that because of axial heat flow, the radial heat flow is not uniform along tube length, as shown in Figure 4.80 as a sample case for $p^* = 0.1$ at $q = 10\ \mathrm{kW/m^2}$. It is observed that some amount of heat flow takes place in axial direction towards the secondary and main structures depending upon the temperature gradient in the tube and hence the radial heat flow reduces in the primary structure and enhances in the secondary and main structures when compared with the average input q as indicated in Figure 4.80. The α-variation versus the tube length has been plotted in Figure 4.80 for reduced pressures of $p^* = 0.1$ and 0.05. A significant variation of α in axial direction has been observed and this variation is symmetric since the surface structures are symmetric to the middle of the tube. It is observed that α is lower in the primary structure and higher in the secondary structure and the main structure because of axial heat flow towards main and secondary structures. This is due to favourable conditions in the secondary structure and the main structure. This variation is as expected, because the axial heat is added to the radial part in the secondary and main structure regions where favourable surface conditions exist. Hence, the experimental α values have to be corrected by

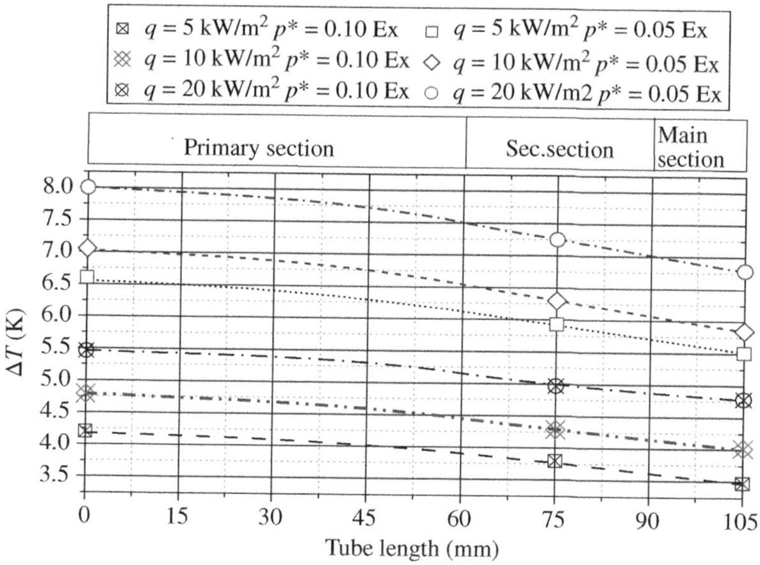

Figure 4.78 ΔT variation versus tube length along with experimental data (symbols) for different heat fluxes : one-dimensional analysis, $p^* = 0.1$ and 0.05.

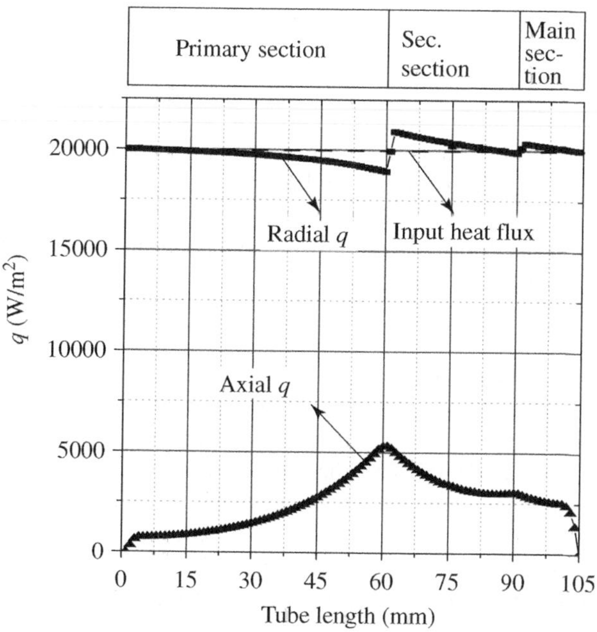

Figure 4.79 Comparison of heat flux (q) distribution in radial and axial directions one- dimensional case, $p^* = 0.1$ and $q = 10\,\text{kW/m}^2$.

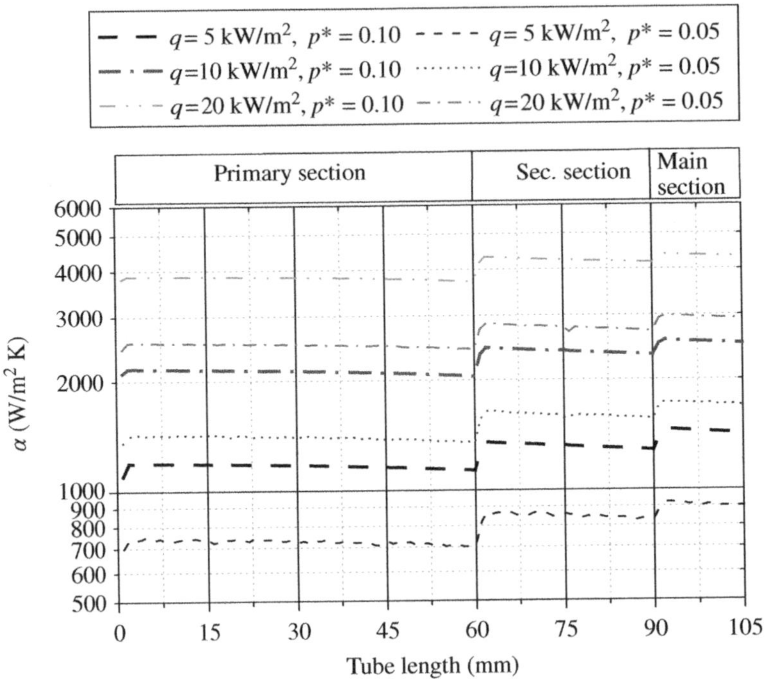

Figure 4.80 α variation in the axial direction for different heat flux (q) values – one-dimensional case for $p^* = 0.1$and 0.05.

taking axial q into account whenever the surface structures are not uniform. Also, it has been observed that the relative amount of the axial heat flow with respect to radial heat flow is higher for low q cases than high q cases, where boiling heat transfer coefficient dominates. However, the relative ΔT variations in radial, axial and azimuthal directions have been listed in a later part of this section in three-dimensional analysis, which will give the exact contributions in all three directions.

4.10.4.2 Two-Dimensional Heat Conduction Case

The azimuthal and axial α variations have been analysed with 2D heat conduction programs (radial-azimuthal and radial-axial). The α variation in axial direction has been shown in Figure 4.81 for cases of $p^* = 0.1$ and 0.05. This diagram is similar to the 1D case but the only difference is the sharp curves become smoother between the structures. The α's versus azimuthal angles are plotted in Figure 4.82 for $p^* = 0.1$ and 0.05 as sample cases. Significant variation of α values is observed in the azimuthal direction as investigated earlier by Kaupmann [56] and Danger et al. [52]. In these figures, the azimuthal angle 0 degree indicates the top of the tube and 180 degrees indicates bottom of the tube. It is noted that α's are higher near the bottom of the tube than at the top of the tube for all cases of q. Also, it is observed that these variations are symmetric with respect to azimuthal angle (# 180°) in general and these α variations are higher for $q = 5$ and $q = 10\ \text{kW/m}^2$ than for $q = 20\ \text{kW/m}^2$.

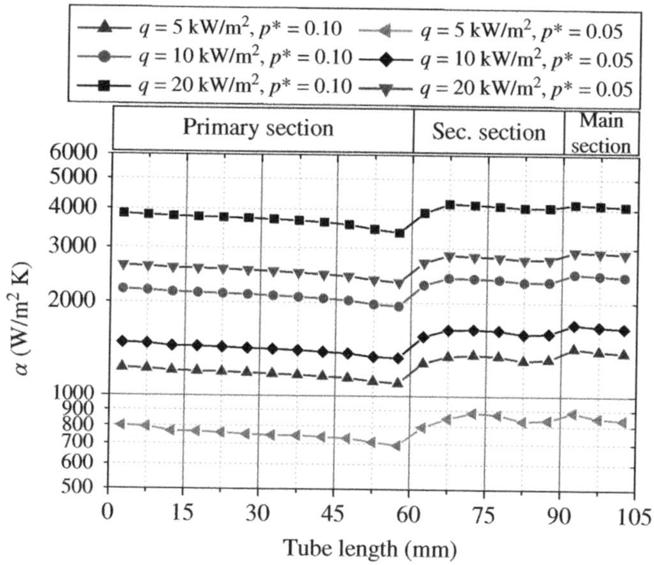

Figure 4.81 α variation in the axial direction for different heat flux (q) values – two-dimensional case for $p^* = 0.1$ and 0.05.

4.10.4.3 Three-Dimensional Heat Conduction Case

Similarly, the results of 3D case are plotted as ΔT vs length, ΔT vs azimuthal angle and ΔT vs tube radius in Figures 4.83–4.85 for a reduced pressure (p^*) of 0.1 to show the three-dimensional temperature distribution. In most of the cases, the ΔT variation in azimuthal direction as well as in axial direction is symmetric. Hence, only a quarter of the tube ($\phi = 0$ to 180° and tube length = 105 mm) has been analysed, to minimize the computing time. Figure 4.83 shows the ΔT variation along the tube length for different azimuthal angles and the Figure 4.84 shows ΔT variation in azimuthal direction for $p^* = 0.1$ at $q = 10\ \text{kW/m}^2$ as a sample case for three-dimensional analysis. These figures show that ΔT values are higher for low azimuthal angles (AA = 0° for top of the tube) than for high azimuthal angles (AA = 180° for bottom of the tube) and ΔT reduces at the middle of the (# $z = 105$ mm) tube caused by the different surface structures.

Similarly, the Figure 4.85 shows ΔT variation with the tube radii for different axial locations for the same case of $p^* = 0.1$ and $q = 10\ \text{kW/m}^2$. This figure indicates that ΔT is higher at the centre of the tube (# $r = 7.05$ mm) than at the outer surface (# $r = 12.7$ mm) of the tube, as expected. It is observed that the tube surface temperature distribution (ΔT variations) are different from the 1D or 2D cases since the temperature distribution is three-dimensional. Figure 4.86 shows the q variation in radial, azimuthal and axial directions of the tube (the left diagram gives q vs azimuthal angle at tube length of 15 mm and the right diagram gives q vs tube length at AA = 71°).

The ΔT variations in the axial direction are shown in Figure 4.87 for different azimuthal angles of the tube. This figure indicates that α increases to the centre of the tube ($z = 105$ mm) and the α is highest in the secondary and main sections. This figure also indicates that α values are higher near the bottom of the tube (AA = 180°) than at the top of the tube (AA = 0°) as reported in earlier investigations.

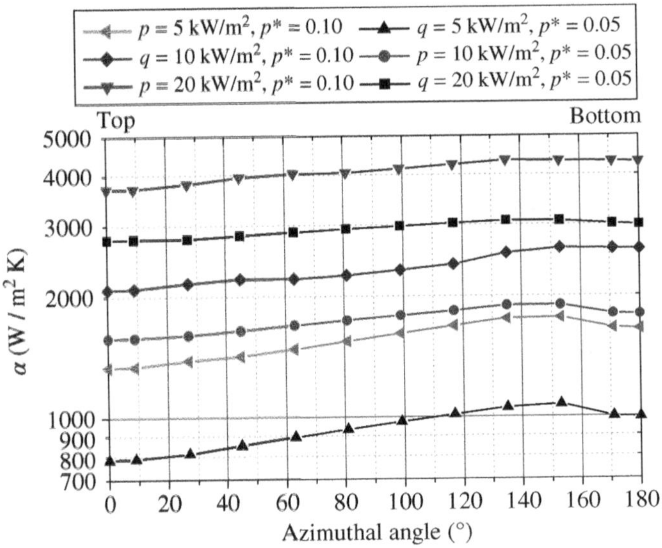

Figure 4.82 α variation in the azimuthal direction for different heat flux values – two-dimensional case $p^* = 0.1$ and 0.05.

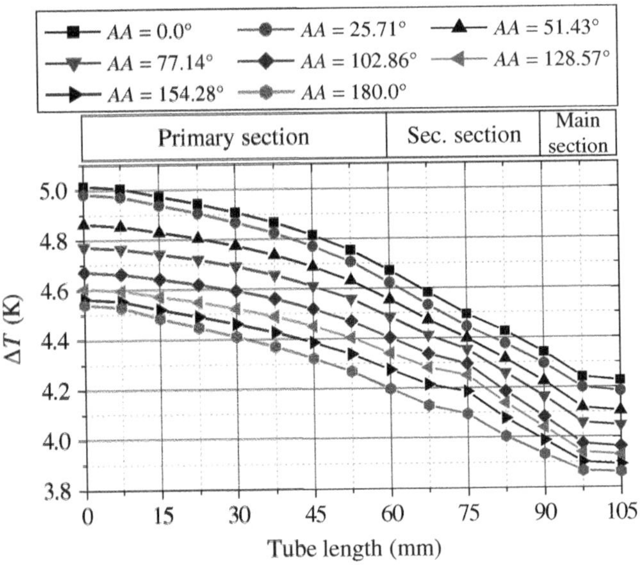

Figure 4.83 ΔT variation along the tube length (# $r = 12.7$ mm) for different azimuthal angles: three-dimensional case for $p^* = 0.1$ and $q = 10$ kW/m^2.

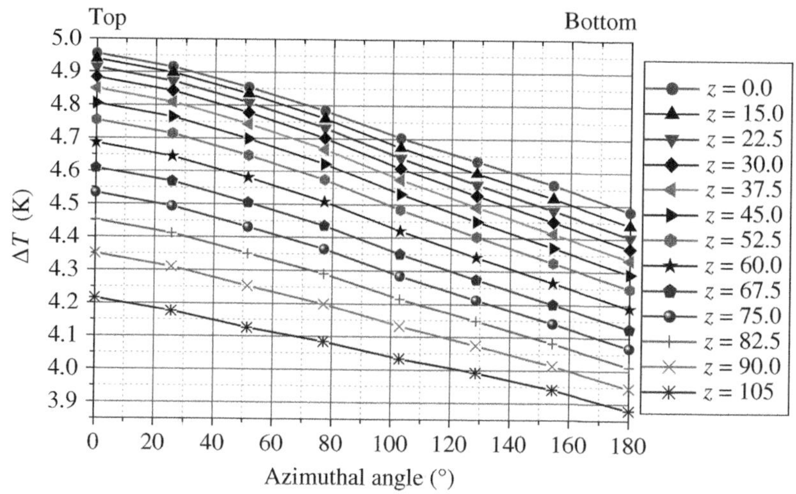

Figure 4.84 ΔT variation in the azimuthal direction (on the surface of the tube, $r = 12.7$ mm) – three-dimensional case for $p^* = 0.1$ and $q = 10$ kW/m^2.

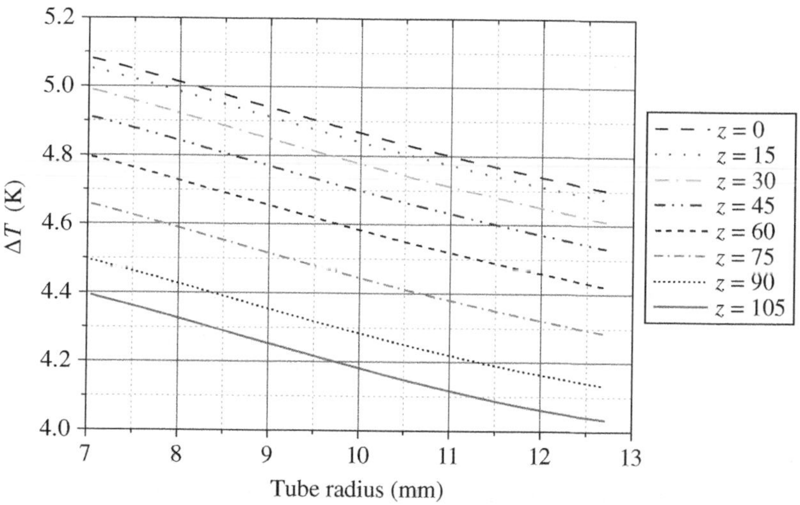

Figure 4.85 ΔT variation with the tube radii for different axial distances (# $AA = 103°$) $p^* = 0.1$ and $q = 10$ kW/m^2 – three-dimensional case.

To compare the amount of heat flux variations in radial, axial and azimuthal directions, the tube temperature variations (ΔT) have been plotted in Figure 4.88 by taking small equivalent lengths (4 mm) in all three directions. This figure has been drawn at the middle of primary section 1 (# axial length, $z = 45$ mm). It can be seen from this figure that as high as 18.39% azimuthal ΔT and 6.36% axial ΔT are observed in three-dimensional analysis for the case with $p^* = 0.1$ and $q = 10$ kW/m^2. Similarly, the ΔT variations for other

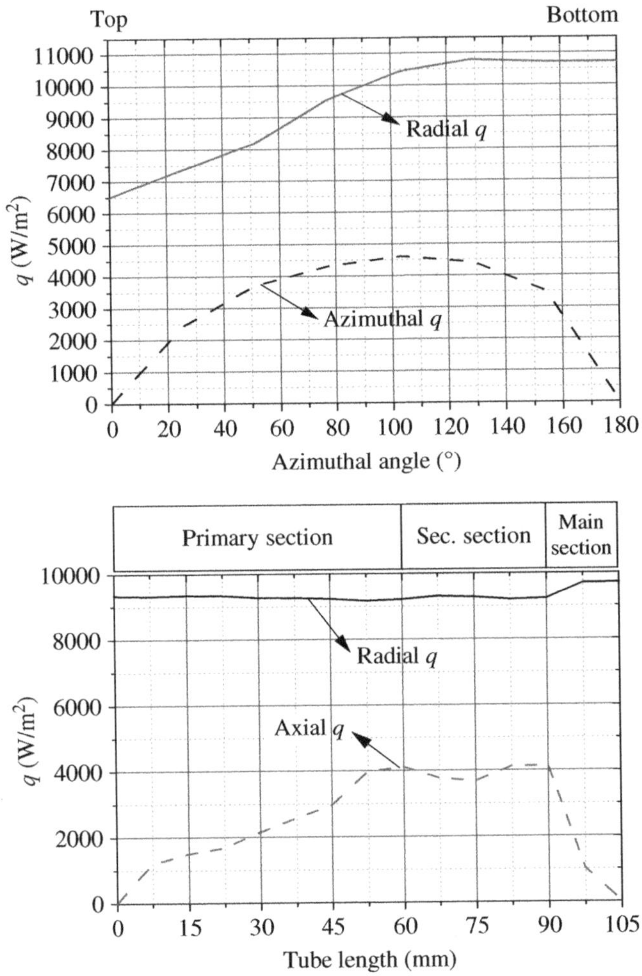

Figure 4.86 q variation in radial, azimuthal and axial directions of the tube, q vs azimuthal angle (left) at $z = 15$ mm and q vs tube length (right) at AA = 71°, $p^* = 0.1$ and $q = 10$ kW/m^2, three-dimensional case.

heat fluxes have been tabulated in Table 4.15 for the $p^* = 0.1$ and 0.05 for the same locations of azimuthal angle and radius. It can be seen from Table 4.15 that the direction cosine in the axial direction may reach 0.532 for $q = 5$ kW/m^2 (with 34.4° radial heat flow tilting angle) and 0.267 in azimuthal direction for $q = 10$ kW/m^2 at $p^* = 0.05$. In the case of $p^* = 0.1$, the value is as high as 0.206 in the axial direction and 0.253 in azimuthal direction. Normally, it is observed that in axial direction, heat flows towards the main section of the tube in all cases. Even though the ΔT variations in the axial direction are up to 0.532 for the direction cosine, the actual α variation is less when compared with α from purely radial heat flux because the heat transfer area available in the axial direction is much less than the heat transfer area for radial flow. However, it can be noted that

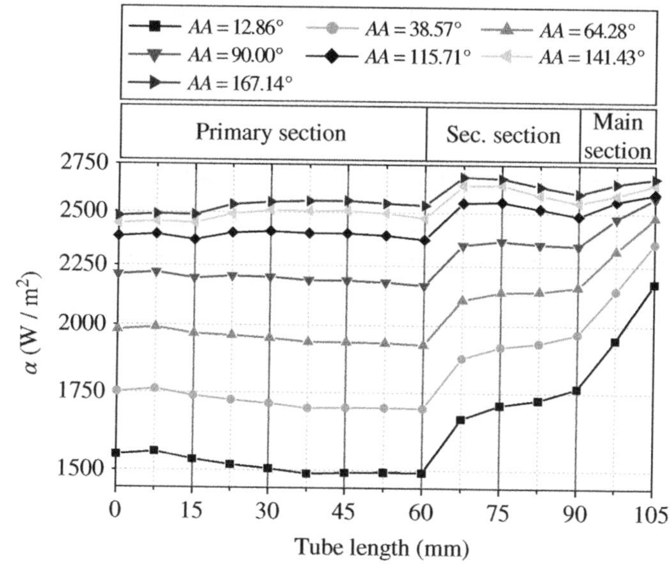

Figure 4.87 Variation of α values along the tube length for different azimuthal angles, $p^* = 0.1$ and $q = 10\ \mathrm{kW/m^2}$, – three-dimensional case.

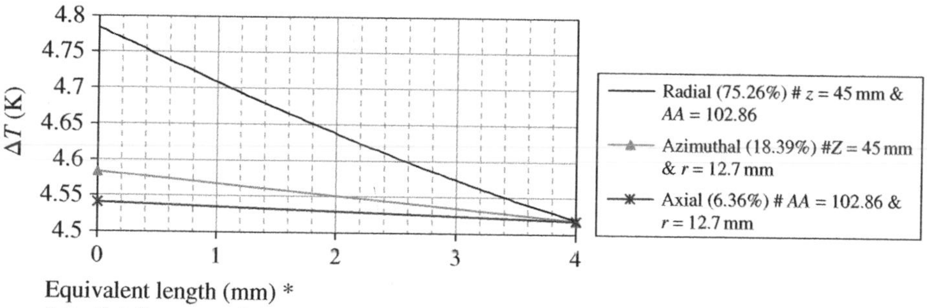

* Radial distance = 8.7–12.7 mm, Azimuthal = 84.81 –102.86 & Axial = 41–45 mm

Figure 4.88 Relative ΔT variation in axial, azimuthal and radial directions in the tube wall near the surface 3D case for $p^* = 0.1$ and $q = 10\ \mathrm{kW/m^2}$.

these variations cannot be neglected where, for example, surface structures are not uniform. The information provided here for ΔT variations is useful for determining such effects and the possible link to different bubble formations. Also, the isotherms over the tube surface have been plotted in Figure 4.89 by taking symmetric values in axial and azimuthal directions of the tube for the same case $p^* = 0.1$ and $q = 10\ \mathrm{kW/m^2}$. This figure shows that heat flows towards the middle (main section) and bottom of the tube which has higher boiling heat transfer coefficient (α) in this region than at the top and ends of the tube (primary sections). These results were also presented by Ranganayakulu and Gerrard Herres [57,58].

Table 4.15 Relative ΔT variation in axial, azimuthal and radial directions of the tube – three-dimensional case for $p^* = 0.1$ and 0.05.

			Direction cosines of ΔT variation and tilting angle of heat flow against radial direction							
			# axial length, $z = 45$ mm				# axial length, $z = 90$ mm			
Sl. No	p^*	q (kW/m²)	Radial	Angle	Azimuthal	Axial	Radial	Angle	Azimuthal	Axial
1	0.05	20.0	0.9652	15.17°	0.2423	0.0987	0.9756	12.68°	0.1956	0.0996
		10.0	0.9317	21.30°	0.2672	0.2461	0.9509	18.03°	0.1908	0.2437
		5.0	0.8251	34.40°	0.1903	0.5320	0.8438	32.46°	0.1327	0.5201
2	0.1	20.0	0.9827	10.67°	0.1744	0.0622	0.9889	8.54°	0.1462	0.0255
		10.0	0.9690	14.31°	0.2337	0.0808	0.9568	16.90°	0.2201	0.1900
		5.0	0.9453	19.03°	0.2531	0.2056	0.9573	16.81°	0.2271	0.1790

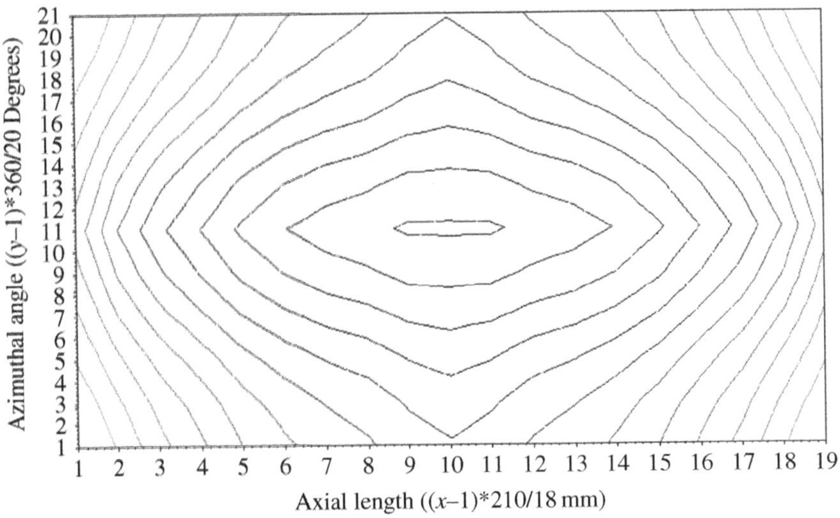

Figure 4.89 Isotherms over the surface of the tube for $p^* = 0.1$ and $q = 10$ kW/m², distance of isotherms = 0.1 K beginning with $\Delta T = 6.1$ K in the centre to 7.2 K at the outermost isotherm.

4.11 Closure

The finite element formulation, presented above, is a very general procedure and is applicable to any situation where a conduction-convection coupled heat transfer takes place. Ravikumar et al. [5] employed a sub-domain finite element concept for the rating of heat exchangers without considering the effects of longitudinal wall heat conduction, inlet flow non-uniformity and temperature non-uniformity. The present formulation with suitable modifications can be used for rating of a network of heat exchangers.

The present model can be used for other similar conventional heat exchangers as well by using appropriate heat transfer coefficients and overall surface characteristics. By using a marching technique, the same model can be used for multipass heat exchangers. With suitable modifications, the model can be used for cooling of gas turbine blades and vanes and pipe flows.

Longitudinal wall heat conduction effects in cross flow plate-fin heat exchangers, and counter flow heat exchangers have been reviewed with a new numerical procedure based on finite element concept. This analysis has been carried out for longitudinal wall heat conduction effects in cross flow plate-fin, counter flow plate-fin, cross flow tube-fin and parallel-flow plate-fin heat exchangers. The relation between the conduction effect factor (τ) and longitudinal wall heat conduction parameter (λ) for magnitude of $C^* = 1.0$ and 0.6 are plotted in Figure 4.9 for cross flow plate-fin, cross flow tube-fin, counter flow plate-fin and parallel-flow plate-fin heat exchangers, to compare the magnitudes of longitudinal wall heat conduction effects with each other. It has been observed that the performance deteriorations of a cross flow plate-fin heat exchanger are higher than the counter flow and parallel-flow heat exchangers for all cases. Also, the performance deteriorations of cross flow tube-fin heat exchanger is less than the cross flow plate-fin heat exchanger. It is expected that the temperature gradient in the tube in a cross flow tube-fin exchanger is one-dimensional.

The thermal performance deteriorations and pressure drop variations in cross flow plate-fin and tube heat exchangers have been reviewed along with effects of flow non-uniformity. A mathematical equation has been developed to generate different types of cross flow non-uniformity models. Using these flow non-uniformity models, the exchanger effectiveness and its deterioration due to flow non-uniformity have been analysed for both balanced and unbalanced flows of cross flow plate-fin and tube-fin heat exchangers. The results obtained, however, are not limited to the proposed models and can also be interpolated to any other similar flow/temperature mal-distribution models. The information presented in this book can provide the designer a means of estimation of effective performance variations when the flow/temperature non-uniformity, longitudinal heat conduction and their combined effects exist in heat exchanger core. A Galerkin finite element formulation is also extended to study the performance of a micro-compact heat exchanger that is used in electronic package cooling.

In addition, a finite element analysis has been carried out on the evaporator tube in pool boiling, taking into account all possible modes (one-dimensional, two-dimensional and three-dimensional) of heat conduction in radial, axial and azimuthal directions. Information obtained in this analysis indicates that the axial and azimuthal heat conduction is significant and cannot be neglected for a tube with thick walls in boiling applications.

Nomenclature

a	elemental heat transfer area, m^2
a'	elemental heat transfer area per unit core area, dimensionless
A	heat transfer area or surface area of the tube, m^2
AA	azimuthal angle, °

A_c	exchanger minimum free flow area, m^2
A_f	fin or extended surface area, m^2
ANN	artificial neural network
av	average values
A_w	total solid area available for longitudinal heat conduction, m^2
c_p	specific heat of the fluid, (Mc_p), J/kg·K
C	flow stream heat capacity rate with a subscript c or h, W/K
C^*	heat capacity rate ratio, C_{min}/C_{max}, dimensionless
C^r	heat capacity rate of storage type heat exchanger, W/°C
D	pressure gradient constant in equation
D_h	hydraulic diameter of flow passage, m
ex	experimental data
FN	flow non-uniformity case
FEM	finite element method
f	mean friction factor
GA	genetic algorithm
G	exchanger flow stream mass velocity, (M/Ac), kg/s·m^2
g	proportionality factor in Newton's second law
Hc	micro-channel height, m
h	convective heat transfer coefficient, W/m^2·K
I	number of divisions in x direction
J	number of divisions in y direction
j	a heat transfer modulus, $St\ Pr^{2/3}$
K	number of divisions in z direction
k	thermal conductivity, W/m·K
L,l	length of the exchanger/channel in the x or y directions respectively, m
LHC	longitudinal heat conduction case
M	mass flow rate, kg/s
N	shape functions
n	number of divisions/channels
NE	number of elements
NTU	number of transfer units, (UA/C_{min}), dimensionless
Nu	Nusselt number
p	cross flow tube-fin exchanger tube perimeter, m
p^*	p_s/p_c, reduced pressure

p_c	critical pressure
Pr	Prandtl number ($\nu c_{p/k}$), dimensionless
Pe	Peclet number, $Re\ Pr$, dimensionless
Q	enthalpy/heat entering or leaving the plate per unit area, W
q	heat flux, kW/m^2
Re	Reynolds number, GD_h/ν, dimensionless
R_{Total}	total thermal resistance, °C/W
r	tube radius, mm
s	local coordinates in x direction
S_t	Stanton number, h/Gc_p, dimensionless
S_f	minimum free flow area, m^2
t	local coordinates in y direction or fin half thickness in the case of microchannel, m
T	temperature of hot or cold fluids and metal wall, K
T_∞	surrounding fluid temperature, K
TN	temperature non-uniformity case
U	overall heat transfer coefficient, $W/m^2{\cdot}K$
V	heat exchanger volume, m^3
V_1, V_2	specific volumes of inlet and outlet fluids, m^3/kg
V_m	mean specific volume of the fluid, m^3/kg
X	axial distance, $x/(D_h\ Pr\ Re)$, dimensionless
W	fluid velocity component
x	coordinate in x direction or flow length along hot fluid side, m
y	coordinate in y direction or flow length along cold fluid side, m, or tube azimuthal length, mm
z	coordinate in z direction or tube axial length, mm
Z	width of the heat sink, m
1D	one-dimensional
2D	two-dimensional
3D	three-dimensional

Greek Symbols

α	flow non-uniformity parameter, or heat transfer coefficient, $W/m^2\ K$
β	temperature non-uniformity parameter

ε_0	exchanger effectiveness without considering the effects of longitudinal wall heat conduction, inlet flow non-uniformity and temperature non-uniformity, dimensionless
ε_L	exchanger effectiveness with the effects of longitudinal wall heat conduction, dimensionless
ε_F	exchanger effectiveness with the effects of inlet flow non-uniformity, dimensionless
ε_T	exchanger effectiveness with the effects of temperature non-uniformity, dimensionless
$\varepsilon_{L,F}$	exchanger effectiveness with the effects of longitudinal wall heat conduction, and inlet flow non-uniformity, dimensionless
$\varepsilon_{L,T}$	exchanger effectiveness with the effects of longitudinal wall heat conduction and temperature non-uniformity, dimensionless
$\varepsilon_{L,F,T}$	exchanger effectiveness with the effects of longitudinal wall heat conduction, inlet flow non-uniformity and temperature non-uniformity, dimensionless
$\varepsilon_{F,T}$	exchanger effectiveness with the effects of inlet flow non-uniformity and temperature non-uniformity, dimensionless
$\Delta\varepsilon$	change in effectiveness, dimensionless
ΔT	$(=T-T_\infty)$, superheat, K, temperature variations in tube material, K
λ	longitudinal wall heat conduction parameter based on the total conduction area, dimensionless
η_f	fin efficiency or temperature effectiveness of fins or an extended surface, dimensionless
η, ϑ	total surface efficiency, dimensionless
τ	performance correction factor defined with subscripts L for longitudinal effect factor, F for flow non-uniformity effect factor, T for temperature non-uniformity correction factor, L,F for the combined effects of longitudinal heat and flow non-uniformity, L,T for the combined effects of longitudinal heat conduction and temperature non-uniformity, F,T for the combined effects of flow non-uniformity and temperature non-uniformity and L,F,T for the combined effects of all the three, dimensionless
$(\eta ha)^*$	the ratio of $(\eta ha)_c/(\eta ha)_h$, dimensionless
σ	ratio of free flow area to frontal area

Subscripts

a	average
b	bottom plate
c	cold fluid side

h	hot fluid side
i	inlet
o	outlet
t	top plate
min	minimum magnitude
max	maximum magnitude
w	wall
1–16	node numbers when used with T or q

References

1 L.J. Segerlind, *Applied Finite Element Analysis*, John Wiley and Sons, 2nd edn, 1982.
2 W.M. Kays and A.L. London, *Compact Heat Exchangers*, 3rd edn, McGraw-Hill, New York, 1984.
3 J.P. Chiou, Thermal performance deterioration in a cross flow heat exchanger due to the flow non-uniformity, *ASME J. Heat Transfer*, vol. 100, pp. 580–587, 1978.
4 J.P. Chiou, The effect of non-uniformity of inlet fluid temperature on the thermal performance of Cross flow heat exchanger, *7th Int. Heat Transfer Conference, Heat Transfer* 1982, pp. 179–184, 1982.
5 S.G. Ravikumar, K.N. Seetharamu and P.A. Aswatha Narayana, Performance evaluation of cross flow compact heat exchangers using finite elements, *Int. J. Heat Mass Transfer*, vol. 32, pp. 889–894, 1989.
6 J.P Chiou, The advancement of compact heat exchanger theory considering the effects of longitudinal heat conduction and flow non-uniformity effects, in *Compact heat exchangers – history, technological advancement and mechanical design problems*, Eds. Shah. R.K et al. ASME, New York, Book no. G00183, HTD, vol. 10, pp. 101–121, 1980.
7 J.P. Chiou, The effect of longitudinal heat conduction in cross flow heat exchangers, *J of Heat Transfer, Trans. of ASME*, vol. 100, pp. 346–351, May 1978.
8 Kroeger P.G., Performance deterioration in high effectiveness heat exchangers due to axial heat conduction effects, *Advances in Cryogenic Engineering*, no. 12, pp. 363–372, 1967.
9 C. Ranganayakulu, K.N. Seetharamu and K.V. Sreevatsan, The effects of longitudinal heat conduction in compact plate-fin and tube-fin heat exchangers using a finite element method, *Int. J of Heat Mass Transfer*, vol. 40, no. 6, pp. 1261–1277, 1997.
10 Ranganayakulu C., Seetharamu K.N., Sreevatsan K.V., The Effect of Longitudinal Heat Conduction in Cross Flow Compact Heat Exchanger, *Proceedings of XI National Heat and Mass Transfer & I ISHMT-ASME Heat and Mass Transfer Conference*, IIT Madras, India, 57–62, 1991.
11 J.P. Chiou, The effect of longitudinal heat conduction on cross flow heat exchanger, *ASME J. Heat Transfer*, vol. 100, pp. 346–351, 1978.

12 G.D. Bahnke and C.P. Howard, The effect of longitudinal heat conduction on periodic-flow heat exchanger performance, *ASME J. Eng. for Power*, vol. 86, pp. 105–121, 1964.

13 R.K. Shah, A correlation for longitudinal heat conduction effects in periodic-flow heat exchangers, *ASME J. Eng. Power*, vol. 97, pp. 453–454, 1975.

14 R.K. Shah, A review of longitudinal wall heat conduction in recuperators, *J. Energy, Heat and Mass Transfer*, vol. 16, pp. 15–25, 1994.

15 Bahnke G.D and Howard C.P., The effect of longitudinal heat conduction on Periodic-flow heat exchanger performance, *J. of Engineering for Power, Trans. of ASME*, pp. 105–120, April 1964.

16 Lee H.W., Teng Y.J., Azid I.A, K.N. Seetharamu, Neuro-genetic optimization of micro compact heat exchanger, *International Journal of Numerical Methods for Heat & Fluid Flow* vol. 17 no. 1, pp. 20–23, 2007.

17 Shah R.K., Flow Maldistribution in Heat Exchangers, in *Handbook of Heat Transfer Applications*, Edited by Rohsenow W.M, Hartnett, New York, pp. 266–280, 1985.

18 C. Ranganayakulu, K.N. Seetharamu and K.V. Sreevatsan, The effects of inlet fluid flow non-uniformity on thermal performance and pressure drops in cross flow plate-fin compact heat exchangers, *Int. J. Heat Mass Transfer*, vol. 40, no. 1, pp. 27–38, 1997.

19 C. Ranganayakulu, K.N. Seetharamu and K.V. Sreevatsan, The performance evaluation of cross flow tube-fin heat exchanger considering the effects of inlet flow non-uniformity and temperature non-uniformity using a finite element method, *Proc. 9th Int. Conference on Numerical Methods in Thermal Problems*, eds. by R.W. Lewis et al, vol. IX, part 2, pp. 1010–1021, 1995.

20 Ranganayakulu C., Panigrahi A., Influence of Header Design on Pressure Drop and Thermal Performance of a Compact Heat Exchanger, *Proceedings of III International Conference on Compact Heat Exchanger for Process Industries*, Davos, Switzerland, 251–258, 2001.

21 Ranganayakulu C., Seetharamu K.N., Sreevatsan K.V., The performance evaluation of cross flow tube-fin heat exchanger considering the effects of inlet flow non-uniformity and temperature non-uniformity using a finite element method, *Proceedings of 9th International Conference on Numerical Methods in Thermal Problems*, Atlanta, USA, IX, 1010–1021, 1995.

22 J.P. Chiou, The effect of airflow non-uniformity on the thermal performance of an automobile air conditioning condenser, SAE Trans., paper no. 830542, pp. 2.587–2.600, 1984.

23 J.P. Chiou, The effect of airflow non-uniformity on the thermal performance of an evaporator of automobile air conditioning system, SAE trans., paper no. 840381, pp. 2.991–2.1005, 1985.

24 J.P. Chiou, Thermal performance deterioration in a cross flow heat exchanger due to the flow non-uniformity on both hot and cold sides, *Proc. 6th Int. Heat Transfer Conference, Heat Transfer*, vol. 4, pp. 279–284, 1978.

25 Fleming R.B., The effects of flow distribution in parallel channels of counter flow heat exchangers, *Advances in Cryogenic Engineering*, vol. 12, pp. 353–362, 1927.

26 J.P. Chiou, The effect of non-uniform inlet air temperature distribution on the sizing of the engine radiator, SAE Trans., Paper no. 820078, pp. 250–260, 1982.

27 Gaddis E.S., Schlunder E.U., Temperature distribution and heat exchange in multipass shell and tube exchangers with baffles, *Heat Transfer Engineering*, vol. 1, no. 1, pp. 43–52, 1979.

28 Ranganayakulu C., Seetharamu K.N., The Effects of Inlet Fluid Temperature Non-Uniformity in Cross Flow Plate-Fin Heat Exchangers, *Proceedings of XVII National Heat and Mass Transfer & VI ISHMT-ASME Heat and Mass Transfer Conference*, IGCAR, Kalpakkam, India, 2004.

29 H. Yamasheta, R. Izumi and S. Yamaguchi, Performance of Cross flow heat exchangers with variable physical properties, *Bulletin of JSME*, no. 20, pp. 1008–1015, 1977.

30 S.G. Ravikumar, K.N. Seetharamu and P.A. Aswatha Narayana, Analysis of compact heat exchangers using finite element method, *Heat Transfer*, Hemisphere publishing Corp., New York, vol. 2, pp. 379–384, 1986.

31 Ranganayakulu C., Seetharamu K.N., The combined effects of wall longitudinal heat conduction and inlet flow non-uniformity on thermal performance of cross flow plate-fin heat exchanger, *International Journal of Heat and Mass Transfer*, 36, 247–256, 2000.

32 Ranganayakulu C., Seetharamu K.N., The combined effects of wall longitudinal heat conduction, inlet flow non-uniformity and temperature non-uniformity on thermal performance of cross flow plate-fin heat exchanger, *International Communications in Heat and Mass Transfer*, 26, 669–678, 1999.

33 Ranganayakulu C., Seetharamu K.N., Sreevatsan K.V., The combined effects of wall longitudinal heat conduction, inlet flow non-uniformity and temperature non-uniformity on thermal performance of cross flow tube-fin heat exchanger, *International Journal of Heat and Mass Transfer*, 42, 263–273, 1999.

34 Ranganayakulu C., Seetharamu K.N., The combined effects of wall longitudinal heat conduction and inlet and fluid flow non-uniformity on thermal performance of cross flow tube-fin heat exchanger – a finite element method, *Proceedings of International Conference on Compact Heat Exchangers for Process Industries*, Snowbird, USA, 425–437, 1997.

35 Ranganayakulu C., Seetharamu K.N., The combined effects of inlet flow non-uniformity and temperature non-uniformity on thermal performance of cross flow plate-fin heat exchanger, *International Journal of Warme und Stoffubertragung*, 32, 375–383, 1997.

36 Ranganayakulu C., Seetharamu K.N., Sreevatsan K.V., The performance evaluation of cross flow tube-fin heat exchanger considering the combined effects of inlet flow non-uniformity and temperature non-uniformity using a finite element method, *Proceedings of XIII National Heat and Mass Transfer & III ISHMT-ASME Heat and Mass Transfer Conference*, REC, Surathkal, India, 713–718, 1995.

37 Dakshina Murthy, C.M., S.G. Ravi Kumar, and K.N. Seetharamu, FEM Applications in the Phase Change Exchangers, *Warme und Stoffubertragung*, 26, 3, 137–140, 1991.

38 Ravikumar S.G, K.N. Seetharamu and P.A. Aswatha Narayana Applications of finite elements in heat exchangers, *Communications in Applied Numerical Methods*, vol. 2, pp. 229–234, 1986.

39 Ravikumar S.G., K.N. Seetharamu and P.A. Aswatha Narayana Finite element analysis of shell and tube heat exchanger, *Int. Communication Heat Mass Transfer*, vol. 15, no. 2, pp. 151–163, 1988.

40 Seetharamu K.N., Finite element analysis of heat exchangers, *Lecture Notes In Engineering*, vol. 37, pp. 773–803, Springer-Verlag, 1988.

41 Quadir G.A., S.L. Beh, K.N. Seetharamu and A.Y. Hassan A transient finite element analysis of microchannel heat exchangers, *International Journal of Heat Exchangers*, vol. 3, no. 1, pp. 21–42, June 2002.

42 Jeevan K., G.A. Quadir, K.N. Seetharamu, I.A. Azid and Z.A. Zainal, Optimization of thermal resistance of stacked micro-channel using genetic algorithms, *Int. J. Num. Methods for Heat & Fluid Flow*, vol. 15, no. 1, pp. 27–42, 2005.

43 Pradeep Hegde, K.N. Seetharamu, G.A. Quadir, P.A. Aswathanarayana, M.Z. Abdullah and Z.A. Zainal Thermal analysis of micro-channel heat exchangers with two phase flow using FEM, *Int. J. Num. Methods for Heat & Fluid Flow* vol. 15, no. 1, pp. 43–60, 2005.
44 Dieter Gorenflo, State of the art in pool boiling heat transfer of new refrigerants, *Int. J. of Refrigeration*, 24, pp. 6–14, 2001.
45 Peter Hübner, Dieter Gorenflo and Andrea Luke, Circumferential temperature distributions on plain and finned tubes in pool boiling, *Proceedings of 3rd Int. Conference on Compact Heat Exchangers*, Davos, Switzerland, pp. 383–390, 2001.
46 Andrea Luke, Dieter Gorenflo, Peter Hübner and Paul Kaupmann, Convective contributions to pool boiling heat transfer, Two phase flow group meeting, Aveiro, Portugal, 2001.
47 V.K. Dhir, Numerical simulations of pool boiling heat transfer, *AIChE Journal*, 47:813–834, 2001.
48 Paul Kaupmann, Dieter Gorenflo and Andrea Luke, Pool boiling heat transfer on horizontal steel tubes with different diameters. *Boiling 2000: Phenomena & Emerging Applications, Proc. Eng. Found. Conf.,* Anchorage, vol. 2, pp. 1999–215, 2000.
49 Andrea Luke and Dieter Gorenflo, Roughness of Evaporator surfaces for joint research project and development of a database for roughness, bubble formation and heat transfer, *Proc. of Int. Conference, IIR Commission B1, Thermo-physical Properties and Transfer Processes of New Refrigerants,* Paderborn Germany, Oct. 3–5, pp. 358–367, 2001.
50 Dieter Gorenflo, Danger E., Andrea Luke, Stephan K., Chandra U. and Ranganayakulu C., Bubble formation with pool boiling on tubes with or without basic surface modification for enhancement, *Proceedings of V Int. Conference of Boiling Heat Transfer*, Jamaica, 2003.
51 Dieter Gorenflo, Elisabeth Danger, Andrea Luke, Stephan Kotthoff, Untung Chandra, Ranganayakulu C., Bubble formation with pool boiling on tubes with or without basic surface modifications for enhancement, *International Journal of Heat and Fluid Flow*, 25, 288–297, 2004.
52 E. Danger, Untung Chandra and Dieter Gorenflo, Basic modification of the heated surface for enhancement of pool boiling heat transfer, *Proc. of Int. Conference, IIR Commission B1, Thermo-physical Properties and Transfer Process of New Refrigerants,* Paderborn Germany, Oct. 3–5, pp. 376–383, 2001.
53 S.S. Rao, *The Finite Element Method in Engineering*, Pergamon press, 1982.
54 G. Comini et al. *Finite Element Analysis in Heat Transfer*, Taylor & Francis, 1994.
55 K.H. Huebner, et al., *The Finite Element Method for Engineers*, 3rd edn., John Wiley & Sons. Inc., 1995.
56 Paul Kaupmann, Durchmessereinfluß und örtlicher Wärmeubergang beim Blasensieden an horizontalen Stahlrohren, Dissertation, Universität Paderborn, 1999.
57 Ranganayakulu C., Gerrard Herres, Heat transfer from horizontal tubes in pool boiling: influence of three-dimensional heat conduction in the wall of the evaporator tube and fouling aspects, *Proceedings of International Conference on Heat Exchanger Fouling*, Kloster Irsee, Germany, 2005.
58 Ranganayakulu C., Gerhard Herres, Heat transfer from horizontal tubes in pool boiling: influence of three-dimensional heat conduction in the wall of the evaporator tube – A finite element analysis, *Heat and Mass Transfer Journal*, Springer-Verlag, 42, 271–278, 2006.

5

Generation of Design Data – Finite Volume Analysis

5.1 Introduction

Increasing awareness for the effective utilization of energy resources, minimizing operating cost and maintenance-free operation have led to the development of efficient heat exchangers such as compact heat exchangers. London and Shah [1] in their elaborate discussion of the classification of heat exchangers have defined the "compact heat exchanger" as one having a surface area density of more than 700 m^2 per m^3. Such compactness is achieved by providing the extended surfaces, that is, fins on the flow passages that work as the secondary heat transfer area.

Various techniques can be used to make heat exchangers more compact. There are three general types of extended surface geometries which can be used to increase gas-side heat transfer coefficients (a) finned-tube heat exchanger with flat fins (b) finned-tube heat exchanger with individually finned tubes and (c) plate-fin heat exchangers. The most common compact heat exchangers are plate-fin heat exchangers. London [2] provided the information about the design methodology of CHEs.

5.2 Plate Fin Heat Exchanger

The plate-fin heat exchanger is a type of compact heat exchanger where the heat transfer surface area is enhanced by providing an extended metal surface interface between the two fluids, referred to as the fins. Out of the various compact heat exchangers, plate-fin heat exchangers are unique due to their construction and performance. They are characterized by high effectiveness, compactness, low weight and moderate cost. A plate-fin heat exchanger is made of a stack of corrugated fins alternating with a nearly equal number of flat separators, known as parting sheets, bonded together to form a monolithic block. A schematic view of such a heat exchanger is shown in Figure 2.13. Appropriate headers are welded to provide the necessary interface with the inlet and the exit streams. Plate-fin heat exchangers can be made in a variety of materials. Aluminium is preferred in cryogenic and aerospace applications because of its low density, high thermal conductivity and high strength at low temperature. The maximum design pressure for a brazed aluminium plate-fin heat exchanger is around 90 bars. At temperatures above ambient,

Compact Heat Exchangers – Analysis, Design and Optimization using FEM and CFD Approach,
First Edition. C. Ranganayakulu and K.N. Seetharamu.
© 2018 John Wiley & Sons Ltd. Published 2018 by John Wiley & Sons Ltd.

most aluminium alloys lose mechanical strength. Stainless steel, nickel and copper alloys have been used at temperatures up to 500 °C.

Figure 2.14 showed an exploded view of two layers of a plate-fin heat exchanger. Such layers are arranged together in a monolithic block to form a heat exchanger. The corrugations serve both as a secondary heat transfer surface and as mechanical support against the internal pressure between layers. The side bars prevent one fluid spilling over and mixing with the other fluid. The fins and side bars are brazed with the parting sheet to ensure a good thermal link and to provide mechanical stability. Plate-fin heat exchangers have been extensively used around the world for several decades, the technologies related to their design and manufacture remaining confined to a few companies in developed countries.

5.3 Heat Transfer Surfaces

A wide variety of plate-fin geometries, as shown in Figure 2.4, have been used to obtain enhanced heat transfer. Engineers are constantly working to develop new and more effective enhanced surfaces. Typical fin spacings are 400 to 1000 fins per metre. Due to their small hydraulic diameter and the low density of gases these surfaces are usually operated in the Reynolds number range $500 < Re < 10000$.

5.3.1 Lance and Offset Fins

Offset fins are very commonly used fins to provide extended surfaces for compact heat exchangers. These fins provide a high heat transfer coefficient when compared to other types of fins. However, they have a high friction factor, which results in increased pumping power. The surface geometries of the fins are described by the fin height (h), transverse spacing (s) and thickness (t). The uninterrupted flow length of offset strip fin is described as the offset length (l). The schematic of offset fins along with a dimensional representation and a photograph are shown in Figure 5.1a and b.

5.3.2 Wavy Fins

Wavy fins are also very commonly used as extended surfaces for compact heat exchangers. These fins provide a similar value of heat transfer coefficient to offset fins, but they have lower friction factors, which results in lower pumping power. The surface geometries of the fins are described by the fin height (h), transverse spacing (s) and thickness (t). Uninterrupted flow length of wavy fins is represented by the pitch of the wave (L). An isometric view of wavy fins is shown in Figure 5.1c and its photograph is shown in Figure 5.1d for clarity. The geometrical features of the wavy fin are shown in Figure 5.1e. The dimensionless representations of these variables are given by ratios of s/h, $2A/s$ and $L/2A$. In the fin designation as denoted by Kays and London [3], the first number indicates the fin density (fins/inch), the second number indicates the fin wave length (L) in inches and the third number indicates the fin thickness in inches as 11.44-3/8 W-0.006.

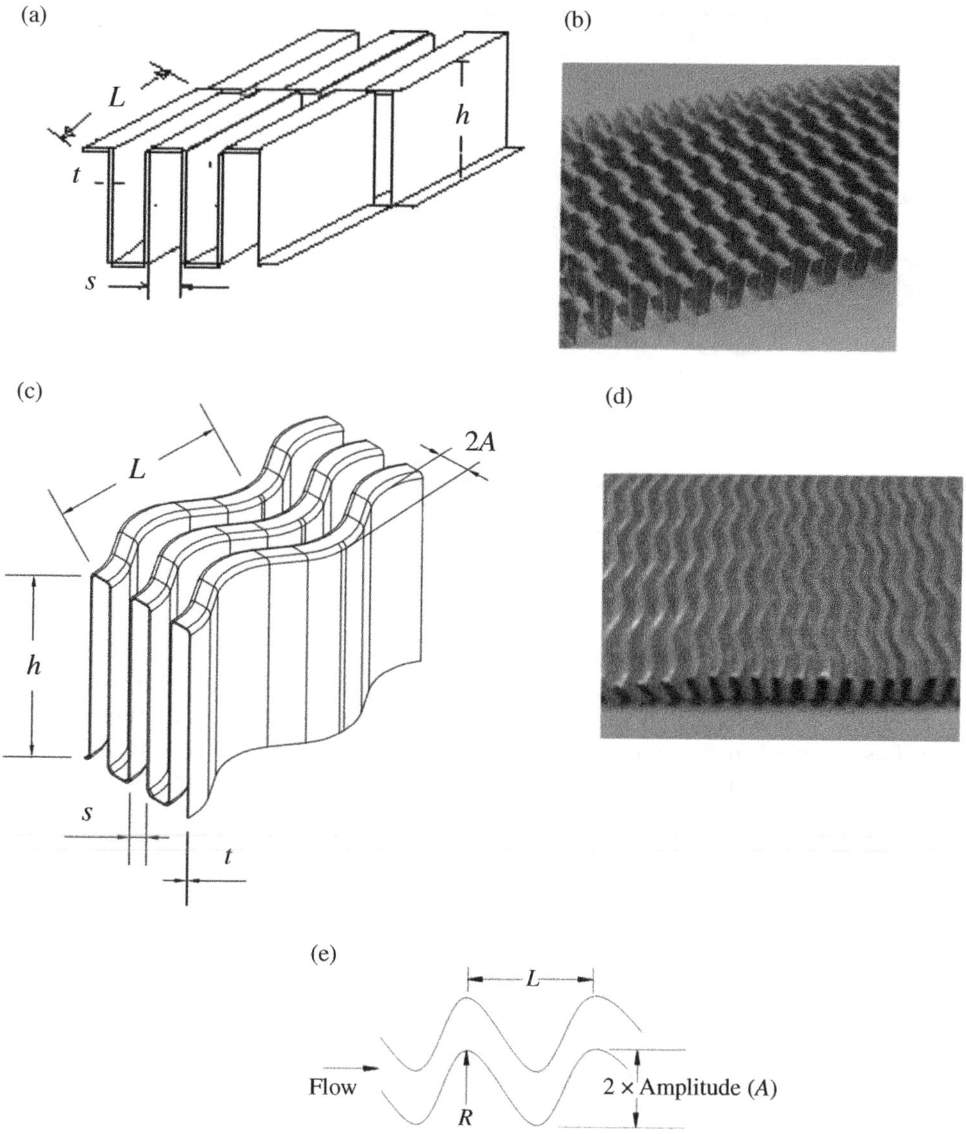

Figure 5.1 Schematics of fin geometry: (a) offset fin (b) offset fin photograph (c) wavy fin (d) wavy fin photograph (e) wavy fin dimensional notations.

5.3.3 Rectangular Plain Fins

Plain fins are the most common of all compact cores or surfaces used in compact heat exchangers. The plain fin surfaces are characterized by long uninterrupted flow passages. Although passages of triangular and rectangular cross-section are more common, any desired shape can be given to the fins, considering only manufacturing constraints. Straight fins in a triangular arrangement can be manufactured at high speeds and hence

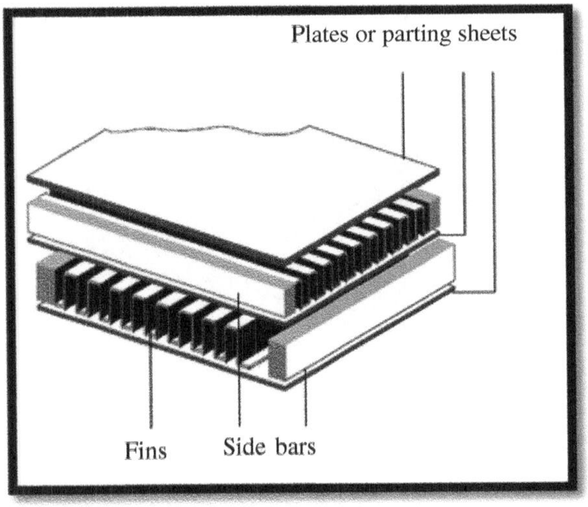

Figure 5.2 Plate-fin heat exchanger with rectangular plain fins.

are less expensive than rectangular fins. However, generally, they are structurally weaker than rectangular fins for the same passage size and fin thickness. They also have lower heat transfer performance compared to rectangular fins, particularly in laminar flow. Plain fins are used in those applications where core pressure drop is critical. A heat exchanger with rectangular plain fins, as shown in Figure 5.2, requires a smaller flow frontal area than that with interrupted fins for a specified pressure drop, heat transfer and mass flow rate. The heat transfer enhancement achieved with plain fins results mainly from increased area density, rather than any substantial rise in the heat transfer coefficient.

5.3.4 Rectangular Perforated Fins

This surface geometry is made by punching a pattern of spaced holes in the fin material before it is formed into flow channels, as shown in Figure 2.4. The channels may be triangular or rectangular in shape with either round or rectangular perforations. The thermal performance of the perforated fin is less than that of a good offset strip fin. Even though a negligible quantity of material is removed for making the holes or perforations, there is an improvement in performance of the heat exchanger due to the interruption in boundary layer and flow mixing. As the aim is to increase the heat transfer coefficient, it cannot be seen as material waste.

Perforated fins are used in the distributor sections of exchangers and also for boiling applications. The perforations in distributors allow for lateral migration of flow (including the vapour phase), without the pressure drop penalty of the offset strip fin. The interruptions given by them give a slight performance improvement per unit surface area, but offset by loss of surface area, and there is generally a higher friction factor.

Shah [4–6] showed that there is a modest improvement of heat transfer for Reynolds numbers greater than 2000 – suggesting that the onset of turbulence is brought forward.

5.3.5 Triangular Plain Fins

The model of the triangular plain fin between the plates is shown in Figure 2.6. The computations are performed in this type of compact heat exchanger with triangular fins, and the passage formed by the plate and the surfaces of the triangular fins is considered as a triangular duct. The heat transfer coefficients and frictional factors are similar to rectangular plain fins.

5.3.6 Triangular Perforated Fins

The perforated triangular fin is similar to the plain triangular fin, the only difference being that the perforated fin has pierced holes along the sides of the fin, which is used to study the effects on heat transfer rate, but it also has its own limitations because the diverged fluid path increases the pumping power needed at the inlet. The channel is in a triangular shape with round perforations, as shown in Figure 2.7. This surface geometry is made by punching a pattern of spaced holes in the fin.

5.4 Performance Characteristic Curves

The basic performance data for an enhanced surface is often shown as curves of the Colburn factor j and the Fanning friction factor f plotted against Reynolds number. One of the oldest and most authoritative sources of experimental j and f data on plate fin surfaces is the monograph *Compact Heat Exchangers* by Kays and London [3]. The plate fin surface performance is characterized by the following two dimensionless groups (see Section 2.6.4.2): Fanning friction factor f for pressure drop and Colburn j factor for heat transfer, in addition to the Reynolds number:

- **Fanning friction factor f** is defined on the basis of an equivalent shear force in the flow per unit heat transfer area, and it is represented by [3]

$$\Delta P = \frac{G^2}{2g} v_1 \left[\left(1 + \sigma^2\right) \left(\frac{v_2}{v_1} - 1\right) + f \frac{A\, v_m}{A_c\, v_1} \right] \tag{5.1}$$

$$\text{In general, } f = (\Delta P) \times \frac{r_h}{L} \tag{5.2}$$

  The friction factor f is computed from the area averaged mean pressure drop (ΔP) over the fin or passage length.

- **Colburn factor j** is defined by

$$j = (S_t)\,(Pr)\,2/3 = \frac{H}{GC_p} Pr^{2/3} \tag{5.3}$$

$$\text{where } S_t = \frac{Nu}{Re\,Pr} \tag{5.4}$$

- **Reynolds number (*Re*):** The characteristic of fluid flow is governed by dimensionless quantity called the Reynolds number:

$$Re = \frac{(\rho u D_h)}{\mu} \tag{5.5}$$

This characterizes the type of flow, whether it is laminar or turbulent flow.

5.4.1 Working Fluids

Studies on fin surfaces using liquid as the working fluid has been reported in Tinaut et al. [7] and Sen and Keith [8]. Sen and Keith [9] compared the experimental data for liquids against air ($Pr = 0.7$) using correlations from Wieting [10] and Joshi and Webb [11]. The results of both models and experiments are obtained for inlet fluid temperature of 10 °C. It is observed that, at the same Reynolds number, the Colburn factor for liquids predicted using air models is approximately twice the Colburn factor obtained from the experimental data for liquids. The air model over-predicted the heat transfer coefficient for liquids. This comparison between air models and experimental results demonstrated that the results for air cannot be accurately applied to liquid applications.

5.5 CFD Analysis

The finite volume based CFD software tool Fluent is used for computing the flow and temperature fields over different types of compact heat exchanger fins. 3D steady Navier–Stokes and energy equations have been solved (see Chapter 3) with appropriate boundary conditions for fin surfaces. In this analysis, the laminar flow model is used up to the Reynolds number 800, and above 1000 the k–ε turbulent models with enhanced wall treatment are used. In this chapter, velocity, pressure and temperature fields have been computed, and j and f factors are determined over appropriate ranges of Reynolds number and geometric dimensions. Owing to the importance of j and f in the design of compact heat exchangers, this work mainly focused on the determination of above factors. Similarly, the fins used in these heat exchangers are characterized for the thermo-hydraulic performance parameters to develop a correlation with the aid of simulation tool Fluent.

Computational fluid dynamics is the analysis of systems involving fluid flow and heat transfer by means of computer-based simulation. The technique is very powerful and spans a wide range of industrial and non-industrial application areas. The availability of affordable high performance computing hardware and the introduction of user-friendly interfaces have led to the development of commercial CFD packages. Today, well-tested commercial CFD packages have not only made CFD analysis a routine design tool in industries, but have also helped the research engineers focus on the physical systems more effectively. All formal CFD software contains three elements: (i) pre-processor (ii) main solver and (iii) post processor.

5.5.1 Pre-Processor

Pre-processing, as the first step of CFD analysis, consists of the input of a flow problem to a CFD program, in which the user provides the following details:

- selection of the phenomena that need to be modelled
- definition of properties
- definition of the geometry of the region, i.e. the computational domain
- specification of appropriate boundary conditions
- grid generation

5.5.2 Main Solver

The solver is the heart of the CFD software; it sets up the equation set according to the options chosen by the user and meshes points generated by the pre-processor, and it then solves them to compute the flow field. The process involves the following tasks:

- integration of the governing equations of fluid flow over all the control volumes of the domain
- discretization – conversion of the resulting integral equations into a system of algebraic equations
- solution of the algebraic equations by an iterative method

5.5.3 Post-Processor

The post processor is the final part of CFD software. It helps the user to examine the results and extract useful data. The results may be displayed as vector plots of velocities, contour plots of scalar variables such as pressure and temperature, streamlines and animation in the case of unsteady simulation. Global parameters such as drag coefficient, lift coefficient, Nusselt number and friction factor may be computed through appropriate formulas. This data from a CFD post-processor can also be exported to visualization software for better display. Several general-purpose CFD packages have been published in the past decade. Prominent among them are Phoenics, Fluent and Star-CD. Most of them are based on the finite volume method. CFD packages have also been developed for special applications, including Flotherm and Icepak for electronics cooling, CFX-Tascflow and Turbo for turbo machinery and Orca for mixing process analysis. Most CFD software packages contain their own grid generators and post-processors. Visualization software used with CFD packages includes Tecplot, Fieldview and CFD Post.

5.5.4 Errors and Uncertainty in CFD Modelling

The application of CFD modelling as an engineering tool can only be justified on the basis of its accuracy and the level of confidence in its results. To address the issue of trust and confidence in CFD, the fraternity has now carried out extensive reviews of the factors influencing simulation results. In the context of trust and confidence in CFD modelling, the following definitions of error and uncertainty have now been widely accepted.

1) **Error:** a recognizable deficiency in a CFD model that is not caused by lack of knowledge
   - numerical errors – round-off errors, iterative convergence errors, discretization errors
   - coding errors – mistakes or bugs in the software
   - user error – human errors through incorrect use of the software
2) **Uncertainty:** a potential deficiency in a CFD model that is caused by lack of knowledge
   - input uncertainty – inaccuracies due to limited information or approximate representation of geometry, boundary conditions, material properties etc.
   - physical model uncertainty – discrepancies between real flows and CFD due to inadequate representation of physical or chemical processes (e.g. turbulence) or due to simplifying assumptions in the modelling process (e.g. incompressible flow)

5.6 CFD Approach

The CFD approach for both pressure drop and heat transfer prediction across the fins is explained in this section. Initially, mesh finalization is done across the computational domain. Different mesh configurations starting with very coarse to very fine are taken at a particular Reynolds number and analysed using Fluent. The grid independence test is carried out for each fin surface and a graph is plotted between the number of elements and the performance parameter, i.e. pressure drop of the fin. The graph is then analysed for the minimum number of elements required for consistent results and hence further analysis is conducted with more than the minimum number of grid elements.

In the flow analysis, in order to overcome the entrance effect, the concept of periodic fully developed flow (Patankar et al. [12,13]) is implemented. The friction factor is calculated from the pressure drop obtained under quasi-isothermal conditions and using the basic equations. The corresponding velocity profile is also obtained, which is used as an input for the thermal analysis. The same procedure is repeated for the range of Reynolds number from 100 to 10,000. This analysis is repeated for the whole fin geometry.

In the thermal analysis, the "velocity inlet" (taken from flow analysis) and "outflow" boundary conditions are used at the inlet and outlet of the fin geometry, respectively. The constant wall temperature boundary condition is employed for walls. After the thermal analysis, post-processing is done for temperatures and pressures at the inlet and outlet over the computational domain using mass-weighted average. The pressure, temperature and velocity profiles are taken at the various sections of the fin for corresponding Reynolds number. This temperature difference between inlet and outlet of the core, in turn, is used for calculating the j factor using the basic equations. The same procedure is repeated for the range of Reynolds numbers from 100 to 10,000. The same approach is used to characterize all different fin geometries.

The following steps are carried out in numerical analysis for the generation of Colburn factor j and Fanning friction factor f for plain and perforated fins:

- modelling of the fin and suitable mesh generation
- calculating the input values for CFD analysis

- analysing the modelled fins using Fluent 12.1
- numerical calculation of j and f factors
- comparison of calculated values with standard values – validation of results
- forming the correlations of j and f factors for extended surfaces

5.6.1 Mathematical Model

The conservation equations of mass, momentum and energy are solved using the finite volume method. There are several turbulence models available in the code. The turbulent flow is calculated by the semi-implicit Simpler as outlined in Versteeg and Malalasekera [14]. A standard κ–ε model as given in Versteeg and Malalasekera [14] and Manabu Horiuchi et al. [15], with enhanced wall treatment, is used to predict turbulent flow in the fin geometry. The Reynolds transport equations can be written in a generalized form (Patankar [16–18] and Anderson [19]):

$$\mathrm{div}(\rho\mu\phi) = \mathrm{div}(\Gamma\,\mathrm{grad}\,\phi) + S_{\emptyset} \tag{5.6}$$

where ϕ stands for a generalized transport variable, which is used for all conserved variables in a fluid flow problem, including mass, momentum and the turbulence variables κ and ε. Γ represents the effective diffusivity (sum of the eddy diffusivity and the molecular diffusivity). S_ϕ is the source term for the respective dependent variable. The solution of the above equation is applied for the prediction of velocity and turbulence levels throughout the domain. The convergent criteria are specified to absolute residuals ($\leq 1.0 \times 10^{-5}$).

5.6.2 Governing Equations

The governing equations of fluid flow represent mathematical statements of the conservation laws of physics:

- mass of a fluid is conserved
- rate of change of momentum equals the sum of the forces on a fluid particle (Newton's second law)
- rate of change of energy is equal to the sum of the rate of heat addition to and the rate of work done on a fluid particle (first law of thermodynamics)

The meaning assigned to ϕ, Γ and S_ϕ for conservation of mass, momentum and energy are listed in Table 5.1. Substitution of the variable ϕ in equation (5.6) by those defined in Table 5.1 leads to the general Equations 5.7–5.9 of fluid dynamics:

Conservation of mass

$$\frac{D\rho}{D\lambda} = -\rho\,(\nabla\,.\,v) \tag{5.7}$$

$$\text{where}\quad \frac{D}{D\lambda} = \frac{\partial}{\partial\lambda} + u\frac{\partial}{\partial x} + v\frac{\partial}{\partial y} + w\frac{\partial}{\partial z} \tag{5.8}$$

$$\text{and}\quad \nabla\,.\,v = \frac{\partial u}{\partial x} + \frac{\partial v}{\partial y} + \frac{\partial w}{\partial z} \tag{5.9}$$

Table 5.1 Description of variables ϕ, Γ and S_ϕ for conservation of mass, momentum and energy.

Conservation law	ϕ	Γ	S_ϕ
Conservation of mass (*Continuity* Equation)	1	0	0
Conservation of momentum in x direction	u	μ	$-\frac{\partial\psi}{\partial x}+\rho g_x+$ fluid friction $+S_u$
Conservation of momentum in y direction	v	μ	$-\frac{\partial\psi}{\partial y}+\rho g_y+$ fluid friction $+S_v$
Conservation of momentum in z direction	w	μ	$-\frac{\partial\psi}{\partial z}+\rho g_z+$ fluid friction $+S_w$
Conservation of energy (temperature)	θ	$\frac{k}{c_p}$	$-\frac{D\psi}{D\lambda}+$ viscous dissipation + other heat sources

Conservation of momentum

$$\rho\frac{D\upsilon}{D\lambda}=-\nabla\rho-[\nabla.\tau]+\rho g \tag{5.10}$$

Conservation of Energy

$$\rho\frac{DU}{D\lambda}=-(\nabla.q)-\psi(\nabla.\upsilon)-(\tau.\nabla\upsilon) \tag{5.11}$$

5.6.3 Assumptions

The following assumptions are made in the CFD simulation for numerical solution of the equation set. While some of these assumptions help in simplifying the numerical scheme, others direct the problem towards practical applications.

- The flow is steady and incompressible. The fluid density is constant throughout the computational domain.
- Air/water/R134a are the working fluids, the fluid properties (viscosity, thermal conductivity and specific heat) being constant throughout the computational domain.
- The flow is periodically developed both hydrodynamically and thermally. Patankar et al. [12] have shown that it is indeed the case in most practical situations.
- The fluid flow meets the Boussinesq assumption.
- No slip boundary condition is considered on the wall.
- The CAD model does not include burrs, sharp edges etc.

5.6.4 Boundary Conditions

All CFD problems are defined in terms of initial and boundary conditions. It is important to specify these correctly and understand their role in the numerical algorithm. The most common boundary conditions that are used in discretized equations of the finite volume method are

- inlet
- outlet

- wall
- prescribed pressure
- symmetry
- periodicity

5.6.4.1 Inlet Boundary Conditions

The distribution of all flow variables needs to be specified at inlet boundaries. The various types of inlet boundary conditions that are predominantly used are

- pressure inlet
- velocity inlet
- mass flow inlet
- inlet vent
- intake fan.

Generally, for all the incompressible flows, the velocity inlet boundary condition can be used, as the velocity variations with respect to the density are zero. For compressible problems, the mass flow inlet boundary condition is more appropriate. The velocity will be calculated at the inlet, based on the mass flow rate, area and specified inlet density. The pressure inlet conditions are suitable for both incompressible and compressible flow calculations. This boundary condition is used when inlet pressure is known, but flow rate and/or velocity is not known. This situation may be best suited for buoyancy-driven flows.

5.6.4.2 Outlet Boundary Conditions

The outlet boundary conditions that are predominantly used are

- pressure outlet
- pressure far field
- out flow
- outlet vent
- exhaust vent.

For the pressure outlet boundary condition the static pressure at the outlet boundary has to be specified. The value of specified static pressure is used only while the flow is subsonic. If the flow becomes locally supersonic, the specified pressure will no longer be used. For incompressible problems the constant static pressure of zero can be defined, but for compressible problems proper static pressure should be defined as it has a role to play in the solution accuracy and convergence.

A pressure far field boundary condition is used in Fluent to model a free stream condition at infinity, with free stream Mach number and static conditions being specified. This condition is often called a characteristic boundary condition since it uses characteristic information to determine flow variables at the boundaries.

Outflow boundary conditions are used in flow situations where the details of flow velocity and pressure are not known prior to solution. No parameter is defined at the outflow boundary condition. The major assumption for this boundary condition is that the flow should be fully developed.

5.6.4.3 Wall Boundary Conditions

In most of the flows, whether internal flow or external flow, solid boundaries are encountered. The wall boundary conditions are then used. The wall boundary conditions play a very significant role, especially whenever the energy equation is switched on and the heat transfer is to be properly determined. This is particularly important for turbulence modelling. The walls should be properly captured, and very fine mesh should be taken at the wall. The conduction, convection or radiation – or any combination of them – and the other parameters can be defined in the wall boundary conditions. For the heat transfer, there are two types of boundary conditions that can be specified at the wall for the fully developed flows:

- constant wall heat flux
- constant wall temperature

In the constant wall heat flux boundary condition, a constant specified heat flux is defined over the length while in the constant wall temperature the wall temperature is defined to be constant.

5.6.4.4 Constant Pressure Boundary Condition

The constant pressure boundary condition is used in situations where exact details of the flow distribution are unknown but the boundary values of the pressure are known. Typical problems where this boundary condition is appropriate include external flows around objects, free surface flows, buoyancy-driven flows such as natural ventilation and fires, and also internal flows with multiple outlets. In applying the fixed pressure boundary, the pressure correction is set to zero at the nodes.

5.6.4.5 Symmetric Boundary Condition

Symmetric boundary conditions can be used when the physical geometry of interest, such as the outlined heat exchanger channel, and the expected pattern of the flow/thermal solution are symmetric. When using this type of boundary condition in such regions no additional boundary conditions are required. Fluent assumes zero flux of all quantities across a symmetric boundary.

5.6.4.6 Periodic Boundary Condition

Since the heat exchanger channel is characterized by a geometry periodically repeating in the flow direction, and the expected flow pattern would have a periodically repeating nature, the use of this type of boundary conditions is valid. Fluent treats the flow at a periodic boundary as though the opposing periodic plane is a direct neighbour to the cells adjacent to the first periodic boundary. Thus, when calculating the flow through the periodic boundary adjacent to a fluid cell, the flow conditions at the fluid cell adjacent to the opposite periodic plane are used. The details of periodicity and periodic boundary conditions are given in Patankar et al. [12].

5.6.5 Turbulence Models

Accuracy of the various turbulence models for the numerical simulation of fins has been discussed in the works of Amano et al. [20], Patel et al. [21], Amano [22] and Manabu Horiuchi et al. [15]. According to Patel et al. [21], the turbulence model of Chen and Patel

[23] appears to capture most of the important physical features of such flows. In addition, the Reynolds stress model (RSM) can take non-isotropic effects into account, which are strong in the recirculation region. Later, Manabu Horiuchi et al. [15] made a numerical simulation of turbulent flow with the Reynolds-averaged Navier–Stokes (RANS) turbulent models. The models are the two layers, the low Reynolds number k–ε, and the V2F models. This reveals that sophisticated models such as the V2F model and the non-linear low k–ε model predicted more precisely the influence of the flow in the recirculation region on the wall.

In turbulent flow, velocity fields are fluctuating. These fluctuations mix with transport quantities such as momentum, energy and species concentration; consequently, the transport quantities are fluctuating as well. The exact governing equation, however, can be time-averaged or ensemble-averaged to cancel the small fluctuations. A modified set of equations are created from this operation. Unknown variables are generated in the modified equations, and these variables are determined as known quantities by using the turbulence model.

The following turbulence models in Fluent can be used:

1) Spalart–Allmaras model
2) κ–ε models:
   a) standard κ–ε model
   b) renormalization group (RNG) κ–ε model
   c) realizable κ–ε model
3) κ–ω models:
   a) standard κ–ω model
   b) shear stress transport (SST) κ–ω model
4) Reynolds stress model (RSM)
5) large eddy simulation model (LES)

For each model, there are advantages and disadvantages. In the present study, standard κ–ε model is selected for these types of extended surfaces. The standard κ–ε model is considered the simplest 'complete model' of turbulence. This model is widely used in industrial flow simulation due to robustness, economy and reasonable accuracy for a wide range of turbulent flows. It is the workhorse of practical engineering flow calculations.

There is no single model that is universally accepted for all classes of problems; the choice of turbulence model depends on considerations such as the physics encompassed in the flow, the established practice for a specific class of problem, the level of accuracy required, the available computational resources and the amount of time available for the simulation. There are a few turbulent flow models such as κ–ε, RANS, DNS and V2F models in the Fluent 12.1 package.

5.7 Numerical Simulation

In this section the numerical simulation through CFD method is applied to steady state and transient analysis for generation of j and f data for compact heat exchanger fins.

5.7.1 Transient Analysis

In addition to the steady state κ–ε model proposed for CFD analysis of various types of fins for generation of *j* and *f* data, a transient method is also presented here for generation of *j* factors for lance and offset fins as an alternate method. Generally the single-blow testing technique will be used for generation of heat transfer coefficient and pressure drop values of any heat exchanger test core. A simple schematic of the single-blow testing technique along with a test core is shown in Figure 5.3 for easy understanding. A typical single transient test rig is an open circuit wind tunnel and consists of a blower, a heating element, a test segment and devices for temperature, flow and pressure drop measurements, similar to references [24–32]. The modified test rig schematic is shown in Figure 5.3, which is the basis for proposed transient numerical analysis.

If a one-dimensional fluid flow with constant properties is assumed to pass through the test core matrix for transient analysis, the model can be represented with three governing equations to describe the single-blow system. From the energy balance between the fluid flow and the test core matrix and the side wall, it yields the dimensionless energy equations for the fluid similar to Chen and Chang [33,34]:

$$\frac{\partial T_f}{\partial \mathrm{x}} + NTU(T_f - T_s) + NTU_w(T_f - T_w) = 0 \tag{5.12}$$

for the test core matrix

$$\frac{\partial T_f}{\partial t} - \lambda \frac{\partial^2 T_s}{\partial x^2} + NTU(T_s - T_f) = 0 \tag{5.13}$$

and for the side wall

$$\frac{\partial T_w}{\partial t} - R_{tc}\lambda_w \frac{\partial^2 T_w}{\partial x^2} + R_{tc}NTU_w(T_w - T_f) = 0 \tag{5.14}$$

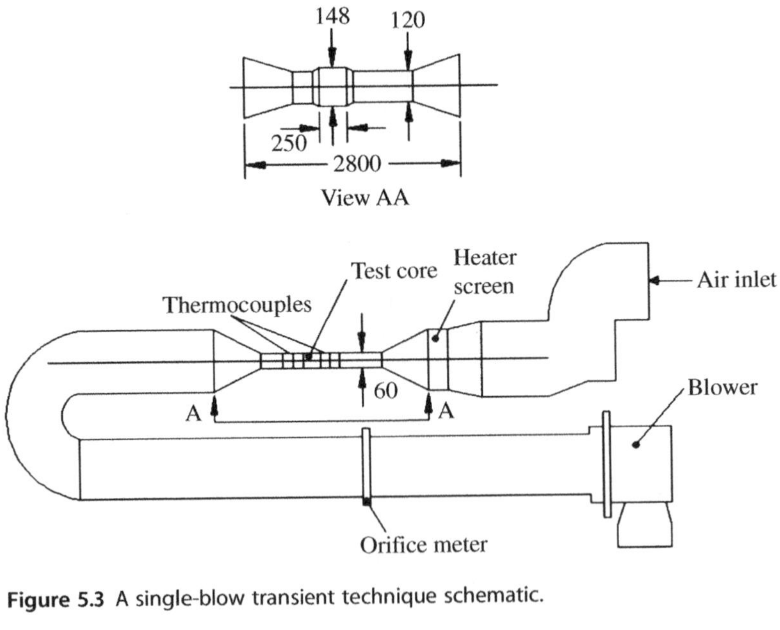

Figure 5.3 A single-blow transient technique schematic.

The dimensional variables or parameters used in Equations 5.12–5.14 are defined as

$$T_f = \frac{T^*_f - T^*_0}{T^*_{fmax} - T^*_0},\ T_s = \frac{T^*_s - T^*_0}{T^*_{fmax} - T^*_0},\ T_w = \frac{T^*_w - T^*_0}{T^*_{fmax} - T^*_0},\ x = \frac{x^*}{L}$$

$$t^* = \frac{t}{(M_s C_s / m_f C_p)},\ R_{tc} = \frac{M_s C_s}{M_w C_w} \qquad (5.15)$$

$$NTU = \frac{HA}{m_f C_p},\ NTU_w = \frac{H_w A_w}{m_f C_p},\ \lambda = \frac{k_s A_s}{m_f C_p L}\ \text{and}\ \lambda_w = \frac{k_w A_{cw}}{m_f C_p L} \qquad (5.16)$$

As compared with an adiabatic side wall, additional non-dimensional groups such as NTU_w, R_{tc} and λ_w can also be included in the model. Equations 5.12–5.14 are subject to the following initial and boundary conditions

$$t^* = 0, T_f(x) = T_s(x) = T_w(x) = 0 \qquad (5.17)$$

$$\left.\begin{aligned} &x = 0,\ T_f = 1 - \exp(-t^*/\tau) \\ &x = 0,\ \frac{\partial T_s}{\partial x} = 0,\ \frac{\partial T_w}{\partial x} = 0 \\ &x = 1,\ \frac{\partial T_s}{\partial x} = 0,\ \frac{\partial T_w}{\partial x} = 0 \end{aligned}\right\}\ \text{for}\ t^* > 0 \qquad (5.18)$$

where τ is the non-dimensional time constant of the suddenly heated inlet fluid temperature response. The above formulation is presented here to understand the basic governing equations for transient analysis. The test section fin details are tabulated in Table 5.2 for transient analysis.

5.7.1.1 Data Reduction and Validation

In the present work, the maximum slope method is used for data reduction. As explained in Pucci et al. [35], for a given *NTU*, the *maximum slope* (using Table 5.3) has a unique value. Thus, if the *maximum slope* can be measured experimentally, the *NTU* can be found.

The maximum slope for a given *NTU* is found by gradually increasing the time parameter, τ. The slope steadily increased until the maximum value is reached. Once the slope began to decrease, the maximum slope for the given *NTU* is known. In addition, the dimensional values which defined *NTU* and τ are substituted into the left side of Equation 5.19 to obtain the slope as defined by Jeffrey L. Wiese [36]:

$$\text{Slope} = \frac{d\left(\dfrac{T^*_f - T^*_0}{T^*_{f\,max} - T^*_0}\right)}{d\left(\dfrac{\tau}{NTU}\right)} = \left(\frac{W_s C_s}{W C_p}\frac{dT_f}{d\phi}\bigg|_{max}\right) \qquad (5.19)$$

$$\text{where}\ \tau = \frac{HAt}{M_s C_p} \qquad (5.20)$$

The maximum slope defined by Equation 5.19 generally can be found from the experimental values of temperature, time, test section mass, mass flow rate and tabulated specific heats. However, a CFD approach is proposed here as an alternate method in the subsequent section. Knowing the maximum slope, the *NTU* could be determined from the *NTU* vs maximum slope of Table 5.3. The next step is to consider the definition of

Table 5.2 Lance and offset and wavy fin details.

Test core/ parameters	Core 1 stainless steel, offset	Core 2 aluminium, offset	Core 3 aluminium, wavy	Core 4 aluminium, wavy	Core 5 aluminium, offset
FPI	20	18	28	15.5	28
t (mm)	0.0762	0.152	0.152	0.0762	0.127
h (mm)	2.79	5.33	9.00	4.09	5.00
l (mm)	3.175	3.175	9.525	9.525	3.175
s (mm)	1.27	1.41	0.907	1.64	0.91
$2A$ (mm)	—	—	0.99	0.99	—
d_h (mm)	1.659	2.026	1.392	2.25	1.345
t_p (mm)	0.254	0.5	0.5	0.5	0.75
N_{layer}	10	10	6	10	10
W (m)	0.167	0.060	0.092	0.132	0.117
Y (m)	0.0307	0.0588	0.0575	0.0464	0.0583
L (m)	0.118	0.250	0.090	0.160	0.125
M_s (kg)	0.749	0.5133	0.2946	0.4576	0.4966
C_s (J/K)	376.75	458.38	263.08	408.7	443.46
A (m^2)	1.128	1.271	1.008	1.310	1.634
A_c (mm^2)	3965	2575	3896	4605	4394

Table 5.3 ***NTU*** **vs maximum slope – Pucci et al. [35].**

Sl. No.	Max slope	*NTU*
1	0.5437	2.2
2	0.5494	2.4
3	0.5766	3
4	0.6208	3.8
5	0.6603	4.5
6	0.7414	6
7	0.8397	8
8	0.9286	10
9	1.121	15
10	1.286	20
11	1.432	25
12	1.565	30

Stanton number, $S_t = H/Gc_p$. From the definition of G and Stanton number S_t, the desired Colburn factor j can be related to NTU.

$$j = S_t Pr^{2/3} = NTU\left(\frac{A_c}{A}\right)Pr^{2/3} \tag{5.21}$$

Since a unique relationship exists between the NTU and the exit fluid temperature response, the NTU value of the test heat exchanger is then evaluated by matching the recorded measured exit fluid temperature history with the predicted one. Different methods have been proposed in the literature for determining the NTU from the transient test data such as (a) maximum slope method (b) curve matching method (c) centroid method and (d) differential fluid enthalpy method. In addition, the curve-matching scheme, a maximum-slope scheme, is also frequently used to evaluate the NTU value of the test heat exchanger. It may be noted that the determined NTU value of the test heat exchanger varies with the model used in the single-blow measurement. For example, the NTU value of the test heat exchanger obtained from the model by including the axial conduction is higher, than that obtained from the model without considering the axial conduction, for $\lambda \times NTU > 0.06$ according to Chen and Chang [33,34].

Even though the one-dimensional fluid flow formulation was presented above, to understand the basic governing equations, a 3D model is considered for CFD analysis using Fluent 6.2 for an estimation of outlet temperature curve using transient model for validation of single-blow experimental results for a typical model. The 3D model of duct was created in Gambit. The three-dimensional computational domain of this model is shown in Figure 5.4.

The model is divided into three zones, the middle zone being the porous zone. Meshing is carried out using quadrilateral grids. The number of mesh cells is optimized by carrying out a grid independency exercise. The appropriate cell size is found to be 2,800,000. Another important parameter in CFD transient analysis is time step. This has been optimized by a trial and error method considering the mass flow rate and size of test core. The following assumptions are made for CFD analysis:

1) The flow is turbulent and the k–ϵ model is used for Reynolds numbers 1000 to 4500.
2) Entrance and exit effects are negligible, i.e. fully developed flow is assumed.
3) Radiation heat transfer effects are ignored.

Figure 5.4 CFD fluid mesh model of duct and core for transient analysis.

4) The fluid flow is incompressible.
5) The physical properties of both solid and fluid phases are constant throughout the temperature range investigated.
6) The matrix is assumed to be an isotropic porous body.
7) The fluid and the matrix are assumed to be at the same temperature.
8) The set of governing equations is solved in a segregated fashion.
9) Momentum and pressure coupling is handled with Simple algorithm.
10) Discretization schemes: pressure: Presto; momentum: first-order upwind; energy: first-order upwind.
11) Under-relaxation factors: momentum: 0.7, pressure-correction: 0.3, energy: 1, density: 1 and body forces: 1.

The porosity (γ) is defined as

$$\gamma = \frac{Total\ volume - Solid\ volume}{Total\ volume} \tag{5.22}$$

In order to compare the numerical prediction with the experimental results, viscous resistance and inertia resistance factors of the porous media are experimentally determined by using the Forcheimer equation.

$$\Delta p = \left(\frac{\mu}{\alpha}v + \frac{C_2\rho}{2}v^2\right)L \tag{5.23}$$

The procedure followed relies on measuring the pressure drop Δp over a porous medium in a rectangular duct of an equivalent radius to that of the porous material at different duct velocities, v,, by fitting a second-order polynomial through these points, as shown in Figure 5.5. The viscous resistance coefficient, $1/\alpha$ and inertial resistance coefficient and C_2 given in Equation (5.23) are determined.

At the inlet section, mass flow rate and temperature are given as input for CFD analysis. For the porous zone, porosity, inertia resistance and viscous resistance factors obtained experimentally are given as inputs. For the outlet section, outflow boundary conditions are given. Initially, room temperature is given as input and solved for steady state condition.

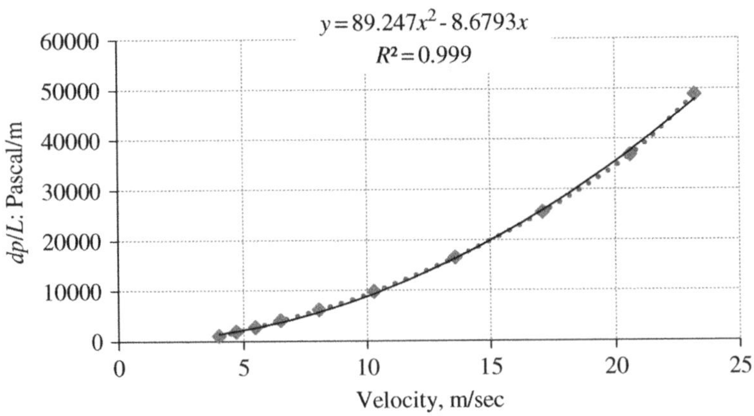

Figure 5.5 Determination of inertial and viscous resistance factors.

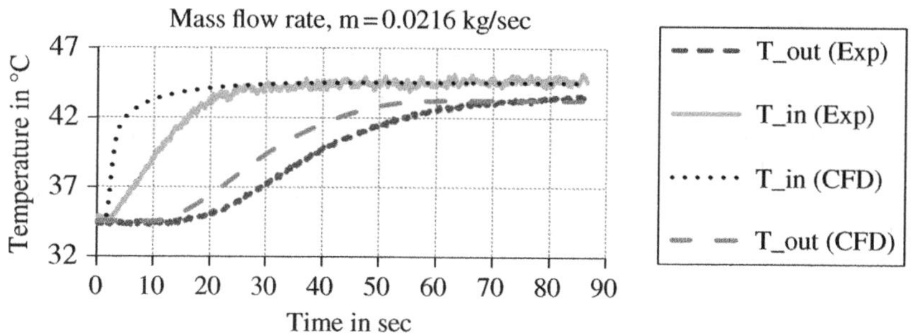

Figure 5.6 Inlet and outlet temperature responses for offset fin core.

Then at the inlet, a step temperature input is made. Then the solver is set for transient conditions with a suitable time step. The temperature response at the outlet is noted. Then using the maximum slope method, *NTU* is obtained. A sample temperature prediction using CFD is compared with experimental data for an offset fin core with mass flow rate of 0.0216 kg/sec, as shown in Figure 5.6 with time step of 0.0001 s.

For the corresponding case, the maximum slope graph is shown below in Figure 5.7. The *NTU* is calculated at max slope value using Equation 5.19 and using Table 5.3 as per the procedure given by Wiese [36]. In this graph, the max slope of the curve is 0.95 at $t^* = 1.1$ and the corresponding *NTU* is 11 from Table 5.3. In Figure 5.7 the experimental value of max slope from the same author is marked, which is around 0.92, and the corresponding *NTU* is around 10 from Table 5.3, for comparison of results. It is observed that the variation between numerical and experimental prediction is around 10%.

Five types of compact heat exchanger fins are tested using the single-blow testing method. The experiments are conducted at different mass flow rates, ranging from 0.017 kg/s to 0.126 kg/s. The single-blow transient test results are presented as *j* factor vs *Re* for all five test cores, as shown in Figure 5.8. In Figure 5.8a, the experimental *j* factors of the stainless steel offset fin test core no. 1 are compared with an available published correlation based on the CFD calculations. The single-blow testing method under-predicts the *j* values by about 20% when compared to the CFD results. This

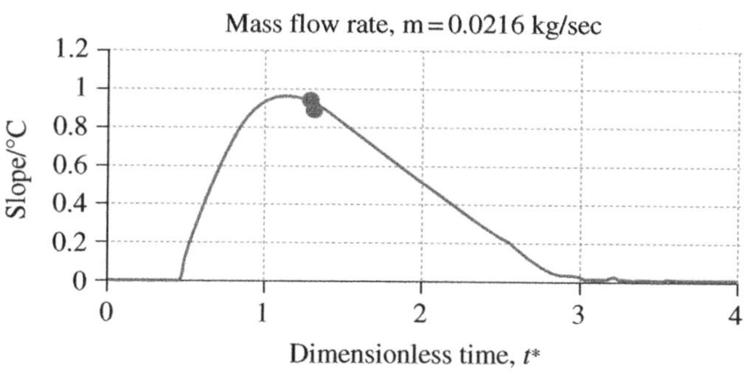

Figure 5.7 Max slope vs time for offset fin core using CFD data.

Table 5.4 Values of *NTU*, R_{tc}, R_{NTU}, λ and λNTU for test cores

Test core/ parameters	Core 1 Stainless steel, offset	Core 2 Aluminium, offset	Core 3 Aluminium, wavy	Core 4 Aluminium, wavy	Core 5 Aluminium, offset
m_f (kg/s)	0.024–0.126	0.019–0.082	0.023–0.125	0.027–0.125	0.022–0.125
NTU	7.3–3.2	11.0–5.4	6.7–3.0	3.7–2.0	12.8–4.6
R_{tc}	96.6	143.3	83.5	99.7	152.9
R_{NTU}	0.007–0.011	0.005–0.007	0.009–0.013	0.014 – 0.019	0.005–0.007
λ	0.005 – 0.0009	0.014 –0.0032	0.029 – 0.0050	0.033 – 0.007	0.072 –0.0125
λ*NTU*	0.035–0.003	0.156–0.017	0.200–0.014	0.12–0.014	0.928–0.058
Comparison of parameters	Moderate R_{tc} and R_{NTU}	High *NTU*, and moderate λ*NTU*	—	—	High *NTU*, high λ and high λ*NTU*
λ_w			3.36×10^{-5} to 1.23×10^{-5}		

deviation is justifiable mainly due to the lower heat capacity ratio R_{tc} and higher *NTU* ratio R_{NTU}, as listed in Table 5.4.

The experimental *j* factors of the aluminium offset fin test core no. 2 are given in Figure 5.8b. It shows that the single-blow testing method under-predicts the *j* values by about 25% compared to the CFD results. This deviation is mainly due to influence of higher λ and λNTU as listed in Table 5.3. This is a bigger core among the aluminium test cores and hence has higher heat capacity ratio R_{tc} and lower *NTU* ratio R_{NTU}. Figure 5.8c shows that the single-blow testing method predicts the *j* values of the aluminium wavy fin test core no. 3 very well compared to the CFD results. This is mainly due to insignificant influence of λ and R_{NTU}. The single-blow testing method predicts the *j* values of the aluminium wavy fin test core no. 4 within 15%

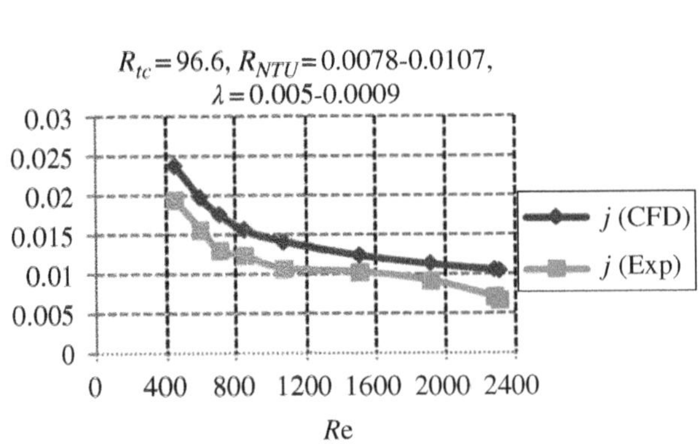

Figure 5.8a *Re* vs *j* for SS offset fin, test core no. 1.

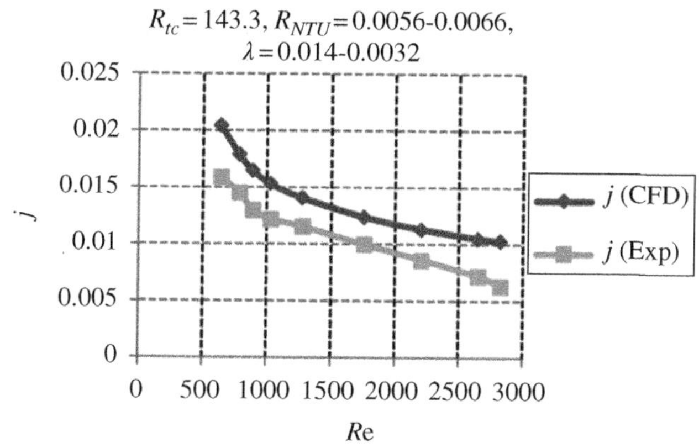

Figure 5.8b *Re* vs *j* for aluminium offset fin test core no. 2.

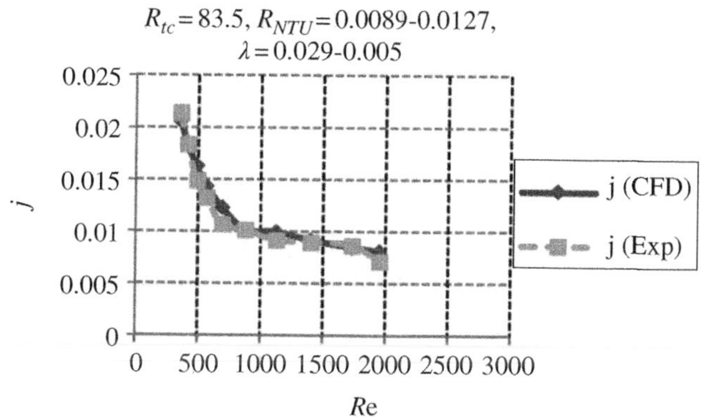

Figure 5.8c *Re* vs *j* for aluminium wavy fin test core no. 3.

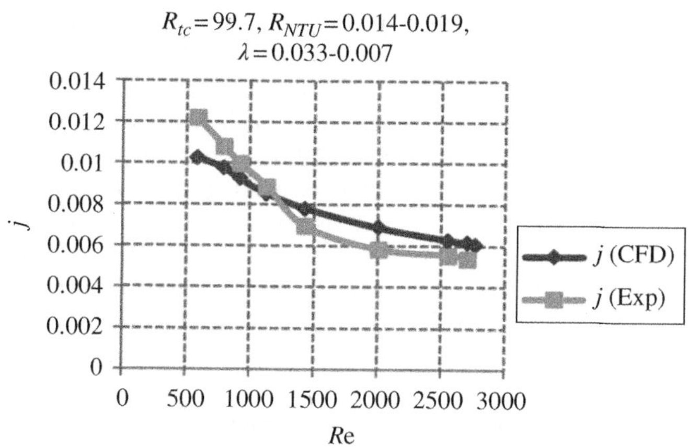

Figure 5.8d *Re* vs *j* for aluminium wavy fin test core no. 4.

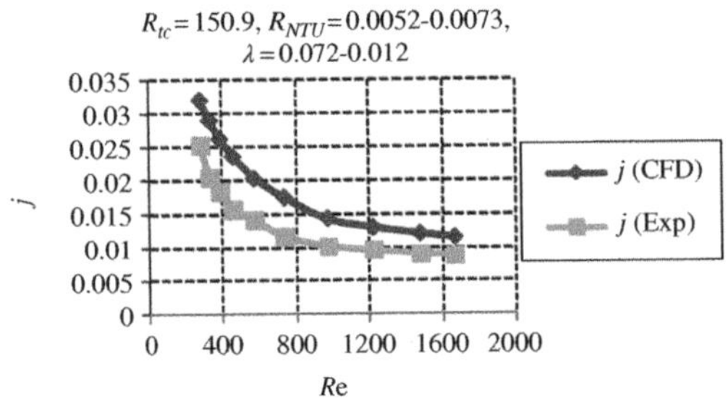

Figure 5.8e *Re* vs *j* for aluminium offset fin test core no. 5.

deviation when compared to the CFD results, as shown in Figure 5.8d. In this case no other parameters are influencing except moderately higher ratio of R_{NTU}.

In Figure 5.8e, the single-blow testing method under predicts the *j* values of the aluminium offset fin test core no. 5 with max variation of 25% when compared to the CFD results. This is mainly due to significant influence of λ and λNTU, which is much higher than 0.06 as indicated in Table 5.4. According to Chen and Chang [33,34], if $\lambda NTU >$ 0.06, the effect of the axial heat conduction is significant.

5.7.2 Steady State Analysis

The steady state analysis is applied to various types of compact heat exchanger fins for generation of *j* and *f* data correlations, which are required for designing compact plate-fin heat exchangers.

5.7.2.1 Wavy Fin

A CFD approach is used to conduct a number of numerical experiments to determine the thermo-hydraulic parameters, with a view to arriving at the optimum parameters for steady state analysis of wavy fin. The computational domain of the wavy fin model is shown in Figure 5.9. The CFD analysis is carried out for different geometric parameters and Reynolds numbers using Fluent 6.2 on a Red Hat Linux OS/HP xw8000 workstation. The analysis is carried out in two phases. In first phase, the wavy fin is taken and characterized for *f* values over a range of Reynolds number. In the second phase, the *j* value is determined for the same range by switching on the energy equation. The mass flow rates are determined for a range of Reynolds number from 100 to 15,000. In order to overcome the entrance effect, the concept of periodic fully developed flow, as suggested by Patankar et al. [12], is implemented for flow analysis. After the analysis, the pressure drop for unit length is multiplied by the actual length to get the total pressure drop for corresponding fins. From the pressure drop, the friction factor is calculated as per Kays and London [3]. Finally, the corresponding two-dimensional fully developed velocity profile is listed out. Similarly, the same procedure is repeated for the range of Reynolds numbers from 100 to 15,000 in order to draw the *f* vs *Re* curve.

The mesh finalization is done before analysing for the *j* and *f* factors. Different mesh configurations, starting with very coarse to very fine, are taken at a particular Reynolds

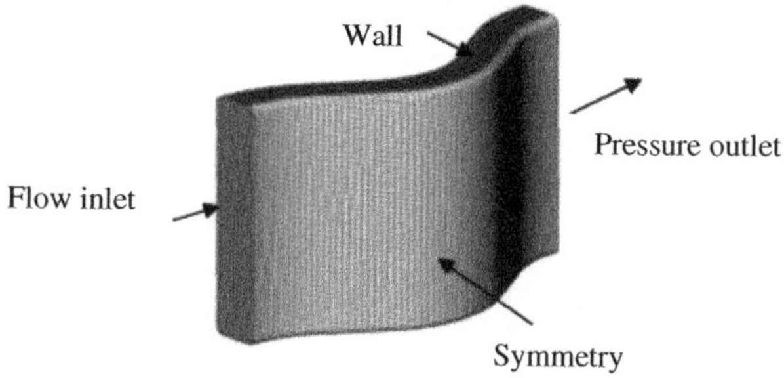

Figure 5.9 Computational domain of a wavy fin.

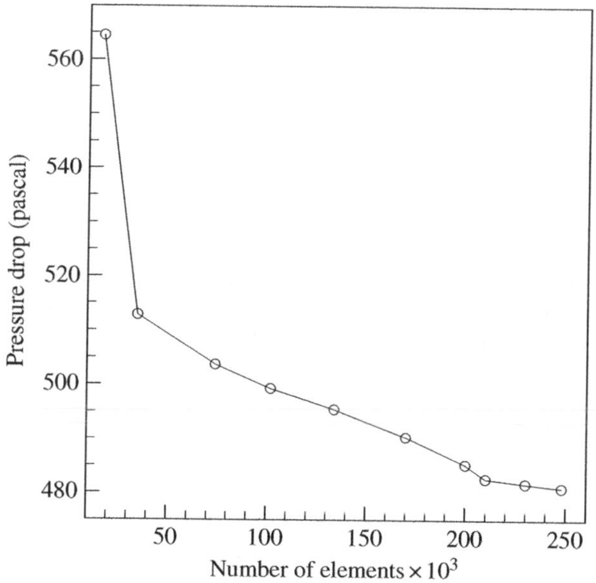

Figure 5.10 Grid independency graph – pressure drop variation.

number and analysed. The grid independency curve is drawn between the number of elements and the pressure drop parameters. Then the mesh size is determined based upon the curve where the slope is almost zero, as shown in Figure 5.10 for pressure drop variations.

The actual mass flow rate is used as the boundary condition. The no-slip boundary condition is used for the walls. The computation time taken for solving the first phase using the 3D segregated k–ε turbulent model with enhanced wall treatment is 30 min. The time taken for solving the second phase using the same model is 20 min. The velocity vector plot for a typical wavy fin computational domains is presented in Figure 5.11 for $Re = 1000$.

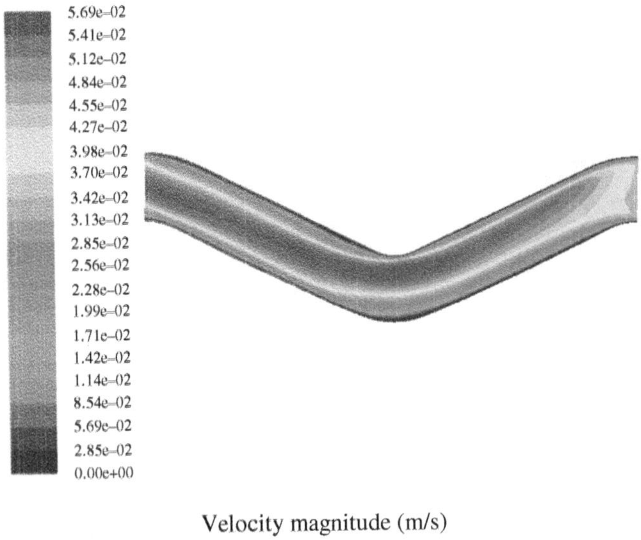

Figure 5.11 Velocity magnitude plot at *Re* =1000 for a typical wavy fin.

5.7.2.1.1 *Wavy Fin Data Comparison*

The results obtained from Fluent for wavy fins in the form of Colburn *j* and Fanning friction *f* factors are compared in Figure 5.12. It is evident that all data of both *j* and *f* factors matched well [38–42], as these are close to each other except Jungi et al. [37] for wavy fins.

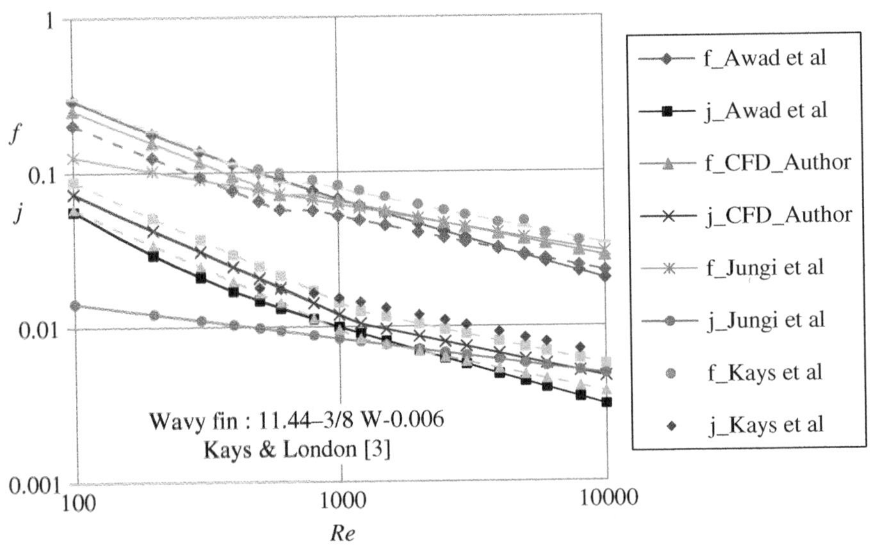

Figure 5.12 Comparison of Kays and London [3] and CFD data of *j* and *f* vs *Re* along with Awad et al. [42] and Jungi et al. [37] data for the wavy fin: 11.44–3/8 W-0.006.

5.7.2.1.2 Correlations of j and f factors for Air

An exhaustive numerical study is carried out on the heat transfer phenomena in plate fin surfaces with wavy fins. The correlations have been expressed in terms of two separate equations over the low and high *Re* regions, along with dimensionless geometric parameters. The power law expressions are used for determining the Colburn factor j and friction factor f as a function of the Reynolds number and dimensionless fin parameters.

The Fanning friction factor f is functionally related to *Re*, h/s, $2A/s$ and $L/2A$ and it can be represented as

$$f = B\,R_e^{a1}\left(\frac{h}{s}\right)^{a2}\left(\frac{2A}{s}\right)^{a3}\left(\frac{L}{2A}\right)^{a4} \tag{5.24}$$

where B, $a1$, $a2$, $a3$ and $a4$ are constants.

The use of these power law expressions is justified because variations in f with *Re*, h/s, $2A/s$ and $L/2A$ follow constant slope log-linear lines in both laminar and fully turbulent flow regions. The f vs *Re* data for wavy fin surfaces shows significant non-linearity over the Reynolds number range $100 \le Re \le 15{,}000$. Therefore, two separate equations have been proposed [45–48] for the low and the high *Re* regions as follows:

$$f = 9.827 R_e^{-0.705}\left(\frac{h}{s}\right)^{0.322}\left(\frac{2A}{s}\right)^{-0.394}\left(\frac{L}{2A}\right)^{-0.603} \tag{5.25}$$

for laminar range ($100 \le Re \le 800$)

$$f = 10.628\,R_e^{-0.359}\left(\frac{h}{s}\right)^{0.264}\left(\frac{2A}{s}\right)^{-0.848}\left(\frac{L}{2A}\right)^{-1.931} \tag{5.26}$$

for turbulent range ($1000 \le Re \le 15{,}000$)

The above correlations correctly predict 99% of the f data for laminar regions and 96% of the f data for the turbulent regions with an RMS value of 0.995. Similarly, two separate equations have been proposed for the low and the high *Re* regions for j factors as

$$j = 2.348\,R_e^{-0.786}\left(\frac{h}{s}\right)^{0.312}\left(\frac{2A}{s}\right)^{-0.192}\left(\frac{L}{2A}\right)^{-0.432} \tag{5.27}$$

for laminar range ($100 \le Re \le 800$)

$$j = 0.242\,R_e^{-0.375}\left(\frac{h}{s}\right)^{0.235}\left(\frac{2A}{s}\right)^{-0.288}\left(\frac{L}{2A}\right)^{-0.553} \tag{5.28}$$

for turbulent range ($1000 \le Re \le 15{,}000$)

The above correlations correctly predict 96% of the j data for laminar regime, and 95% of the j data for the turbulent regions, with RMS value of 0.987.

5.7.2.1.3 j and f Factors for Water and R134a

All the studies carried out above are used for air as the coolant medium. Only a few studies have been performed using liquid as coolant. Even though liquid cooling (using these fins) has been used for many years in applications, adequate data is not available in the available literature. Since they have better heat transfer coefficient than the air, liquids are preferred to air in some applications. Hu and Herold [43,44] carried out studies both analytically and experimentally using liquid water and polyalphaolefin for Prandtl number ranges from 3 to 150 on an offset fin surface. They have showed that the Prandtl number had a significant effect on the heat transfer factor *j* of the fin. They also observed that air models over-predict the *j* factor for liquids. However, the Prandtl number is found to have little effect on the Fanning friction factor. Ranganayakulu et al. [45–48] performed numerical analysis on offset fins and wavy fins using a CFD approach. They found that the *j* of water is lower by about 30% when compared to air, whereas there is no significant deviation in *f* values. Even though *j* is lower for water, the heat transfer coefficient (h) is much higher when compared to air. Kim and Sohn [49] carried out experiments using water and R113 as a test fluid on offset strip fins by considering the effect of fin geometry. In their experimental findings, they observed that there is no significant variation of *f* factor. However, the Colburn factor measured in their study is about 25% smaller than the prediction of Manglik and Bergles [50] for $Re < 1000$. For $Re > 1000$, the difference between the experimental data and the prediction of Manglik and Bergles [50] is reduced as the Reynolds number increased.

Most of the studies performed analysis on offset fins, and there are no specific correlations for the *j* and *f* characteristics of wavy fin surfaces using liquid as coolant. The experimental results of Kays and London [3] on three wavy fins are the only data available in open literature. But they used air as a medium. Maiti [51] carried out numerical analysis on wavy fins using air as a medium and proposed correlations. An extensive numerical study is extended to heat transfer phenomena and pressure drop in plate fin surfaces with wavy fins for water and R134a. In this analysis, the following ranges of fin geometries and Reynolds numbers are taken for *j* and *f* data generation with water and R134a fluids:

- $2.83 \le h/s \le 11.24$
- $1.13 \le 2A/s \le 2.53$
- $4.14 \le L/2A \le 8$ and
- $100 \le Re \le 1000$ for laminar region
- $1000 \le Re \le 15{,}000$ for turbulent region

5.7.2.1.4 Validation of Data – Wavy Fins

To validate the CFD results, offset fin geometry from Kim and Sohn [49] has been taken for the CFD analysis and modelled, since experimental results with wavy fin for liquid water/refrigerant R134a as working medium are not available in the open literature. The offset fin geometry considered for the analysis is as follows:

Fin height, h (mm) 2.8, fin length, l (mm) 1.5, lateral fin spacing, s (mm) 3.5, fin thickness, t (mm) 0.2 and hydraulic diameter, D_h (mm) 2.84

The above fin geometry is modelled, and generated in a suitable grid and the analysis carried out using ANSYS CFD. The results are analysed and compared with experimental results of Kim and Sohn [49] and plotted in Figure 5.13. The CFD results are found to be in good agreement with experimental results. The variation is found to be less than 5% in both *f* and *j* values. Uniform wall temperature boundary conditions are used for the

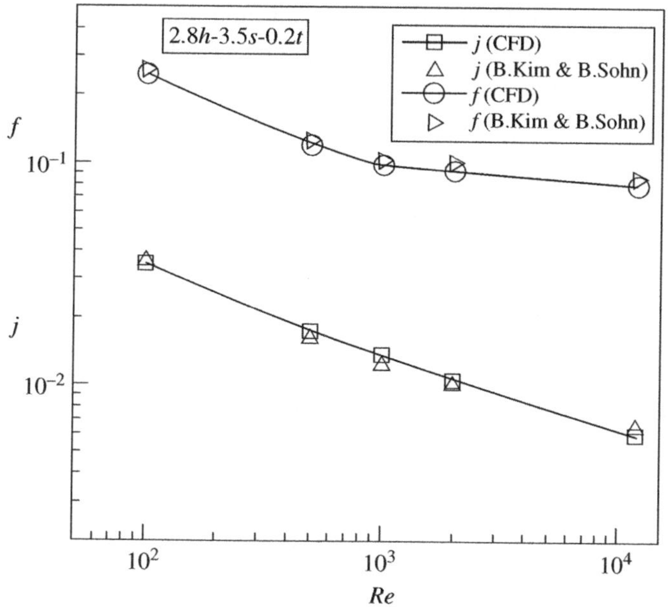

Figure 5.13 Validation with experimental results for the fin OSF: 1/9 inch-7.3 FPI-0.008 inch thick.

analysis. Before carrying out the analysis, a grid independence check is also carried out to optimize the mesh size.

5.7.2.1.5 j and f Correlations for Water and R134a

Similar to the above, the use of these power law expressions is justified because variations in *f* and *j* with *Re*, *h*/*s*, 2A/s, L/2A follow constant slope log-linear lines in both laminar and fully turbulent flow regions. The *j* vs *Re* data for wavy fin surfaces show significant non-linearity over the Reynolds number range $100 \le Re \le 15{,}000$. Hence, two separate equations have been proposed below for the low and high *Re* regions of *j* data for water and liquid R134a:

$$j = 2.989Re^{-0.54241}(h/s)^{-0.72276}(2A/S)^{-0.83914}(L/2A)^{-0.7588} \quad \text{for } 100 \le Re \le 1000 \tag{5.29}$$

$$j = 3.245Re^{-0.66388}(h/s)^{-0.53614}(2A/S)^{-0.80626}(L/2A)^{-0.63465} \quad \text{for } 1000 \le Re \le 15000 \tag{5.30}$$

$$j = 1.154Re^{-0.65938}(h/s)^{-0.96698}(2A/S)^{0.176702}(L/2A)^{0.288785} \quad \text{for } 100 \le Re \le 1000 \tag{5.31}$$

$$j = 0.323Re^{-0.69341}(h/s)^{-0.9602}(2A/S)^{0.633246}(L/2A)^{0.889252} \quad \text{for } 1000 \le Re \le 15000 \tag{5.32}$$

Equations 5.29 and 5.30 predicts the *j* factor for R134a and Equation 5.31 and Equation 5.32 predicts for the water. The above correlations predict 99% of the *j* data for laminar regions and 96% of the *j* data for the turbulent regions.

Similarly, the f vs Re data for wavy fin surfaces shows significant non-linearity over the Reynolds number range $100 \le Re \le 15000$ and hence, two separate equations have been proposed for the low and high Re regions:

$$f = 18.607 Re^{-0.59381} (h/s)^{-0.088954} (2A/S)^{-0.46976} (L/2A)^{-0.92621} \quad \text{for } 100 \le Re \le 1000 \tag{5.33}$$

$$f = 24.413 Re^{-0.46532} (h/s)^{-0.226497} (2A/S)^{-0.94256} (L/2A)^{-1.70937} \quad \text{for } 1000 \le Re \le 15{,}000 \tag{5.34}$$

The deviation between f data of water and liquid R134a is less than 5%. Hence, single correlations have been proposed for both fluids. The Prandtl number is found to have little effect on the Fanning friction factor. The above correlations correctly predict 99% of the f data for laminar regions and 96% of the f data for the turbulent regions.

5.7.2.2 Offset Fins

Similar to wavy fins, the CFD approach is extended to offset fins. The computational domain of the offset fin model is shown in Figure 5.14. Similarly, the same procedure is repeated for offset fins in the range of Reynolds numbers from 100 to 10,000 in order to draw the f vs Re curve.

The mesh finalization is carried out before analysing for the j and f factors. Different mesh configurations, starting with very coarse to very fine, are taken at a particular Reynolds number and analysed using Fluent. The grid independency curve is drawn between number of elements and the pressure drop parameters. Then the mesh size is determined based upon the curve where the slope is almost zero, as shown in Figure 5.15.

The actual mass flow rate is used as the boundary condition. The no-slip boundary condition is used for the walls. The computation time for solving first phase using a 3D segregated k–ε turbulent model with enhanced wall treatment is 30 min. The time taken for solving the second phase, using the same model is 20 min. The velocity vector plot for typical offset fins computational domains are presented in Figure 5.16 for $Re = 10{,}000$.

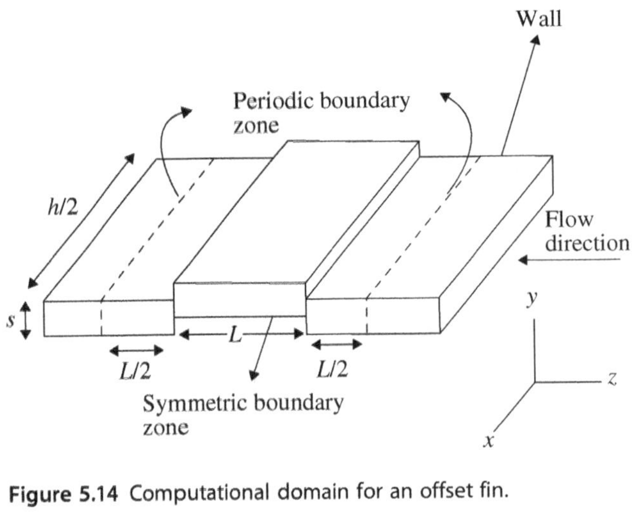

Figure 5.14 Computational domain for an offset fin.

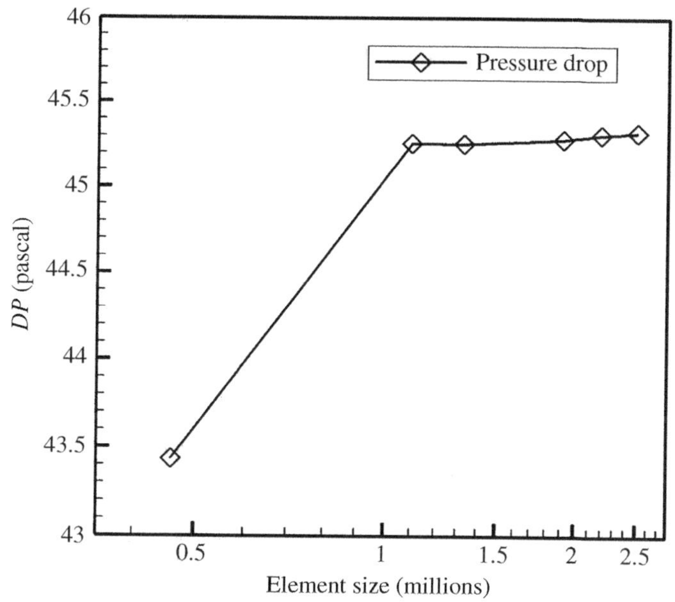

Figure 5.15 Grid independency for offset fin.

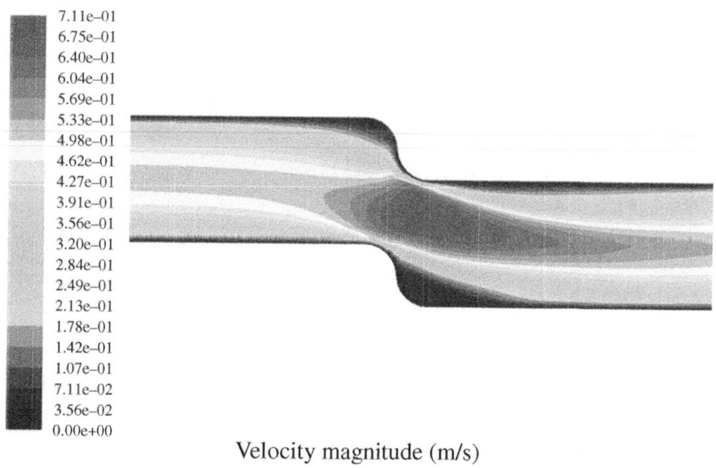

Figure 5.16 Velocity magnitude plot at $Re = 10{,}000$ for a typical offset fin.

5.7.2.2.1 Offset Fins – Steady State Data Comparison

In this section, the offset fins are analysed for both laminar and κ–ε turbulent model for the Reynolds numbers ranging from 300 to 15,000 and the variations in j and f are compared with various published data [52–58], as shown in Figure 5.17. It is observed that the Fluent predictions are matched well with available literature data except Manson [58], which is far away from the other data. In the laminar region of f factors, all the correlations are predicted well with the experimental data of Kays and London [3], except Manson [58], correlation as given by Manglik and Bergles [50] which over-predicted

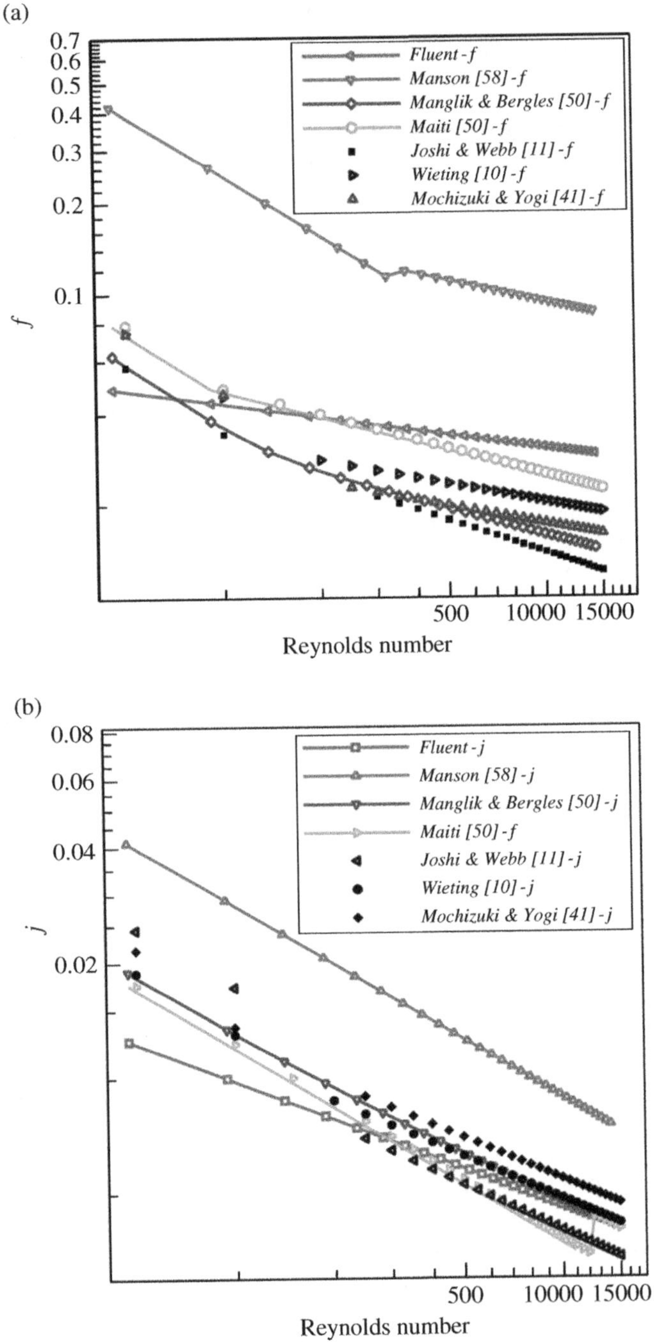

Figure 5.17 Comparison of offset fin: 3.0S-28.5-0.0762 (a) friction factor *f* vs *Re* data (b) Colburn *j* factor vs *Re* data.

by twice or three times. In the turbulent region of f factors, Manglik and Bergles [50] and Maiti [51] predicted well, but Joshi and Webb [11] under-predicted more than 15%. Wieting [10] over-predicted by 30% and Manson [58] over-predicted twice. Where the j factor is concerned for both regions, except for Wieting [10], all the others under-predicted by about 20%. Manson [58] over-predicted by almost twice the experimental data of Kays and London [3]. Giving exact reasons for variation of these factors may not be possible due to the involvement of so many parameters such as manufacturing aspects and testing conditions.

5.7.2.2.2 *Offset Fins – j and f Correlations*

The heat transfer and pressure drop correlations in the form of j and f for the offset strip fins are reported in Manson [58], London and Shah [2], Wieting [10], Joshi and Webb [11], Manglik and Bergles [50], Maiti [51] and Ranganayakulu et al. [47]. These correlations are compared. Based on the above survey, it is understood that there is limited data available in the literature for different geometries of offset strip fins. The CFD simulation eliminates the experimental setup and testing cost to a large extent. Similar to the above, the power law expressions for offset fins [47] are presented in two separate equations for the low and the high Re regions of f and j factors:

$$f = 10.882(Re)^{-0.79}(s/h)^{-0.359}(t/s)^{-0.187}(t/l)^{0.284} \quad \text{for laminar range } (300 \le Re \le 800) \tag{5.35}$$

$$f = 2.237(Re)^{-0.236}(s/h)^{-0.347}(t/s)^{-0.151}(t/l)^{0.639} \quad \text{for turbulent range } (1000 \le Re \le 15000) \tag{5.36}$$

$$j = 0.661(Re)^{-0.651}(s/h)^{-0.343}(t/s)^{-0.538}(t/l)^{0.305} \quad \text{for laminar range } (300 \le Re \le 800) \tag{5.37}$$

$$j = 0.185(Re)^{-0.396}(s/h)^{-0.178}(t/s)^{-0.403}(t/l)^{0.29} \quad \text{for turbulent range } (1000 \le Re \le 15000) \tag{5.38}$$

5.7.2.3 Rectangular Plain Fin

A numerical model of rectangular plain fin has been developed by considering the fin thickness (t), fin height (h), fin length (l) and fin spacing (s), as shown in Figure 5.18.

The thermal performance of a plain fin surface depends on the fin thickness for a given length, spacing of the fin and fin height. The plain fin surface is often treated as a set of rectangular conduits in parallel, and heat transfer correlations to appropriate non-circular ducts are used for predicting performance. The rectangular plain fin passage is modelled by CFD and j and f correlations are generated, instead of using those for fully developed flow in ducts. Earlier authors such as Wieting [10], Kays and London [3] and Joshi and Webb [11] have used the hydraulic diameter D_h as the characteristic dimension in expressing heat transfer and flow friction data.

5.7.2.3.1 *Computational Domain*

The computational domain of a plain fin model, as shown in Figure 5.19 (front view) and Figure 5.20 (isometric view), is analysed by using the CFD simulation tool Fluent. The CFD analysis is carried out for different geometric parameters and Reynolds numbers

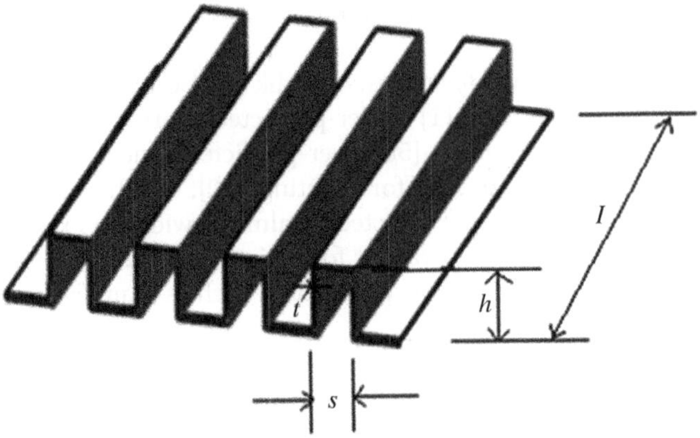

Figure 5.18 Plain fin geometry.

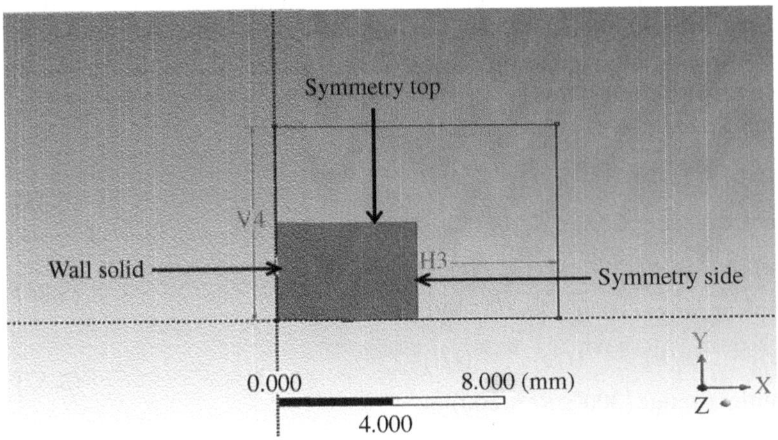

Figure 5.19 Computational domain of a plain fin – front view.

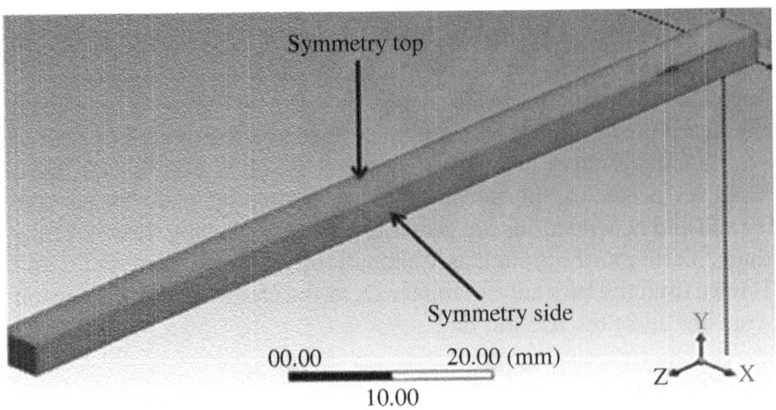

Figure 5.20 Computational domain of a plain fin – isometric view.

using Fluent 12.1. The analysis is carried out in two phases for plain fin. In the first phase, a plain fin is taken and characterized for *f* values over a range of Reynolds numbers under isothermal conditions. In second phase, the *j* values are determined for the same range by switching on the energy equation. The domain takes full advantage of geometrical symmetries of the system. Half the fin spacing in *x* direction, half the fin height in *y* direction and the full length (l) in *z* direction constitute the computational domain.

The optimum plain fin surfaces are taken for parametric study. A total of 144 different fin surfaces are modelled and CFD analysis is carried out for developing correlations. A numerical procedure has been used to find out *j* and *f* factors for the plain fin surfaces within the range of dimensionless parameters. The ranges of dimensionless geometric parameters and Reynolds number are as follows:

- $l = 100$ mm
- $s = 2, 4, 6$ mm
- $h = 2, 4, 6, 8, 10$ mm
- $t = 0.254$ mm
- $1 \leq h/s \leq 5$
- $04 \leq t/s \leq 0.14$
- $100 \leq Re \leq 1000$ for laminar region
- $1000 < Re \leq 7500$ for turbulent region

5.7.2.3.2 Grid Independency

The mesh finalization is done before analysing for the *j* and *f* factors. Different mesh configurations – starting with very coarse to very fine – are taken at a particular Reynolds number and a grid independence graph is plotted against the number of elements vs pressure drop, as shown in Figure 5.21. This figure shows, after 252,000 cells that there is not much variation in the pressure drop. Hence, the minimum grid size greater than 252,000 cells maintained throughout the analysis. The computational domain is meshed in such a way that very fine mesh is used near the wall and coarse mesh is used for the region away from the wall.

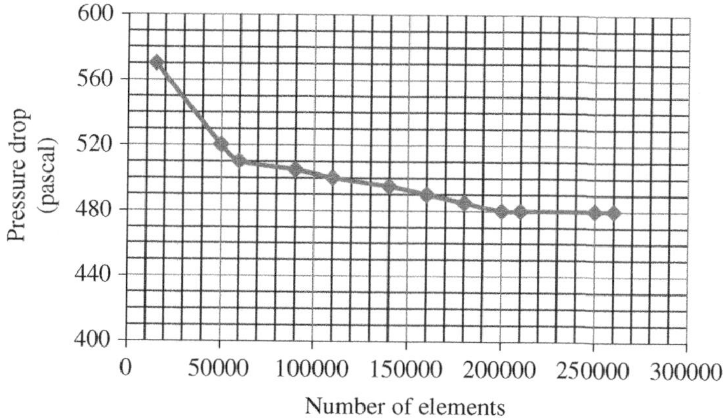

Figure 5.21 Grid independency graph.

The mass flow rates are determined for a range of Reynolds number 100 to 7500. In order to overcome the entrance effect, the concept of periodic fully developed flow as suggested by Patankar et al. [12,13] is implemented for flow analysis. After the analysis, the pressure drop for unit length is one of the outputs, and that is multiplied by the actual length to get the total pressure drop for corresponding fins. Finally, the corresponding two-dimensional fully developed velocity profile is listed. The same procedure is repeated for the range of Reynolds numbers from 100 to 7500 in order to draw the f vs Re curve. Pressure, temperature and velocity profiles are taken at the various sections of the fins for corresponding Reynolds numbers. This temperature difference between inlet and outlet of the fin is used for calculating j factor using Kays and London [3] methodology. Similarly the same procedure is repeated for the range of Reynolds numbers 100 to 1000 for laminar flow and $Re > 1000$ for turbulent flow in order to draw the j vs Re characteristics curves. The actual mass flow rate is used as the boundary condition. A no-slip boundary condition is used for walls.

Figure 5.22 shows the static pressure distribution at Reynolds number 2500 based on the ANSYS output data. This figure indicates the variation of static pressure from 10 to 100 pascal. Figure 5.23 indicates the temperature distribution at Reynolds number 2500 based on the ANSYS output data. This figure gives the variation of static temperature from 300 K to 310 K.

Figure 5.24 Indicates the velocity contour at Reynolds number 2500 based on the ANSYS output data. This figure gives the variation of velocity from 3 to 25 m/s.

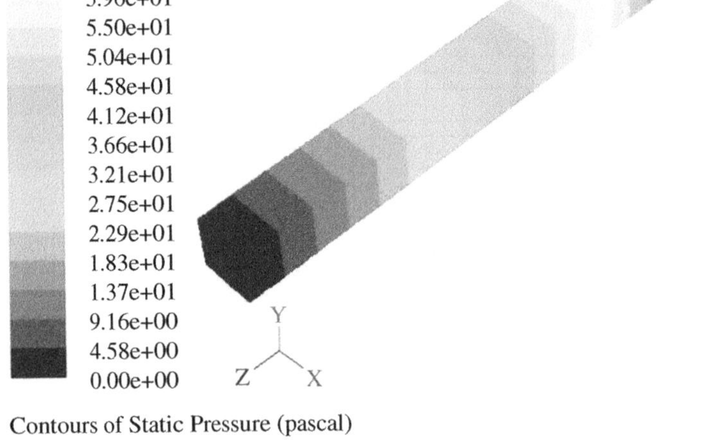

Figure 5.22 Static pressure graph at $Re = 2500$.

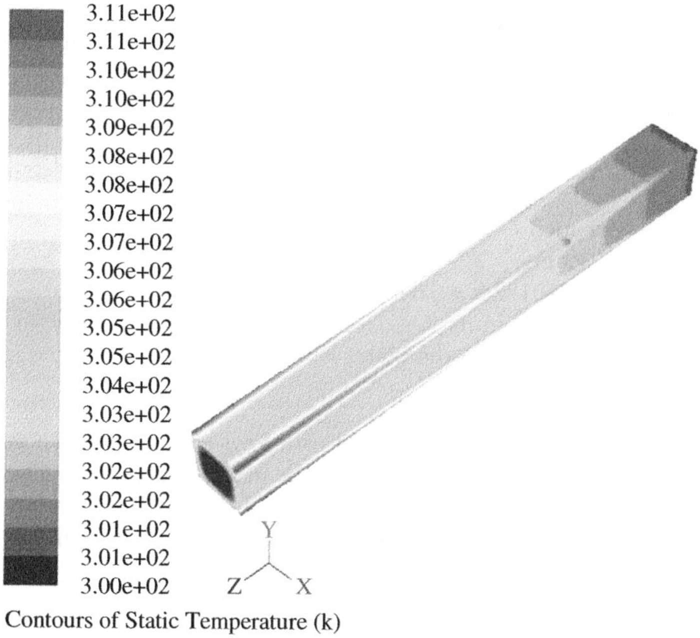

Figure 5.23 Static temperature graph at $Re = 2500$.

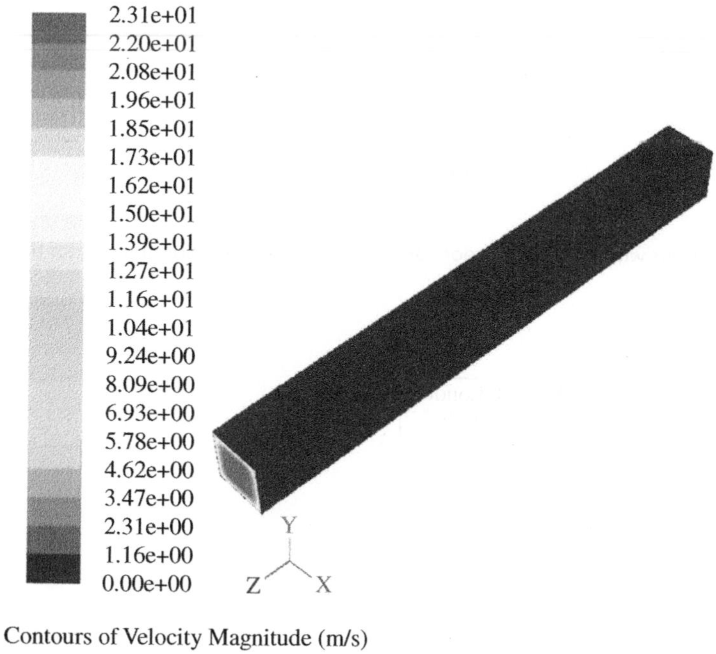

Figure 5.24 Velocity contour at $Re = 2500$.

5.7.2.3.3 Validation of Results

The rectangular fin channel has been analysed by considering a uniform wall temperature boundary condition. For the validation of numerical analysis conducted in the present study, the following case of the rectangular plain fin from Kays and London [3] has been analysed.

Fin specification:

- FPI (or) fin density = 6.2 fins per inch
- plate spacing = 0.405 inch = 10.29 mm
- flow passage hydraulic diameter = 0.0182 ft = 5.54 mm
- fin metal thickness = 0.010 inch = 0.254 mm
- total heat transfer area per volume between the plates = 669.3 m^2/m^3
- fin area/total area = 0.728

The CFD results for rectangular plain fin are compared with Kays and London [3], as shown in Figures 5.25 and 5.26 for j and f data, and found to be in good agreement. The CFD results are in good agreement with results given by Kays and London [3] for the low and high Reynolds number region. The variations are found to be 1–7% in f and 2–9% in j values.

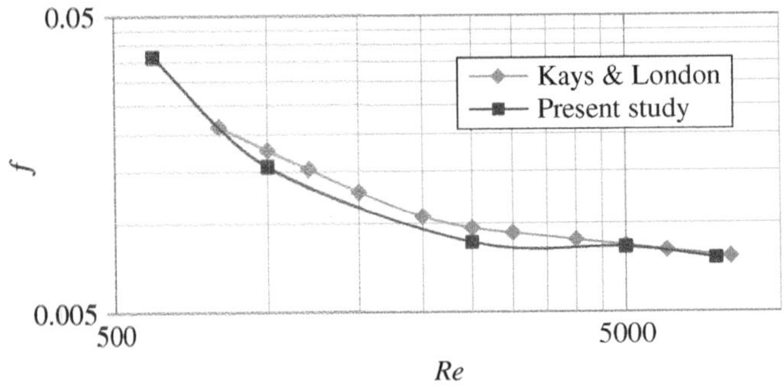

Figure 5.25 Comparison of results with Kays and London [3] friction factor vs Reynolds number, for the fin density of 6.2.

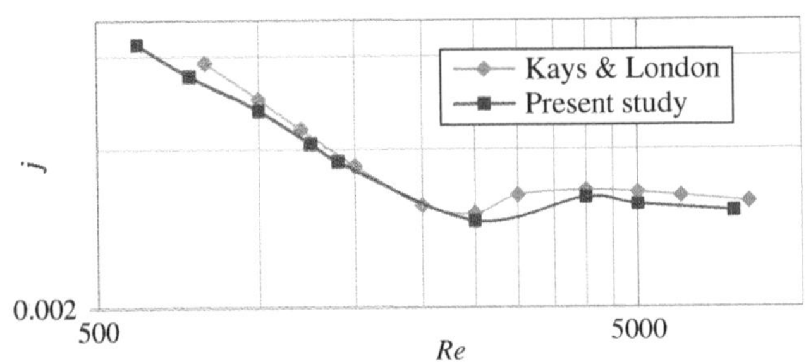

Figure 5.26 Comparison of results with Kays and London [3], Colburn factor vs Reynolds number, for the fin density of 6.2.

5.7.2.3.4 Correlations for f and j Factors

The geometrical features and computational domains of the three-dimensional rectangular plain fin are shown in Figures 5.18–5.20. The optimum plain fin surfaces are taken for parametric study. A total of 144 different fin surfaces are modelled, and CFD analysis is carried out for developing correlations. The ranges of dimensionless geometric parameters and Reynolds number are as follows:

- $1 \le h/s \le 5$
- $0.04 \le t/s \le 0.14$
- $100 \le Re \le 1000$ for laminar region
- $1000 < Re \le 7500$ for turbulent region

An exhaustive numerical study has been carried out on the heat transfer phenomena in plate fin surfaces with rectangular fins. The curves f vs Re and j vs Re show significant non-linearity, as shown in Figures 5.25 and 5.26. The power law expressions have been used for determining the Colburn factor j and friction factor f as a function of the Reynolds number and dimensionless fin parameters. The use of the power law expressions is justified because variations in f and j with Re, h/s and t/s follow constant slope log-linear lines in both laminar and turbulent flow regions. The Fanning friction factor f is functionally related to Re, h/s and t/s, and two separate equations are proposed, one for the low and the other for high Re regions:

$$f = 12.892(Re)^{-1.229}(h/s)^{0.452}(t/s)^{-0.198} \quad \text{for laminar range } (100 \le Re \le 1000) \tag{5.39}$$

$$f = 3.133(Re)^{-1.285}(h/s)^{0.247}(t/s)^{-0.181} \quad \text{for turbulent range } (1000 < Re \le 7500) \tag{5.40}$$

The above correlations correctly predict 97% of the f data for laminar regions and 96% of the f data for the turbulent regions, with RMS value of 0.985.

Similarly, the Colburn factor j are functionally related to Re, h/s and t/s, and it can be represented as

$$j = 0.454(Re)^{-0.977}(h/s)^{0.435}(t/s)^{-0.227} \quad \text{for laminar range } (100 \le Re \le 1000) \tag{5.41}$$

$$j = 0.166(Re)^{-1.011}(h/s)^{0.228}(t/s)^{-0.366} \quad \text{for turbulent range } (1000 < Re \le 7500) \tag{5.42}$$

The above correlations correctly predict 96% of the j data for laminar regions and 95% of the j data for the turbulent regions with RMS value of 0.99.

5.7.2.3.5 Error Analysis

Error analysis is carried out to evaluate the efficacy of these correlations by considering 144 data points. Figures 5.27 a and b show f and j factors obtained from the correlations and numerical models for laminar and turbulent regimes.

It is observed that most of the data points in Figure 5.27 lie near the 45° line. The maximum absolute divergence is found to be less than 13% for f factor and 12% for j factor. Therefore these correlations can be adopted conveniently to predict the

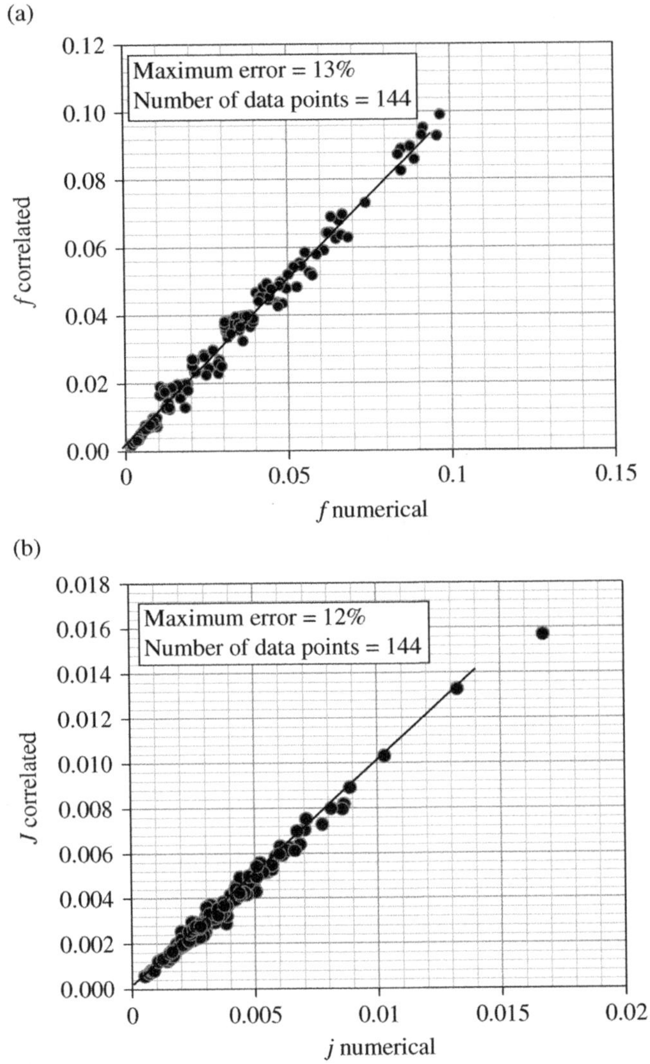

Figure 5.27 Plots showing *f* and *j* factors obtained from the correlations and numerical models for laminar and turbulent regimes: (a) *f* correlated vs *f* numerical and (b) *j* correlated vs *j* numerical.

Fanning friction factor *f* and Colburn *j* factor for the rectangular plain fins in both laminar and turbulent regimes.

5.7.2.4 Rectangular Perforated Fin

Numerical model of rectangular perforated fin has been developed by considering the fin thickness (t), fin height (h), fin length (l), fin spacing (s), diameter of the perforation (d), perforation spacing and number of perforations per inch length. The model of the perforated fin is shown in Figure 5.28.

Figure 5.28 Model of perforated fin.

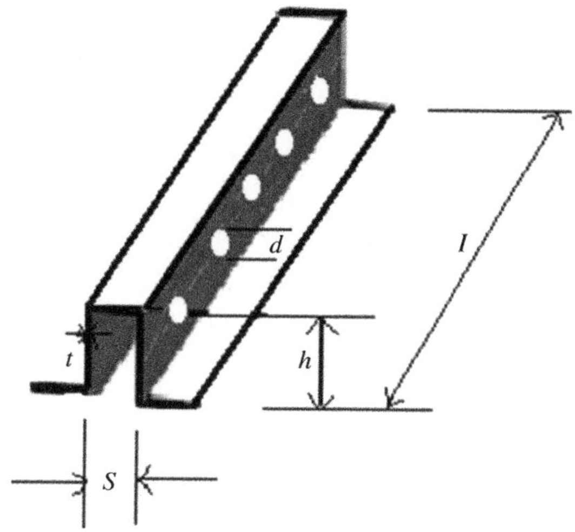

The fluid path is modelled by using CFD and the fine mesh has been applied to get more accurate values. A fin is considered, as shown in Figure 5.28 and the fluid travels in the punched hole space. By changing the diameter of the punched hole, spacing and height, the iterations are taken and then the graphical results of static pressure, temperature and velocity could be drawn. The geometry of the perforated fin is varied continuously by using the parameter set. The appropriate nodes have been given. The fluid selected is air and the material used is aluminium.

5.7.2.4.1 Computational Domain

The computational domain of the perforated fin model is shown in Figure 5.29. It is analysed using CFD simulation tool Fluent 12.1 for different geometric parameters and Reynolds numbers. The analysis is carried out in two phases for the perforated

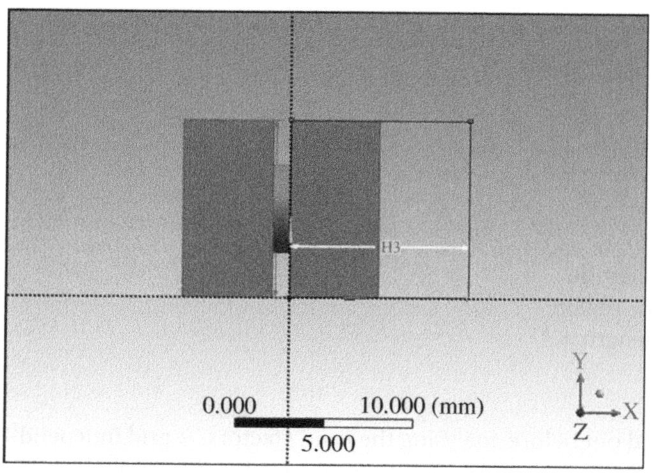

Figure 5.29a Computational domain of a perforated fin – front view.

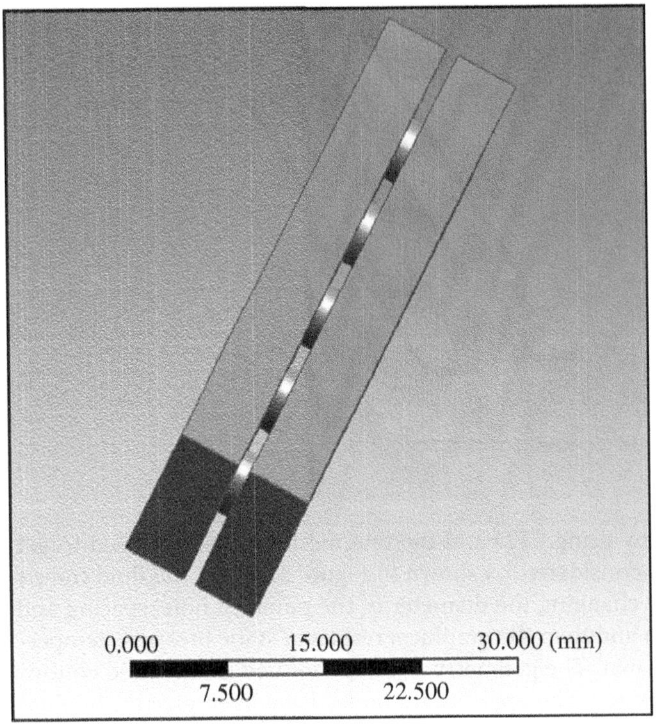

Figure 5.29b Computational domain of a perforated fin – isometric view.

fin under isothermal condition. In the first phase, a rectangular perforated fin is taken and characterized for f values over a range of Reynolds numbers. In second phase, the j values are determined for the same range by switching on the energy equation.

The ranges of dimensionless geometric parameters and Reynolds number are as follows:

- $l = 100$ mm
- $s = 2, 6, 8, 10, 12$ mm
- $h = 5, 9, 10$ mm
- $t = 0.305$ mm
- $d = 2, 4, 5$ mm
- $0.4 \leq h/s \leq 5$
- $0.025 \leq t/s \leq 0.15$
- $100 \leq Re \leq 1000$ for laminar region
- $1000\ Re \leq 7500$ for turbulent region
- Number of holes per inch length = 2

5.7.2.4.2 Grid Independency

The mesh finalization is carried out before analysing the j and f factors. A grid independence test is carried out for the same fin and a graph is plotted against the number of elements vs pressure drop, as shown in Figure 5.30. This figure shows that after 260,000

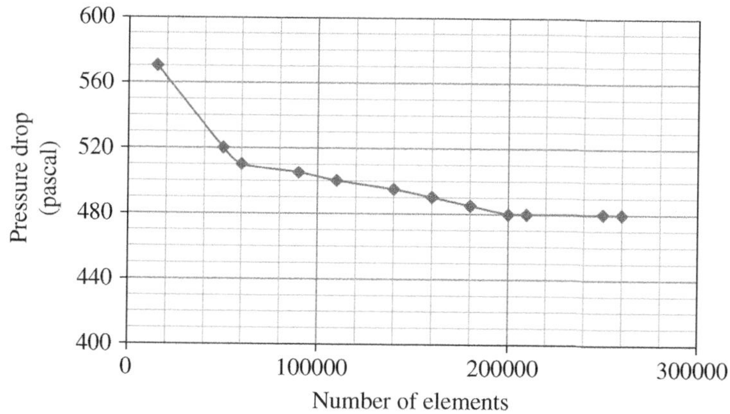

Figure 5.30 Grid independency graph.

cells there is not much variation in the pressure drop. Hence, the minimum grid size maintained throughout the analysis is greater than 260,000 cells.

The same methodology and procedure is adopted for CFD simulation studies of paragraph 5.7.2.3.2.

Pressure, velocity and temperature profiles are taken at the various sections of the fin for corresponding Reynolds number of 2500, as shown in Figure 5.31.

5.7.2.4.3 Validation of Results

The CFD data of rectangular perforated channel is analysed and compared with available published data. The results are found to be in good agreement with Kays and London [3] for the case of 13.95 fins per inch (or) fin density, 5.08 mm fin height and 0.305 mm fin thickness.

Figure 5.31a Static pressure graph at $Re = 2500$.

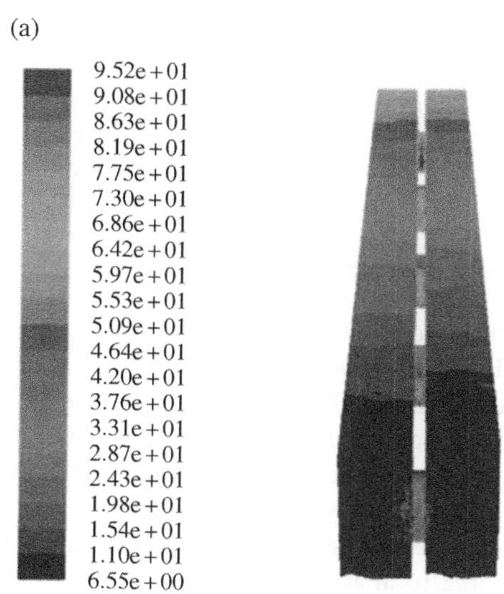

(b)

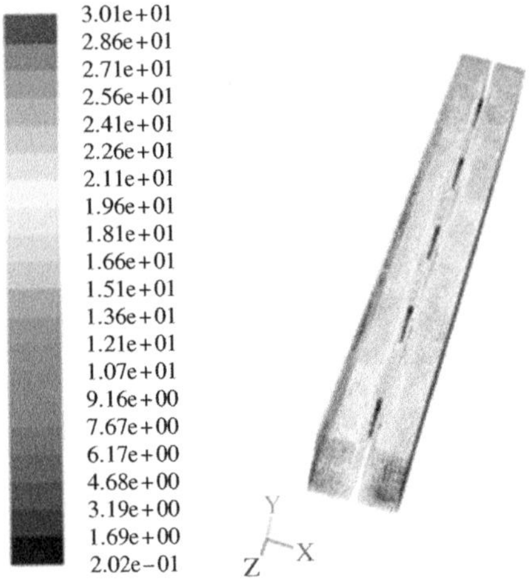

Figure 5.31b Velocity contour at *Re* = 2500.

(c)

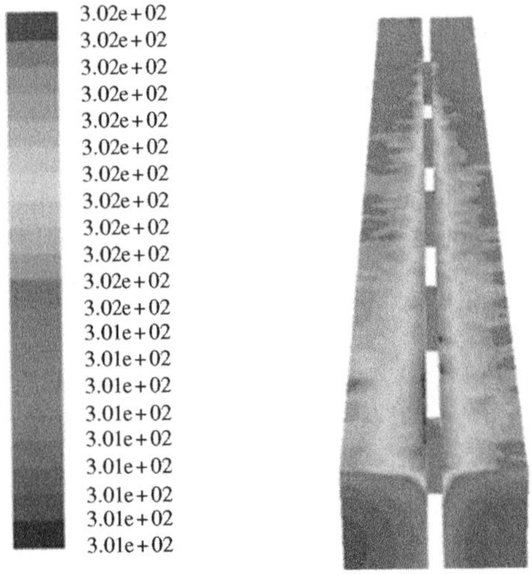

Figure 5.31c Static temperature graph at *Re* = 2500.

The fin specifications are:

- surface designation = 13.95(P)
- FPI (or) fin density = 13.95
- plate spacing = 0.0167 inch = 5.08 mm
- flow passage hydraulic diameter = 0.00822 ft = 2.50 mm
- fin metal thickness = 0.012 inch = 0.305 mm
- hole diameter = 0.079 inch = 2 mm
- total heat transfer area per volume between the plates = 1250 m^2/m^3
- fin area/total area = 0.705

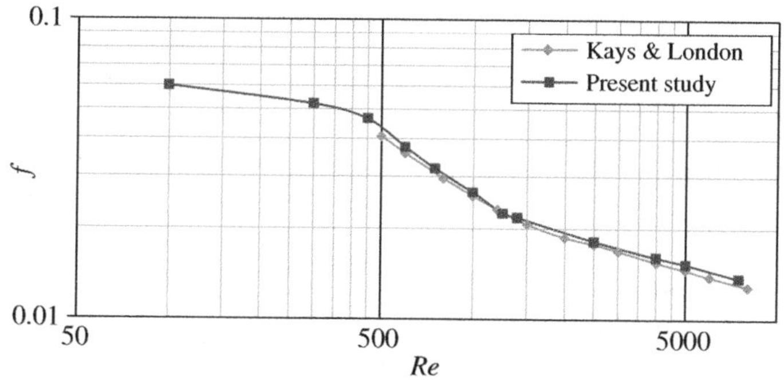

Figure 5.32 Comparison of numerical results with Kays and London [3], friction factor vs Reynolds number, for the fin density of 13.95 and height of the fin 5.08 mm.

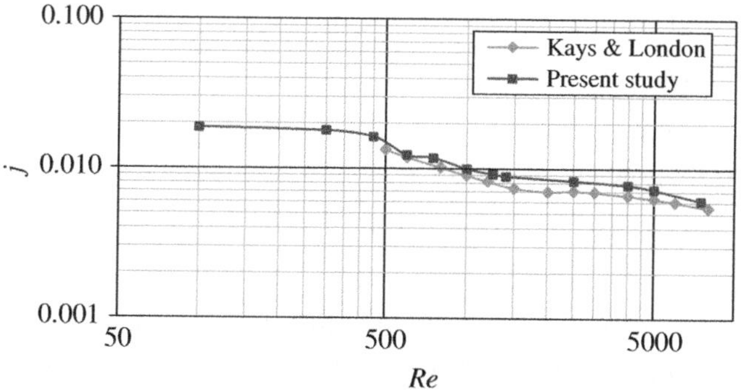

Figure 5.33 Comparison of numerical results with Kays and London [3], Colburn factor vs Reynolds number, for the fin density of 13.95 and height of the fin 5.08 mm.

The CFD results for rectangular perforated fin are compared with Kays and London [3], as shown in Figure 5.32 and 5.33. The CFD results are in good agreement with experimental results given by Kays and London [3] for the low and high Reynolds number region. The maximum variations are found to be 2–7% in f and 3–6% in j values.

5.7.2.4.4 Correlations for f and j Factors

The ranges of dimensionless geometric parameters and Reynolds number are as follows:

- $d = 2, 4, 5$ mm
- $0.4 \leq h/s \leq 5$
- $0.025 \leq t/s \leq 0.15$
- $100 \leq Re \leq 1000$ for laminar region
- $1000 < Re \leq 7500$ for turbulent region
- number of holes per inch length = 2

The f vs Re and j vs Re curves show significant non-linearity, as shown in Figures 5.32 and 5.33. The power law expressions are used for determining the Colburn factor j

and friction factor f as a function of the Reynolds number and dimensionless fin parameters.

$$f = 0.7127(Re)^{-1.8858}(h/s)^{0.4196}(t/s)^{-1.4826} \tag{5.43}$$

for laminar range ($100 \le Re \le 1000$)

$$f = 0.4345(Re)^{-1.3029}(h/s)^{0.3725}(t/s)^{-1.3178} \tag{5.44}$$

for turbulent range ($1000 < Re \le 7500$)

The above correlations correctly predict 95% of the f data for laminar regions and 97% of the f data for the turbulent regions, with RMS value of 0.985.

Similarly, the Colburn factor j is functionally related to Re, h/s and t/s, and it can be represented as

$$j = 0.121(Re)^{-2.2920}(h/s)^{2.75}(t/s)^{-1.830} \tag{5.45}$$

for laminar range ($100 \le Re \le 1000$)

$$j = 11.71(Re)^{-2.311}(h/s)^{2.144}(t/s)^{-1.9237} \tag{5.46}$$

for turbulent range ($1000 < Re \le 7500$)

The above correlations correctly predict 96% of the j data for laminar and turbulent regions, with RMS value of 0.98.

5.7.2.5 Triangular Plain Fin Surface

A numerical model of triangular plain fin has been developed by considering fin thickness (t), fin height (h), fin length (l) and fin spacing (s). The model of triangular plain fin is shown in Figure 5.34.

5.7.2.5.1 Computational Domain

The CFD analysis is carried out for different geometric parameters and Reynolds numbers using Fluent 12.1. The analysis is carried out in two phases. In first phase, a triangular plain fin is taken and characterized for f values over a range of Reynolds numbers. In second phase, the j values are determined for the same range by switching on the energy equation. The computational models of triangular plain fin are shown in Figure 5.35.

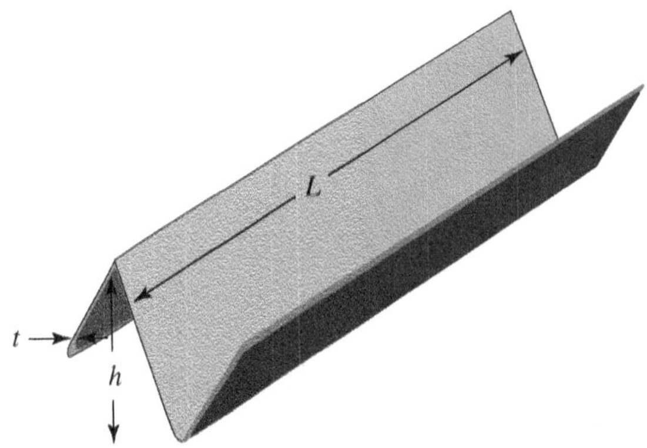

Figure 5.34 Model of triangular plain fin.

(a)

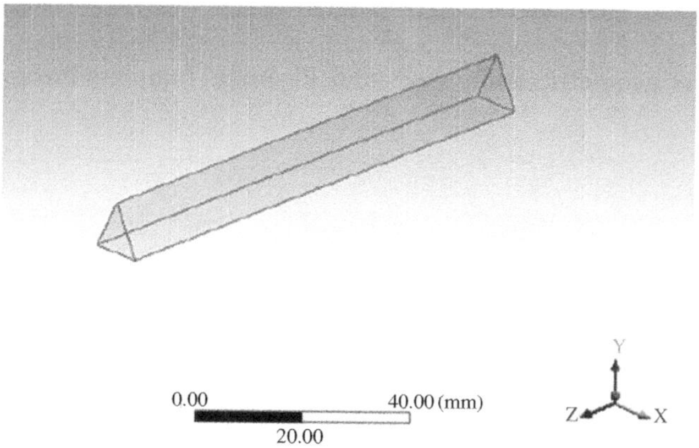

Figure 5.35a Isometric view of triangular plain fin.

(b)

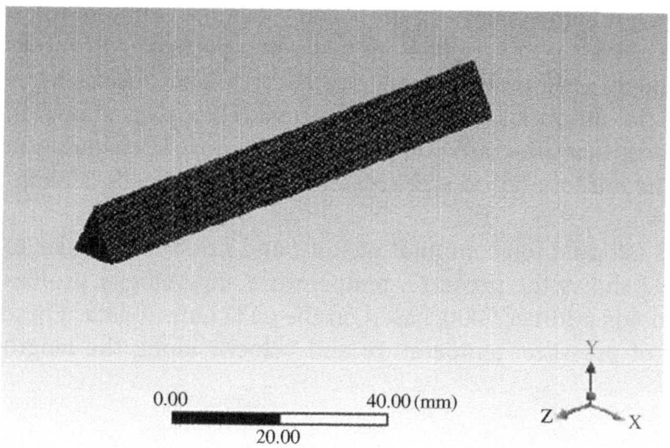

Figure 5.35b Isometric view of meshed model.

(c)

Figure 5.35c Orthogonal view of meshed model.

The ranges of dimensionless geometric parameters and Reynolds numbers are as follows:

- $l = 100$ mm
- $s = 2, 4, 6, 8, 10$ mm
- $h = 2, 4, 6, 8, 10$ mm
- $t = 0.152$ mm
- $0.20 \le h/s \le 5$
- $015 \le t/s \le 0.08$
- $100 \le Re \le 1000$ for laminar region
- $1000 < Re \le 10000$ for turbulent region

The total pressure drop or Fanning friction factor f for flow across the heat exchanger and the dimensionless Colburn factor j are calculated as per the Equations 5.1 and 5.3.

5.7.2.5.2 Grid Independency

The mesh finalization is carried out before analysing the 'j' and 'f' factors. Different mesh configurations starting with very coarse to very fine are taken at a particular Reynolds number and analysed using Fluent. Grid independence test is carried out for the same fin and graph is plotted with the number of elements vs pressure drop, as shown in Figure 5.36. This figure indicates that after 235,000 cells, there is not much variation in the pressure drop. Hence, the minimum grid size greater than 235,000 cells is maintained throughout the analysis.

The CFD simulation studies adopted for computations of f and j data are similar to paragraph 5.7.2.4.2. Figure 5.37 shows the pressure, temperature and velocity profiles of a triangular plain fin at Reynolds number 5000, based on the CFD output data. These figures indicate the variation of pressure, temperature and velocity along the length of the fin.

5.7.2.5.3 Validation of Results

The CFD data of a triangular plain fin channel has been analysed and compared with available published data. For the validation of numerical analysis conducted in the present study, the following triangular plain fin from Kays and London [3] was analysed: 12.00 T fins per inch (or) fin density, 6.35 mm fin height and 0.152 mm fin thickness.

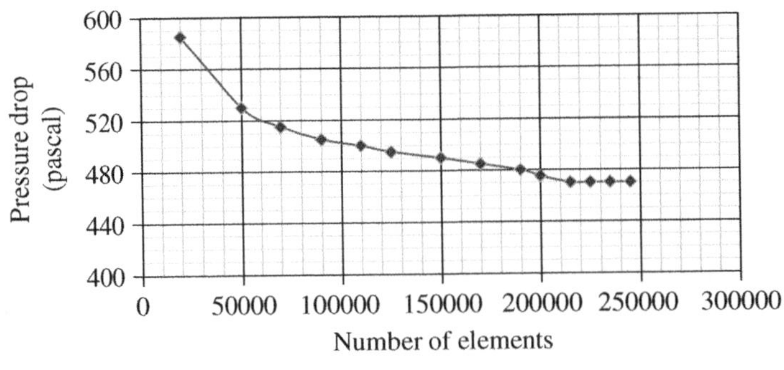

Figure 5.36 Grid independency graph.

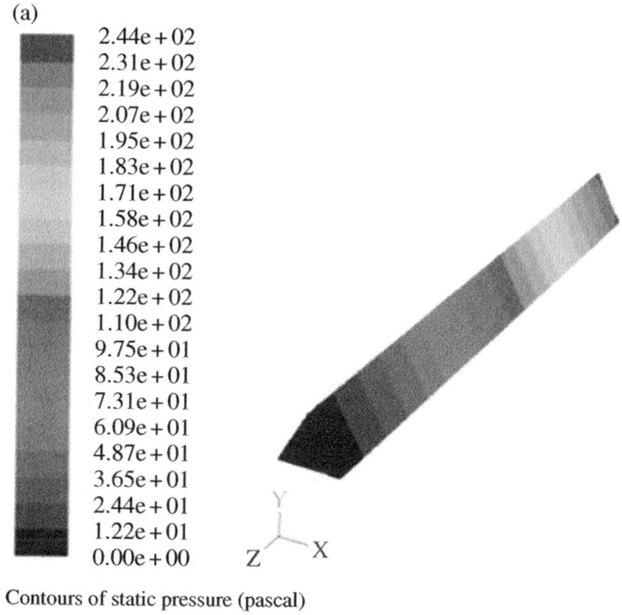

Figure 5.37a Pressure profile of triangular plain fin at $Re = 5000$.

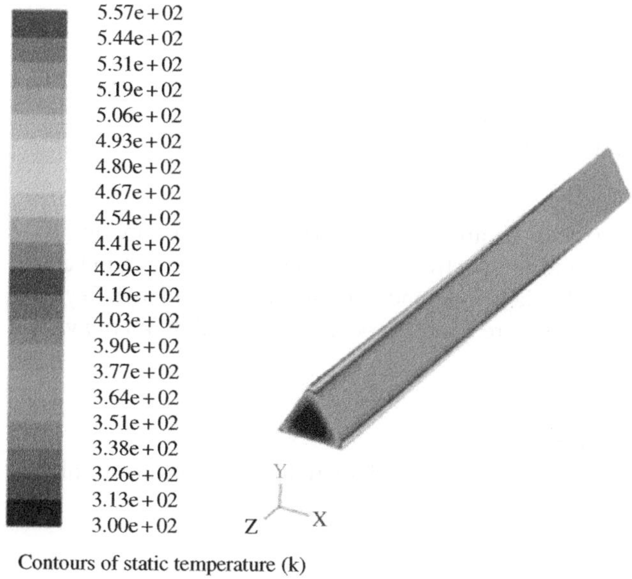

Figure 5.37b Temperature profile of triangular plain fin at $Re = 5000$.

(c)

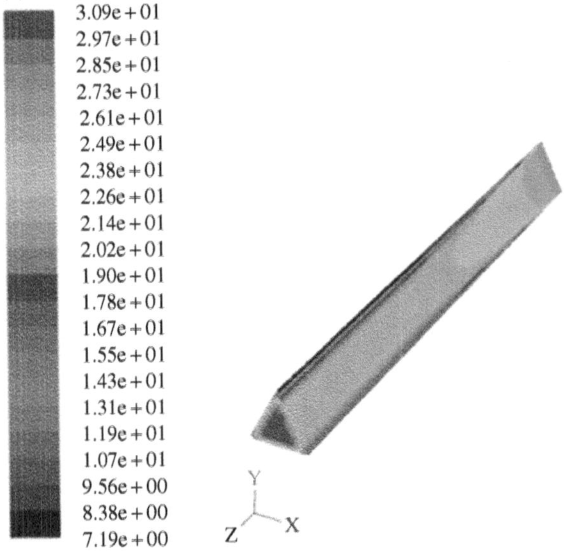

Velocity vectors colored by velocity magnitude (m/s)

Figure 5.37c Velocity profile of triangular plain fin at $Re = 5000$.

Fin specifications, as per Kays and London[3] surface geometry- plate-fin surfaces and triangular plain fin:

- surface designation = 12.00 T
- fpi (or) fin density = 12.00
- plate spacing = 0.0208 ft = 6.35 mm
- flow passage hydraulic diameter = 0.009412 ft = 2.87 mm
- fin metal thickness = 0.006 inch = 0.152 mm
- total heat transfer area per volume between the plates = 1288.4 m^2/m^3
- fin area/total area = 0.773

The CFD results for a triangular plain fin are compared with Kays and London [3], as shown in Figures 5.38 and 5.39 for f and j respectively. The CFD results are in good agreement with experimental results given by Kays and London [3] for the low and high Reynolds number region. The maximum variations are found to be 2–9% in f and 3–4% in j values.

5.7.2.5.4 Correlations for f and j Factors

The ranges of dimensionless geometric parameters and Reynolds numbers are as follows:

- $0.2 \le h/s \le 5$
- $0.015 \le t/s \le 0.08$
- $100 \le Re \le 1000$ for laminar region
- $1000 < Re \le 10000$ for turbulent region

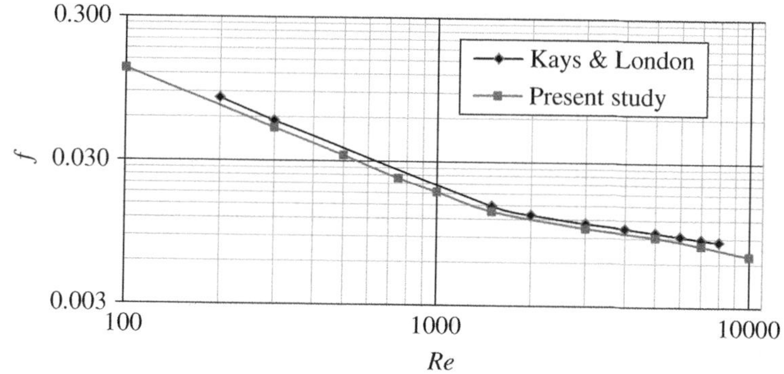

Figure 5.38 Comparison of results with Kays and London [3], friction factor *f* vs Reynolds number, for the fin density of 12.00.

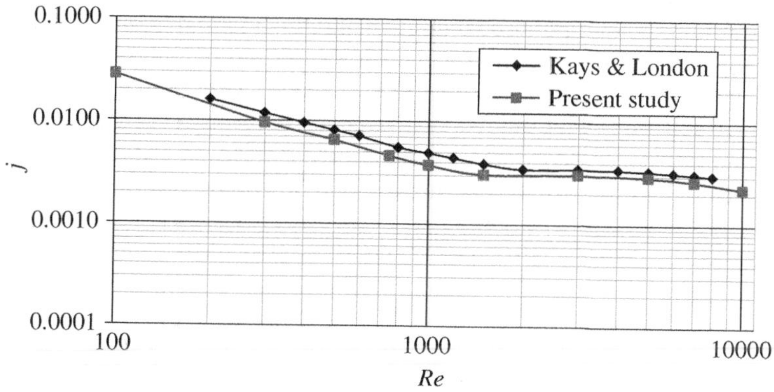

Figure 5.39 Comparison of results with Kays and London [3], Colburn *j* factor vs *Re*, for the fin density of 12.00.

The power law expressions have also been used for determining the Colburn factor j and friction factor f as a function of the Reynolds number and dimensionless fin parameters. The f vs Re and j vs Re data for triangular plain fin surfaces show significant non-linearity over the Reynolds number range $100 \le Re \le 10000$. Therefore, two separate equations have been proposed for the low and the high Re regions:

$$f = 3.12(Re)^{-0.852}(h/s)^{0.156}(t/s)^{-0.184} \tag{5.47}$$

for laminar range ($100 \le Re \le 1000$)

$$f = 2.69(Re)^{-0.918}(h/s)^{0.355}(t/s)^{-0.175} \tag{5.48}$$

for turbulent range ($1000 < Re \le 10000$)

The above correlations correctly predict 96% of the f data for laminar regions and 97% of the f data for the turbulent regions, with RMS value of 0.98.

Similarly, the Colburn factor j are functionally related to Re, h/s and t/s, and it can be represented as

$$j = 0.718(Re)^{-0.625}(h/s)^{0.765}(t/s)^{-0.765} \quad (5.49)$$

for laminar range ($100 \le Re \le 1000$)

$$j = 0.789(Re)^{-1.1218}(h/s)^{1.235}(t/s)^{-0.764} \quad (5.50)$$

for turbulent range ($1000 < Re \le 10000$)

The above correlations correctly predict 97% of the j data for laminar regions and 96% of the j data for the turbulent regions, with RMS value of 0.99.

5.7.2.6 Triangular Perforated Fin Surface

A numerical model of triangular perforated fin is developed by considering the fin thickness (t), fin height (h), fin length (l), fin spacing (s), diameter of the perforation (d), number of perforations per inch length and perforation spacing. The triangular perforated fin is shown in Figure 2.7 in Chapter 2.

5.7.2.6.1 *Computational Domain*

The CFD approach describes the computational domain and the boundary conditions to analyse using the Fluent tool and its validation with available experimental data. As the initial step, the fluid path is modelled in the ANSYS package. After modelling, it is imported to Fluent 12.1 software. The models of triangular perforated fin with three holes and eight holes on triangular perforated fin are shown in Figures 5.40 and 5.41.

The ranges of dimensionless geometric parameters and Reynolds numbers are as follows:

- $l = 100$ mm
- $s = 5, 10$ mm
- $h = 6, 8$ mm
- $d = 2, 3$ mm
- $t = 0.152$ mm
- $0.6 \le h/s \le 1.6$
- $0.015 \le t/s \le 0.03$
- $300 \le Re \le 1000$ for laminar region
- $1000 < Re \le 7500$ for turbulent region
- Number of holes per inch length = 2, 3 and 4

The total pressure drop or Fanning friction factor f for flow across the heat exchanger and the dimensionless Colburn factor j are calculated as per Equations 5.1 and 5.3.

5.7.2.6.2 *Grid Independency*

Different mesh configurations starting with very coarse to very fine are taken at a particular Reynolds number and analysed. In addition, a grid independence test is carried out for the same fin and a graph is plotted is the number of elements vs pressure drop, as shown in Figure 5.42. This figure shows that after 220,000 cells there is not much variation in the pressure drop. Hence, a minimum grid size of 244,000 cells is maintained throughout the analysis.

Figures 5.43, 5.44 and 5.45 represent the pressure, temperature and velocity profile of a triangular perforated fin at Reynolds number 5000 based on the Fluent 12.1 output data. In Figure 5.43, it is observed that local pressure variations are seen near perforations.

(a)

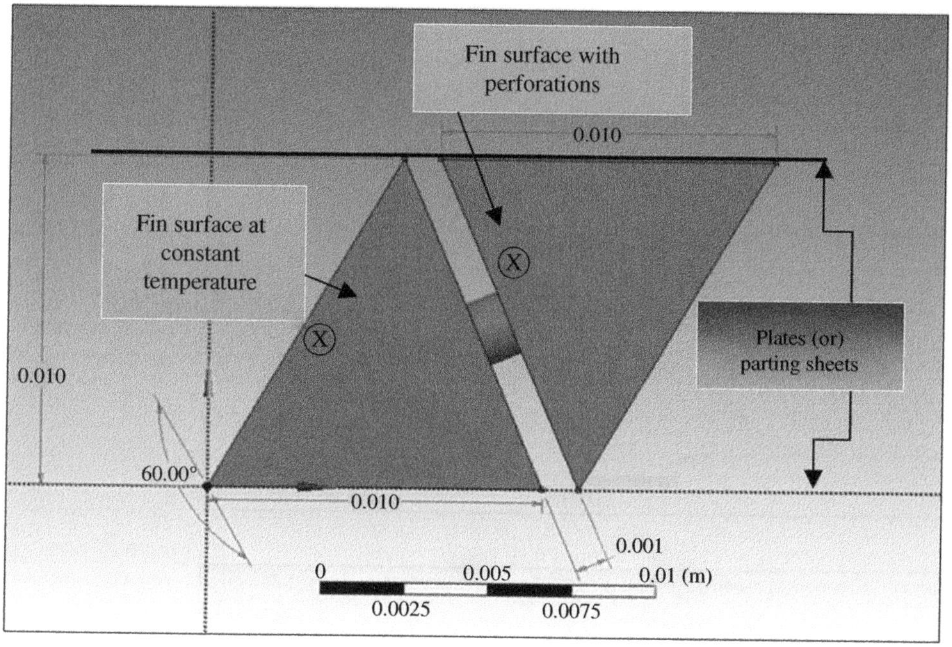

(b)

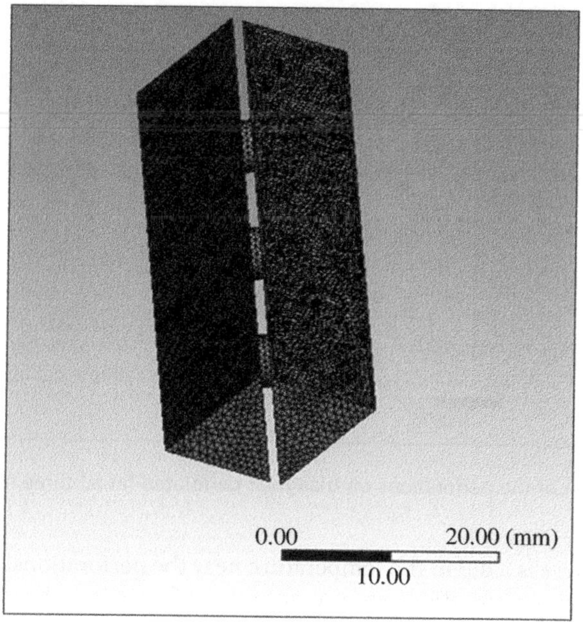

Figure 5.40 Model of triangular perforated fin (a) orthogonal view (b) isometric view.

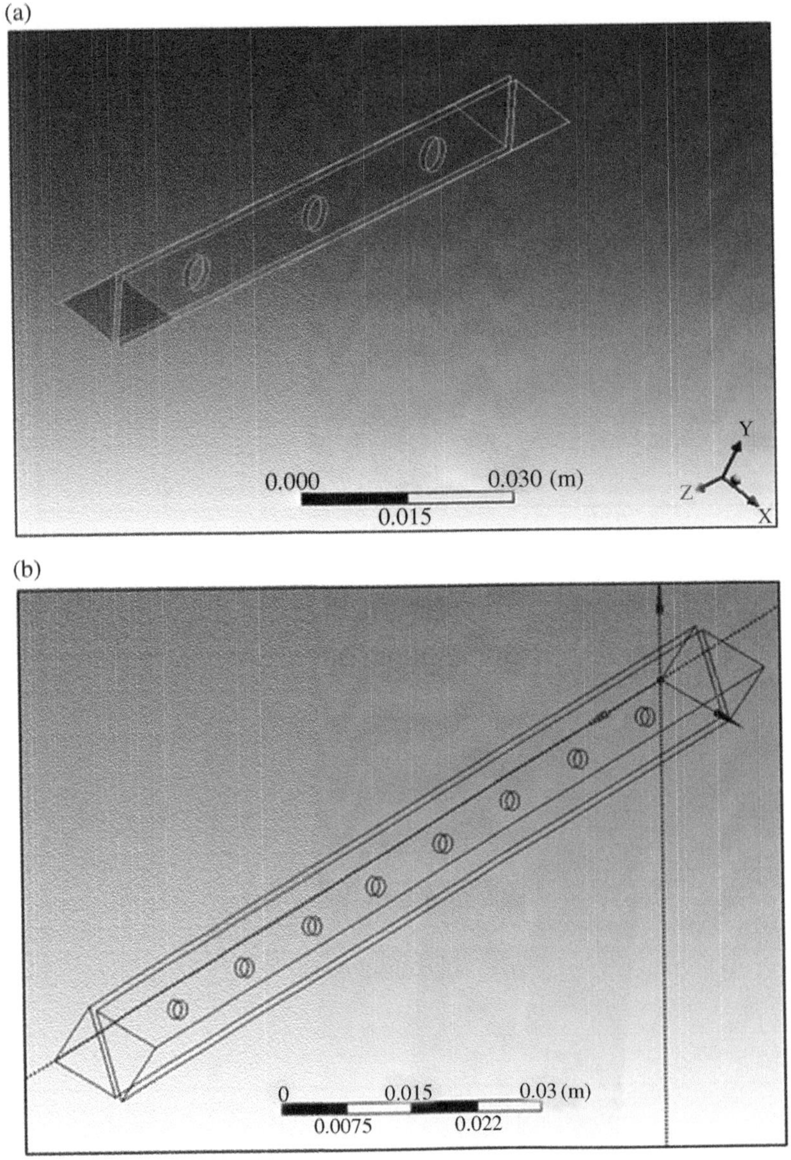

Figure 5.41 Position of the perforations on triangular perforated fin (a) three holes (b) eight holes.

In Figure 5.44, there is a dip in the temperature near the perforations. In Figure 5.45, it is observed that where perforations are present on the fin surface the direction of velocity vectors are disturbed.

5.7.2.6.3 Validation of Results

Simulation studies based on the methodology outlined have been carried out for triangular plain fins with and without perforations. The *j* and *f* factors obtained are found to be

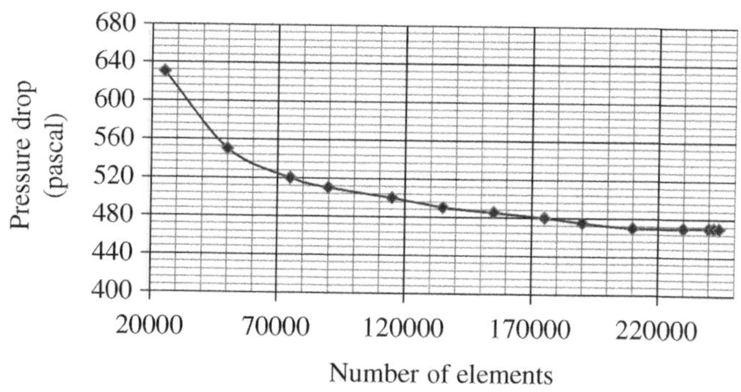

Figure 5.42 Grid independency graph.

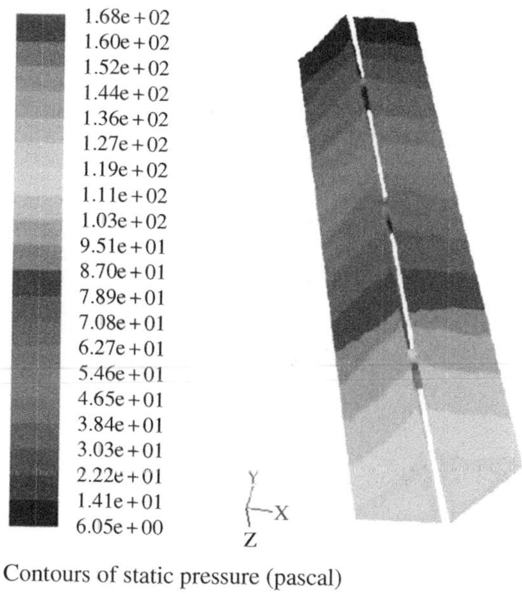

Figure 5.43 Pressure profile of triangular perforated fin at $Re = 5000$.

in good agreement with the figures of Kays and London [3] Triangular plain fin results also match satisfactorily. For the validation of present study, the triangular plain fin from Kays and London [3] has been analysed for the following case: 11.94 T fin density, 6.325 mm fin height and 0.152 mm fin thickness.

Fin specifications:

- FPI (or) fin density (or) fin pitch = 11.94
- plate spacing = 0.0207 ft = 6.325 mm
- flow passage hydraulic diameter = 2.87 mm
- fin metal thickness = 0.006 inch = 0.152 mm
- hole diameter = 0.079 inch = 2 mm
- total heat transfer area per volume between the plates = 1289 m^2/m^3

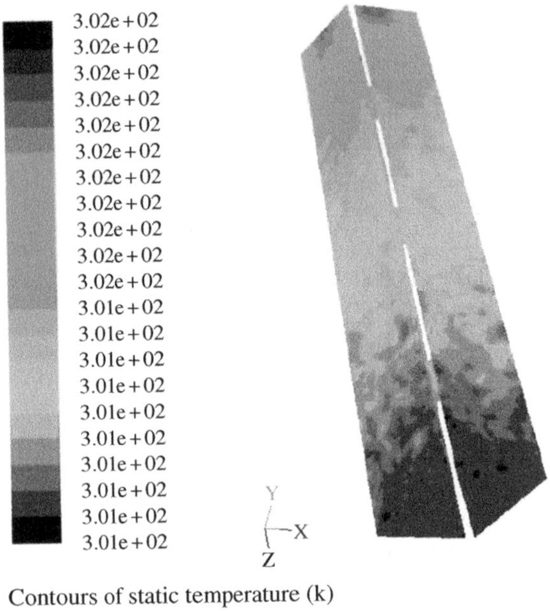

Figure 5.44 Temperature profile of triangular perforated fin at $Re = 5000$.

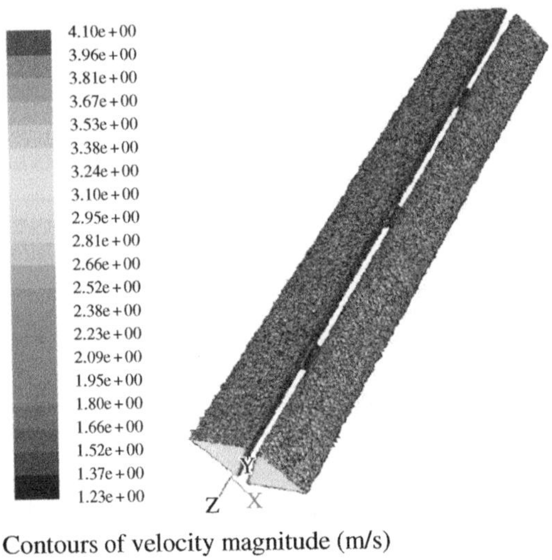

Figure 5.45 Velocity profile of triangular perforated fin at $Re = 5000$.

Figure 5.46 represents *f* vs *Re* relative comparison of triangular perforated fin of FPI 5.08 and FPI 11.94 with the data of Kays and London [3] of triangular plain fin of FPI 11.94 T, which is well matched with the FPI 11.94 of present study, where it is shown that higher FPI will result in higher *f* values. The maximum variations are found to be 6–8% in *f* and 5–7% in *j* values. Figure 5.47 represents Colburn factor *j* vs Reynolds number relative comparison of triangular perforated fin of FPI 5.08 and 11.94 with the data of Kays and London [3] of triangular plain fin of FPI 11.94 T, which is well matched with FPI 11.94 of the present study, where it is shown that higher FPI will result in higher *j* values.

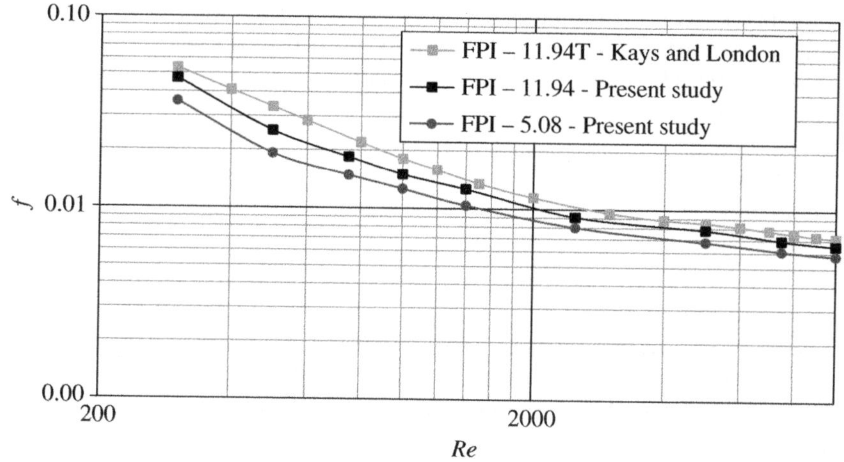

Figure 5.46 *f* vs *Re* comparison between triangular perforated fin present study values and triangular plain fin Kays and London [3] data FPI-11.94 T.

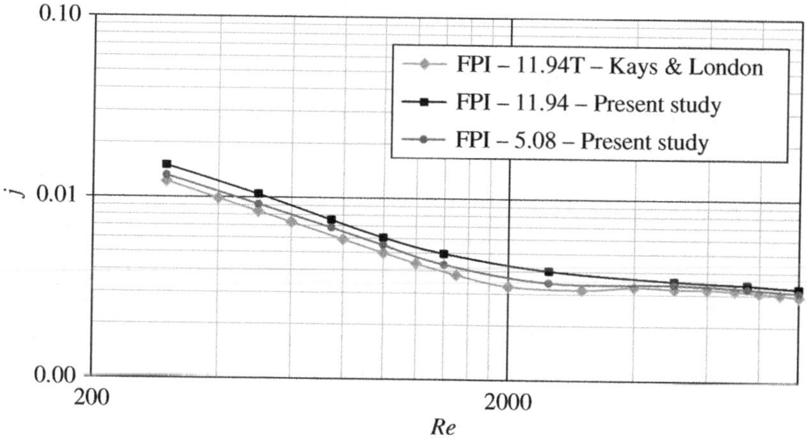

Figure 5.47 *j* vs *Re* comparison between triangular perforated fin present study values and triangular plain fin Kays and London [3] data FPI-11.94 T.

5.7.2.6.4 Correlations for f and j Factors

The ranges of the various fin geometric parameters are as follows:

- $d = 2$ and 3 mm
- $0.6 \le h/s \le 1.6$
- $0.015 \le t/s \le 0.03$
- $300 \le Re \le 1000$ for laminar region
- $1000 < Re \le 7500$ for turbulent region
- number of perforations per inch length = 2, 3 and 4

An exhaustive numerical study is carried out on the heat transfer phenomena in plate fin surfaces with triangular perforated fins. The power law expressions are used for correlating the Colburn factor *j* and Fanning friction factor *f* as a function of the Reynolds number and dimensionless fin parameters h/s and t/s. The f vs Re and j vs Re data for perforated triangular fin surfaces shows significant non-linearity over the Reynolds number range $300 \leq Re \leq 7500$. Therefore, two separate equations have been proposed for the low and high Re regions.

$$f = 10.127(Re)^{-1.588}(h/s)^{0.778}(t/s)^{-0.868} \tag{5.51}$$

for laminar range ($300 \leq Re \leq 1000$)

$$f = 1.685(Re)^{-0.798}(h/s)^{0.447}(t/s)^{-0.276} \tag{5.52}$$

for turbulent Range ($1000 < Re \leq 7500$)

The above correlations correctly predict 97% of the f data for laminar regions and 98% of the f data for the turbulent regions, with RMS value of 0.985.

Similarly, the Colburn factor j are functionally related to Re, h/s and t/s, and it can be represented as

$$j = 0.544(Re)^{-1..673}(h/s)^{2.278}(t/s)^{-1.589} \tag{5.53}$$

for laminar range ($300 \leq Re \leq 1000$)

$$j = 7.579(Re)^{-1..626}(h/s)^{1.185}(t/s)^{-1.689} \tag{5.54}$$

for turbulent range ($1000 < Re \leq 7500$)

The above correlations correctly predict 96% of the j data for laminar regions and 97% of the j data for the turbulent regions, with RMS value of 0.99.

5.7.2.6.5 f Factor Performance Comparison

Figure 5.48 shows the f factor performance comparison between Kays and London [3] data and the present study values of rectangular plain, rectangular perforated, triangular plain and triangular perforated fin for the same fin density. The graph indicates that the rectangular perforated fin gives the highest f factor while triangular perforated fin gives the lowest, which closely matches with the data presented by Kays and London [3].

5.7.2.6.6 Colburn j Factor Performance Comparison

Figure 5.49 shows, the Colburn j factor performance comparison between Kays and London [3] data and the present study values of rectangular plain, rectangular perforated, triangular plain and triangular perforated fin for the same fin density. The graph indicates that rectangular perforated fin gives the highest Colburn j factor while triangular perforated fin gives the lowest, which closely matches with the data presented by Kays and London [3].

5.7.3 Flow Non-Uniformity Analysis

A typical plate-fin heat exchanger, as shown in Figure 5.50, is analysed in order to study flow mal-distribution effects. This is made of stainless steel and is a cross flow heat exchanger with one pass on the cold air side and two passes on the hot air side. It has 31 layers on the hot air side for each pass and 32 layers on the cold air side.

This plate-fin heat exchanger is manufactured with a combination of offset strip fins on the hot fluid side and wavy fins on the cold fluid side. Based on earlier studies by

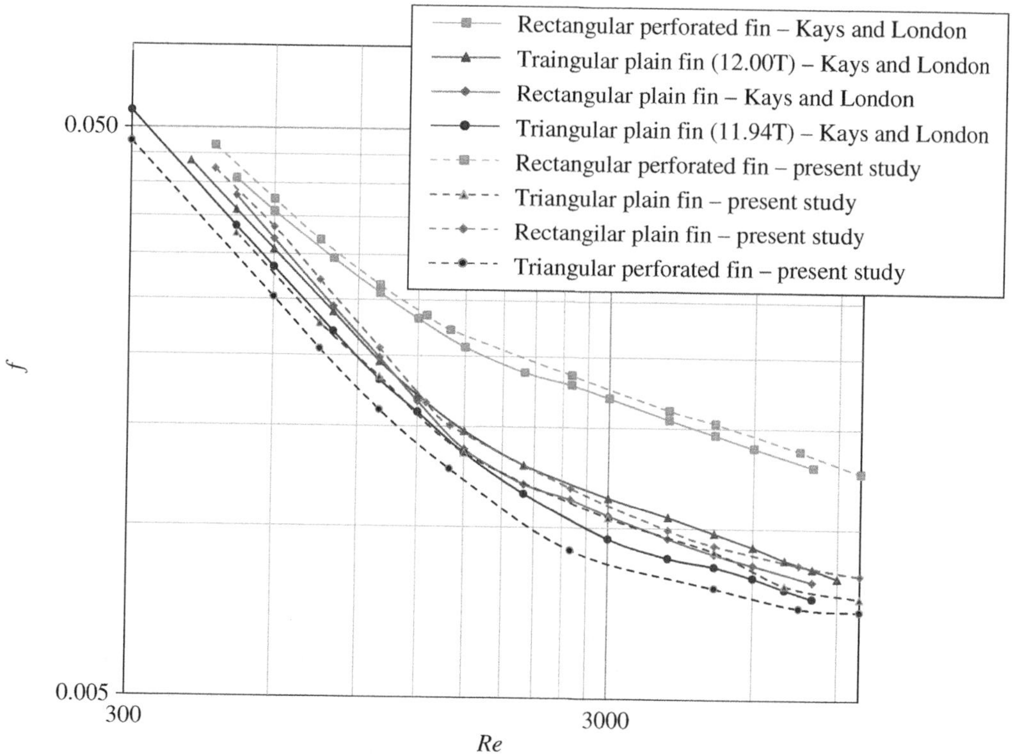

Figure 5.48 *f* vs *Re* comparison between rectangular plain fin, rectangular perforated fin, triangular plain fin and triangular perforated fin of the present study values with Kays and London [3] data.

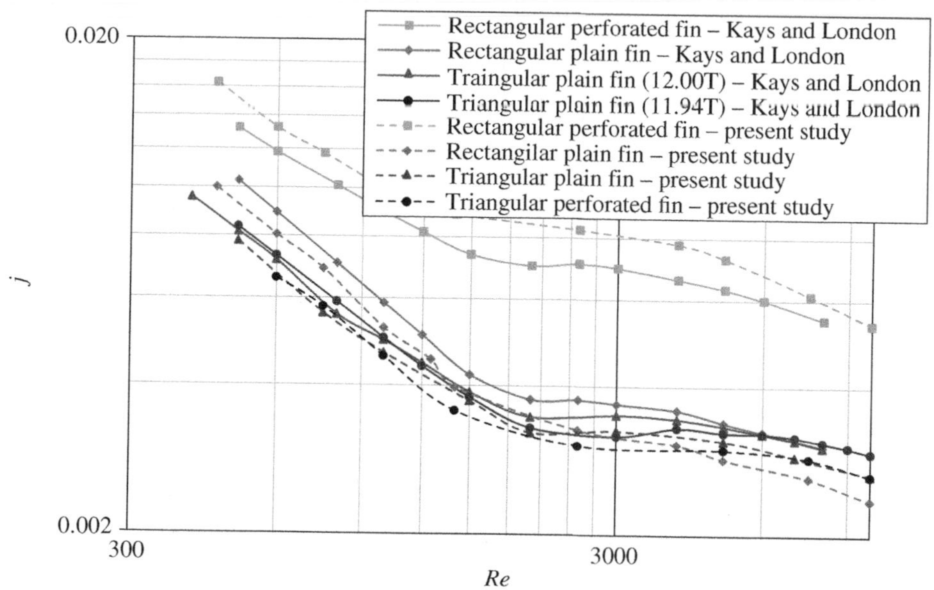

Figure 5.49 *j* vs *Re* comparison between rectangular plain fin, rectangular perforated fin, triangular plain fin and triangular perforated fin of the present study values with Kays and London [3] data.

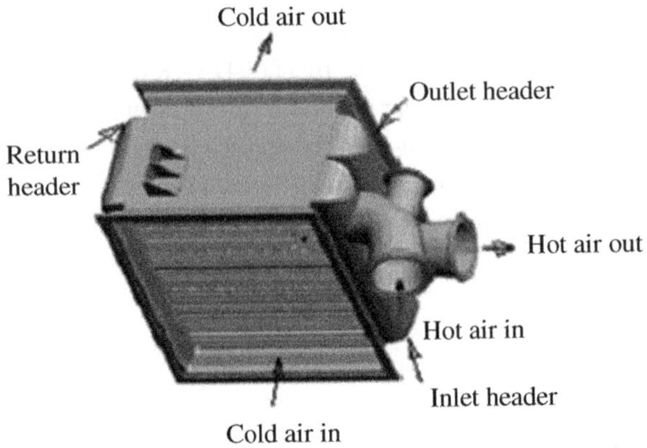

Figure 5.50 Compact Heat Exchanger.

Ranganayakulu et al. [52–57], proper header design is an essential part of the solutions to reduce flow mal-distribution and improve performance. The header is modified by placing a baffle plate in the inlet header and the effects of flow non-uniformity are studied. The study of the effects of entrance configuration on fluid flow mal-distribution is very important for optimizing the configuration and making the fluid flow more uniformly in the heat exchanger.

The typical improvised header configuration of the heat exchanger is shown in Figure 5.51a and b. It consists of a punched baffle plate placed at the distance of 85 mm

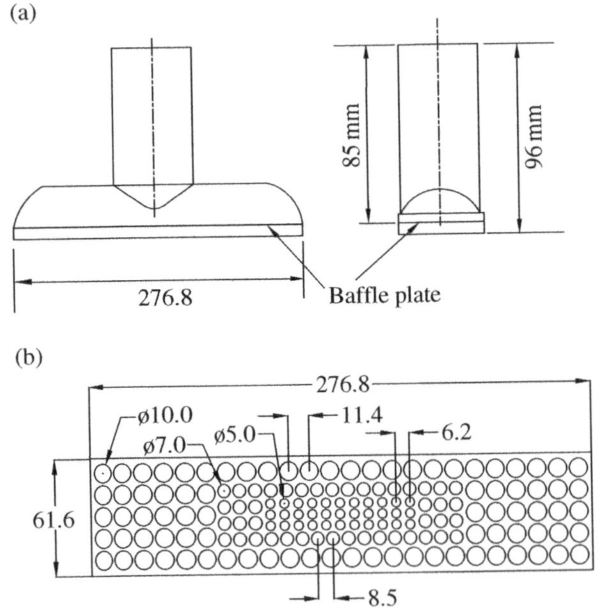

Figure 5.51 Inlet header configuration of heat exchanger (a) header with a baffle (b) details of punched baffle plate (all dimensions are in mm).

from the inlet of the pipe to the header. The baffle plate consists of a plate with holes of varying dimensions, arranged such that the flow is made uniform throughout; otherwise most flow from the centred inlet pipe will go through the center region of the baffle plate. The holes at the center are small, whereas the holes away from the center are larger in diameter. A baffle plate with hole diameters of 10, 7 and 5 mm is shown in Figure 5.51b. As more flow is passing through the center of exchanger core than through the sides/corners, 5 mm holes are selected for the center region of the baffle plate. Accordingly, hole diameters are increased to 7–10 mm in the peripheral regions of the baffle plate. These hole diameters are optimized such that more or less uniform flow enters the exchanger core. With the chosen hole pattern, the free flow area is reduced by 46.7%.

The ideal case header is nothing but perfectly uniform flow entering the inlet of the core, i.e. completely uniform velocity throughout the core. Practically this is not possible due to the space limitations of the aerospace industry. The pressure drop performances of a heat exchanger with and without baffle plate at the inlet header are compared numerically to find out the variations in the flow distributions at the entry of the core. The pressure drops for the real case and ideal case are compared in Table 5.5. In this table, the CFD results of (a) real case with existing header (b) ideal case by assuming perfect flow distribution and (c) with baffle plate provided in inlet header are compared with experimental one where total pressure is measured for the old configurations. It can be seen that CFD estimation of total pressure drop (1880.6 mbar) is close to measured pressure drop of 1890 mbar. There is a reduction of about 256 mbar in the pressure drop that is about 16% higher in the real case when compared with the ideal case. Moreover, as presented earlier in Chapter 4, there will be an increase in thermal performance if uniform flow is maintained in the exchanger core. This is primarily because of the greater flow uniformity throughout all channels as compared to the ideal case, where flow mal-distribution is predominant. The pressure drop change is nearly 13 mbar (about 1%) in the baffle plate case, when compared with the ideal case. The velocity profiles for the ideal and baffle plate cases are shown in the Figures 5.52 and 5.53, respectively. These figures show that core velocity

Table 5.5 Comparison of pressure drop of heat exchanger – CFD with experimental data.

Description CFD/ experimental results	Inlet to core (mbar)	Core (1st pass) (mbar)	Return header (mbar)	Core (2nd pass) (mbar)	Core to outlet (mbar)	Total pressure drop (mbar)
Real case with existing header	178.1	578.6	76.5	590.6	456.8	1880.6
Ideal case by assuming uniform flow	105.7	492.3	57.8	588.5	380.4	1624.7
With baffle plate provided in inlet header	175.3	492.3	58.0	588.5	323.8	1637.9
Experimental data at **0.575** kg/s	–	–	–	–	–	1890*

*Total pressure drop including headers and nozzles

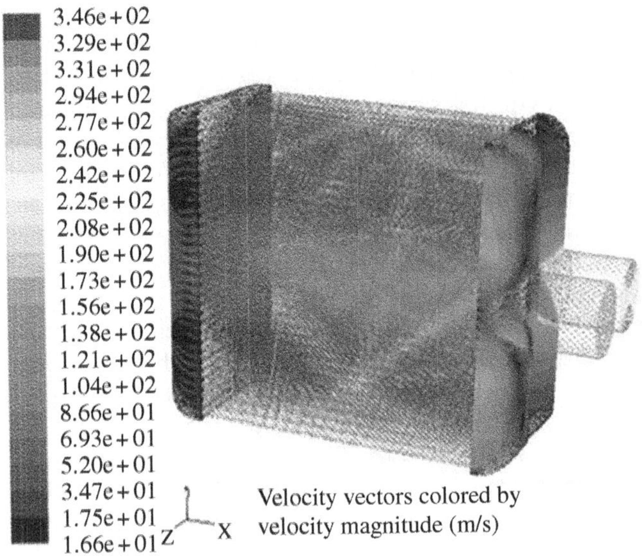

Figure 5.52 Sectional view of velocity distribution in a typical heat exchanger.

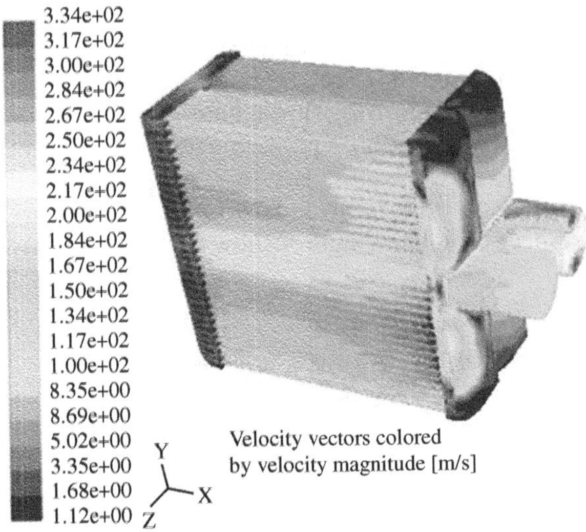

Figure 5.53 Velocity distribution with baffle plate in a typical heat exchanger.

vectors without baffle plate and with baffle plate give more or less uniform flow in the exchanger core.

5.7.4 Characterization of CHE Fins for Two-Phase Flow

In this section two-phase flow phenomena of compact heat exchanger fins are presented for both condensation and boiling applications. In phase change, the heat transfer of a compact heat exchanger, in addition to fluid properties and geometrical parameters, the heat flux, mass flux and vapour quality also affect the heat transfer coefficients. Extensive experimental data and empirical correlations are obtained for the condensation heat transfer and pressure drop characteristics of refrigerant R134a and other refrigerants on the inside helicoidal pipe and herringbone fins by Yi-Yie Yan et al. [59]; Kang et al. [60]; Longo and Gasparella [61, 64], Barbaara Watel [62], Longo [65, 66], Amir Jokar et al. [67] and Wang et al. [63]. The mechanism of condensation in compact heat exchangers is essentially that of gravity-controlled film condensation, for which a Nusselt-type theory applies, and a shear-controlled process in which the convective mechanism dominates [67]. Compact heat exchangers have an advantage again, in general, in that the condensate drainage process is augmented by surface tension drawing the liquid into the corners of the surface, which act as preferred drainage paths and thin the liquid layer on the active surface [67]. In this section, the average condensation heat transfer coefficients and frictional pressure drop measured during R134a inside small brazed compact plate-fin heat exchanger are presented. The effects of hydraulic diameter, refrigerant mass flux, saturation temperature (pressure), average heat transfer coefficient and frictional pressure drops are investigated. The correlation for average refrigerant heat transfer coefficient and frictional pressure drop are developed for the selected serrated fin surfaces.

5.7.4.1 Experimental Set-Up

The experimental test facility is established to study the condensation and boiling of R134a in different types of serrated fins used in compact plate-fin heat exchangers. A schematic diagram of the experimental test facility is shown in Figure 5.54; it has five main loops and a SCADA system. Specifically, the system consists of a refrigerant loop, condenser loop, evaporator loop, superheater loop and de-superheater loop.

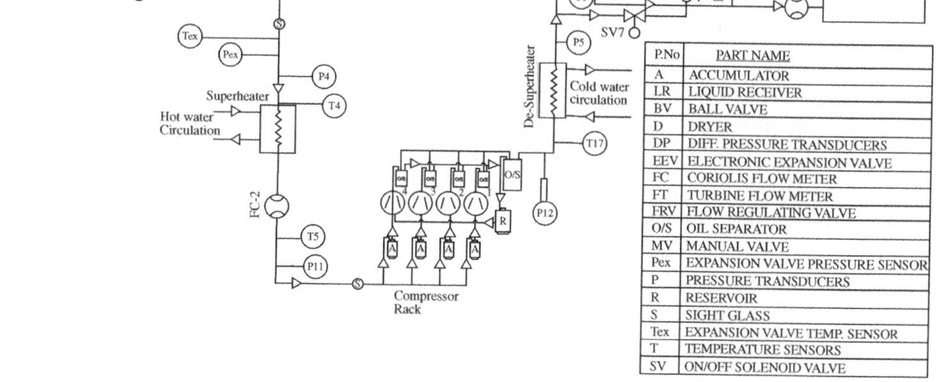

Figure 5.54 Schematic view of the experimental test rig.

Refrigerant R134a is circulated in the refrigerant loop, whereas de-mineralized (DM) water is used in all other loops. The condenser and evaporator loops are capable of supplying DM water flow at a constant temperature in the range of 5 °C to 60 °C with stability within ± 0.5 K. In order to obtain different conditions of R134a in the test heat exchanger, it is necessary to control the temperature and flow rate of the working fluids in the other loops.

The refrigerant loop contains the compressor rack, accumulators, oil separators, de-superheater, condenser, test condenser (brazed compact plate-fin heat exchanger), liquid receiver, filter drier, sight glasses, electronic expansion valve, evaporator, superheater and Coriolis flow meter. The refrigerant loop is a basic vapour compression refrigeration system, consisting of four parallel semi-hermitically sealed variable speed reciprocating compressors, having different capacity, test condenser, condenser, electronic expansion valve and evaporator. The variable speed reciprocating compressor rack is designed to control the flow rate between 0.005 kg/s to 0.06 kg/s in steps of 0.005 kg/s using a variable frequency drive. The flow rate was measured by a mass flow meter (Micro motion DN4) installed after the test condenser with an accuracy of ±0.05%. The de-superheater installed in the downstream of compressor rack, controls the inlet conditions to the test condenser by the water cooling system. After the condenser, the liquid refrigerant flows back to the receiver. The filter drier provides high water adsorption at low and high condensing temperatures, as well as at low and high degrees of humidity. The liquid line filter drier protects the refrigeration system from moisture and solid particles. The Coriolis flow meter is installed in the circuit after the evaporator and condenser to measure the refrigerant flow rate. The system also has sight glasses for physical verification of the state of the refrigerant R134a. The standard brazed plate heat exchanger is installed in the main circuit for the calibration of the test rig. After attaining the required conditions, the condenser is bypassed by the test heat exchanger which is installed as a separate test section. This test condenser has a water circuit to provide the cooling fluid, where the water flow rate can be varied between 0.08 kg/s and 0.7 kg/s. The pressure transducers and temperature sensors are located at the entrance and exit of the test condenser.

5.7.4.2 Brazed Test Core

The test cores are a brazed compact plate-fin heat exchanger with serrated/wavy fin surfaces for condensation and boiling effects. The test cores are basically a cross flow arrangement with two passes on the water side and one pass on the R134a side. The schematic representation of the test core is shown in Figure 5.55. It is made of aluminium alloy because of its low density, high thermal conductivity and high strength at low temperature. The geometric data of fins used in the test cores is given in Table 5.6. The fin surfaces, side-bars, parting sheets and cap sheets are held together in a jig under a predefined load, placed in a furnace and brazed to form the plate-fin heat exchanger block. The vacuum brazing process is used for brazing of the test core. The brazed test core with inserted wires is shown in Figure 5.55a and its photograph is shown in Figure 5.55b. The headers and nozzles are then welded to the block, taking care that the brazed joints remain intact during the welding process. In all three test cores (TC1, TC2 and TC3), the same type of fin is used on the DM water side. In R134a circuit of test cores TC1, TC2 and TC3 the hydraulic diameters of serrated fin are 1.1894 mm, 1.345 mm and 1.7461 mm respectively.

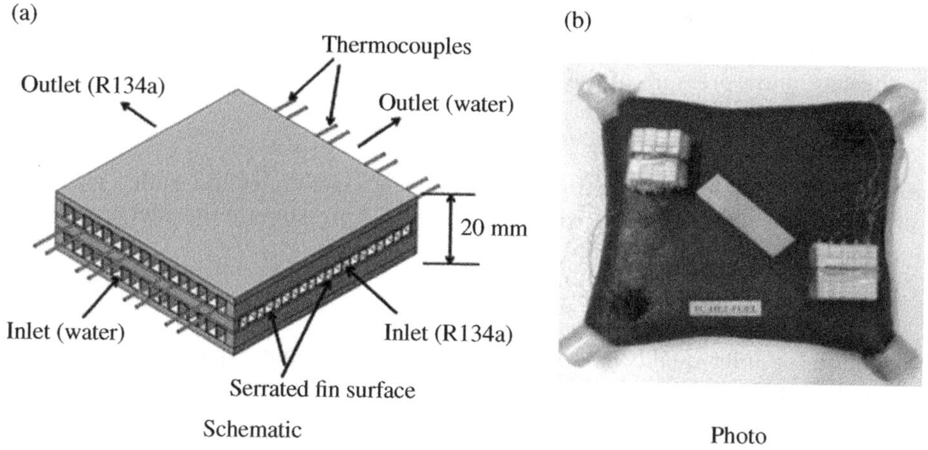

Figure 5.55 Test core.

Table 5.6 Geometric data of fins in test core (TC).

		Common		TC2	
Parameter	**Unit**	**DM water side**	**TC1**	**Refrigerant side**	**TC3**
Type			Serrated (lanced and offset)		
Material			Aluminium		
Fin frequency	FPI	28	30	28	18
Fin height	mm	5	3.05	5	3.8
Fin thickness	mm	0.127	0.1016	0.127	0.254
Serration length	mm	3.175	3.175	3.175	1.588
No of layers	no.	2	1	1	1
Hydraulic diameter	mm	1.3450	1.1894	1.345	1.7461

The temperatures of the refrigerant and water at the inlet and outlet of the test condenser and evaporator are measured by using resistance thermometers (RTD). The refrigerant pressures at the inlet of the test condenser and the evaporator are measured by a strain gauge pressure transducer; the test condenser pressure drop is measured by a strain-gauge differential transducer. The refrigerant mass flow rate is measured by Coriolis effect mass flow meter. Water flow rates through the test condenser and evaporator are measured by turbine flow meters. A data acquisition system is used, to collect the data of pressure, temperature, differential pressure, water flow rate, refrigerant flow rate at different test points, and the repeatability of experimental data is ensured. It includes an industrial PC with 187-channel Proficy SCADA, which gives the measured data in the form of digital output. The pressure transducer, differential pressure transducer, turbine flow meter and Coriolis flow meter need a power supply of 4–20 mA electric current to drive the output.

5.7.4.3 Boiling Heat Transfer Coefficient

The evaporation temperature is the saturation temperature, and the evaporation pressure is the saturation pressure of the refrigerant:

$$T_{ev} = T_{sat}; \quad P_{ev} = P_{sat} \tag{5.55}$$

The specific enthalpies of the refrigerant at the evaporator inlet and outlet, H_{evi} and H_{evo}, are calculated with the measured temperatures and pressures at the inlet and outlet of the evaporator.

$$H_{evi} = f(T_{evi}, P_{evi}); \quad H_{evo} = f(T_{evo}, P_{evo}) \tag{5.56}$$

The evaporator inlet quality is controlled by the electronic expansion valve. The vapour quality at the inlet and outlet of the test channel is given by state variables, enthalpy and pressure.

$$(x_{evo} - x_{evi}) = \frac{\dot{m}_w c_p (T_{wi} - T_{wo})}{\dot{m}_r h_{fg}} \tag{5.57}$$

$$x_{evi} = \frac{H_{expo} - (H_{evi})_{sat}}{(H_{evo} - H_{evi})_{sat}} \tag{5.58}$$

The local enthalpy and vapour quality inside the test section at every position z from the channel inlet to the channel outlet are calculated from the equation

$$x_z = x_{evi} + \frac{Qz}{\dot{m} h_{fg}} \tag{5.59}$$

where

$$Q = \dot{m}_w c_p (T_{wi} - T_{wo}) \tag{5.60}$$

Energy balance in the test section is verified by calculating the heat lost by the water and comparing it with the latent heat of vaporization of the refrigerant.

$$\dot{m}_w c_p (T_{wi} - T_{wo}) = (x_{evo} - x_{evi}) \, \dot{m}_r h_{fg} \tag{5.61}$$

The pressure drop due to flow friction and acceleration is the evaporation pressure, and this will not be constant in the test section (test evaporator). To simplify the problem, it is assumed that the evaporation temperature is constant in the test section.

The two-phase heat transfer coefficient on the refrigerant is given by

$$H_{tp} = \frac{Q}{\left(A_p + \eta_f A_f\right)(T_{wall} - T_{ev})} \tag{5.62}$$

$$\text{fin efficiency}, \eta_f = \frac{\tanh(ml)}{ml} \tag{5.63}$$

$$m = \left[\frac{2H_{tp}(t + l)}{k_s t l}\right]^{0.5} \tag{5.64}$$

T_{wall} is the mean wall temperature of two adjacent wall temperatures, measured at each axial position in the refrigerant flow direction in the test section.

$$T_{wall} = \frac{TC_1 + TC_2}{2} \tag{5.65}$$

$T_{ev} = T_{sat}$ = inlet temperature of the test section.

Heat is transferred from the water to the refrigerant liquid by nucleate boiling, as well as by two-phase convective mechanisms. The boiling heat transfer coefficient on the refrigerant, H_{tp}, is the sum of the convective term H_{cb} and the nucleate boiling term H_{nb}, as proposed by Barbara Watel [62].

$$H_{tp} = H_{cb} + SH_{nb} \tag{5.66}$$

The convective heat transfer coefficient H_{cb} is related to the single-phase heat transfer coefficient H_l by the Reynolds number factor F.

$$H_{cb} = F\,H_l \tag{5.67}$$

$$F = \frac{H_{cb}}{H_l} \tag{5.68}$$

where H_l can be obtained from single-phase experimental heat transfer data. The Reynolds number factor has generally been postulated to be a function of Lockhart and Martinelli parameter X

$$X = \left(\frac{(dp/dz)_{F,l}}{(dp/dz)_{F,v}}\right)^{0.5} = \left(\frac{\Delta P_l}{\Delta P_v}\right)_F^{0.5} = \frac{1-x}{x}\left(\frac{f_l\, v_l}{f_v\, v_v}\right)^{0.5} \tag{5.69}$$

where f_l is the liquid friction factor and f_v is the vapour friction factor.

The general form is given by Hesselgreaves [68] as

$$X = \left(\frac{1-x_g}{x_g}\right)^{\left(1-\frac{n}{2}\right)} \left(\frac{\rho_g}{\rho_l}\right)^{0.5} \left(\frac{\mu_l}{\mu_g}\right)^{\frac{n}{2}} \tag{5.70}$$

where n is the index of Re number in the single-phase frictional correlation $f = CRe^{-n}$ for the surface considered. Thus for turbulent flows of both components, the index n will be approximately 0.2, while for fully developed laminar flow $n = 1$.

The nucleate boiling heat transfer coefficient will be contributed for both the primary and secondary surface areas. The nucleate boiling contribution to the local boiling heat transfer coefficient for the primary and secondary surface area is given by Ranganayakulu et al. [69].

$$H_{nb} = \frac{.00916\Delta\, T_{sat}^{0.24}\,\Delta\, P_{sat}^{0.75}\, C_{pl}^{0.45}\, \rho_l^{0.49}\, k_l^{0.79}}{\sigma^{0.5}\, h_{fg}^{0.24}\, \mu_l^{0.29}\, \rho_g^{0.24}} \tag{5.71}$$

The nucleate boiling component H_{nb}, of the two-phase heat transfer coefficient is predicted using the suppression factor S. More recently, experiments have been carried out on different compact heat exchanger fins for boiling heat transfer and pressure drop values, presented by Amaranatha Raju et al. [70,71], which are also presented here for data collection and for the design of two-phase compact heat exchangers.

The suppression factor (S) proposed by Butterworth [72] can be directly extended to the flow boiling in a channel with offset strip fins in the present study.

Table 5.7 Summary of *F* factor correlations developed for offset and wavy fins.

Type of fin surface	*Two*-phase Reynolds number factor, *F*
OSF1	$F=\left[1+\frac{33.3}{X}\right]^{0.5}$
OSF2	$F=\left[1+\frac{30}{X}\right]^{0.5}$
Combined equation of OSF1 and OSF2	$F=\left[1+\frac{32}{X}\right]^{0.5}$
WF1	$F=\left[1+\frac{21}{X}\right]^{0.5}$
WF2	$F=\left[1+\frac{20}{X}\right]^{0.5}$
Combined equation of WF1 and WF2	$F=\left[1+\frac{20.5}{X}\right]^{0.5}$

$$S=\frac{1}{1+2.53\times10^{-6}Re_s^{1.17}} \tag{5.72}$$

where Re_s is the two-phase Re given by $Re_s = Re_l F^{1.25}$ (5.73)

The summary of F correlations developed by Amaranatha Raju et al. [70,71] for two offset strip and two wavy fins are presented in Table 5.7.

For single-phase heat transfer coefficient calculations, the following equations, which are developed recently by Ramanamurthy et al. [73,74] for water on OSF can be used:

$$H_w=0.049\left(\frac{k_w}{D_h}\right)Re_w^{0.69}Pr^{\frac{1}{3}}\quad \text{for laminar flow} \tag{5.74}$$

$$H_w=0.016\left(\frac{k_w}{D_h}\right)Re_w^{0.85}Pr^{\frac{1}{3}}\quad \text{for turbulent flow} \tag{5.75}$$

Using the above equations the augmented convective coefficient (Equation 5.67) and then the total boiling heat transfer coefficient (Equation 5.66) can be calculated.

5.7.4.3.1 *Two-Phase Boiling Pressure Drop Coefficient*

The two-phase pressure gradient in the boiling channel mainly consists of the frictional, acceleration and gravitational pressure gradients. The sum of the other form losses in the channel due to inlet/outlet header, expansion and contraction are also included. It is assumed to be fully liquid at the inlet of the test section and fully vapour at the exit of the test section for the entrance loss and exit loss calculation. Neglecting the entrance and exit effects, the overall pressure drop in the test section for a vapour–liquid mixture over a length L is estimated as

$$\Delta P_{tp}=\Delta P_{IH}+\Delta P_F+\Delta P_A+\Delta P_G+\Delta P_{OH} \tag{5.76}$$

The acceleration pressure gradient can be expressed as Kim [49]

$$\Delta P_A = \dot{G}^2 \left[\frac{x^2 v_g}{\alpha} + \frac{(1-x)^2 v_l}{1-\alpha} - 1\right] \tag{5.77}$$

Assuming the flow as a homogenous equilibrium model, the void fraction α is expressed in terms of Carey [75]

$$\alpha = \frac{1}{1 + \left(\frac{1-x}{x}\right)\left(\frac{\rho_v}{\rho_l}\right)} \tag{5.78}$$

The gravitational pressure gradient can be expressed as Carey [75]

$$\Delta P_G = \left[\frac{\alpha}{v_g} + \frac{(1-\alpha)}{v_l}\right] gL \tag{5.79}$$

Inlet header enlargement pressure gradient can be expressed as Kays and London [3]

$$\Delta P_{IH} = -\frac{\dot{G}^2 v_l}{2}\left[2\sigma_A(1-\sigma_A)\right] \tag{5.80}$$

Outlet header contraction pressure gradient can be expressed as Kays and London [3]

$$\Delta P_{OH} = \frac{\dot{G}^2 v_g}{2}\left[1.5\left(1-\sigma_A{}^2\right)\right] \tag{5.81}$$

The two-phase frictional pressure gradient can be expressed as Carey [75]

$$\Delta P_F = \frac{2 v_l \dot{G}^2 (1-x)^2 f \phi_f{}^2 L}{D_h} \tag{5.82}$$

where ΔP_{IH} is the expansion loss in the inlet header, ΔP_{OH} is the exit header contraction pressure loss, $\dot{G}$ is the mass velocity, v is the specific volume, f is the fluid friction factor, α is the local void fraction, x is the local quality and g denotes the gravitational acceleration.

The two-phase frictional pressure gradient is usually expressed in terms of the single-phase pressure gradient for the liquid phase flowing alone in the channel. The two-phase frictional multiplier in round tubes, ϕ_f^2 is uniquely correlated as a function of a parameter X by Lockhart and Martinelli [76], where

$$\phi_f{}^2 = 1 + \frac{C}{X} + \frac{1}{X^2} \tag{5.83}$$

where X is the Martinelli parameter, X is calculated using Equations 5.69 or 5.70 and constant C depends upon the flow characteristics of each phase. The effects of flow parameters (mainly q, x, G) on the two-phase frictional pressure drops are analysed for both offset strip and wavy fins in this chapter. The summary of frictional multiplier ϕ_f^2 developed by Amaranatha Raju et al. [70,71] for two offset strip and two wavy fins are presented in Table 5.8.

Using Table 5.8 and the Equation 5.76, the total pressure drop, ΔP_{tp} can be estimated for OSF and wavy fin surfaces.

Table 5.8 Summary of correlations developed for offset and wavy fins.

Type of fin surface	Two-phase frictional multiplier,
OSF1	$\phi_f^2 = 1 + \frac{30.1}{X} + \frac{4}{X^2}$
OSF2	$\phi_f^2 = 1 + \frac{28}{X} + \frac{3.3}{X^2}$
Combined equation of OSF1 and OSF2	$\phi_f^2 = 1 + \frac{29}{X} + \frac{3.5}{X^2}$
WF1	$\phi_f^2 = 1 + \frac{25}{X} + \frac{3}{X^2}$
WF2	$\phi_f^2 = 1 + \frac{25}{X} + \frac{3.2}{X^2}$
Combined equation of WF1 and WF2	$\phi_f^2 = 1 + \frac{25}{X} + \frac{3.1}{X^2}$

5.7.4.4 Two-Phase Condensation

Hu and Herold, [77,78] have published the effects of the Prandtl number on Colburn j factor for water and polyalphaolefin (PAO) fluids and claimed that air models over-predict the j factor for liquids and found that the heat transfer coefficient of liquids is approximately two times that of air. Hence, a detailed analysis has been carried out using ANSYS Fluent tool to estimate the j and f factors for serrated fins and reported in Ramana Murthy et al. [79].

The correlation for water side Nu and friction factor for the above selected serrated fin surface in accordance with Ramana Murthy et al. [73,74] in laminar range and turbulent range are as follows:
for laminar range ($100 \le Re \le 800$)

$$Nu_w = 0.049\, Re_w^{0.69} Pr^{\frac{1}{3}} \tag{5.84}$$

$$f_w = 5.189\, Re_w^{-0.596} \tag{5.85}$$

and for turbulent range ($1000 \le Re \le 15000$)

$$Nu_w = 0.016\, Re_w^{0.85} Pr^{\frac{1}{3}}, \tag{5.86}$$

$$f_w = 0.4533\, Re_w^{-0.24} \tag{5.87}$$

The steady state conditions for the test condenser are determined by checking the heat balance between the released heat from R134a and the heat absorbed by the cooling water. The amount of heat released from R134a can be calculated as

$$Q_r = m_r \Delta H \tag{5.88}$$

where Q_r, m_r and ΔH are the heat load, mass flow rate and enthalpy drop of R134a across the test condenser respectively. The criterion used for the steady state condition during the test is

$$\frac{|Q_r - Q_w|}{Q_{ave}} \leq 5\% \tag{5.89}$$

where Q_{ave}, is the average value of heat exchanged between R134a and the cooling water and is determined by

$$Q_{ave} = \frac{1}{2}(Q_r + Q_w) \tag{5.90}$$

5.7.4.4.1 Pressure Drop Measurements

The frictional pressure drop on the refrigerant side is computed in accordance with Yan et al. [80] by subtracting the pressure losses at the test condenser inlet and exit manifolds and ports (ΔP_c) then adding the momentum pressure rise (deceleration, ΔP_a) and the gravity pressure rise (elevation, ΔP_g) to the total pressure drop measured (ΔP_t).

$$\Delta P_f = \Delta P_t - \Delta P_c + \Delta P_a + \Delta P_g \tag{5.91}$$

The momentum and gravity pressure drops are estimated by the homogeneous model for two-phase gas-liquid flow as per Collier [81]:

$$\Delta Pa = G^2(v_g - v_l)|\Delta x| \tag{5.92}$$

$$\Delta Pg = g\rho_m L \tag{5.93}$$

where v_g and v_l are the specific volume of liquid and vapour phase, $|\Delta x|$ is the absolute value of the vapour quality change between inlet and outlet and

$$\rho_m = X_m/\rho_g + (1 - X_m)/\rho_l^{-1} \tag{5.94}$$

is the average two-phase density between inlet and outlet, calculated by the homogeneous model at the average vapour quality X_m between inlet and outlet. The pressure drop in the inlet and outlet manifolds and ports is empirically estimated, in accordance with Shah and Focke [82] as

$$\Delta Pc = 1.5\,G^2/(2\rho_m) \tag{5.95}$$

5.7.4.4.2 Condensation Correlations for Pressure Drop

Similarly, the two-phase serrated fin surface frictional pressure drops are functionally related to (*KE*/*V*) and it can be represented as:

$$\Delta P_f = C(KE/V)^{b1} \tag{5.96}$$

where C and $b1$ are constants.

Therefore, the proposed equation for this particular serrated fin surface is

$$\Delta P_f = 12.49(KE/V)^{0.546} \tag{5.97}$$

More recently, experiments have been carried out on different CHE fins for condensation heat transfer and pressure drop values, presented by Ramana Murthy et al. [73,74], which are also presented here for data collection and for design of two-phase CHE. The two-phase serrated fin surface frictional pressure drops are functionally related to (*KE*/*V*) and equations are proposed for particular serrated fin surfaces as listed in Table 5.9.

Table 5.9 Condensation pressure drop correlations for serrated fin surfaces.

Unit no.	ΔP_f (kPa) correlations
TC1	$\Delta P_f = 0.8828\ (KE/V)^{0.62}$
TC2	$\Delta P_f = 2.389\ (KE/V)^{0.283}$
TC3	$\Delta P_f = 1.654\ (KE/V)^{0.478}$
TC4	$\Delta P_f = 0.8875\ (KE/V)^{0.5867}$

The correlations are applicable for equivalent Reynolds number range 600–1800 for refrigerant saturated conditions of serrated fins used for experiments. The correlations of condensation refrigerant frictional pressure drop are accurate to ±15%.

In addition, the correlations are explored for serrated fin surface hydraulic diameter ranges from 1.1894 mm to 1.8667 mm and fin density ranges from 10 FPI (393.7 fins per metre) to 30 FPI (1181 fins per metre) as:

$$\Delta P_f = 1.2556(KE/V)^{00.4987} \tag{5.98}$$

The above ΔP_f correlations correctly predict 87% of the ΔP_f data for equivalent Reynolds number range of 600 to 1800, and the root-mean-square error is 4.3%.

5.7.4.4.3 *Condensation Correlations for Heat Transfer*

The correlations have been expressed in terms of equivalent Reynolds number, liquid Prandtl number, liquid thermal conductivity and fin hydraulic diameter. The power law expressions have been used for determining the two-phase refrigerant heat transfer coefficient and two-phase refrigerant friction factor as a function of the equivalent Reynolds number and dimensionless fin parameters. The two-phase refrigerant condensation heat transfer coefficient (H_r) is functionally related to (K_L/D_h), Re_{eq}, Pr_L and T_{sat} and it can be represented as

$$H_r = B(K_L/D_h)^{a1} Re_{eq}^{a2} Pr_L{}^{a3} \tag{5.99}$$

where B, $a1$, $a2$ and $a3$ are constants, and Re_{eq} can be calculated as

$$Re_{eq} = \frac{G\left[(1-x) + x\left(\dfrac{\rho_l}{\rho_g}\right)^{0.5}\right] D_h}{\mu_l} \tag{5.100}$$

Experiments have been carried out by Ramana Murthy et al. [73,74] on different CHE fins for condensation heat transfer, and they presented the equations shown in Table 5.10 for serrated fin surfaces:

The correlations are applicable for equivalent Reynolds number range of 600–1800 for refrigerant saturated conditions of serrated fins used for experiments. The correlations of condensation refrigerant heat transfer coefficients are accurate within ±20%. In addition, the correlations are explored for serrated fin surface hydraulic diameter ranges from 1.1894 mm to 1.8667 mm and fin density ranges from 10 FPI (393.7 fins per metre) to 30 FPI (1181 fins per metre) as

Table 5.10 Heat transfer correlations for condensation.

Unit no.	H_r (w/m²k) correlations
TC1	$H_r = 6.5139*10^{-4}(K_l/D_h)^{0.08826}(Re_{eq})^{1.5747}(Pr_l)^{3.7787}$
TC2	$H_r = 6.909*10^{-28}(K_l/D_h)^{12.174}(Re_{eq})^{1.594}(Pr_l)^{8.024}$
TC3	$H_r = 2.2284*10^{-4}(K_l/D_h)^{0.1465}(Re_{eq})^{1.6917}(Pr_l)^{3.7301}$
TC4	$H_r = 3.1072*10^{-22}(K_l/D_h)^{9.9534}(Re_{eq})^{1.6491}(Pr_l)^{7.2095}$

$$H_r = 1.786*10^{-4}(K_L/D_h)^{0.271}(Re_{eq})^{1.5947}(Pr_l)^{4.105} \quad (5.101)$$

The above H_r correlations correctly predict 94% of the H_r data for equivalent Reynolds number range of 600–1800, and the root-mean-square error is 3.8%.

The above correlations correctly predict 88% of the ΔP_f data for equivalent Reynolds number range of 600–1800, and root-mean-square error is 5.5%. No other experimental data is available in the open literature on the condensation average saturated heat transfer coefficient and pressure drops for R134a for serrated fin surface.

5.7.5 Estimation of Endurance Life of Compact Heat Exchanger

Compact heat exchangers (CHE) used in aircraft will be subjected to arduous operating conditions during flight. They experience high temperatures and are also subjected to thermal cycles during operation [83]. In this section, computational prediction of endurance life of compact heat exchangers is presented. In general, compact heat exchanger is subjected to combined pressure, temperature and flow cycling. However, in this section, estimation of endurance life of CHEs is presented numerically using both CFD and FEM and is compared with experimental results. A combined thermal and structural analysis is carried out to predict the endurance life of compact heat exchanger computationally, considering the effects of fatigue and creep. The computational results are in close agreement with experimental data. The valid computational approach is used for various case studies to understand the endurance life of compact heat exchangers.

For aerospace applications where a premium is placed on high performance with minimum weight and low footprint, compact heat exchangers are found suitable. The compact heat exchangers used in combat aircraft operate at temperatures in excess of 550 °C and pressures of 7.0 bar (g) and are subjected to varying operating conditions during the course of a flight. As the heat exchangers operate at high temperatures for a considerable duration, creep phenomena play a major role. The cyclic operation experienced during take-off and landing and during combat manoeuvres can accelerate the creep phenomena. This cycling operation also causes thermal fatigue. When creep and fatigue are combined, it reduces the endurance life of the compact heat exchanger.

When new heat exchangers are developed for such rigorous aerospace applications, it has to be evaluated for thermal performance and endurance life during the design stage. It is mandatory from the airworthiness point of view to evaluate the endurance life of CHE.

CFD analysis is carried out to obtain the temperature distribution in the heat exchanger along the flow direction. This is used as an input in the finite element analysis to calculate the stress distribution in the CHE which is then utilized to calculate the endurance life of the CHE in terms of creep and associated fatigue life. A typical compact heat exchanger shown in Figure 5.50 is used for estimation of the endurance life of a CHE.

5.7.5.1 Computational Analysis

The endurance life of a heat exchanger is predicted using computational methods [84] by considering the effect of fatigue and creep. The analysis is carried out for the experimental conditions as given in paragraph 5.7.5.2. The heat exchanger is made up of various parts such as headers, shell, side bars and the core consisting of separating sheets, wavy fins, lanced and offset fins. The failure of the separating sheet, which transfers the heat from hot to cold sides is termed the failure of the heat exchanger, as hot and cold streams would mix with each other. Hence, for analysis, only the core of the heat exchanger is considered.

The core of the heat exchanger in the present study has a complex geometry and consists of alternate layers of cold air and hot air streams flowing in fins, separated by thin sheets. The broad details of the core of heat exchanger are as follows:

- hot side: 31 layers, lanced and offset type fins of 3.048 mm height, 20.5 fins per inch with a lancing length of 2.032 mm
- cold side: 32 layers, wavy type fins of 5.207 mm height and 13.5 fins per inch
- separating sheets: 64 numbers, 0.254 mm thick

This exchanger has both wavy fins and offset fins, as shown in Figures 2.4c and 2.4d respectively in Chapter 2. The maximum stress to which the heat exchanger is subjected is found by finite element analysis (FEA) for the given operating conditions. The solid model is generated in the Ideas software and meshed in Hyper Mesh. FEA is carried out in the ANSYS software [85]. Due to symmetry in the geometry and flow, a geometric model consisting of three layers of cold air streams and three layers of hot air streams separated by seven numbers of separating sheets is considered. For FEA, a model of 24 mm (cold air entry face) × 28 mm (hot air entry face) is considered.

Two assumptions are considered in the analysis:

- simplification of lanced and offset fins in the hot side to plain fins
- better connectivity of fin with separating sheets using coupling elements

The FE model considered for analysis is shown in Figure 5.56, the FE model grid consisting of 30 million elements. The fin surfaces are modelled as "3D shell" elements and separating sheet as "3D solid" elements.

5.7.5.2 CFD Analysis of CHE

CFD analysis [85] is carried out to find out the temperature distribution of hot and cold fluid in the heat exchanger. The computational domain considered for CFD purposes is a model consisting of three layers of cold air side fins and two layers of hot air side fins separated by six thin sheets, because the heat exchanger geometry is symmetrical in length and height on both hot and cold air sides. Gambit software is used for modelling the geometry and generating the grids. This grid has approximately 3 million cells.

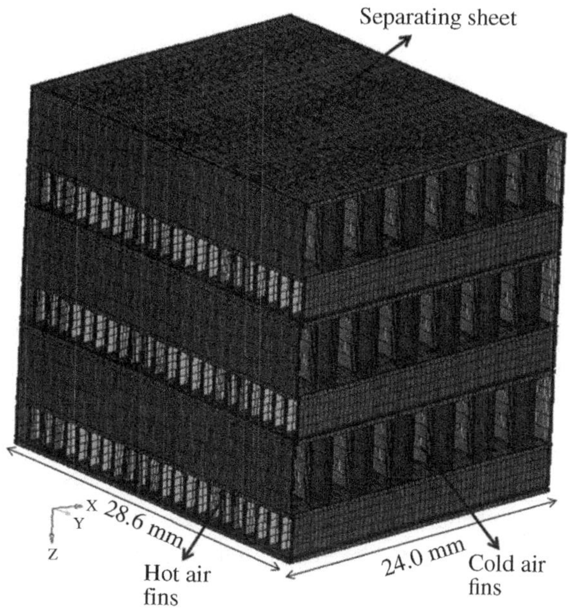

Figure 5.56 Compact heat exchanger – FE model.

Figure 5.57 shows the CFD model of the CHE considered for analysis. The boundary conditions are specified and the solution is obtained by running the CFD code, Fluent solver [84]. The turbulence model K–epsilon, realizable with standard wall treatment functions is considered with 5% turbulent intensity.

Fluid temperature distribution for the hot and cold streams is obtained for the operating conditions. As the operating envelope is cyclic, having a maximum and a minimum condition, CFD analysis is carried out to obtain the fluid temperature distribution for both the conditions. Temperature distribution along the flow length direction for both hot and cold air streams for maximum and minimum conditions are shown in Figures 5.58 and 5.59 respectively.

It is observed that during maximum condition, the hot air enters the CHE at 600 °C and is cooled by cold air and leaves CHE at 327 °C, while the cold air enters the CHE at 120 °C and exits at 144 °C. During the minimum condition, the hot air enters the

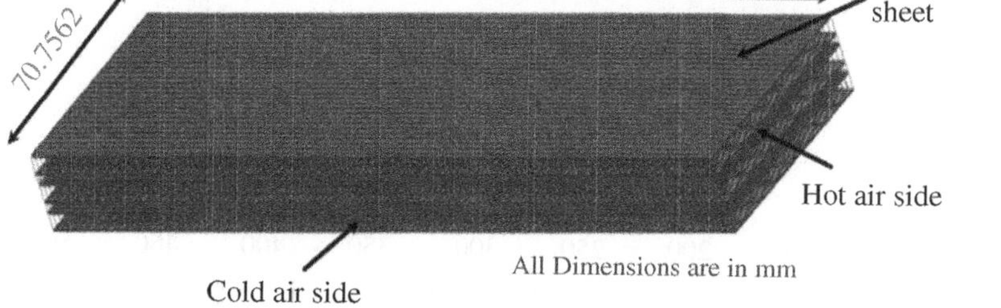

Figure 5.57 CFD model of heat exchanger.

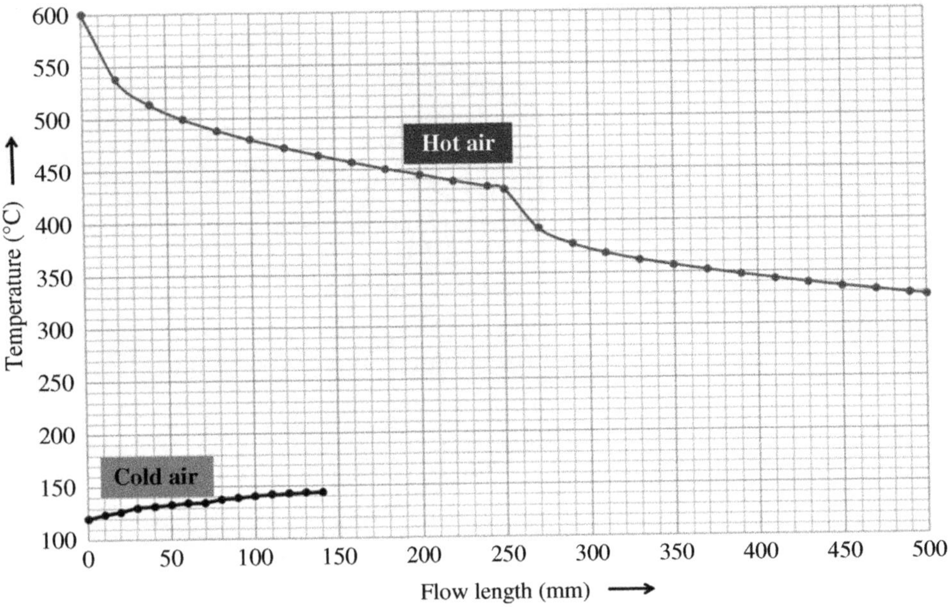

Figure 5.58 Temperature distribution of hot and cold air stream for maximum condition.

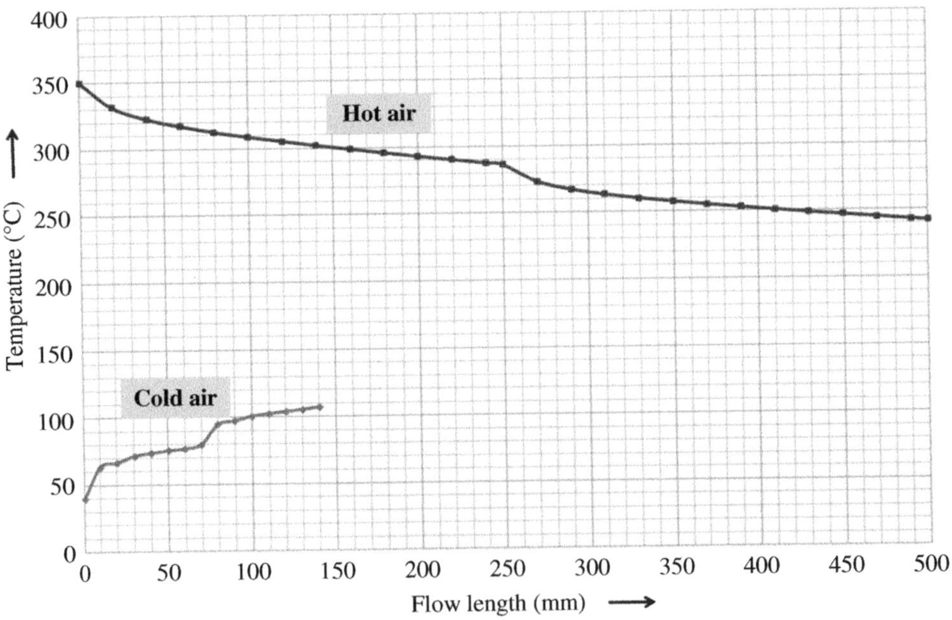

Figure 5.59 Temperature distribution of hot air and cold air stream for minimum condition.

CHE at 350 °C and leaves at 242 °C, with an average temperature drop of 108 °C, while the cold air enters CHE at 40 °C and leaves at 107 °C with an average temperature rise of 67 °C. It is also seen that for maximum operating condition, the gradient of temperature drop in the hot air stream is high in the first 50 mm from the entry along the flow direction. Also, the hot air temperature drop during maximum condition is very high compared to the minimum condition. This is due to the fact that the cold air mass flow during maximum condition (300 kg/min) is almost 10 times higher than during minimum condition (32 kg/min).

5.7.5.2.1 *Thermal Analysis*

The separating sheet transfers heat from the hot side to the cold side by conduction and convection, resulting in a temperature gradient across the separating sheet, which in turn produces thermal load. Bulk fluid temperature for both hot and cold air flows is obtained for the given operating conditions. The heat transfer coefficient for the hot and cold flows near the separating sheet is also calculated for the given operating conditions. These inputs are then used for defining thermal loading on the separating sheet in FEM. Using the fluid temperature distribution and by applying convective boundary conditions on the FE model for the hot and cold sides of the separating sheet surface, thermal analysis is carried out using ANSYS software [85] to obtain the temperature distribution on the separating sheet. Figure 5.60 shows the temperature distribution obtained on the separating sheet for maximum and minimum conditions.

5.7.5.2.2 *Structural Analysis*

Linear static analysis is performed using ANSYS software [85] on the FE model for the maximum and minimum operating conditions, and the stress distribution on the

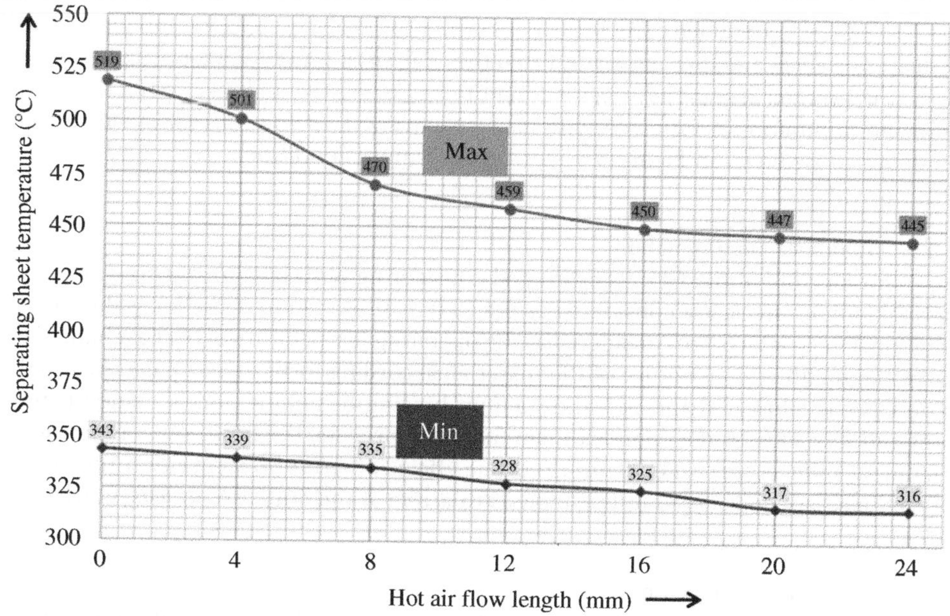

Figure 5.60 Temperature distribution on the separating sheet for maximum and minimum condition.

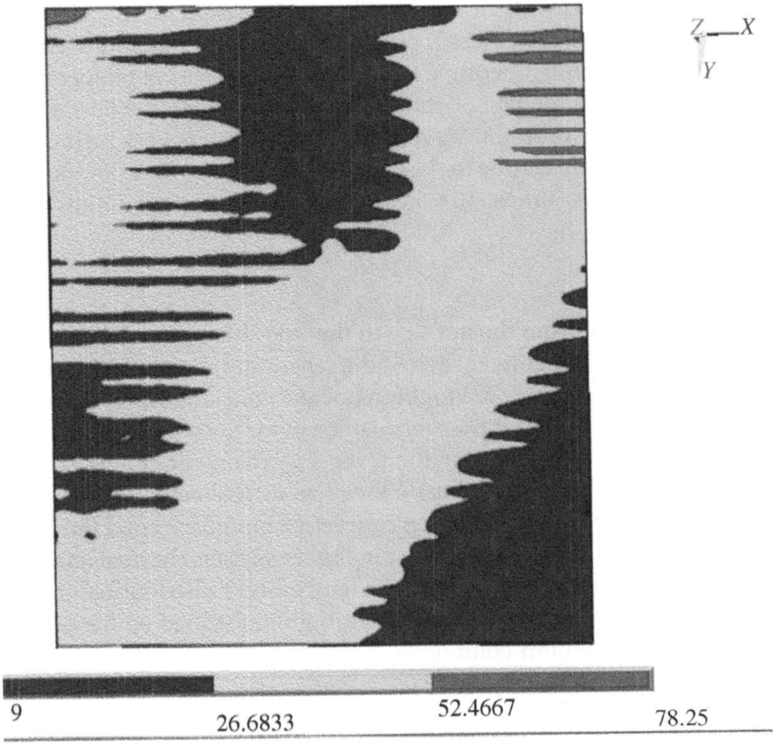

Figure 5.61 Von Mises stress distribution on the separating sheet for maximum condition.

separating sheet is obtained. The loads considered for analysis are differential pressure on separating sheet and thermal loads. From the analysis, it is found that the highest Von Mises stress on the separating sheet is found to be 78.25 MPa and 39.25 MPa for maximum and minimum operating conditions, respectively. The Von Mises stress distribution obtained is shown in Figure 5.61 for the maximum condition.

5.7.5.3 Endurance Life Estimation

The endurance life of the CHE is estimated considering both fatigue and creep effects. Thermal fatigue effects are considered as the CHE is subjected to cyclic loads during its operation. Creep effects are considered as the operating temperature of heat exchanger is 30% of the melting temperature of the material, according to Bazu and Bajenescu [86].

The CHE is fabricated through the vacuum brazing technique and is made of thin plates and fins. Due to the complex manufacturing process, the fin to separating sheet attachment induces stress concentration. The effects of stress concentration (Peterson [87]) induced are considered in the analysis to arrive at the critical Von Mises stresses.

5.7.5.4 Fatigue Life Estimation

Fatigue life is calculated using the Coffin–Manson (universal slope) equation. For load cycles where stress ratio, $R \neq -1.0$, the Coffin–Manson equation is

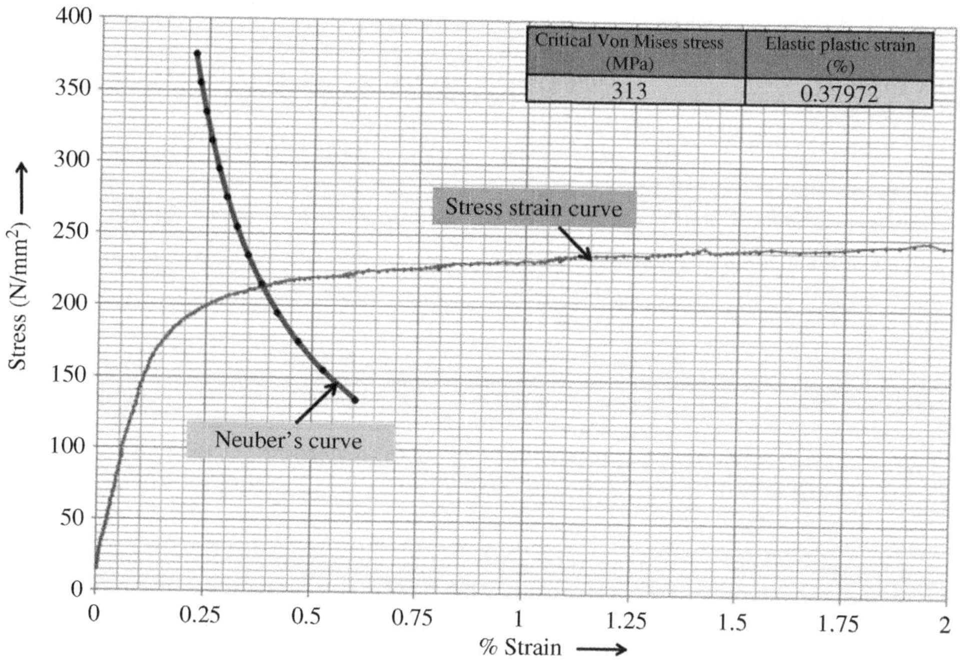

Figure 5.62 Neuber's hyperbola for maximum operating condition.

$$\Delta\varepsilon = \frac{3.5(UTS - mean)}{E} N_f^{-0.12} + D^{0.6} N_f^{-0.6} \tag{5.102}$$

$$D = -\ln(1 - RA) \tag{5.103}$$

$$\infty_{mean} = (\infty_{max} + \infty_{min})/2 \tag{5.104}$$

For maximum operating condition, the critical Von Mises stress is found to be greater than the proof strength of the separating sheet material. To obtain the actual strain, Neuber's hyperbola [88] is constructed, intersected with the real stress–strain curve of material by Balram Gupta [89], as shown in Figure 5.62. It is found that for the given operating conditions the heat exchanger has a fatigue life of 65,550 cycles, which is shown in Table 5.11.

5.7.5.5 Effect of Creep

The CHE experiences creep due to exposure to high temperature and high stresses. The rate of creep is highly dependent on both stress and temperature. The heat exchanger is subjected to a maximum temperature of 600 °C for 23 seconds and minimum temperature of 350 °C for 23 seconds. However, in the present work, only the maximum temperature of 600 °C is considered for creep effects.

The Larson–Miller parameter [90] predicts the lifetime of material vs time and temperature which is expressed as

$$LMP = T^{*}(C + \log t)/1000 \tag{5.105}$$

where $C = 20$, for SS 347.

Table 5.11 Number of cycles for failure by fatigue.

Max Von Mises stress (MPa)			Critical Von Mises stress (MPa)				
Max	**Min**	**SCF**	**Max**	**Min**	**Mean**	**Delta total strain**	**Number of cycles**
78.25	39.25	4.0	313	157	235	0.0029	65550

Table 5.12 Creep life of the compact heat exchanger.

Gross stress (MPa)	SCF	Peak stress (MPa)	Larson–Miller parameter	Creep life (hrs)
78.25	4.00	313	32.85	8

For the peak stress of 313 MPa, the Larsen–Miller parameter (32.85) is obtained from the material properties of SS. Substituting for operating temperature, the creep life in hours is obtained and tabulated in Table 5.12.

The endurance life of CHE is estimated using the Miner–Robinson Equation (5.106) by combining the fatigue and creep life, as per Wieslaw Beres [91].

$$1/\textit{Endurance life} = \{1/\textit{Fatigue life} + 1/\textit{Creep life}\} \tag{5.106}$$

To calculate the endurance life, the equivalent number of creep cycles is calculated by considering the dwell time in maximum condition only. Bhaskar et al. [92,93] explained the qualification test facility, endurance tests and endurance life estimation for the CHE. In the present case, dwell time in maximum condition is 23 s. Thus the creep life of 8 hours translates to 1247 equivalent creep cycles. The total life of the heat exchanger is computed using the above equation and is found to be 1224 cycles.

5.7.5.6 Results of Endurance Life

The heat exchanger is subjected to high pressure and high temperature in a cyclic manner to experimentally evaluate endurance life. It is found that the CHE has withstood 1100 cycles. The endurance life of compact heat exchanger using computational methods is found to be 1224 cycles and is in close agreement with the experimental results. The computational results also predicted the failure region of CHE. This region is located at the junction of hot air entry with cold air exit, which is the actual location of failure in the experimental tests. In this region, the CHE experiences maximum metal temperature and maximum metal temperature gradient, which causes maximum stress. Case studies are carried out using computational methods to estimate the life of the CHE with varying conditions. The inlet temperature of the hot air is varied and the results are graphically illustrated in Figure 5.63. It is seen that creep plays a more important role in the failure of CHE than fatigue at higher temperatures.

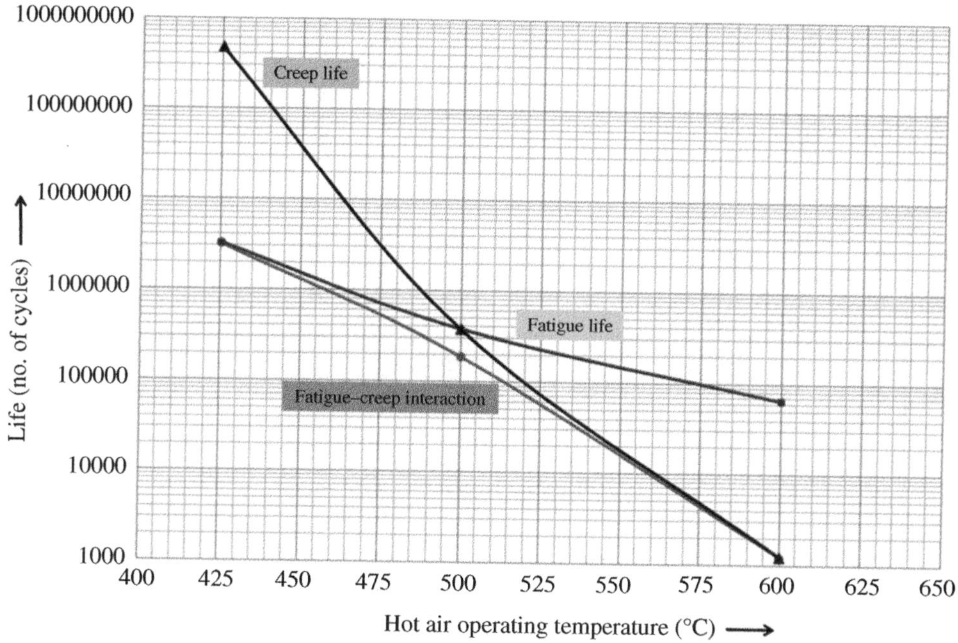

Figure 5.63 Effect of variation of inlet temperatures on fatigue, creep and endurance life of the CHE.

5.8 Closure

Several promising developments are taking place in the field of compact heat exchangers, but the technology has remained proprietary. In the present study, a small contribution has been made to the technology of plate-fin heat exchangers. Issues related to manufacturing techniques and design approaches remain crucial to widespread application of plate-fin heat exchangers, which are discussed in Chapter 7. Heat transfer and flow friction characteristics of plate-fin surfaces will play the most vital role in its success. An extensive survey has been carried out on compact plate-fin heat exchangers and the problems faced by various researchers are reviewed.

- Introduced a new approach to develop heat transfer and flow friction correlations by computational methods. Experiments on heat transfer over plate-fin surfaces are expensive. Results generated by reputed international laboratories have remained almost totally proprietary.
- Presented the heat transfer and friction coefficient correlations for wavy fins, offset fins, rectangular plain fins, rectangular perforated fins, triangular plain fins and triangular perforated fins. CFD analysis has been carried out using Fluent software, and design data has been generated for different fin geometries and the results are validated with open literature.
- Developed the correlations for the Fanning friction factor f and Colburn j factor as a function of Re and geometrical parameters such as fin height h, fin spacing s, fin thickness t, lance length l, in case of offset fins and wavy length L, in case of wavy fins. These correlations can be used effectively for design activities of compact heat exchanger.

Table 5.13 Fin surfaces *f*, *j* and *h* correlations.

S. no.	Type of fin surface	Design data	Correlations	RMS	Fluid flow	Range of applicability
1	Wavy fins	f	$9.827R_e^{-0.705}\left(\frac{h}{s}\right)^{0.322}\left(\frac{2A}{s}\right)^{-0.394}\left(\frac{L}{2A}\right)^{-0.603}$	0.995	laminar	$100 \le Re \le 800$
		f	$10.628\,R_e^{-0.359}\left(\frac{h}{s}\right)^{0.264}\left(\frac{2A}{s}\right)^{-0.848}\left(\frac{L}{2A}\right)^{-1.931}$		turbulent	$1000 \le Re \le 15000$
		j	$2.348\,R_e^{-0.786}\left(\frac{h}{s}\right)^{0.312}\left(\frac{2A}{s}\right)^{-0.192}\left(\frac{L}{2A}\right)^{-0.432}$	0.995	laminar	$100 \le Re \le 1000$
		j	$0.242\,R_e^{-0.375}\left(\frac{h}{s}\right)^{0.235}\left(\frac{2A}{s}\right)^{-0.288}\left(\frac{L}{2A}\right)^{-0.553}$		turbulent	$1000 \le Re \le 15000$
2	Wavy fin R134a	j	$2.989Re^{-0.54241}\frac{h}{s}^{-0.72276}\frac{2A}{s}^{-0.83914}\frac{L}{2A}^{-0.7588}$	0.98	laminar	$100 \le Re \le 1000$
		j	$3.245Re^{-0.66388}\left(\frac{h}{s}\right)^{-0.53614}\frac{2A}{s}^{-0.80626}\frac{L}{2A}^{-0.6346}$	0.98	turbulent	$1000 \le Re \le 15000$
3	Wavy fin water	j	$1.154Re^{-0.65938}\frac{h}{s}^{-0.96698}\frac{2A}{s}^{0.176702}\frac{L}{2A}^{0.288785}$	0.98	laminar	$100 \le Re \le 1000$
4	Wavy fin water and R134a	f	$18.607Re^{-0.59381}\frac{h}{s}^{-0.088954}\frac{2A}{s}^{-0.46976}\frac{L}{2A}^{-0.92621}$	0.99	laminar	$100 \le Re \le 1000$
		f	$24.413Re^{-0.46532}\alpha\left(\frac{h}{s}\right)^{-0.226497}\left(\frac{2A}{s}\right)^{-0.94256}\left(\frac{L}{2A}\right)^{-1.70937}$	0.99	turbulent	$1000 \le Re \le 15000$
5	Offset fin for water		$H_w = 0.049\left(\frac{K_w}{D_h}\right)Re_w^{0.69}Pr^{\frac{1}{3}}$	0.96	laminar	$100 \le Re \le 800$
			$f_w = 5.189Re_w^{-0.596}$	0.95		
			$H_w = 0.016\left(\frac{K_w}{D_h}\right)Re_w^{0.85}Pr^{\frac{1}{3}}$	0.96	turbulent	$1000 \le Re \le 15000$
			$f_w = 0.4533Re_w^{-0.24}$	0.95		

6	Offset fins for air	f	$10.882(Re)^{-0.79}(s/h)^{-0.359}(t/s)^{-0.187}(t/l)^{0.284}$	0.987	laminar	$300 \le Re \le 800$
		f	$2.237(Re)^{-0.236}(s/h)^{-0.347}(t/s)^{-0.151}(t/l)^{0.639}$		turbulent	$1000 \le Re \le 15000$
		j	$0.661(Re)^{-0.651}(s/h)^{-0.343}(t/s)^{-0.538}(t/l)^{0.305}$	0.982	laminar	$300 \le Re \le 800$
		j	$0.185(Re)^{-0.396}(s/h)^{-0.178}(t/s)^{-0.403}(t/l)^{0.29}$		turbulent	$1000 \le Re \le 15000$
7	Offset fins for R134a	j	$0.01197 \times R_e^{-0.5375} \times (s/h)^{0.1212} \times \delta(t/s)^{-0.1906} \times (t/l)^{-0.8473}$	0.982	laminar	$100 \le Re \le 1000$
		j	$0.0099 \times R_e^{-0.3744} \times (s/h)^{-0.4331} \times (t/s)^{-0.5475} \times (t/l)^{-0.6931}$		turbulent	$1000 \le Re \le 15000$
		f	$0.33648 \times R_e^{-0.5909} \times (s/h)^{-0.1275} \times (t/s)^{-0.2356} \times (t/l)^{-0.6108}$		laminar	$100 \le Re \le 1000$
		f	$0.00632 \times R_e^{-0.2163} \times (s/h)^{0.2253} \times (t/s)^{-0.3799} \times (t/l)^{-0.9331}$		turbulent	$1000 \le Re \le 15000$
8	Rectangular plain for air	f	$12.892(Re)^{-1.2299}(h/s)^{0.452}(t/s)^{-0.198}$	0.985	laminar	$100 \le Re \le 1000$
		f	$3.133(Re)^{-1.285}(h/s)^{0.247}(t/s)^{-0.181}$		turbulent	$1000 < Re \le 7500$
		j	$0.454(Re)^{-0.977}(h/s)^{0.435}(t/s)^{-0.227}$	0.99	laminar	$100 \le Re \le 1000$
		j	$0.166(Re)^{-1.011}(h/s)^{0.228}(t/s)^{-0.366}$		turbulent	$1000 < Re \le 7500$
9	Rectangular perforated fin for air	f	$0.7127(Re)^{-1.8858}(h/s)^{0.4196}(t/s)^{-1.4826}$	0.985	laminar	$100 \le Re \le 1000$
		f	$0.4345(Re)^{-1.3029}(h/s)^{0.3725}(t/s)^{-1.3178}$		turbulent	$1000 < Re \le 7500$
		j	$0.121(Re)^{-2.2920}(h/s)^{2.75}(t/s)^{-1.830}$	0.98	laminar	$100 \le Re \le 1000$
		j	$11.71(Re)^{-2.311}(h/s)^{2.144}(t/s)^{-1.9237}$		turbulent	$1000 < Re \le 7500$
10	Triangular Plain Fin for Air	f	$3.12(Re)^{-0.852}(h/s)^{0.156}(t/s)^{-0.184}$	0.98	laminar	$100 \le Re \le 1000$
		f	$2.69(Re)^{-0.918}(h/s)^{0.355}(t/s)^{-0.175}$		turbulent	$1000 < Re \le 10000$
		j	$0.718(Re)^{-0.628}(h/s)^{0.765}(t/s)^{0.765}$	0.99	laminar	$100 \le Re \le 1000$
		j	$0.789(Re)^{-1.1218}(h/s)^{1.235}(t/s)^{-0.764}$		turbulent	$1000 < Re \le 10000$
11	Triangular perforated fin for air	f	$10.127(Re)^{-1.588}(h/s)^{0.778}(t/s)^{-0.868}$	0.985	laminar	$300 \le Re \le 1000$
		f	$1.685(Re)^{-0.798}(h/s)^{0.447}(t/s)^{-0.276}$		turbulent	$1000 < Re \le 7500$
		j	$0.544(Re)^{-1.673}(h/s)^{2.278}(t/s)^{-1.589}$	0.99	laminar	$300 \le Re \le 1000$
		j	$7.579(Re)^{-1.626}(h/s)^{1.185}(t/s)^{-1.689}$		turbulent	$1000 < Re \le 7500$

- Provided the expressions for the Colburn factor j and friction factor f for all values of Reynolds number, including the laminar and turbulent regions. The generalized correlations are developed by taking all geometrical parameters into account.
- Presented a new set of correlations of f and j factors for wavy fins, offset fins, rectangular plain, rectangular perforated, triangular plain and triangular perforated fin surfaces used in a compact plate-fin heat exchanger. Expected that industry will find this information and the approach productive and useful.
- Developed f and j correlations for present study fin surfaces both laminar and turbulent flow regimes by using the power law expressions and the results are compared with open literature. All these correlations are summarized in Table 5.13.
- Shown that the values obtained from these expressions are in agreement with the literature data when compared with available experimental data. The above correlations can be used by the heat exchanger designers and can reduce the number of tests and modifications of the prototype to a minimum for similar applications.
- Extended the CFD analysis for flow non-uniformity effects and compared with measurements for a typical compact plate-fin heat exchanger. Almost uniform flow distribution is achieved in the core by providing suitable baffle plate in the inlet header of a typical CHE.

Nomenclature

A	test core total heat transfer area, m^2
A_c	exchanger minimum free flow area, m^2
A_f	fin area, m^2
A_s	conduction area of test core, m^2
A_w	heat transfer area of side wall, m^2
A_p	primary surface area, m^2
A_{cw}	conduction area of side wall, m^2
CHE	compact heat exchanger,
CFD	computational fluid dynamics
C_p	specific heat of air, J/kg°C
C_s	specific heat capacity test core, J/kg°C
C_w	specific heat capacity of side wall, J/kg°C
C_2	inertial resistance
d	hole diameter, m
D	fracture ductility
d_h or D_h	hydraulic diameter, m
E	Young's modulus
f	Fanning friction factor

FEM	finite element method
FPI	fins per inch
G	fluid mass flux, m_f / A_c, kg/(m^{2}s)
g	gravitational constant
H	convective heat transfer coefficient, W/m^2·K
H_{evi}	specific enthalpy at evaporator inlet
H_{evo}	specific enthalpy at evaporator outlet
H_w	convective heat transfer coefficient for side wall, W/m^2·K
h	height of the fin, m
h_{fg}	latent heat evaporation
j	Colburn factor, dimensionless
k	thermal conductivity, W/m·K
kE	kinetic energy, J
L	length of test core or wavy fin length or lance length of offset fin, m
LMP	Larsen Miller parameter
l	fin length, m
l_w	pitch of wave for wavy fin
M	mass, kg
M_s	mass of matrix, kg
M_w	mass of side wall, kg
m or $\dot{m}$	fluid mass flow rate, kg/s
NTU	number of heat transfer units, $HA/m_f c_p$
NTU_w	number of heat transfer units, $H_w A/m_f c_p$
Nf	number of fatigue cycles for failure
Nu	Nusselt number
N_{layer}	number of layers
OSF	offset strip fin
P	pressure, Pa or bar
P_{sat}	saturation pressure, Pa or bar
ΔP	pressure drop of test core, Pa
Pr	Prandtl number
q	heat flux, kW/m^2
Q	heat load, kW
R	wavy fin radius, degrees
Re	Reynolds number

R_{tc}	heat capacity ratio, M_sC_s / M_wC_w,
R_{NTU}	*NTU* ratio, NTU_w / *NTU*
r_h	hydraulic radius, m
s	fin spacing, m
S_t	Stanton number, H/GC_p
SCF	stress concentration factor
T	temperature, °C
TC	test core for boiling and condensation cases
T^*	temperature, °R
T_f	dimensionless temperature of fluid flow
T^*_f	fluid outlet temperature, °C
$T^*_{f\,max}$	Max heated fluid outlet temperature, °C
T^*_s	test core temperature, °C
T_s	dimensionless temperature of test core
T_w	dimensionless temperature of side wall
T^*_w	temperature of side wall, °C
T^*_0	inlet temperature of fluid flow at time = 0
T_{sat}	saturation temperature,°C
t	fin thickness. m
t^*	dimensionless time
t	time, sec, hours in equation 5.105
tp	plate thickness
UTS	ultimate tensile strength, Pa
u	velocity, m/sec
v	mean fluid velocity, m/sec
V	specific volume, m^3/kg, velocity, m/sec
X	Martinelli parameter
x	quality in case two-phase flows
x^*	flow length, m
x, *y*,*z*	coordinates
W	test core width, m
WF	wavy fin
Y	test core height, m
2*A*	wavy fin amplitude, m

Greek Symbols

Δp	pressure drop of test core in transient analysis
ΔP	pressure drop in two-phase flow cases
ΔT	superheat
ΔH	enthalpy drop
$\Delta \varepsilon$	strain
γ	porosity factor
λ	heat conduction effect factor,
μ	dynamic viscosity, N/s·m^2
ρ	density, kg/m^3
τ	non-dimensional time constant
α	void fraction
∞	stress, MPa
λ	dimensionless longitudinal heat conductivity, wave length
σ	ratio of free flow area to frontal area
η_f	fin efficiency
ϕ_f	Two-phase frictional multiplier

Subscripts

1...4	number 1...4
1	inlet
2	outlet
ave	average
cb	convective boiling
eq	equivalent
ev	evaporation
evi	evaporation inlet
evo	evaporation outlet
f	fluid, fin frictional
fin	fin
g	gas
i	inlet
l	liquid
m	mean

min	minimum
max	max/final
nb	nucleate boiling
r	refrigerant
s	test core solid material
tp	two-phase flow
o	outlet
0	initial value at time $t^* = 0$
w	wall or water side
wall	wall
wi	water inlet
wo	water outlet

References

1 London, A.L. and Shah, R.K., Offset rectangular plate-fin surfaces – heat transfer and flow friction characteristics, *Journal of Engineering for Gas Turbines and Power*, vol. 90, no. 3, pp. 218–228, 1968.

2 London, A.L., Compact Heat Exchangers – Design Methodology in Kakac, S., Shah, R.K. and Bergles, A.E. (Eds), *Low Reynolds Number Flow Heat Exchangers*, Hemisphere publishing Corp., Washington DC, pp. 21–27, 1983.

3 Kays W.M. and London, A.L., *Compact Heat Exchangers*, 1st edition, McGraw Hill, New York, 1964.

4 Shah, R.K., Advances in science and technology of compact heat exchangers, *Heat Transfer Engineering*, vol. 27, no. 5, pp. 3–22, 2006.

5 Shah, R.K., Perforated heat exchanger surfaces: Part 2 – Heat transfer and flow friction characteristics, ASME Paper no. 75 – WA/HT-9, 1975.

6 Shah, R.K., Compact heat exchanger surface selection methods in heat transfer, *In Heat Transfer, 4th Ed.*, Hemisphere Publishing Co, New York, pp. 193–199, 1978.

7 Tinaut, F.V., Melgar, A and Rahman Ali, A.A., Correlations for heat transfer and flow friction characteristics of compact plate-type heat exchangers, *International Journal of Heat and Mass Transfer*, vol. 35, pp. 1659–1665, 1992.

8 Sen Hu and Keith Herold Prandtl number effect on offset fin heat exchanger performance – predictive model for heat transfer and pressure drop, *Journal of Heat and Mass Transfer*, vol. 38, pp. 1043–1051, 1995a.

9 Sen Hu and Keith Herold, Prandtl number effect on offset fin heat exchanger performance – Experimental results, *Journal of Heat and Mass Transfer*, vol. 38, pp. 1053–1061, 1995b.

10 Wieting A.R., Empirical correlations for heat transfer and flow friction characteristics of rectangular offset-fin plate-fin heat exchangers, *ASME, Journal of Heat Transfer* 97, pp. 480–490, 1975.

11 Joshi H.M., Webb R.L., Heat transfer and friction in the offset-strip fin heat exchangers, *International Journal of Heat and Mass Transfer* 30 (1), pp. 69–84, 1987.

12 Patankar S.V., Liu C.H., Sparrow E.M., Fully developed flow and heat transfer in ducts having stream wise periodic variations of cross-sectional area, *ASME, J. Heat Transfer*, vol. 99, pp. 80–186, 1977.

13 Patankar, S.V. and Spalding, D.B., A calculation procedure for heat, mass and momentum transfer in three-dimensional parabolic flows, *Int. J. Heat Mass Transfer*, vol. 15, pp. 1787–1806, 1972.

14 Versteeg, H.K. and Malalasekera, W., *An Introduction to Computational Fluid Dynamics: The Finite Volume Method*, Addison Wesley, 1996.

15 Manabu Horiuchi, Shinji Nakagawa and Masaru Ishizuka, Numerical simulation of turbulent flow over a wavy wall, *Proceedings of 17th International Symposium on Transport Phenomena*, Toyama, 2006.

16 Patankar S.V., *Numerical Heat Transfer and Fluid Flow*, 1st edition, Hemisphere Publishing Corporation, USA, pp. 79–86, 1980.

17 Patankar, S.V., Numerical Prediction of Flow and Heat Transfer in Compact Heat Exchanger Passages, in *Compact Heat Exchangers* edited by Shah, R.K., Kraus, A.D. and Metzger, D., Hemisphere Publishing Corp., Washington DC, pp. 191–204, 1990.

18 Patankar, S.V., A calculation procedure for two-dimensional elliptic situations, *J. Num. Heat Transfer*, vol. 4, pp. 409–425, 1981.

19 Anderson, J.D., *Computational Fluid Dynamics, the Basics with Applications*, McGraw-Hill Companies, Inc., New York, 1995.

20 Amano, R.S., Bagherlee, A., Smith, R.J. and Niess, T.G., Turbulent heat transfer in corrugated-wall channels with and without fin, Transaction of *ASME, Journal Heat Transfer*, vol. 109, pp. 62–67, 1987.

21 Patel, V.C., Tyndall, Chon, J., Yoon, J.Y., Turbulent flow in a channel with a wavy walls, Transactions of ASME, *Journal Fluids Engineering*, vol. 113, pp. 579–586, 1991.

22 Amano, R.S., A numerical study of laminar and turbulent heat transfer in a channel with bends, *Proceedings of the 23rd National Heat Transfer Conference* Denver, ASME Paper no. 85-H.T.-20, 2000.

23 Chen, H.C., Patel, V.C., Near-wall turbulence models for complex flows including separation, *Journal of AIAA*, vol. 26, pp. 641–648, 1988.

24 Liang C.Y, Yang W.J., Modified single blow technique for performance evaluation on heat transfer surfaces, *Trans. ASME J. Heat Transfer* 97, pp. 16–21, 1975.

25 Xing Luo, Wilfried Roetzel, The single-blow transient testing technique for plate-fin heat exchangers, *International J. of Heat and Mass Transfer*, 44, pp. 3745–3753, 2001.

26 Xing Luo, Wilfried Roetzel, Ulrich Ludersen, The single-blow test technique considering longitudinal core conduction and fluid dispersion, International *J. of Heat and Mass Transfer*, 44, pp. 121–129, 2001.

27 Krishnakumar K., Venkatarathnam G., Transient testing of perforated plate matrix heat exchangers, *Cryogenics* 43, pp. 101–109, 2003.

28 Krishnakumar K., Venkatarathnam G., On the use of time at maximum slope in determining the heat transfer coefficients in complex surfaces using the single blow transient test method, *Int. J. Heat Exch.* VIII, pp. 31–44, 2007.

29 Krishnakumar K., John Anish K., Venkatarathnam G., A review on transient test techniques for obtaining heat transfer design data of compact heat exchanger surfaces, *Experimental Thermal and Fluid Science* 35, pp. 738–743, 2011.

30 Nair Anjan R., Krishnakumar K., Numerical modelling of single blow transient testing of a wire screen mesh heat exchanger, *10th National Conference on Technological Trends* (NCTT09), pp. 6–7, 2009.
31 W. Roetzel, X. Luo, Extended temperature oscillation measurement technique for heat transfer and axial dispersion coefficients, in: *Compact Heat Exchangers for the Process Industries*, R.K. Shah (ed.), Begell House, New York, pp. 381–388, 1997.
32 K. Zhou, U. Ludersen, X. Luo, W. Roetzel, Transient impulse test technique for heat transfer and axial dispersion coefficient, in: *Proceedings of Int. Conference on Heat Exchangers for Sustainable Development*, Lisbon, Portugal, pp. 645–652, 1998.
33 Chen Ping-Hei, Chang Zei-Chi, An improved method for the single-blow measurement including the non-adiabatic side wall effect, *Int. Comm. Heat Mass Transfer* 23, pp. 55–68, 1996.
34 P.-H. Chen, Z.-C. Chang, Measurements of thermal performance of cryo-cooler regenerators using an improved single-blow method, *Int. J. Heat Mass Transfer*, 40(10) 2341–349, 1997.
35 Pucci P.F., Howard C.P., Piersall Jr C.H., The single-blow transient testing technique for compact heat exchangers, *J. Eng. Power* 89, pp. 29–40, 1967.
36 Wiese Jeffrey L., Experimental analysis of heat transfer characteristics and pressure drop through screen regenerative heat exchangers, Thesis: AFIT/GAE/ENY/93D-30. 1993.
37 Junqi D., Jiangping C., Zhijiu C., Yimin Z., Wenfeng Z., Heat transfer and pressure drop correlations for the wavy fin and flat tube heat exchangers, *Applied Thermal Eng.*, vol. 27, no. 11–12, pp. 2066–2073, 2007.
38 Ranganayakulu C., Steady state and transient analysis of Compact plate fin heat exchanger fins for generation of design data using CFD, *Int. J of Numerical Methods for Heat and Fluid Flow*, vol. 26, Issue 2, pp. 440–460, 2016.
39 Manglik R. M., Zhang J. Muley A., Low Reynolds number forced convection in three-dimensional wavy-plate-fin compact channels fin density effects, *Int. J of Heat Mass Transfer*, 48, pp. 1439–1449, 2005.
40 Jose Fernandez-Seara, Ruben Diz, Francisco J. Uhia, Pressure drop and heat transfer characteristics of a titanium brazed plate-fin heat exchanger with offset strip fins, *Applied Thermal Engineering*, 51, Issues 1–2, pp. 502–511, 2013.
41 Mochizuki S., Yagi S., Heat transfer and friction characteristics of strip fins, *International Journal of Refrigeration* 50, pp. 36–59, 1975.
42 Awad M.M., Muzychka, Y.S., Models for pressure drop and heat transfer in air cooled compact wavy fin heat exchangers, *Journal of Enhanced Heat Transfer*, vol. 18, no. 3, pp. 191–207, 2011.
43 Hu S. and Herold K. E., Prandtl number effect on offset fin heat exchanger performance: predictive model for heat transfer and pressure drop, *Int. J. Heat Mass Transfer*, V138, no. 6, pp. 1043–1051, 1995.
44 Hu S., Herold K.E., Prandtl number effect on offset fin heat exchanger performance: experimental results. *Int J. Heat Mass Transfer*, 38–1053-1061, 1995.
45 Ranganayakulu C., Ismail L.S., Vasudeva Rao V., Rajeshwar S., Optimization of wavy fin parameters in compact heat exchangers, *Proceedings of XIX National Heat and Mass Transfer and VIII ISHMT-ASME Heat and Mass Transfer Conference, JNTU*, Hyderabad, India, 2008.

46 Sheik Ismail L., Ranganayakulu C., Shah Ramesh K., Numerical study of flow patterns of compact plate-fin heat exchangers and generation of design data for offset and wavy fins, *International Journal of Heat and Mass Transfer*, 52, pp. 3972–3983, 2009.

47 Pallavi P., Ranganayakulu C., Development of heat transfer coefficient and friction factor correlations for Offset fins using CFD, *Int. Journal of Numerical Methods for Heat and Fluid Flow*, vol. 21, no. 8, pp. 935–951, 2011.

48 Bala Sundar Rao R., Ranganath G., Ranganayakulu C., Development of Colburn 'j' factor and Fanning friction factor 'f' correlations for compact heat exchangers plain fins by using CFD, *Journal of Heat Mass Transfer*, vol. 49(7), pp. 991–1000, 2013.

49 Kim, B., Sohn, B., An experimental study of flow boiling in a rectangular channel with offset strip fins, *International Journal of Heat and Fluid Flow*, 27, pp. 514–521, 2006.

50 Manglik, R.M. and Bergles, A.E., Heat transfer and pressure drop correlations for the rectangular offset strip fin compact heat exchanger, *Int. J. Exp Thermal Fluid Science*, vol. 10, pp. 171–180, 1995.

51 Maiti, D.K., Heat transfer and flow friction characteristics of plate-fin heat exchanger surfaces – A numerical study, Ph.D. Thesis, Indian Institute of Technology, Kharagpur, India, 2002.

52 J.P. Chiou, Advancement of compact heat exchanger theory considering longitudinal heat conduction and flow non-uniformity effect, in: *Compact Heat Exchanger: History, Technological Advancement and Mechanical Design Problems*, R.K. Shah et al. (ed.), ASME New York, pp. 101–121, 1980.

53 C. Ranganayakulu, K.N. Seetharamu, K.V. Sreevatsan, The effects of longitudinal heat conduction in compact plate-fin and tube-fin heat exchangers using a finite element model, *Int. J. Heat Mass Transfer*, 40 pp. 1261–1277, 1997.

54 J.P. Chiou, Thermal performance deteriorations in a cross flow heat exchanger due to the flow non-uniformity, *J. Heat Transfer*, 100 pp. 580–587, 1978.

55 C. Ranganayakulu, K.N. Seetharamu, K.V. Sreevatsan, The effects of inlet fluid flow non-uniformity on thermal performance and pressure drops in cross-flow plate-fin compact heat exchangers, *Int. J. Heat Mass Transfer*, 40, pp. 27–38, 1997.

56 J.P. Chiou, The effect of non-uniformity of inlet fluid temperature on the thermal performance of cross flow heat exchanger, 7th International Heat Transfer Conference, Munich, Germany, pp. 179–184, 1982.

57 C. Ranganayakulu, K.N. Seetharamu, K.V. Sreevatsan, The combined effects of wall longitudinal heat conduction, inlet flow non-uniformity and temperature non-uniformity on thermal performance of cross flow tube-fin heat exchanger, *Int. J. Heat Mass Transfer*, 42, pp. 263–273, 1999.

58 Manson, S.V., Correlations of heat transfer data and of friction data for interrupted plate fins staggered in successive rows, NACA Technical Note 2237, National Advisory Committee for Aeronautics, Washington D.C. 1950.

59 Yi-Yie Yan, Hsiang-Chao Lio, Tsing-Fa Lin, Condensation heat transfer and Pressure drop of refrigerant R134a in a plate heat exchanger, *Int. J. Heat and Mass Transfer* 42 pp. 993–1006, 1999.

60 H.J. Kang, C.X. Lin, M.A. Ebadian, Condensation of R134a flowing inside helicoidal pipe, *Int. J. Heat and Mass Transfer* 43 pp. 2553–2564, 2000.

61 G.A. Longo, A. Gasparella, R. Sartori, Experimental heat transfer coefficients during refrigerant vaporization and condensation inside herringbone-type plate heat exchangers with enhanced surfaces, *International Journal of Heat and Mass Transfer* 47 pp. 4125–4136, 2004.

62 Barbara Watel, Review of saturated flow oiling in small passages of compact Heat exchangers, *International journal of Thermal Sciences*, 42, pp. 107–140, 2003.

63 L.K. Wang, B. Sunde´n, Q.S. Yang, Pressure drop analysis of steam condensation in a plate heat exchanger, *Heat Transfer Engineering* 20(1) pp. 71–77, 1999.

64 Longo, G.A., Gasparella, A., Refrigerant R134a vaporization heat transfer and pressure drop inside a small brazed plate heat exchanger, *International Journal of Refrigeration* 30 pp. 821–830, 2007.

65 Giovanni A. Longo, Refrigerant R134a condensation heat transfer and pressure drop inside a small brazed plate heat exchanger, *International Journal of Refrigeration* 31 pp. 780–789, 2008.

66 Giovanni A. Longo, R410A condensation inside a commercial brazed plate heat exchanger, *Experimental Thermal and Fluid Science* 33 pp. 284–291, 2009.

67 A. Jokar, Mohammad H. Hosni, Steven J. Eckels, Dimensional analysis on the evaporation and condensation of refrigerant R134a in mini-channel plate heat exchangers, *Applied Thermal Engineering* 26 pp. 2287–2300, 2006.

68 J.E. Hesselgreaves, *Compact Heat Exchangers: Selection, Design and Operation*, Elsevier Science and Technology Books, 2001.

69 Ranganayakulu C., Stephan Kabelac, Boiling of R134a in a compact plate- fin heat exchanger having offset strip fins. *ASME Journal of Heat Transfer*, No. H.T.-14-1235, 2015.

70 Amaranatha Raju Muppala, M.E., Ashok Babu T.P., Chennu Ranganayakulu, Boiling heat transfer and pressure drop analysis of R134a in a brazed heat exchanger with offset strip fins, *J. of Heat and Mass Transfer*, Springer, accepted, Ref. HAMT-D.-16-00515R1, 2017.

71 Amaranatha Raju Muppala, M.E., Ashok Babu T.P., Chennu Ranganayakulu, Investigation of flow boiling heat transfer and pressure drop of R134a in a rectangular channel with wavy fin, *International Journal of Refrigeration*, 2017. [under review].

72 Butterworth D., Hewitt G.F., Two-phase flow and heat transfer, *International J. of Energy Research*, vol. 2, Issue 4, 1978, pp. 407–408, Oxford University Press(Harwell Series), 1977.

73 K.V. Ramana Murthy, Mahesh Bondhu, C. Ranganayakulu. T.P. Ashok Babu, Condensation heat transfer and pressure drop of R134a saturated vapour inside a brazed plate-fin heat exchanger with serrated fin, Paper DOI 10.1007/s00231-016-1827-0, Online in *J. of Heat and Mass Transfer*, 2016.

74 K.V. Ramana Murthy, C. Ranganayakulu, T.P. Ashok Babu, "Influence of hydraulic diameter on heat transfer and pressure drop during condensation of R134a refrigerant in a brazed plate-fin heat exchangers" *Int. J. Engineering and Innovative Technology* (IJEIT), vol. 5, Issue 7, ID: IJEIT201512-4802, 2016.

75 Carey V.P., Mandrusiak Annular Film flow boiling of liquids in a partially heated vertical channel with offset strip fins, *Int. J. Heat Mass Transfer*, 29(6), pp. 927–939, 1996.

76 Lockhart, R.W. and Martinelli, R.C., Proposed correlation of data for isothermal two-phase two-component flow in a pipe, *Chem. Eng. Prog.*, 45, 1949.

77 Sen Hu S., Keith E. Herold, Prandtl number effect on offset fin heat exchanger performance: predictive model for heat transfer and pressure drop. *Int J. Heat Mass Transfer*, 38: pp. 1043–1051, 1995.

78 Sen Hu S., Keith E. Herold, Prandtl number effect on offset fin heat exchanger performance: experimental results. *Int J. Heat Mass Transfer*, 38: pp. 1053–1061, 1995.

79 Ramana Murthy K.V., Ranganayakulu C., Ashok Babu T.P., Development of heat transfer coefficient and friction factor correlations for serrated fins in water medium using CFD. *Journal of Physical Science and Application*, 5(3): pp. 238–248, 2015.

80 Yi-Yie Yan, Hsiang-Chao Lio, Tsing-Fa Lin, Condensation heat transfer and pressure drop of refrigerant R.-134a in a plate heat exchanger, *Int. J. Heat and Mass Transfer* 42 pp. 993–1006, 1999.

81 J.G. Collier, *Convective Boiling and Condensation* 2nd ed., McGraw Hill, 1982.

82 R.K. Shah, W.W. Focke, Plate heat exchangers and their design theory, *Heat Transfer Equipment Design*. Hemisphere, Washington, pp. 227–254, 1988.

83 Military Specification MIL-A-83116A (USAF), 31-March-1971.

84 Carter. P., Carter. T.J., Viljoen. A., Failure Analysis and life prediction of a large complex plate-fin heat exchanger, *Engineering Failure Analysis*, vol. 3, no. 1, pp. 21–43, 1996.

85 ANSYS Inc., *Fluent 12 User Manual, User's Guide*, ANSYS Mechanical, Release 12.

86 Marius Bazu and Titu Bajenescu, *Failure Analysis – A Practical guide for Manufacturers of Electronic Components and Systems*, Wiley Series Quality and Reliability Engineering, John Wiley and Sons Ltd., UK, 2011.

87 R.E. Peterson, *Stress Concentration Factors*, A Wiley-Inter science publication, 1974.

88 Miodrag Janković, The application of some approximate solutions of stress and strain concentration for life estimation in the low cycle fatigue region, *The Scientific Journal FactA Universitatis* Series: Mechanical Engineering vol. 1, no. 8, pp. 1025–1032, UDC 620.193.91, 2001.

89 Balram Gupta, V. Gopalakrishna, J.S. Yadav, *Aerospace Materials, Vol. 2*, Aeronautics Research and Development Board, S. Chand Publications, 1996.

90 Larson, F.R., Miller, James: A time-temperature relationship for rupture and creep stresses. Trans. *ASME, Journal of Heat Transfer* vol. 74, pp. 765 – 775, 2007.

91 Wieslaw Beres, Mechanics of materials failure, in Chapter 5 of *Recommended practices for monitoring gas turbine engine life consumption*, Centre for Aerospace information (CASI), RTO TR 28, ISBN 92-837-1032-0, April 2000.

92 Bhaskar A.C., Ranganayakulu C., Paul Vizhian S., Mohan Rao T., Design of qualification test facility and endurance tests of compact heat exchanger, *Proceedings of 13th International Symposium on Air Breathing Engines* (ISABE-13) Beijing, China, September 2007.

93 A.C. Bhaskar, M. Palanikumar, C. Ranganayakulu, S. Paul Vizhian, Endurance life estimation of aircraft compact heat exchanger, *Int. Journal of Aerospace Science and Technologies*, vol. 64, no. 4, pp. 248–258. Nov. 2012.

6

Thermal and Mechanical Design of Compact Heat Exchanger

6.1 Introduction

This chapter gives detailed information about compact heat exchanger design, and in particular how the size and shape of the exchanger are affected by the thermal and allowable pressure drop requirements. Most compact heat exchangers are formed of layers of plates or finned channels of fixed length and width. The surface shape of the plate or its fin determines the surface performance, which is described by non-dimensional heat transfer coefficient and friction factor relationships. Thus a feature of design is to determine a channel size and then find the number of layers to meet the duty. For the specialized cases of plate and frame exchangers and other welded plate products, the plate sizes are already specified in ranges by the manufacturers. The design problem is then confined to selecting the most appropriate size and the number of plates. This selection process is readily achieved by adaptation of the procedure given here. In general, it is assumed for the sake of brevity that the fluid properties are constant throughout the exchanger, and are evaluated for each stream at an appropriate mean temperature. In the case of the numerical method designs described in Chapter 4, local fluid properties can be applied for a more accurate design process. The main methods of heat exchanger design and analysis are those of the logarithmic mean temperature difference (LMTD) method, the effectiveness-number of transfer units (ε-*NTU*) method and the P-*NTU* method. The methods can be shown to be mathematically equivalent to each other. In the description here for single-phase flows attention is mainly confined to the ε-*NTU* method as it has distinct advantages in certain aspects of design and analysis, and is also physically easier to understand and interpret. An outline of the LMTD method is also given.

Boiling and condensing flows in compact passages are the subject of increasing research attention, stimulated partly by their long term application in cryogenic applications, and partly by the outstanding success of the brazed plate heat exchanger refrigeration systems. A summary of the design method and recommended correlations are given in Chapter 5. The methods described here will be adequate for most practical design purposes. For rapid design, and for comparison purposes, use of single power law relationships of Colburn *j* factor and friction factor *f* with Reynolds number and geometrical parameters of various fins are presented.

Compact Heat Exchangers – Analysis, Design and Optimization using FEM and CFD Approach,
First Edition. C. Ranganayakulu and K.N. Seetharamu.
© 2018 John Wiley & Sons Ltd. Published 2018 by John Wiley & Sons Ltd.

A heat exchanger by definition exchanges heat, or energy, between two or more streams given by Sekulic [1]. In process applications with two streams, the process stream may be heated or cooled by a service or utility stream. The process stream flow rate, its upstream and downstream temperatures (or heat load) and allowable pressure drop are often specified closely. In these cases, the service stream may only have its inlet temperature specified, for example, and part of the design problem is to find the most economic value for its flow rate and for the heater/cooler. More often, heat is to be exchanged between two process streams of given flow rates and given inlet temperatures, and the minimum approach temperature is an important design criterion. Pressure drop of either or both streams may be specified. In some cases there may be a pressure drop in a stream, arising from process requirements, which can be "used up" by the heat exchanger. More often the allowable pressure drop is determined by system constraints such as the available pump head. For gas streams this can be critical, and the corresponding pumping power is often a limiting factor; this is rarely the case with liquid streams.

6.2 Basic Concepts and Initial Size Assessment

When the thermal and pressure drop specification has been fixed, it is possible to determine quickly whether the most appropriate flow configuration is that of cross flow, counter flow or often multipass overall counter flow arrangements. We will examine first the effectiveness – NTU design approach (Kays and London [2]) and then the LMTD approach.

6.2.1 Effectiveness Method

For each stream, the rate equation is, for hot and cold streams respectively,

$$Q = m_h c_{p.h}(T_{h.out} - T_{h.in}) \quad \text{or} \quad Q = m_c c_{p.c}(T_{c.out} - T_{c.in}) \tag{6.1}$$

The product mc_p is often denoted by C for convenience, the highest and lowest between the two streams being C_{max} and C_{min}. The ratio of stream capacity rates is denoted by C^*, given by

$$C^* = \frac{C_{min}}{C_{max}} \tag{6.2}$$

Typical temperature distributions for common arrangements are shown in Figure 6.1. Note that the curvature changes according to whether C_h or C_c is the C_{min} stream. If C^* is unity, the exchanger is said to be balanced, a condition often approached in gas turbine recuperators, and the distributions would be straight and parallel, provided that the heat transfer coefficients are constant along the length. Representative surface temperature distributions are also shown, and these will be closer to the hot stream temperatures if the heat transfer coefficient is higher for the hot stream, a point which will be further developed later.

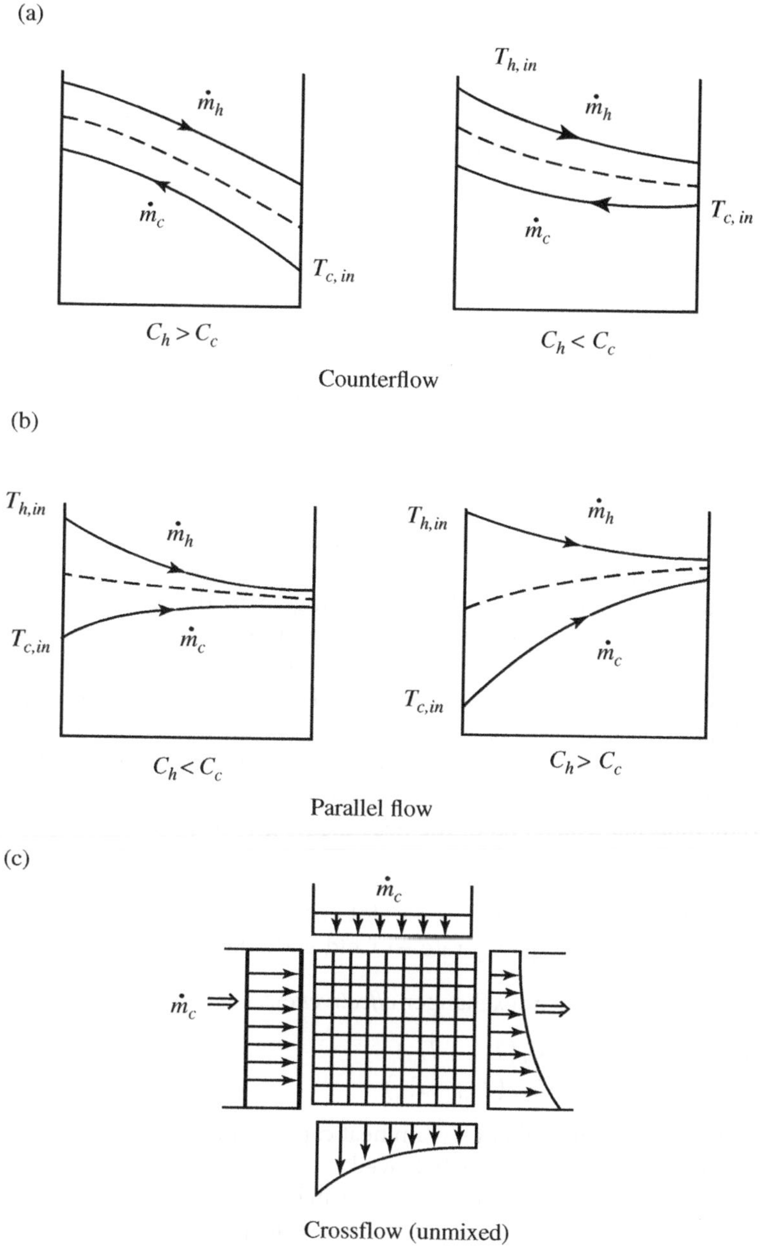

Figure 6.1 Typical temperature distributions in heat exchangers.

The maximum possible heat exchanged between the streams Q_{max}, is given by

$$Q_{max} = C_{min}(T_{h,i} - T_{c,i}), \tag{6.3}$$

which represents the idealized performance with infinite surface area. This is best described physically with the aid of the idealized temperature distribution diagrams taken from Hesselgreaves [3] shown in Figure 6.2.

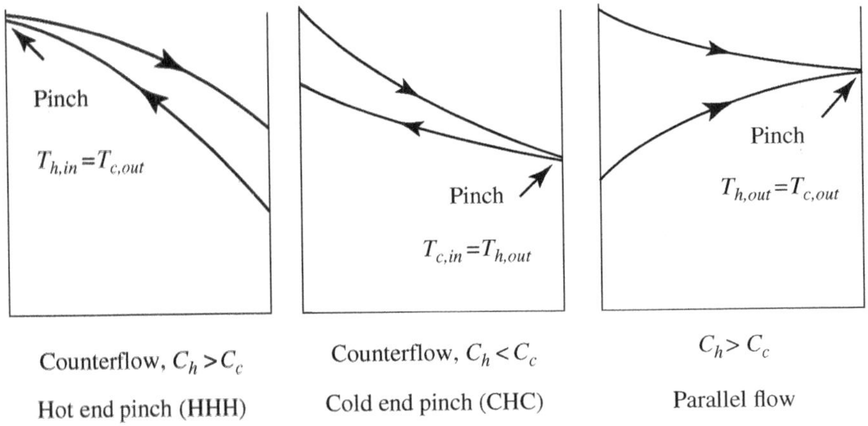

Figure 6.2 Idealized temperature distributions showing the pinch. (Hesselgreaves [3])

Clearly the maximum heat transferred is obtained when the stream of lowest heat capacity rate has an outlet temperature equal to the inlet temperature (for a counter flow configuration) of the other stream. For the parallel flow arrangement, the state is reached when both streams attain the same temperature at outlet. In this case, the state corresponds to a theoretically infinite surface area. The point of equal temperatures is called the pinch point or simply the pinch. In this ideal ease it corresponds to zero temperature difference, but in general, especially in the process industries, the pinch refers to the minimum temperature difference in the heat exchanger, also referring to the minimum in a heat exchanger network. It is important for energy targeting.

An acronym for remembering which end of the heat exchanger the pinch is at can be given as

HHH = Hot fluid has Highest heat capacity rate → Hot end pinch
CHC = Cold fluid has Highest heat capacity rate → Cold end pinch

From this definition of Q_{max} the heat exchanger *effectiveness*, ε, can be given as

$$\varepsilon = \frac{Q}{Q_{max}} = \frac{C_h(T_{h,i} - T_{h,o})}{C_{min}(T_{h,i} - T_{c,i})} = \frac{C_c(T_{c,o} - T_{c,i})}{C_{min}(T_{h,i} - T_{c,i})} \tag{6.4}$$

where C_c and C_h are the cold stream and hot stream heat capacity rates respectively, one of which will be C_{min} unless the exchanger is balanced.

Thus the *effectiveness* of the exchanger to be designed can be determined directly from the terminal temperatures, if these are given, and the stream parameters. The other major relationship needed for this stage is that of the heat exchange equation. This is given by

$$Q = (UA_s)\Delta T_m \tag{6.5}$$

where ΔT_m is a suitably averaged temperature difference between streams, usually expressed as the LMTD multiplied by a correction factor which depends on the flow configuration, and UA_s is the product of overall heat transfer coefficient and reference surface area. This product is often called the heat transfer conductance.

We define the number of transfer units (NTU) for the exchanger as

$$NTU = \frac{UA_s}{C_{min}}, \tag{6.6}$$

which is a measure of the thermal length of the exchanger. It can be shown that the effectiveness ε is a function of NTU and C^*, the form of function depending on the flow configuration. The most common are counter flow and cross flow (both streams unmixed) and these are shown below.

For computer-aided design purposes, algebraic relationships are needed, and a selection for the most commonly used configurations provided by Hesselgreaves [3] is given in Table 2.2 in Chapter 2. Figure 6.3 gives a graphic illustration of how the NTU as in Kays and London [2], giving the product of required area and heat transfer coefficient, is related to the effectiveness, determined by the specified temperatures, and the capacity rate ratio C^*. It is clear from these figures that the configuration makes little difference to the effectiveness if either C^* is low, ($C^* < 0.25$ for example) or NTU is low (e.g. $NTU < 1$) the latter implying of course that effectiveness is low. Because of its simplicity, a cross flow design is most appropriate for these cases, most often applying to liquid/gas exchangers, for which the gas side heat transfer coefficient is low and dominates the overall UA_S. If C^* is higher than about 0.25, and especially if the required effectiveness is high ($\varepsilon > 0.8$), a counter flow configuration will usually give the most economic design. Figure 6.3c shows that for multipass cross–counter flow the pure counter flow value of effectiveness is closely approached for three or more passes, and this is often preferred over the pure counter flow configuration because porting is easier and cheaper.

6.2.2 Inverse Relationships

In inverse calculations, the terminal temperatures are often specified, and these directly determine the effectiveness ε. It is then helpful to be able to express NTU explicitly in terms of ε instead of the vice-versa relationships above. This then gives the overall conductance UA_S, and is directly equivalent to the LMTD method discussed next. Some expressions for simple flow configurations provided by Hesselgreaves [3] are given in Table 2.3 in Chapter 2.

6.2.3 LMTD Method

This approach utilizes the commonly available specification of the terminal temperatures. It is easily shown that it is mathematically compatible with the effectiveness-NTU method, the working equations of one being derivable from those of the other. The method includes design and rating options. These basic analytical approach, are fully treated by Kays and Crawford [4] and Smith [5]. Only a summary will be presented here.

The design or sizing problem involves the selection of the heat load Q, the inlet temperatures and the two mass flow rates (of the specified fluids), or a combination, which allows the derivation of all four terminal temperatures, as shown in Figure 6.1. At this stage the configuration is not important. The method is predominantly used for the design of shell and tube exchangers.

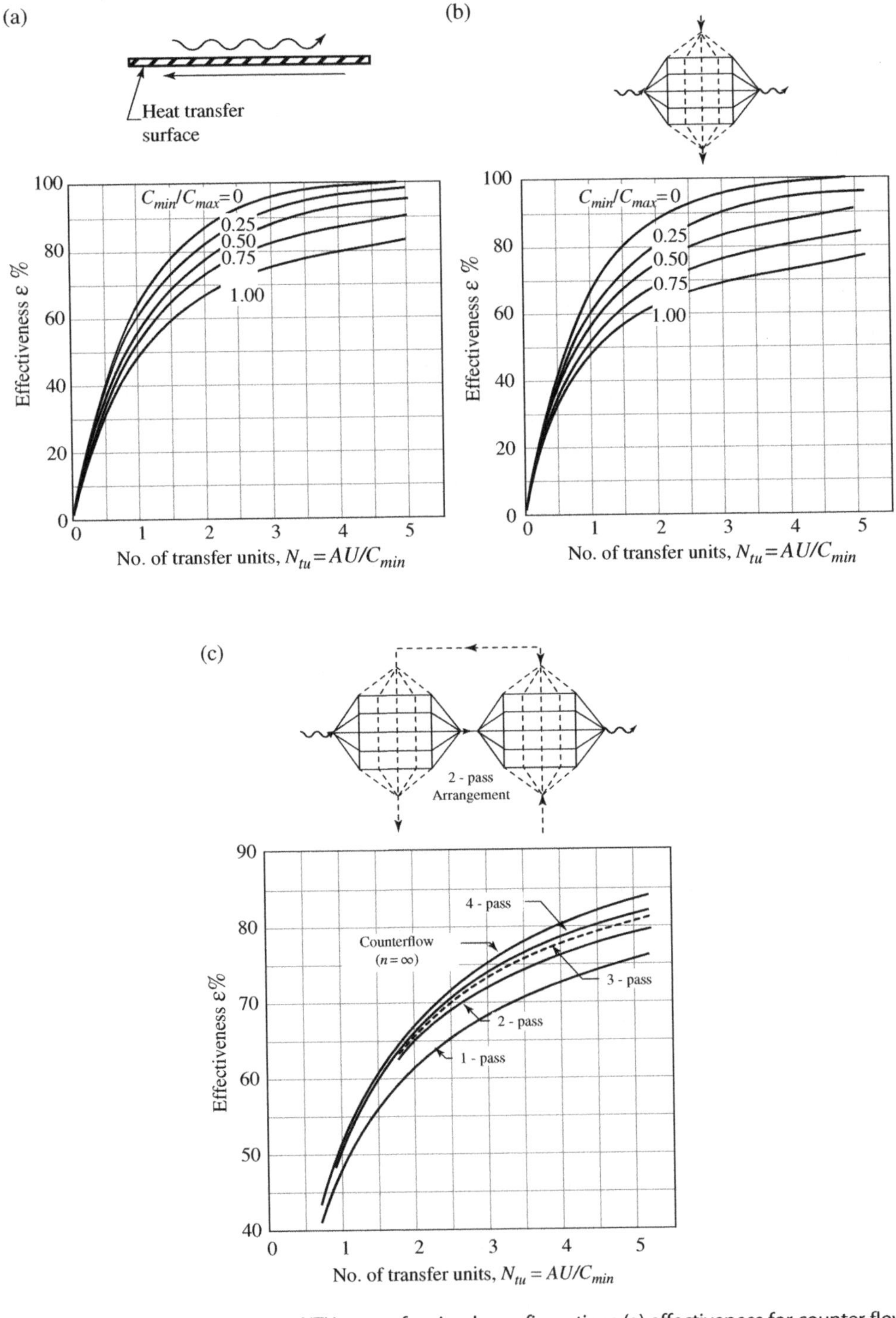

Figure 6.3 Effectiveness versus *NTU* curves for simple configurations (a) effectiveness for counter flow (b) effectiveness for cross flow (both streams unmixed) (c) effectiveness for multipass cross–counter flow, $C* = 1$, unmixed streams within passes, both streams mixed between passes. (Hesselgreaves [3])

The essential equation is the heat transfer, Equation 6.5. The assumption made, as for the effectiveness method, is that the heat transfer coefficients of both sides are constant along the exchanger (or a chosen segment of it if necessary). By consideration of heat flows along the exchanger length it is easily shown that the appropriate mean temperature difference in this equation is

$$\Delta T_{lm}(=LMTD)=\frac{\Delta\vartheta_1-\Delta\vartheta_2}{\ln(\Delta\vartheta_1/\Delta\vartheta_2)},\tag{6.7}$$

the $\Delta\vartheta_S$ being the terminal temperature differences, regardless of configuration, as shown in Figure 6.4.

For simple exchangers (pure counter flow, pure co-current flow), the required conductance then follows directly:

$$UA_s=\frac{Q}{\Delta T_{lm}}\tag{6.8}$$

For other configurations, the LMTD is still used and is calculated as if for pure counter flow, but with a correction factor F to reflect the departure from counter flow:

$$Q=F(UA_s)\Delta T_{lm}\tag{6.9}$$

Note that F does not represent the effectiveness of the exchanger, so that a high value (close to unity) is not necessarily indicative of a high effectiveness, but merely shows how close the performance is to that of a counter flow exchanger for the specified operating conditions – Shah [6]. Although analytical expressions for F for some configurations are available in Shah [6], it is usual to use charts.

The use of F for shell and tube design necessitates the introductions of two new variables, which, in contrast to the effectiveness approach, distinguish between shell side and tube side streams. The heat capacity rates are denoted C_S and C_T for shell and tube sides correspondingly, and an effectiveness parameter P is defined as

$$P=\frac{C_{min}}{C_T}\varepsilon\tag{6.10}$$

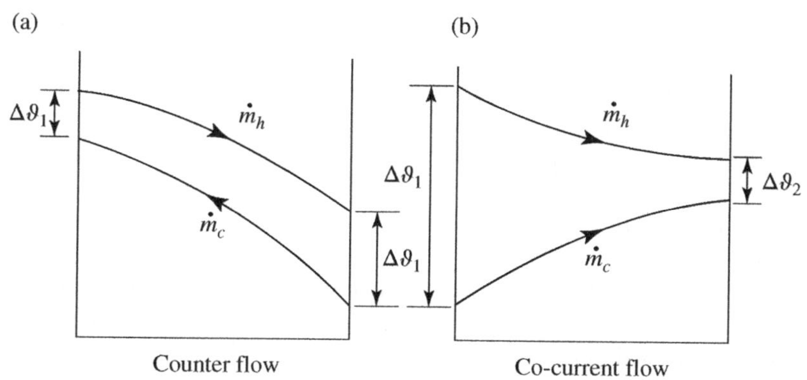

Figure 6.4 Temperature profiles.

so that $P = \varepsilon$ for $C_T = C_{min}$,
and $P = \varepsilon\, C^*$ for $C_T = C_{max}$.

The other parameter needed is the heat capacity rate ratio R, defined as

$$R = \frac{C_T}{C_S} \tag{6.11}$$

so that $R = C^*$ for $C_T = C_{min}$,
and $R = 1/C^*$ for $C_T = C_{max}$.

It is also convenient to define an additional number of transfer units, NTU_T, as the overall conductance divided by the tube side heat capacity rate:

$$NTU_T = \frac{UA_s}{C_T} \tag{6.12}$$

Thus the relationship between NTU_T and NTU based on C_{min} is

$$NTU_T = NTU\frac{C_{min}}{C_T}, \tag{6.13}$$

and $NTU_T = NTU$ for $C_T = C_{min}$,
$NTU_T = NTU\, C^*$ for $C_T = C_{max}$.

Then F can be expressed in terms of R and P by

$$F = \frac{1}{NTU_T(1-R)}\ \ln\left[\frac{1-RP}{1-P}\right] \text{for} R \neq 1 \tag{6.14}$$

$$F = \frac{P}{NTU_T(1-P)} \text{for} R = 1, \tag{6.15}$$

If Equations 6.8–6.11 are used for parallel flow it is necessary to use the following expression for F, since for this configuration the LMTD is defined in terms of the inlet and outlet terminal temperature differences of each stream:

$$F = \frac{R+1}{R-1}\frac{\ln\left[(1-RP)/(1-P)\right]}{\ln\left[1-(1+R)P\right]} \tag{6.16}$$

In terms of effectiveness ε and C^*, using equations 6.9, 6.14 and 6.15,

$$F = \frac{NTU_{cf}}{NTU} \tag{6.17}$$

where NTU_{cf} is the corresponding value for pure counter flow,

$$F = \frac{1}{NTU\,(1-C^*)}\ ln\left(\frac{1-C^*\varepsilon}{1-\varepsilon}\right) \text{for} C^* \neq 1 \tag{6.18}$$

$$F = \frac{\varepsilon}{NTU(1-\varepsilon)} \text{for} C^* = 1 \tag{6.19}$$

Thus in the design for most configurations other than pure counter flow and parallel flow, the required NTU, giving UA_S, can only be found implicitly by solving iteratively for F: for a cross flow exchanger with one fluid mixed and for 1–2 parallel counter flow with shell fluid mixed an explicit relation between NTU and ε allows a corresponding direct solution for NTU in terms of P and R. Figure 6.5 shows this F-P relationship. Shah [6]

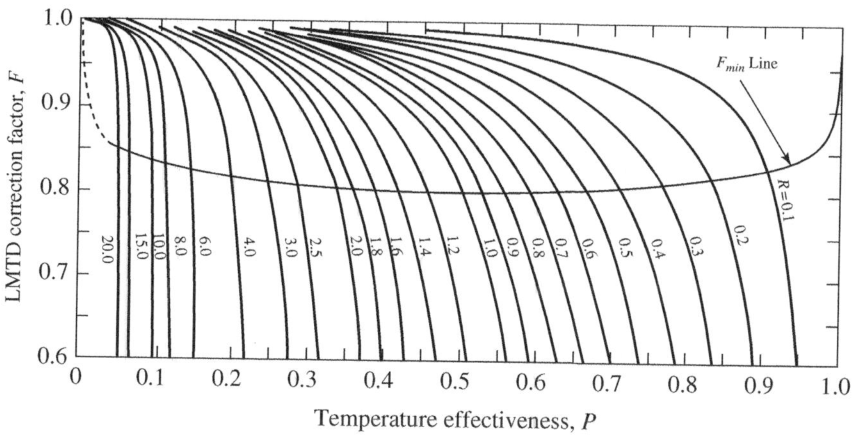

Figure 6.5 Correction factor (F) versus P curves for a1-2 parallel-counter flow exchanger. (Hesselgreaves[3])

recommends that exchanger designs should avoid the steep (high *NTU*) regions of the graph, for good utilization of surface area, and in addition that *F* should not be less than the value F_{min} superimposed on the figure in order to avoid a temperature cross in this arrangement.

6.3 Overall Conductance

For either effectiveness or LMTD approaches it is now necessary to expand on the formation of overall conductance UA_S. This is defined in terms of the individual heat transfer coefficients H and areas A, as

$$\frac{1}{UA_S} = \frac{1}{(\eta_o HA_S)_h} + \frac{1}{(\eta_o H_f A_S)_h} + \frac{1}{\lambda_m A_w} + \frac{1}{(\eta_o HA_S)_c} + \frac{1}{(\eta_o H_f A_S)_c}, \tag{6.20}$$

which is the sum of the reciprocals of the hot stream, wall and cold stream, and fouling or scale conductances on both sides, or simply the sum of the component heat transfer resistances. It is convenient to write

$$R_{f.h} = \frac{1}{(\eta_o H_f A_S)_h} \quad \text{and} \quad R_{f.c} = \frac{1}{(\eta_o H_f A_S)_c} \tag{6.21}$$

for the fouling resistances, related to the fouling thicknesses by

$$R_{f.h} = \left(\frac{t_f/k_f}{\eta_o A_S}\right)_h \quad \text{and} \quad R_{f.c} = \left(\frac{t_f/k_f}{\eta_o A_S}\right)_c \tag{6.22}$$

where the t_f's and k_f's are the fouling thicknesses and conductivities respectively. Note that there is an implicit assumption here that the fouling layers are uniform on the surface(s), which may be far from reality. However, the instances in which a highly non-uniform distribution is expected may well be known to the experienced designer, and can therefore be taken into account. Normally the error incurred in the assumption is relatively small.

If there are no fouling layers Equation 6.20 simplifies to

$$\frac{1}{UA} = \frac{1}{(\eta_o HA_S)_h} + R_w + \frac{1}{(\eta_o HA_S)_c} \tag{6.23}$$

where the wall resistance is

$$R_w = \frac{1}{k_m A_w} \text{ for a plane wall, and} \tag{6.24a}$$

$$R_w = \frac{ln\,(d_o - d_i)}{2\pi k_w L_T N_T} \text{ for a circular tube} \tag{6.24b}$$

where the d_o, d_i are the outer and inner tube diameters, L_T is the tube length and N_T is the number of tubes in the exchanger. For thin-walled tubes a satisfactory approximation to this case is

$$R_w = \frac{t_w}{k_m A_m} \tag{6.24c}$$

where t_w is the tube wall thickness $(= (d_o - d_i)/2)$ and A_m is the area evaluated at the arithmetic mean diameter $(Am = \pi L_T N_T (d_o + d_i)/2)$.

It is clear that the resistances operate in a similar way to electrical resistances in series, the temperature difference being the driving potential, directly analogous to the voltage, and the heat flow, constant through each resistance, analogous to the current, as shown in Figure 6.6.

Note that only in the case of one-dimensional heat flow, that is, with constant area conductive path, is the heat *flux* constant, an important point for heat exchangers but not normally for the electrical analogy.

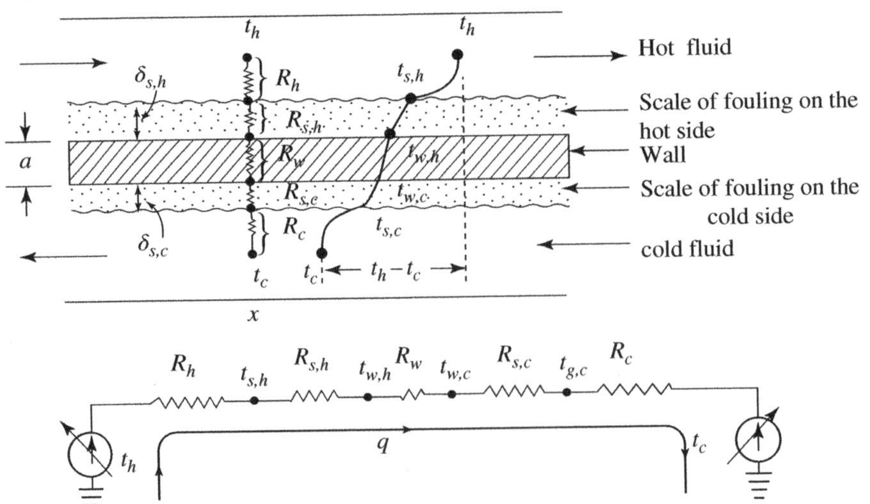

Figure 6.6 Heat transfer resistance analogy. (Hesselgreaves[3])

In many cases the wall resistance is negligible, but with increasing use being made of stainless steels, titanium and nickel alloys, all of which have low conductance, in process exchangers, it is becoming increasingly significant and should be included in the overall *NTU* calculation. It can be the dominating resistance for polymer exchangers. The surface effectivenesses η_o are typically of the order of 0.8, and this value can be used for the first scoping calculation.

The resistance of the largest stream dominates the overall resistance. As mentioned above, if one side is a gas stream, its conductance is likely to be lowest and the resistance highest. This is because typical heat transfer coefficients (*H*) for low pressure gases are in the region 50–100 W/m^2·K, whereas liquid side coefficients are typically 2000–5000 W/m^2·K. Further details of the relative influence of resistances are given in Bemtsson et al. [7]. A low coefficient *H* is usually compensated for by an increase in surface area A_S, on that side, in order to obtain a better balanced design, that is, to have closer to equal resistances. This helps to achieve an economic overall design, and explains the practice of combining a plain tube (inside) for liquid flows with a finned or extended surface on the air side of air conditioning and related exchangers. Note that the balancing referred to here is not the same as the balance of heat capacity rates, which is only a function of specified stream variables.

Equation 6.20 can be expressed in terms of the individual side *NTU*s (N_h or N_c) by

$$NTU = \frac{1}{C_{min}}\left[\frac{1}{1/(\eta_o HA_S)_h + R_{f.h} + R_w + R_{f.c} + 1/(\eta_o HA_S)_c}\right] \quad \text{or} \tag{6.25}$$

$$\frac{1}{NTU} = \frac{1}{N_h(C_h/C_{min})} + \frac{1}{N_c(C_c/C_{min})} \tag{6.26}$$

neglecting the wall term, where the N_h's and N_c's are

$$N_{h,c} = \frac{\eta_o H_{h,c} A_{h,c}}{C_{h,c}} \tag{6.27}$$

6.3.1 Fin Efficiency and Surface Effectiveness

Fin efficiency is defined as the ratio of actual heat flow of the fin to that which would be obtained with a fin of constant temperature uniformly equal to the base surface temperature, that is, with one of infinite thermal conductivity. It is readily found by consideration of the heat flow within the fin, together with that from the fin surface (an example of a conjugate problem). The fin efficiency η_f for a fin of constant rectangular section (Figure 6.7) and with constant heat transfer coefficient over the surface is given by

$$\eta_f = \frac{\tanh(ml)}{ml} \tag{6.28}$$

where m is a dimensional $(1/l)$ function given by

$$m = \sqrt{\frac{2H}{k_m t_f}\left(1 + \frac{t_f}{l_f}\right)} \tag{6.29}$$

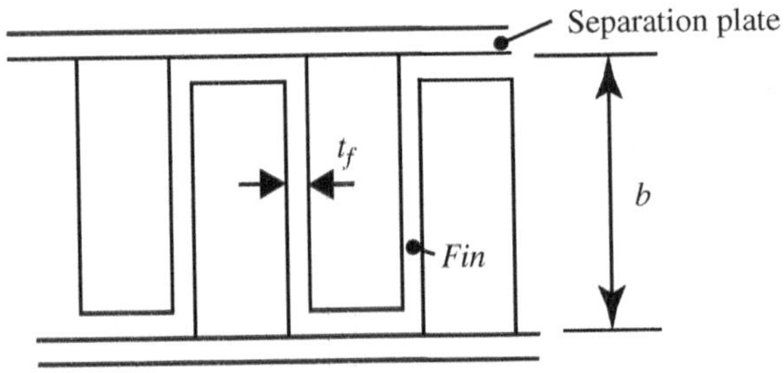

Figure 6.7 Schematic of rectangular fin. (Hesselgreaves [3])

l is the fin length (especially for a triangular fin) or height, l_f is the fin (flow) length and t_f is fin thickness.

In the case of a plate-fin (PFHE) surface with a rectangular fin, the fin height l used is, from symmetry considerations, the half-height of the effective gap, which is the plate separation b minus the fin thickness t_f.

$$l = 0.5\left(b - t_f\right) \tag{6.30}$$

If the ratio of fin surface area to total area is given as A_f/A_s, the surface effectiveness η_o is defined as the ratio of actual heat flow on the total (that is, primary plus secondary) to the ideal heat flow obtained if both primary and secondary surfaces are at the same temperature. It is easily shown that η_o is given by

$$\eta_o = 1 - \left(1 - \eta_f\right)\frac{A_f}{A_s} \tag{6.31}$$

6.4 Pressure Drop Analysis

In this section both single-phase pressure drop and two-phase pressure drop estimation procedure are presented.

6.4.1 Single Phase Pressure Drop

Different types of pressure loss components are explained below:

1) **Single-phase frictional losses**: The most significant pressure loss in single-phase flow is usually the frictional loss through the plate-fin heat exchanger distributors and heat transfer core.
2) **Single-phase momentum losses**: Momentum losses occur where the streams change velocity in magnitude or direction, and may sometimes be significant. The most significant momentum losses occur at entry to the distributors where velocities are highest.
3) **Gravity losses**: Gravity losses or gains may be significant in tall plate-fin heat exchangers.
4) **Fitting losses**: Losses in pipes, bends, nozzles, headers and fittings can be estimated if required by methods published by ESDU [8] or any good fluid mechanics textbook. Within the distributors the geometry varies to suit each duty and each manufacturer's standard practice. In most cases the pressure losses in fittings contribute no more

than 10–15% of the overall pressure loss. Mal-distribution may also be reduced by the use of an artificial resistance, such as a "hard way" fin at the entrance to, exit from or within the heat transfer section.

5) **Header tanks**: For flow from a header tank into a distributor the fittings loss, K_d is taken as 0.5. Hence, assuming the header tank velocity to be zero as indicated by Taylor [9]:

$$\Delta P_i = 1.5\left(\frac{m'^2}{2\rho}\right), \qquad \text{(Pa)} \tag{6.32}$$

At the outlet from a distributor to a header tank there is no velocity head recovery, i.e. $m' = 0$, $K_d = 1$ and $\Delta P_o = 0$.

6) **Gaps between distributors and fins and between fin rods**: In each gap between fin pads, either between the distributor and heat transfer section or within the core itself, there will be a change in momentum as the stream enters and leaves the gap. The change in static pressure as the stream enters the gap depends on the upstream fin void age, ε as indicated by Taylor [9]:

$$-\Delta P_{gap,\ i} = 2\varepsilon(1-\varepsilon)\left(\frac{m'^2}{2\rho}\right), \qquad \text{(Pa)} \tag{6.33}$$

where

$$\varepsilon = \left(1 - \frac{t}{b}\right)(1 - nt) \tag{6.34}$$

t = fin thickness, (m)
n = fin density, (1/m)
b = gap between partition sheets, (m)

The loss in static head as the stream leaves the gap is 1.5 times downstream velocity heads as given in the header tanks Equation 6.32.

7) **Single-phase gravitational losses**: This can be calculated from

$$\Delta P = \rho g \Delta h \qquad \text{(Pa)} \tag{6.35}$$

where

g = acceleration due to gravity (m/s^2)
Δh = change in height of stream (m)

Estimation of the pressure loss in a plate-fin heat exchanger is usually undertaken after the thermal rating. If the calculated pressure loss is either too high, i.e. outside specification, or too low, i.e. uneconomical use of transfer surface, the geometry of the plate-fin should be adjusted and the complete thermal and hydraulic rating repeated. Usually it is sufficient to iterate on the mass velocity chosen at the start by pro-rating and convergence is usually achieved in one iteration.

$$m'_{new} = m'_{old}\sqrt{\left(\frac{\Delta P_{required}}{\Delta P_{calculated}}\right)} \qquad (\text{kg/m}^2\,\text{s}) \tag{6.36}$$

The total pressure drop ΔP consists of entrance effect, acceleration, core friction and exit effect:

$$\Delta P = \frac{1}{2} G^2 v_i \left[\left(k_c + 1 - \sigma^2\right) + 2\left(\frac{v_o}{v_i} - 1\right) + f \frac{A}{A_c} \frac{v_m}{v_i} - \left(1 - k_e - \sigma^2\right) \frac{v_o}{v_i} \right] \tag{6.37}$$

Since any measured pressure drop data is cumulative of all these four effects (ΔP_{entry}, $\Delta P_{acceleration}$, $\Delta P_{core\ friction}$ and ΔP_{exit}), the individual effects can be estimated as follows: entrance and exit pressure drops are calculated by taking k_c and k_e values recommended by Hesselgreaves [3]. This approach, which is shown to be of good accuracy for offset strip fin surfaces, is recommended because it avoids the use of graphs – a feature of Kays and London [2] method.

The entry loss coefficient is $k_c = 0.5\,(1 - \sigma)$ (6.38)

and the exit loss coefficient is $k_e = (1 - \sigma)^2$ (6.39)

8) **Entry and Exit Losses**: In the application of friction factor data for surfaces it is implicitly assumed that the pressure drop of each flow is entirely within the matrix. In reality the flows experience abrupt contractions at entry and expansions at exit to the core, from the headers or ducts. These give rise to net increases in pressure drop which are usually small, being of the order of 10–20% for a long duty. For "short fat" duties (i.e. low pressure drop gas duties) these extra losses should be accounted for, and Kays and London [2] gives calculation processes based on systematic tests of a wide range of cores with different porosities. The basic model of pressure development through a general core is shown in Figure 6.8.

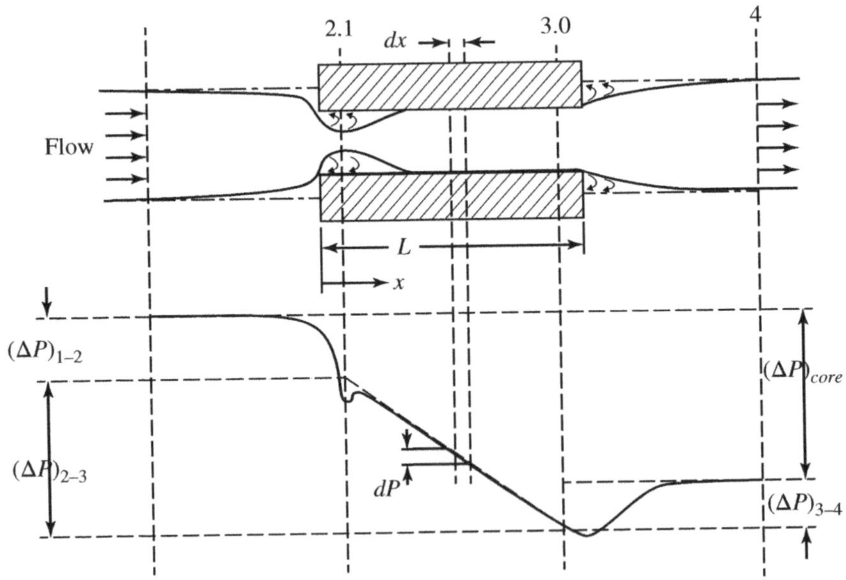

Figure 6.8 Pressure drop development in one core passage of a plate-fin heat exchanger. (Hesselgreaves [3])

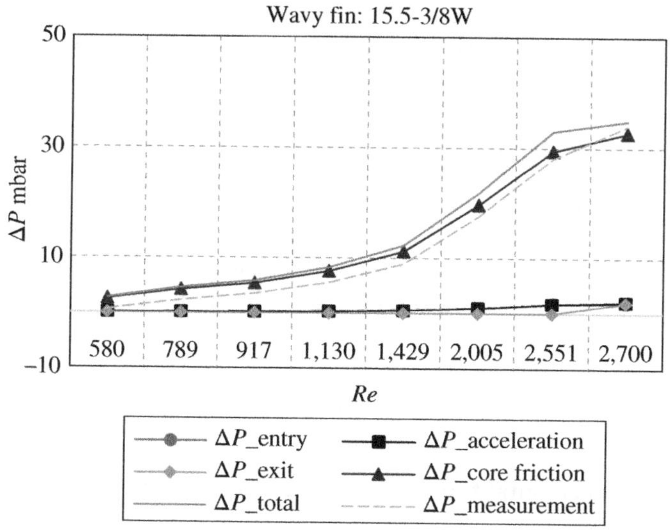

Figure 6.9 Pressure drop comparison in wavy fin test core, 15.5–3/8 W.

Equation 6.40 allows for density changes, which are important for long gas duties but are small for liquids and short gas duties such as in air conditioning and related applications. The mean density is given by

$$\left(\frac{1}{\rho}\right)_m = \frac{1}{2}\left(\frac{1}{\rho_i} + \frac{1}{\rho_o}\right) \tag{6.40}$$

This will apply for gas duties with a linear temperature distribution ($C^* = 1$), and for liquid duties. For gases with other values of C^* a log-mean temperature T_{lm} should be used together with the mean pressure and the perfect gas law to derive the (log mean) density:

$$\left(\frac{1}{\rho}\right)_m = \frac{R}{\rho_{ave}} T_{lm} \tag{6.41}$$

Thus the friction factor f can be obtained by Equation 6.37. The core friction factors are also calculated with the correlations provided by [10–21] and are compared with experimental measurements for a typical small test core of wavy fin, as shown in Figure 6.9 for relative comparison with measured total pressure drop values.

Figure 6.9 indicates that the measured pressure drops are well comparable with the cumulative components of pressure loss values in a small offset fin test core using the correlations based on the CFD simulations within acceptable error.

6.4.2 Two-Phase Pressure Loss

The methods presented here are for calculating the local pressure gradient in a channel of a plate-fin heat exchanger. To calculate the overall pressure change along a complete channel, it is necessary to divide the length into sufficient increments such that the stream quality and physical properties may safely be considered constant over each increment. Where the mass quality changes significantly within the plate-fin heat exchanger, it will be necessary to carry out a simultaneous heat transfer calculation for each increment.

Plate-fin heat exchangers have been used for two-phase horizontal flow but this orientation leads to mal-distribution and thermal inefficiencies, especially in duties requiring high thermal effectiveness. Condensing duties are usually designed for downward flow and boiling duties for upward flow. There are components of pressure loss – friction, momentum and gravity – among them the gravitational and momentum losses are significantly high in two-phase flows. Therefore it is likely that a proper rating can only be carried out by integrating along the length of the plate-fin heat exchanger.

6.4.2.1 Two-Phase Frictional Losses

The following methods apply to two-phase flow in the horizontal plane (no gravitational component) or downward condensation or upward boiling.

Vertically upward (boiling) or horizontal flow: Friedel's [22] method for wetted wall flow at constant quality is as follows:

1) Use the total, liquid-plus-gas, mass velocity to calculate the all-liquid and all-gas *Re* numbers:

$$Re_{Tl} = \frac{m' d_h}{\mu_l} \tag{6.42}$$

$$Re_{Tg} = \frac{m' d_h}{\mu_g} \tag{6.43}$$

where

m' = mass velocity, (kg/m^3·s)
μ_l = liquid viscosity, (Pa·s)
μ_g = gas viscosity, (Pa·s)

2) For each *Re* number determine the Fanning friction factors, f_{Tl} and f_{Tg}.
3) Calculate the single-phase, liquid only, pressure gradient:

$$\left(\frac{dp}{dz}\right)_{fl} = 4\left(\frac{f_{Tl}}{d_h}\right)\left(\frac{m'^2}{2\rho_l}\right), \quad \text{(Pa/m)} \tag{6.44}$$

4) Determine

$$E = (1-q)^2 + q^2 \frac{(\rho_l f_{Tg})}{(\rho_g f_{Tl})} \tag{6.45}$$

$$F = q^{0.78}(1-q)^{0.24} \tag{6.46}$$

$$H = \left(\frac{\rho_l}{\rho_g}\right)^{0.91}\left(\frac{\mu_g}{\mu_l}\right)^{0.19}\left(\frac{1-\mu_g}{\mu_l}\right)^{0.7} \tag{6.47}$$

where

q = mass fraction vapour, (kg$_g$/kg$_{g+l}$)
ρ = stream density, (kg/m^3)

μ = stream viscosity, (Pa·s)
l = liquid subscript
g = vapour subscript

5) Calculate the homogeneous two-phase density, ρ_{tph}

$$\rho_{tph} = \frac{1}{\left(\frac{q}{\rho_g} + \frac{(1-q)}{\rho_l}\right)}, \quad (\text{kg/m}^3) \tag{6.48}$$

6) Determine the two-phase Froude number

$$Fr_{tp} = \frac{m'^2}{\left(g_n d_h \rho_{tph}^2\right)} \tag{6.49}$$

where
g_n = acceleration due to gravity (m/s^2)

7) Determine the two-phase Weber number

$$We_{tp} = m'^2 \frac{d_h}{\left(\rho_{tph}\sigma\right)} \tag{6.50}$$

where
σ = surface tension (N/m)

8) Calculate

$$\emptyset_{lo}^2 = E + 3.24F \frac{H}{\left(F\,r_{tp}^{0.045}\ W\,e_{tp}^{0.035}\right)}, \tag{6.51}$$

9) Calculate the two-phase pressure gradient

$$\left(\frac{dp}{dz}\right)_f = \emptyset_{lo}^2 \left(\frac{dp}{dz}\right)_{fl} \quad (\text{Pa/m}) \tag{6.52}$$

The two-phase pressure gradient must be combined with the gradients due to acceleration and gravity and integrated over the boiling length, using the correct value of average quality for each increment.

A considerable simplification can be made if the following assumptions hold true:

- The friction factor is that for all-liquid and remains constant.
- Flow is homogenous and fully turbulent.
- Quality increases linearly with distance through plate-fin heat exchanger.

These assumptions are reasonably valid for most plate-fin heat exchangers in vaporizing duties. Then the enhancement to the liquid-only frictional pressure gradient is

$$\emptyset_{lo,h}^2 = 1 + q_{av}\left(\frac{\rho_l - \rho_g}{\rho_g}\right) \tag{6.53}$$

where q_{av} = average quality through the length, z
the pressure drop over full length, z, then becomes

$$\Delta P_f = \emptyset_{lo,h}^2 4 f_{Tl}\left(\frac{z}{d_h}\right)\left(\frac{m'}{2\rho_l}\right) \quad (\text{Pa}) \tag{6.54}$$

For conditions where the walls are dry (e.g. for post dry-out in evaporating flow), the frictional pressure gradient is taken as that of the vapour assumed to be flowing alone in the channel.

More recently, experiments have been carried out on different CHE fins for boiling pressure drop correlations presented by Amaranatha Raju et al. [23,24], which can also be used from Section 5.7.4.3 and Table 5.8 for design of a two-phase compact evaporator.

Vertically downward flow (condensation): For condensation in downward flow at low mass velocities, the liquid will exit as a falling film on the channel's surfaces and will have little influence on the behaviour of the vapour core. The frictional component of pressure gradient may therefore be estimated by ignoring the presence of the liquid film and evaluating only the pressure change for the local vapour flow.

More recently, experiments are carried out on different CHE fins for condensation pressure drop correlations presented by Ramana Murthy et al. [25,26] can also be used from Section 5.7.4.4 and Table 5.9 for design of a two-phase compact condenser.

6.4.2.2 Two-Phase Momentum Losses – Change of Quality

As the quality of the flow changes so will its velocity and momentum. The overall momentum head resulting from such a change is

$$\Delta P_{acc} = m'^2 \left(\frac{1}{\rho_{tp,out}} - \frac{1}{\rho_{tp,in}} \right) \text{(Pa)} \tag{6.55}$$

where

$\rho_{tp,in}$ = two-phase density at inlet
$\rho_{tp,out}$ = two-phase density at outlet

If it can be assumed that flow is homogenous and if the inlet quality is zero, as in boiling, above equation simplifies to:

$$\Delta P_{acc} = m'^2 q_{out} \left(\frac{1}{\rho_g} - \frac{1}{\rho_l} \right) \qquad \text{(Pa)} \tag{6.56}$$

where

ρ_l = outlet density of liquid (kg/m^3)
ρ_g = outlet density of vapour (kg/m^3)

6.4.2.3 Two-Phase Gravitational Losses – Upward Flow (Boiling)

The pressure change due to change in height is

$$\frac{dp_{gr}}{dz} = g_n \sin(t) f \rho_{tp} \qquad \text{(Pa)} \tag{6.57}$$

where

g_n = acceleration due to gravity (m/s^2)
t = inclination of plate-fin heat exchanger
dz = incremental length of plate-fin heat exchanger
ρ_{tp} = local two-phase density over dz

$$= \rho_l(1-\varepsilon_g) + \rho_l\varepsilon_g \qquad (\mathrm{kg/m^3}) \tag{6.58}$$

$$\varepsilon_g = \frac{V_R}{(V_R - u_r)} \tag{6.59}$$

$$V_R = \frac{\rho_l}{\rho_g}\frac{q}{(1-q)} \tag{6.60}$$

The two-phase density ρ_{tp} thus depends on the slip ratio u_r between gas and liquid. However, this method is developed for adiabatic flow, However, it may not be entirely applicable to boiling.

If it can be assumed that

- flow is homogenous throughout
- quality increases linearly with distance through the plate-fin heat exchanger
- stream is all liquid at inlet where $z = 0$, the above equation integrates to

$$\Delta P_{gr} = \frac{g_n \rho_l z \ln\left[1 + \left(\frac{\rho_l}{\rho_g} - 1\right) q_o\right]}{\left(\frac{\rho_l}{\rho_g} - 1\right) q_o} \qquad (\mathrm{Pa}) \tag{6.61}$$

where

ρ_l = average liquid density through plate-fin heat exchanger, kg/m^3
ρ_g = average vapour density through plate-fin heat exchanger, kg/m^3
q_o = exit quality

The assumption of homogeneous flow is unlikely to introduce serious error for serrated wavy or perforated finning, but for plate-fin heat exchangers using plain straight fins, there may be sufficient distance for slip to take place, in which case the two-phase density will be under-predicted as will the gravitational pressure loss.

6.4.2.4 Downward Flow (Condensation)

The liquid will normally exit as a falling film; the resulting small addition to the gravitational component can conservatively be ignored. References [9,27–31] describe this effect as it is observed for condensing nitrogen.

6.5 Two-Phase Heat Transfer

The mechanisms of fin efficiency in evaporation or condensation are different from those for single-phase heat transfer. The fluid distribution is always an important consideration when designing plate-fin heat exchangers of high thermal effectiveness and is especially important with two-phase streams. Generally, condensing duties are performed downwards and evaporation duties upwards. Most manufacturers use their own in-house proprietary methods, but references to evaporating and condensing in plate-fin heat exchanger fins are very limited in the open literature. For users without proprietary methods, the following procedure can be used.

6.5.1 Condensation

For condensing streams, an approximate heat transfer coefficient may be calculated from the *j* curves, using the total two-phase mass velocity with the properties of the condensate. The heat transfer coefficient thus obtained is then multiplied by the square root of the volume expansion factor to obtain the condensing coefficient.

6.5.1.1 All Liquid Heat Transfer Coefficient

$$Re_{Tl} = \frac{m' d_h}{\mu_l} \tag{6.62}$$

where

m' = mass velocity for total stream quantity, kg/s m^2
d_h = hydraulic diameter as for single phase
μ_l = dynamic viscosity of the liquid

Determine j_{Tl} from Re_{Tl} as for single phase. Calculate Prandtl number using all-liquid properties:

$$Pr_l = Cp_l \frac{\mu_l}{k_l} \tag{6.63}$$

where

C_{pl} = liquid specific heat at constant pressure, J/kg·K
k_l = liquid thermal conductivity, W/m·K

Calculate the single-phase heat transfer coefficient from

$$\mathrm{H_{Tl}} = j_{Tl} m' \frac{C_{pl}}{P r_l^{2/3} \quad \mathrm{W/m^2{\cdot}K}} \tag{6.64}$$

6.5.1.2 Correction for the Vapour Volume

Correct the single-phase coefficient, H_{tl} for the volume of the vapour:

$$H_{cond} = j_{Tl} \sqrt{(F_v)} \quad \mathrm{W/m^2 \cdot K} \tag{6.65}$$

where

$$\mathrm{F_v} = \left(x_g \rho_l / \rho_g \right) + 1 - x_g \tag{6.66}$$

x_g = quality of vapour as mass fraction
ρ_l = density of liquid, kg/m^3
ρ_g = density of vapour, kg/m^3

More recently, experiments have been carried out on different CHE fins for condensation heat transfer correlations presented by Ramana Murthy et al. [25,26], which can also be used for design of two-phase compact condensers. These correlations are tabulated in Section 5.7.4.4.3 and Table 5.10 for ready use.

6.5.1.3 Correction for the Multicomponent Streams

For multicomponent streams, add the resistance to heat transfer of cooling the vapour from the dew point to the bubble point:

$$Re_g = \frac{m' x_g d_h}{\mu_g} \tag{6.67}$$

where

μ_g = dynamic viscosity of the vapour.

Determine j_g from Re_g as for single-phase. Calculate Prandtl number using all vapour properties:

$$Pr_g = C_{pg} \frac{\mu_g}{k_g} \tag{6.68}$$

Where

C_{pg} = vapour specific heat at constant pressure, J/kg·K
k_g = vapour thermal conductivity, W/m·K

Calculate single-phase heat transfer coefficient from

$$Hg = j_g m' x_g \frac{C_{pg}}{P\, r_g^{2/3}} \quad \text{W/m}^2\text{·K} \tag{6.69}$$

Calculate: $Z = x_g C_{pg} \left(\frac{dh}{dt}\right)$

where

h = enthalpy of condensing mixture, J/kg
T = temperature of condensing mixture, K
$\frac{dh}{dt}$ = slope of heat release curve, J/kg·K

Calculate the multicomponent heat transfer coefficient

$$Hg = \frac{1}{\frac{1}{\alpha_{cond}} + \frac{Z}{\alpha_g}} \quad \text{W/m}^2\text{·K} \tag{6.70}$$

In calculating H, it is assumed that liquid and vapour are in equilibrium and that the fall in saturation temperature due to pressure drop is negligible. The effect of pressure drop on heat release can be calculated incrementally, but normally it is only significant for low pressure streams.

6.5.2 Evaporation

For evaporating streams, it can be assumed that the liquid-only region exists until the bulk of the stream reaches its boiling point, if sub-cooled nucleate boiling is neglected.

In the two-phase region the heat transfer coefficient increases with increasing gas phase mass fraction. Two-phase heat transfer coefficient in plate-fin heat exchangers is usually dominated by forced convective boiling, which tends to suppress nucleate boiling component. The nucleate boiling contribution is therefore usually neglected. The evaporation coefficient may be estimated from the single-phase coefficient times the convective boiling enhancement factor.

As evaporation proceeds, the amount of liquid present decreases so there is a reduction in the single-phase coefficient, although the boiling enhancement factor continues to increase. At very high qualities, typically above 95%, mist flow boiling is possible, then the coefficient is calculated as for single-phase gas flow, based on the gas phase flow (not the total flow).

6.5.2.1 Reynolds Number Calculation

Calculate Reynolds number for the total flow with liquid properties:

$$Re_{Tl} = \frac{m' d_h}{\mu_l} \tag{6.71}$$

where

m' = mass velocity for total stream quantity, kg/s m^2
d_h = hydraulic diameter as for single phase, m
μ_l = dynamic viscosity of the liquid, P$a{\cdot}s$

Calculate the Reynolds number for the flow of liquid alone:

$$Re_l = Re_{Tl}\left(1 - x_g\right) \tag{6.72}$$

where
x_g = quality of vapour i.e. mass fraction
Calculate Reynolds number for the flow of vapour alone:

$$Re_g = \frac{m' x_g d_h}{\mu_g} \tag{6.73}$$

where
μ_g = dynamic viscosity of the vapour

6.5.2.2 Determine *j* and *f* Factors

Determine j_{Tl} from Re_{Tl} as for single-phase.
Determine f_{l} from Re_l as for single-phase.
Determine f_{g} from Re_g as for single-phase.

6.5.2.3 Heat Transfer Coefficient Calculation for Quality between 0 and 0.95

Calculate Prandtl number using all liquid properties:

$$Pr_l = C_{pl} \frac{\mu_l}{k_l} \tag{6.74}$$

where

C_{pl} = liquid specific heat at constant pressure, J/kg·K
k_l = liquid thermal conductivity, W/m·K

Calculate single-phase heat transfer coefficient from:

$$H_{Tl} = j_{Tl} m' (1 - x_g)^{0.8} \frac{C_{pl}}{Pr_l^{\frac{2}{3}}} \quad \text{W/m}^2\text{·K} \tag{6.75}$$

Correct the single-phase coefficient, H_{Tl}, by the convective boiling enhancement factor, F.

$$H_c = H_{Tl} F \qquad \text{W/m}^2 \cdot \text{K} \tag{6.76}$$

where

$$F = 2.35\left(\frac{1}{X} + 0.213\right)^{0.736}$$

$$X = q\left\{\frac{f_l \rho_g}{f_g \rho_l}\right\}^{0.5} \frac{1 - x_g}{x_g} \tag{6.77}$$

ρ_l = density of liquid, kg/m^3
ρ_g = density of vapour, kg/m^3

6.5.2.4 Heat Transfer Coefficient for High and Low Values of Quality

If $x_g > 0.95$ calculate the heat transfer coefficient for the flow of vapour alone:

$$H_g = j_g m' x_g \frac{C_{pg}}{P\, r_g^{2/3}} \quad \text{W/m}^2 \cdot \text{K} \tag{6.78}$$

where j_g is determined from Re_g as for single-phase flow.

Use H_g for all values of $x_g > 0.95$, whether convective or mist flow.

Note that as x_g becomes 1.0 the equations become those for single-phase gas. Similarly as x_g becomes 0, in sub-cooled boiling for instance, the equations simplify to those for single-phase liquid.

More recently, experiments have been carried out on different CHE fins for boiling heat transfer. Correlations are presented by Amaranatha Raju et al. [23,24], which can also be used for design of two-phase compact evaporators. This information is available in Section 5.7.4.3 and Table 5.7.

6.6 Useful Relations for Surface and Core Geometry

Certain geometrical relations are necessary in the application of the basic heat transfer and flow-friction data to the design problem. Few types of surfaces can have a set of these relations formed for them that prove to be convenient, which can be referred to Tables 9.3 and 9.4 of Kays and London [2].

These dimensions are, in effect, part of the basic design data for the surface selected as indicated in the Figure 6.10 methodology.

where

b = plate spacing (plate-fin surfaces only) (m)
r_h = flow-passage hydraulic radius $A_c L/A$ (m)
β = ratio of total transfer area of one side of exchanger to volume between plates of that side (plate-fin surfaces only), (m^2/m^3)

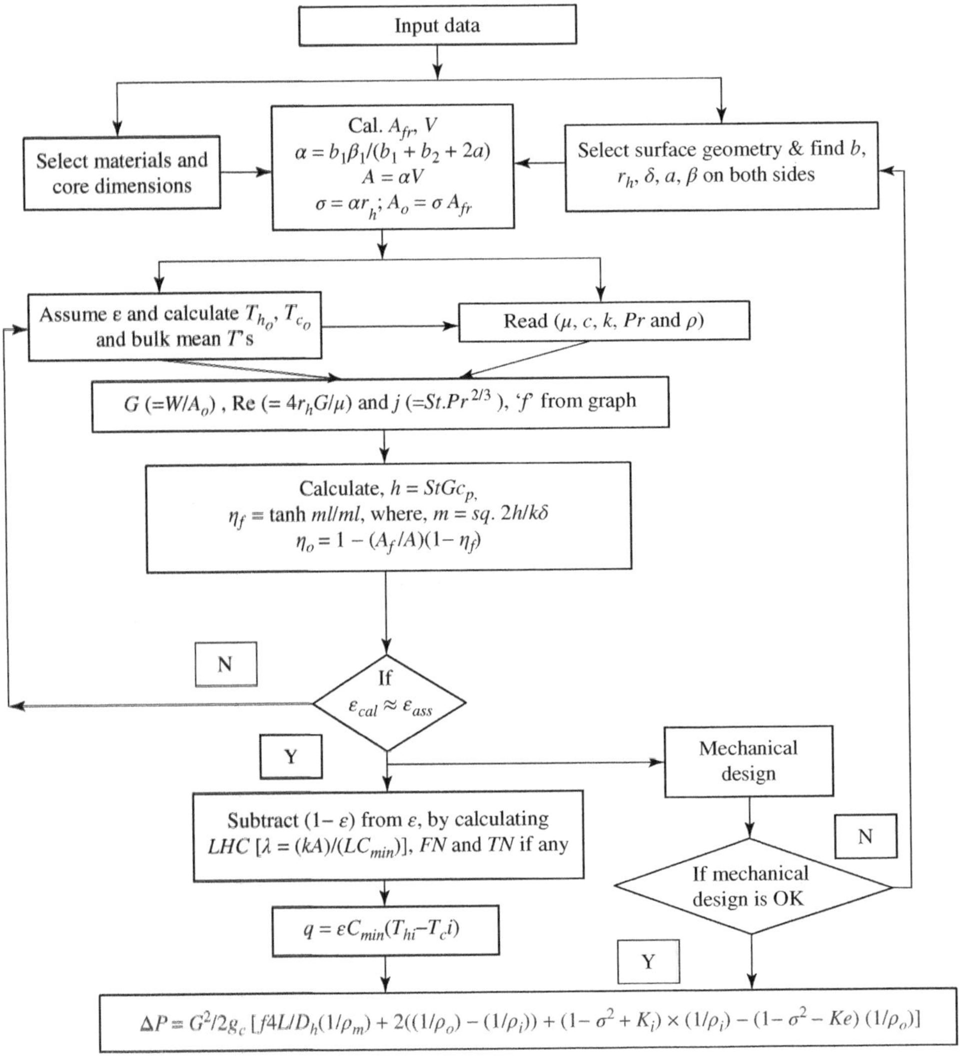

Figure 6.10 Design procedure of compact heat exchanger.

α = ratio of total transfer area of one side of exchanger to total exchanger volume (given only for matrix, tubular, and finned-type surfaces) (m^2/m^3)
p = porosity, void volume total volume (matrix surfaces only)
ε_{cal} = effectiveness calculated
ε_{ass} = effectiveness assumed

In addition, for the plate-fin surfaces, the plate thickness a, (m), must be separately specified. The following geometrical factors are required as a design result for each of the two sides of the complete heat exchanger core:

A = total transfer area of one side of exchanger (m^2)
A_c = free flow area of one side (m^2)

A_f = frontal area of one side (m^2)
L = flow length on one side (m)
V = total exchanger volume (m^3)
α = ratio of total transfer area of one side of exchanger to total exchanger volume (this is already part of given basic data for matrix tubular and finned tube surfaces) (m^2/m^3)
σ = ratio of free-flow to frontal area of one side of exchanger

The equations taken from Kays and London [2], give the relations between surface and core factors for one side of the exchanger. Subscript 1 refers to any one side, and 2 refers to the other side, and factors without a subscript are common to both sides. The same relations apply to side 2 with the exchange of subscripts 1 for 2 and 2 for 1.

$$r_h = L\left(\frac{A_e}{A}\right) \tag{6.79a}$$

$$\sigma_1 = \left(\frac{A_e}{A_{fr}}\right)_1 = \left(\frac{Ar_h}{LA_{fr}}\right)_1 = \frac{(Ar_h)_1}{V} = (\alpha r_h)_1 \quad \text{(except for matrix surfaces)} \tag{6.79b}$$

$$\sigma_1 = \frac{b_1\beta_1 r_{h1}}{b_1 + b_2 + 2a} \quad \text{(plate – fin surfaces only)} \tag{6.79c}$$

$$\alpha_1 = \frac{A_1}{V} = \left(\frac{A}{LA_{fr}}\right)_1 = \left(\frac{\sigma}{r_h}\right)_1 \quad \text{(except for matrix surfaces)} \tag{6.79d}$$

$$\alpha_1 = \frac{b_1\beta_1}{b_1 + b_2 + 2t} \quad \text{(plate-fin surfaces only)} \tag{6.79e}$$

$$A_{e1} = (\sigma A_{fr})_1 = \left(\frac{Ar_h}{L}\right)_1 = \left(\frac{A\sigma}{L\alpha}\right)_1 \quad \text{(except for matrix surfaces)} \tag{6.79f}$$

$$\left(\frac{r_h}{L}\right) = \left(\frac{A_e}{A}\right) = \left(\frac{\sigma}{L\alpha}\right)_1 \quad \text{(except for matrix surfaces)} \tag{6.79g}$$

$$p = \frac{A_e}{A_{fr}} = \alpha r_h \quad \text{(matrix surfaces only)} \tag{6.79h}$$

$$\alpha = \frac{p}{r_h} \quad \text{(for matrix surfaces only)} \tag{6.79i}$$

$$A_c = pA_{fr} = \alpha r_h A_{fr} = A\frac{r_h}{L} \quad \text{(matrix surfaces only)} \tag{6.79j}$$

$$A = \frac{\alpha}{p}LA_c \quad \text{(matrix surfaces only)} \tag{6.79k}$$

$$A = \alpha LA_{fr} \tag{6.79l}$$

The following relations apply for square-mesh crossed-rod matrices only:

$$p = 1 - \frac{\pi}{4x_t} \tag{6.79m}$$

$$\sigma = \frac{(x_t - 1)^2}{{x_t}^2} \tag{6.79n}$$

$$ad = \frac{\pi}{x_t} \tag{6.79o}$$

$$\frac{r_h}{d} = \frac{x_t}{\pi} - \frac{1}{4} = \frac{1}{4}\frac{Re}{Re_d} \tag{6.79p}$$

6.7 Core Design (Mechanical Design)

At the thermal rating stage, it is important that the thermal designer has some means of calculating fin and parting sheet thicknesses. Normally, where the manufacturer carries out both thermal and mechanical design, or where an experienced user carries out the thermal rating, correct thicknesses will be available to the designer by in-house application of the appropriate pressure vessel code for the various fin characteristics. In other situations, where the thermal rating is carried out for budgeting purposes or in feasibility studies, the rating engineer must know at least the approximate minimum thicknesses for fins, sheets and bars referred to by Taylor [9].

6.7.1 Fins

A simple one-dimensional model approach for the fins treats them as tension members restraining the two attached parting sheets, mainly in the direction of the plane of the fins (see Figure 5.1 or Figure 6.7).The tensile stresses in the fin limbs treated as straight members can be arranged to be less than a given allowable stress for the chosen material at the desired temperature by providing sufficient fins with adequate cross-sectional area.

6.7.2 Separating/Parting Sheets

The separating sheet acts as a flat plate with shear loads applied along the edges due to pressure on the side-bars. Additional loads are applied where the header tanks are attached. In severe applications the additional stress on the parting sheet arising from the effects of pipe loads on the nozzles should be considered.

6.7.3 Cap Sheets

The cap sheets are usually thicker than the parting sheets, to give protection to block against physical damage. Discontinuity stresses occur in the cap sheets where they meet the side-bars. These discontinuity stresses arise because of the dissimilar rigidity of the side-bars and the fins. The stresses in the edge bars can be obtained by estimating the shear forces in the joint between the side-bars and the parting sheets.

6.7.4 Headers

Headers can be treated as semi-cylindrical pressure vessels, to obtain a rough approximation of the tensile stress within the chosen radius. Thick cylinder design may be required for small headers with a small radius-to-thickness ratio. The ends of the headers will be sealed by plates called 'end caps' and these will add discontinuity stresses at the joins. Further discontinuity stresses will arise at the penetration of the nozzles into the headers. In practice the header tank thickness and nozzle compensation requirements are determined using appropriate pressure vessel design code formulae for cylinders and nozzles.

6.7.5 Supports

The most common ways of supporting plate-fin heat exchangers are

- brackets on cap sheets
- brackets on side bar face
- shear plates welded to cap sheets with angle supports
- foot supports welded to cap sheets at the end of the heat exchanger.

Supports must be provided with oversize or slotted holes to accommodate the change in block dimensions or its position at operating temperature. The foot type support is sometimes given sufficient length to allow for expansion and contraction of the block. Angle supports welded to the side-bar face should not extend for the full width of a composite exchanger. Layers in which water may condense during operation or shutdown, should be designed so that they drain freely into the header tanks. This is normally achieved by an internal diagonal bar or a sloping bottom bar.

The following equations are taken from the Japanese 'high pressure gas control law' as presented by Taylor [9] and, used in isolation, represent an ideal situation, and they do take into account many of the restrictions mentioned in the previous section. Alternatively, the ASME codes can be used for calculations on the mechanical design of plate-fin heat exchangers. In practice, the Japanese code requires confirmation of the calculated thicknesses by a hydraulic test on a representative sample plate-fin heat exchanger to four times the design pressure. Without such confirmation these equations must only be used for first approximations and are not a substitute for proper mechanical design.

6.7.6 Fin Minimum Thickness

$$t = \left(\frac{P}{n_f \sigma_f}\right) \times \left\{\frac{p_h}{(p_h - d_{hole})}\right\} \tag{6.80}$$

where

P = relevant stream design pressure, Pa
n_f = number of fins per meter, 1/m
p_h = pitch of holes in corrugation if fin is perforated, m
d_{hole} = diameter of holes if fin is perforated
σ_f = allowable tensile stress for fin material at design temperature, Pa

The second term on the right-hand side of Equation 6.80 becomes unity for fins which are not perforated. The allowable tensile stress for aluminium fins can be derived approximately from the material's ultimate tensile stress divided by 4 or its 0.2% proof stress divide by 1.5.

6.7.7 Parting/Separating and Cap Sheets Minimum Thickness

The maximum value of 't' from the following equations should be used.

- **Simple tensile stress (t_1)**

$$t_1 = \mathbf{b}' \frac{P_m}{\sigma_s} \quad \text{(m)} \tag{6.81}$$

where
σ_s = allowable tensile stress for sheet material at design temperature (Pa)
and for cap sheets:
$\mathbf{b}'$ = gap between cap sheet and first or last parting sheet (m)
P_m = design pressure of streams in first or last layers (Pa)

and for parting/separating sheets:

$$\mathbf{b}' = \frac{(b_n + b_{n+1})}{2} \quad \text{(m)} \tag{6.82}$$

$$P_\text{m} = \frac{(P_n b_n + P_{n+1} b_{n+1})}{(b_n + b_{n+1})} \quad \text{(Pa)} \tag{6.83}$$

P_n = design pressure of nth layer (Pa)
P_{n+1} = design pressure of (n+1)th layer (Pa)
b_n = height of nth layer between parting sheets (m)
b_{n+1} = height of (n+1)th layer between parting sheets (m)

- **Flexural stress**

$$t_2 = \left(\frac{P_m}{\left(2\, n_f^2 \sigma_s\right)} \right)^{0.5} \quad \text{(m)} \tag{6.84}$$

$$n_f = \left(n_{fn} n_{fn+1}\right)^{0.5} \tag{6.85}$$

- **Shear stress:**

$$t_3 = \frac{P_m}{(2 n_f \tau_s)} \quad \text{(m)} \tag{6.86}$$

where τ_s = allowable shear stress for sheet material at design temperature (Pa)

6.7.8 Side-Bar Minimum Thickness

$$t_b = b_b \left(\frac{1.25P}{\sigma_b} \right)^{0.5} \quad \text{(m)} \tag{6.87}$$

where

b_b = gap between sheets (m)
σ_b = allowable tensile stress for bar material at design temperature
P = relevant stream design pressure (Pa)

6.7.9 Headers Minimum Thickness

As per ASME Section VIII, Division 1, as presented by Taylor [9], for thin spherical or hemispherical headers the minimum thickness, t, can be calculated as

$$t = \frac{PR}{2SE - 0.2P} \tag{6.88}$$

where

P = max. allowable working pressure (kgf/m^2)
R = radius or arc of header (m)
S = max. allowable working stress (kgf/m^2)
E = joint efficiency ~ 0.85

6.8 Procedure for Sizing a Heat Exchanger

Two broad categories of problem specification are as follows. Given the core geometry, the flow rates and the entering fluid temperatures, what is the rating of the heat exchanger? That is, what heat transfer rate and exchanger effectiveness is predicted and what are the resulting outlet temperatures? The second category, which is the major subject of this section, is termed the *sizing problem* as distinct from the *rating problem.* The sizing problem asks the following question: What is the size of the core, given the flow rates and their entering and *leaving* temperatures? These in turn establish the desired heat transfer rate and exchanger effectiveness. Clearly, one problem is the inverse of the other. The sizing problem is a *design problem,* while the rating problem is a *performance* prediction for a specified design.

The complete design of a heat exchanger involves a whole set of considerations, as indicated by Figure 6.10. It is assumed that not only are the performance characteristics completely established but also the general type of heat exchanger has been selected, the flow arrangement (counter flow, cross flow, etc.) has been chosen and the heat transfer surface configurations for the two fluid sides have been selected. The problem then is to determine the dimensions of the overall heat exchanger. Note that if a plate-fin type of heat exchanger is chosen, the designer can in principle select the surface configurations for the two fluid sides completely independently. This, in fact, is one of the virtues of the plate-fin construction. As an example, one might select surface number of offset fin 30.33 from Figure 6.11 for one fluid side, and surface of wavy fin number 17.8–3/8 W from Figure 6.12.

In any case, for present purposes it is assumed that the surfaces have been established, as well as the flow arrangement and the desired operating parameters. Under these conditions, it is possible to size simple cross flow heat exchanger effectiveness and independently prescribed pressure drops on *both* fluid sides. Similarly, it is possible to size a rotating periodic-flow heat exchanger for prescribed effectiveness and two independently prescribed pressure drops. On the other hand, for a simple counter flow heat exchanger, it is possible to size for prescribed effectiveness and for the pressure drop on only one of the two fluid sides. The pressure drop on the other side becomes a dependent variable. It can be changed only by selecting a different surface for that side.

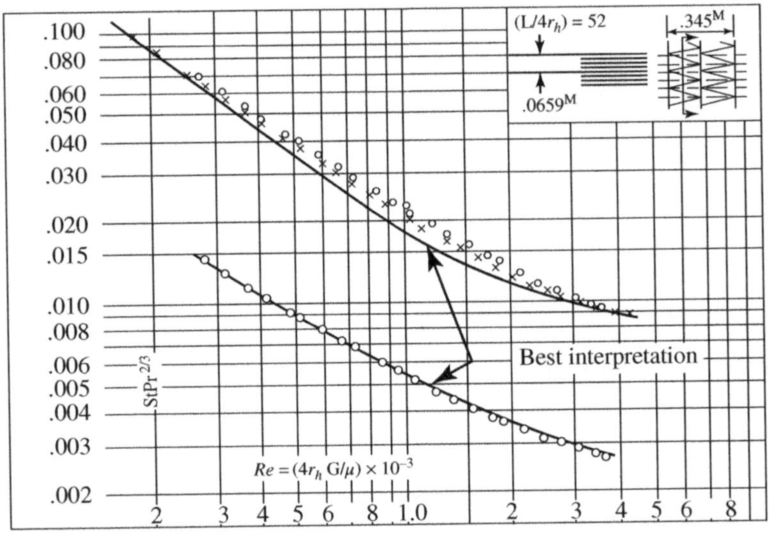

Fin pitch = 30.33 per in =1194 per m
Plate spacing, b = 0.345 in = 8.763 × 10^{-3} m
Splitter symmetricaltly located
Fin length flow direction = 2.50 in = 63.5 × 10^{-3}m
Flow passage hydraulic diameter, $4r_n$ 0.004009 *ft* = 1.222 × 10^{-3}m
Fin metal thickness = 0.004 in, aluminum = 0.101 × 10^{-3} m
Splitter metal thickness = 0.006 in = 0.152 × 10^{-3}m
Total heat transfer area/volume between plates, β = 812.51 ft^2/ft^3 = 2,666 m^2/m^3
Fin area (including splitter)/total area = 0.928

Figure 6.11 Sample *j* and *f* data curves for offset fin number 30.44. (Hesselgreaves [3])

Note then that the counter flow heat exchanger does not provide quite as much design flexibility as either the simple or multipass cross flow exchanger, although it does in general yield a higher effectiveness for a given amount of surface area (i.e. a given *NTU*).

The procedure for sizing any one of these heat exchangers will almost inevitably be an iterative one and thus it lends itself very conveniently to computer implementation, as presented by Kays and London [2]. To illustrate such a procedure, a simple cross flow arrangement will be considered. It will then be shown how a multipass cross flow heat exchanger can be easily derived from a single-pass cross flow design. A procedure for designing a counter flow heat exchanger can then be easily inferred, subject to the restriction that only one fluid pressure drop can be independently prescribed. This procedure is also very easily adapted to the design of a rotating periodic-flow heat exchanger.

The first step, after having chosen the two surfaces, is to assemble the geometric characteristics of the surface pair. The next question is what to use for an iteration variable. A very convenient variable is the mass velocity G for each side of the heat exchanger, since it is easy to make a reasonable initial estimate of both G's, and the design will converge rapidly by a simple adjustment of the G's. Note that G is the mass flow rate W divided by the flow cross-sectional area A_c. A_c is in turn related to the frontal area A_{fr}

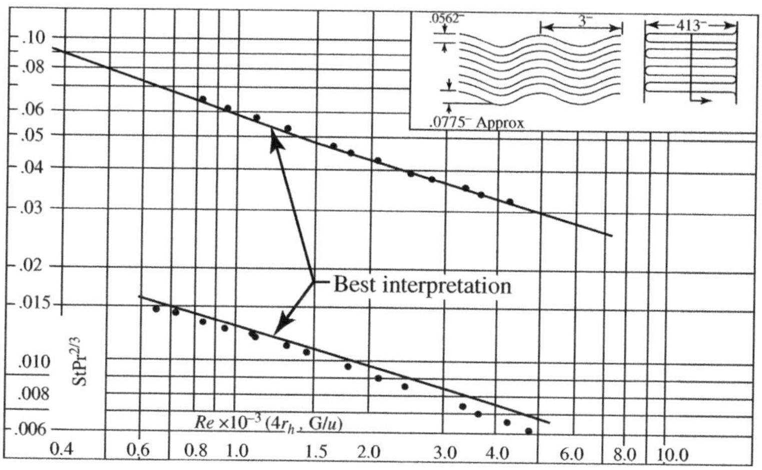

Fin pitch = 17.8 per in = 701 per m
Plate spacing, $b = 0.413$ in $= 10.49 \times 10^{-3}$ m
Flow passage hydraulic diameter, $4r_n = 0.00696\,ft = 2.123 \times 10^{-3}$m
Fin metal thickness = 0.006 in, aluminum $= 0.152 \times 10^{-3}$ m
Total heet transfer area/volume between plates, $\beta = 514\,ft^2/ft^3 = 1{,}686$ m^2/m^3
Fin area/total area = 0.892
Note: Hydraulic diameter based on free-flow area normal to mean flow

Figure 6.12 Sample *j* and *f* data curves for wavy fin number 17.8–3/8 W. (Hesselgreaves [3])

through σ. Thus, when G_h and G_c are specified, the frontal areas are fixed. Alternatively the outlet temperatures can be assumed initially for estimation of heat load and can be iterated until the assumed temperatures meet the heat load requirement. The complete dimensions of the heat exchanger are then established when the volume V is determined.

A first estimate of G_h and G_c can be made using the following approximate relation:

$$G \approx \sqrt{\left(\frac{StPr^{2/3}}{f}\right)\left(\frac{\Delta P}{NTU}\right)\frac{g_c}{v_m Pr^{2/3}}} \tag{6.89}$$

This relation is derived for the core-velocity equation of G/A_c as per Kays and London [2] by treating η_o as unity (neglecting fin ineffectiveness) and taking

$$NTU_{side} \approx 2NTU \tag{6.90}$$

In this last approximation it is implied that, in Equation 6.23, the wall resistance is negligible, and the convective heat transfer resistances are approximately equally distributed on the two sides of the heat exchanger. The ratio $StPr^{2/3}/f$ tends to be reasonably constant over a 2:1 Reynolds number range for most surfaces and does not vary greatly from surface to surface – Kays and London [2]. It can be estimated from the basic data curves for the surfaces in question, but a value of 0.3 provides a good starting point in any case, since this is merely the initial estimate for an iterative procedure. ΔP is the desired pressure drop for the side in question, and NTU is determined for the desired heat exchanger effectiveness and the C_{min}/C_{max}, derived from the given flow rates.

With initial estimates of G_h and G_c, the following variables can be evaluated in succession – Kays and London [2]:

$$\begin{array}{llll}
1) & Re_h & & Re_c \\
2) & (StPr^{2/3})_h & & (StPr^{2/3})_c \\
3) & f_h & & f_c \\
4) & h_h & & h_c \\
5) & \eta_{f_h} & & \eta_{f_c} \\
6) & \eta_{o_h} & & \eta_{o_c} \\
7) & (UA)_h & \text{or} & (UA)_c \\
8) & A_h & \text{or} & A_c \\
9) & & V & \\
10) & A_{fr,h} & & A_{fr,c} \\
11) & A_{c,h} & & A_{c,c} \\
12) & (K_c, K_e)_h & & (K_c, K_e)_c \\
13) & (\Delta P)_h & & (\Delta P)_c
\end{array} \tag{6.91}$$

These pressure-drop results are then compared with those specified for the design, and new estimates of G_h and G_c are made according to the following approximation, which comes directly from Equation 6.37:

$$G \propto \sqrt{\Delta P} \tag{6.92}$$

The procedure is then repeated until the two pressure drops are within some predetermined fraction of the specified pressure drops. The convergence is generally quite rapid, often no more than three or four iterations. It can be done by hand calculator but obviously lends itself very well to a computer program.

A multipass cross-counter flow heat exchanger can be designed using a simple modification of the same procedure if both the passes have the same volume and therefore the same amount of surface area and if the surface configurations are the same. Both have the same frontal areas and therefore the same free-flow areas. If the flow rates are the same, both have the same G's. The total flow length for each fluid is the same, and except for any pressure drop associated with the turning of the cold fluid between passes, the pressure drops for each fluid must then be the same. In fact, the only difference is the effectiveness, which is higher for the two-pass case, since the applicable effectiveness solution from Kays and London [2]. It is apparent, then, that a multipass exchanger can be designed as single-pass exchanger, but using the multipass value of NTU. Alternatively, the outlet conditions of the first pass can be estimated first and using these outlet conditions the second pass outlet conditions can be estimated and then overall effectiveness of two-pass heat exchanger is calculated finally.

6.9 Design Procedure of a Typical Compact Heat Exchanger

The design procedure for a typical compact heat exchanger is outlined below, utilizing the information and equations from Chapters 4, 5 and 6 of this book. The following specification parameters and design procedure steps are listed below for design of a CHE (Kays and London [2]):

Operating conditions and the heat transfer surfaces should be specified as follows: Operating fluids and its conditions: types of fluids, mass flow rates, and humidity content in case of air, inlet pressures, inlet temperatures and heat loads are to be considered.

The objective of this problem is to determine, for the specified conditions and the basic heat transfer and flow-friction characteristics of the surface, (a) the exchanger effectiveness and (b) the pressure drops for both the fluids. The steps in the analysis require the determination of the following factors:

1) Surface characteristics: these should be selected from a known source. Experimental data is available in Kays and London [2] for some types of fins. However, general correlations can be used from Chapter 5, which is CFD-based data for different types of fins.
2) Calculation of heat transfer and free-flow areas: Taking heat exchanger core dimensions, calculate frontal areas, A_{fr}, on both sides, and volume, V. Select material based on inlet temperatures and pressures. Assume a plate thickness, t, between fluids, which can be verified later based on mechanical design.

   Heat transfer and free flow areas from Equation 6.79c:

$$\alpha_1 = \frac{A_1}{V_{total}} = \frac{b_1\beta_1}{b_1 + b_2 + 2t} \tag{6.93}$$

   The total heat transfer area on each side is given by

$$A = \alpha V \tag{6.94}$$

   The ratio of free-flow area to frontal area is given by

$$\sigma = \frac{A_c}{A_{fr}} = \alpha r_h \tag{6.95}$$

   The free-flow areas are then given by

$$A_c = \sigma A_{fr} \tag{6.96}$$

3) Fluid properties from standard tables (Table A. 2 of Appendices):

   Assume as a first approximation, to be verified later, the heat exchanger effectiveness. This approximation need not be of a high order of accuracy, since it is used only to estimate an average temperature for the evaluation of fluid properties.

   From the definition of effectiveness Equation 6.4 for given C_c and C_h, calculate outlet temperatures, $t_{c.\,out}$ and $t_{h.\,out}$ and thus the bulk average temperatures, $t_{c.\,av}$ and $t_{h.\,av}$.

   Arithmetic averages are satisfactory here because of the approximately linear variations of the fluid temperatures with flow length. From standard tables, take viscosity, μ, thermal conductivity, k, Prandtl number, Pr, specific heat, C_p, at these bulk temperatures from Table A.3 (in the Appendices), or more accurately from other available tables. The specific heat on the air side should be corrected for humidity, in case of air as fluid, $\chi_{a,\omega}$.

   Assume as a first approximation a 2% pressure drop on both the fluid sides, which will be verified later, then calculate outlet pressures. The inlet and outlet air and gas specific volumes can then be evaluated from the perfect gas equation of state, employing the correction factors for humidity.

Suitable density correction factor, $\chi_{d,\omega}$ can be used for specific volume. It should be noted that these correction factor apply to the density ρ, the reciprocal of the specific volume v.

The specific volumes are computed as

$$v_1 = \frac{1}{\rho_{\sigma.1}} = \frac{(R/M)T_{a.1}}{\chi_{d,\omega}P_{a.1}} \tag{6.97}$$

As a fair approximation, the mean specific volume will be evaluated as the arithmetic average of the inlet and outlet specific volumes

$$v_m \approx \frac{v_1 + v_2}{2} \tag{6.98}$$

4) Calculation of Reynolds numbers, using equations as

$$G = \frac{m}{A_c} \quad \text{and} \quad Re = \frac{4r_h m}{\mu} \tag{6.99}$$

5) Calculation of $StPr^{2/3}$(j factor) and f factor from the basic characteristics of the surface from Chapter 5 using CFD correlations for surfaces selected in step 1 above. For example, in case of wavy fin turbulent flows use Equations 5.26 and 5.28 as

$$f = 10.628\, R_e^{-0.359} \left(\frac{h}{s}\right)^{0.264} \left(\frac{2A}{s}\right)^{-0.848} \left(\frac{L}{2A}\right)^{-1.931} \tag{6.100}$$

$$j = 0.242\, R_e^{-0.375} \left(\frac{h}{s}\right)^{0.235} \left(\frac{2A}{s}\right)^{-0.288} \left(\frac{L}{2A}\right)^{-0.553} \tag{6.101}$$

6) Calculation of heat transfer coefficients on both sides as

$$St = \frac{j}{Pr^{\frac{2}{3}}} \tag{6.102}$$

and

$$h = St\, G\, c_p \tag{6.103}$$

The two-phase heat transfer coefficients for both boiling and condensation applications have to be corrected taking two-phase Reynolds number factor, F as per Table 5.7 for boiling and two-phase condensation as per Table 5.10 for condensation heat transfer coefficient.

7) Calculation of fin effectiveness on both sides using Equations (6.28–6.30).
8) Calculation of surface effectiveness on both sides using Equation 6.31.
9) Calculation of overall coefficient (U) of heat transfer as

$$\frac{1}{U} = \frac{1}{\eta_{o\cdot}H_c} + \frac{1}{(A_h/A_c)\eta_{o.h}H_h} \tag{6.104}$$

10) Calculation of NTU and exchanger effectiveness:

Calculate heat capacity rates and capacity ratio as

$$C_c = m_c C_{p.c} \tag{6.105}$$

$C_h = m_{hC_{p,h}}$ and (6.106)

C_{min}/C_{max} (6.107)

Finally NTU as, $NTU = \dfrac{UA}{C_{min}}$

From Kays and London [2] for a cross flow exchanger with both fluids unmixed, the effectiveness can be calculated.

11) Subtraction of ineffectiveness values due to longitudinal heat conduction (LHC), flow non-uniformity (FN) and temperature non-uniformity (TN) effects, which cannot be neglected, especially plate-fin compact heat exchangers as estimated by Ranganayakulu et al. [32–37], Chiou [38–40] and Kroeger [41]. Thus, in the presence of LHC, FN and TN effects, the cross flow plate-fin heat exchanger effectiveness equation (4.81) will get changed to a function of the following groups:

$$\varepsilon = \phi(NTU, C^*, \lambda_h, \lambda_c, (\eta hA)^*, FN, TN \; and \; flow \; arrangement) \tag{6.108}$$

Hence, the effectiveness, ε, has to be adjusted by considering LHC, FN, TN ineffectiveness values, and this can be iterated until the assumed value is matched. Both fluid outlet temperatures can then be computed from the definition of effectiveness from Equation 6.108.

12) Calculation of pressure drops on both sides using the complete equation for the pressure drop Equation (6.37).

The approximate entrance and exit loss coefficients can be calculated from Equations 6.38 and 6.39.

From previous calculations,

$$\left(\frac{A}{A_c}\right)_{c \text{ or } h} = \left(\frac{L}{r_h}\right)_{c \text{ or } h} \tag{6.109}$$

Finally pressure drop ΔP or loss in % can be calculated as $\dfrac{\Delta P}{P_1}$

In case of the two-phase pressure drop calculations or both boiling and condensation applications have to be estimated, taking two-phase frictional multiplier ϕ_f^2 as per Table 5.8 for boiling pressure drop and two-phase condensation as per Table 5.9 for condensation pressure drop correlations.

These calculations illustrate a procedure for predicting the performance of a given heat exchanger core. No allowance has been made for additional heat transfer resistance due to fouling or for increased pressure drop due to header design and poor flow distribution. The choice of surfaces and core dimensions does not necessarily represent an optimum design.

13) Sizing of core (matrix) parts: The following equations have to be used for finding out the minimum thicknesses for the following parts:
    a) Calculate minimum thickness of fins using Equation 6.80
    b) Calculate minimum thickness of parting and cap sheets using Equations 6.82–6.86
    c) Calculate minimum thickness of side bars using Equation 6.87
    d) Calculate minimum thickness of headers using Equation 6.88

14) Calculation of exchanger weight: After estimation of minimum thicknesses of all parts, the dimensions can be fixed by taking a suitable factor of safety and then

estimate the overall weight of detailed parts and assembly of the unit. There is a need to add suitable allowances in weight estimation for mounting brackets, nozzles, weld materials and insulation if required.

6.10 Worked Examples

In this section, the following four types of heat exchanger examples are worked out for easy understanding of the design procedures.

6.10.1 Example 1: Direct Transfer Heat Exchanger

Thermal design: This example is taken from Kays and London [2], Appendix B, Figure B.3, with modified calculations taking account of ineffectiveness values of LHC and FN and extended for mechanical design of the same core geometry of a gas turbine regenerator for a 5000-shp (3.7 MW) open-cycle gas-turbine plant. Operating conditions and the heat transfer surfaces are specified as follows:

Operating conditions:

air-flow rate, humid	24.3 kg/s
air humidity 0.015	(kg/kg dry air)
fuel–air ratio 0.015	(kg/kg)
fuel–fuel oil H/C = 0.15	(kg/kg)

Air side:

louvred plate-fin surface	$\frac{3}{8} - 6.06$
entering pressure	9.1×10^5 N/m^2
entering temperature	347 °F = 807 °R (175 °C) (448 K)

Gas side:

plain plate-fin surface	11.1
entering pressure	103×10^3 N/m^2
entering temperature	805 °F = 1265 °R (430 °C) (703 K)

A detailed illustration of this procedure follows:
Surface characteristics are as follows:.

Parameters	Air side	Gas side
Surface	$\frac{3}{8}$-6.06	11.1
Plate spacing b		**6.35** × 10^{-3} m

Hydraulic radius r_h	**6.35** × 10^{-3} m	**7.71** × 10^{-3} m
Fin thickness δ	**1.11** × 10^{-3} m	**0.15** × 10^{-3} m
Transfer area/volume between plates β	**0.15** × 10^{-3} m	1204 m^2/m^3
Fin or extended area/total area	0.640	0.756

The heat exchanger core dimensions are given as

$$A_{fr.a} = L \times h = 2.09\text{m}^2 \qquad A_{fr.g} = L \times B = 4.18\text{m}^2 \qquad V = L \times B \times h = 3.82\text{m}^3$$

Assume a plate thickness between air and gas sides

$$a = 0.3 \times 10^{-3}\text{m}$$

and assume a fin material of high-temperature alloy steel with a thermal conductivity $k = 20.8\ \text{W/m}\cdot\text{K}$

Heat transfer and free flow areas from Equation 6.79e, $\alpha_1 = \dfrac{A_1}{V_{total}} = \dfrac{b_1\beta_1}{b_1 + b_2 + 2a}$

Air side $\quad \alpha_a = 400\text{m}^2/\text{m}^3$

Gas side $\quad \alpha_R = 574\text{m}^2/\text{m}^3$

The total heat transfer area on each side is given by $A = \alpha V$
Consequently, $A_a = 1531\ \text{m}^2$ and $A_g = 2193\ \text{m}^2$

The ratio of free-flow area to frontal area is given by, $\sigma = \dfrac{A_c}{A_{fr}} = \alpha rh$

$$\sigma_a = 400 \times 0.00111 = 0.444$$

$$\sigma_g = 574 \times 0.00771 = 0.443$$

The free-flow areas are then given by $A_c = \sigma A_{fr}$

$$A_{c.a} = 0.444 \times 2.09 = 0.929\,\text{m}^2$$

$$A_{c.b} = 0.443 \times 4.18 = 1.85\,\text{m}^2$$

Fluid properties:

Assume as a first approximation, to be verified later, a regenerator effectiveness of 75%. This approximation need not be of a high order of accuracy, since it is used only to estimate an average temperature for the evaluation of fluid properties.

From the definition of effectiveness Equation 6.4 for $C_c < C_h$ and $\varepsilon = 75\%$,

$$t_{a.out} \approx 366\,^\circ\text{C}\ (639\,\text{K})$$

$$t_{g.out} \approx 239\,^\circ\text{C}(512\,\text{K})$$

Thus the bulk average temperatures are

$$t_{a.av} = \frac{691 + 347}{2} = 246^\circ\text{C}(519\,\text{K})$$

$$t_{g.av} = \frac{805 + 461}{2} = 360^\circ\text{C}\ (633\,\text{K})$$

For fluid properties at these bulk temperatures from Table A.3 (from Appendices).

Property	Air side	Gas side
μ	2.55×10^{-5} Pa·s	3.02×10^{-5}Pa·s
Pr	0.67	0.67
$Pr^{2.3}$	0.766	0.766
C_p for dry air	**1.04** kJ/kg·K	**1.05** kJ/kg·K

The specific heat on the air side, as given above, should be corrected for humidity, and the specific heat on the gas side, as given above, should be corrected for both the humidity and the effects of combustion of the fuel. From data tables for a humidity of 0.015 $lb_m H_2O/\ lb_m$ (kg/kg) dry air and a fuel–air ratio of 0.015 $lb_m/\ lb_m$ (kg/kg)

$$\chi_{a,\omega} = 1.013 \qquad \chi_{e,f} = 1.019$$

Then,

$$C_{p.o} = \chi_{c.\omega} C_p = 1.05\,\mathrm{kJ/kg \cdot K}$$
$$C_{p.g} = \chi_{c.\omega} \chi_{c.f} C_p = 1.08\,\mathrm{kJ/kg \cdot K}$$

Assume as a first approximation a 2% pressure drop on both the air and gas sides. The air outlet pressure is then $8.89 \times 10^5\ \mathrm{N/m^2}$ and the gas outlet pressure is $1.01 \times 10^5\ \mathrm{N/m^2}$. The inlet and outlet air and gas specific volumes can then be evaluated from the perfect gas equation of state, employing the correction factors for humidity and the effect of the fuel.

From data tables:

$$\chi_{d,\omega} = 0.992 \qquad \chi_{d,f} = 1.001$$

It should be noted that these correction factors apply to the density ρ, the reciprocal of the specific volume v.

The specific volumes are computed as

Air side,

$$v_{a,1} = \frac{1}{\rho_{\sigma,1}} = \frac{(R/M)T_{a,1}}{\chi_{d,\omega} P_{a,1}} = 0.142\,\mathrm{m^3} \text{ and } v_{a,2} = \frac{1}{\rho_{\sigma.2}} = 0.208\,\mathrm{m^3/kg}$$

As a fair approximation, the mean specific volume will be evaluated as the arithmetic average of the inlet and outlet specific volumes

$$v_{a,m} \approx \frac{v_{a,1} + v_{a,2}}{2} = 0.175\,\mathrm{m^3/kg}$$

Gas side,

$$v_{g,1} = \frac{1}{\rho_{g,1}} = \frac{RT_{g,1}/M}{\chi_{d,\omega}\chi_{d,f} P_{g,1}} = 1.98\mathrm{m^3/kg} \text{ and } v_{g,2} = \frac{1}{\rho_{g,2}} = = 1.47\,\mathrm{m^3/kg}$$

Similarly, $v_{g,m} = 1.73\,\mathrm{m^3/kg}$

Reynolds numbers:

Air side, $\quad G_a = \dfrac{W_a}{A_{c.a}} = 26.18\,\mathrm{kg/s \cdot m^2}$

$$Re_a = \frac{4r_{h.a}G_a}{\mu_a} = 4090$$

Gas side, $W_R = (1 + f)W_a = 24.7\,\text{kg/s}$

$$G_g = \frac{W_g}{A_{c,g}} = 13.36\,\text{kg/s.m}^2$$

$$Re_g = 13,643$$

***StPr*$^{2/3}$ and *f*:**

Air side (from Figure 10.38 Kays and London [2]):

$$StPr^{2/3} = 0.0073\, f = 0.0375$$

$$St = 0.00954$$

Gas side (from Figure 10.26, Kays and London [2])

$$StPr^{2/3} = 0.0045\, f = 0.0155$$

$$St = 0.00588$$

Heat transfer coefficients:

Air side $H_a = StGc_p = 262\,\text{W}/(\text{m}^2 \cdot \text{K})$

Gas side $H_g = 85\,\text{W}/(\text{m}^2\text{K})$

Fin effectiveness (Equations 6.28, 6.29)

Air side, $m = 407\ \text{m}^{-1}$

$$l = \frac{b}{2} = 0.0032\,\text{m}$$

$ml = 1.29$ then $\eta_{f.a} = 0.665$

Gas side, $m = 231\ \text{m}^{-1}$

$ml = 0.733$ then $\eta_{f,g} = 0.850$

Surface effectiveness from Equation (6.31)

$$\eta_o = 1 - \frac{A_f}{A} = \left(1 - \eta_f\right)$$

Air side, $\eta_{o,a} = 0.786$

Gas side, $\eta_{o.g} = 0.887$

Overall coefficient of heat transfer *U*:

From Equation (6.20), neglecting the very small wall resistance

$$\frac{1}{U_a} = \frac{1}{\eta_{o,a}H_a} + \frac{1}{(A_g/A_a)\eta_{o,g}H_g}$$

$$U_a = 70.9\,\text{W/m}^2 \cdot \text{K}$$

***NTU* and exchanger effectiveness:**
Air-side capacity rate

$$C_a = W_{a}c_{p.c} = 25.5\,\text{kW/K}$$

Gas side capacity rate

$$C_g = 26.7\,\text{kW/K}$$

Capacity ratio

$$\frac{C_{min}}{C_{max}} = 0.955$$

From Equation 6.6, the number of heat transfer units is

$$NTU_{max} = \frac{A_{a}U_a}{C_{min}} = 4.25$$

Number of plates = 2.29/0.00635 = 360 (approx.)
Conduction plate area on air side = 0.91 × 0.0003 × 360 = 0.09828 m^2

$$\lambda_c = kA_a/LC_a = 20.8 \times 0.098284/1.83 \times 25.5 = 0.0438$$

Depending on the parameters LHC, FN and TN, the ineffectiveness values have to be read from available equations or tables from the available literature [32–41]. Hence, from the LHC graph for approximate $\frac{C_{min}}{C_{max}} = 1.0$ and $\lambda_c = 0.05$, the ineffectiveness is 0.09 for assumed FN model of A5 [38]. Subtracting this ineffectiveness value from assumed effectiveness, we get exchanger effectiveness is 0.75–0.09 = 0.66.Taking this effectiveness, there is a need to iterate the assumed effectiveness by correcting fluid properties at actual bulk mean temperatures.

Then gas and air outlet temperatures can be computed from the definition of effectiveness (at 0.66) as

$$t_{a.out} = 343.3\,^{\circ}\text{C} \quad t_{g.out} = 269.2\,^{\circ}\text{C}$$

One more iteration may be required by taking fluid properties at revised bulk mean temperatures, whose effects may be negligible.

Pressure drops:

The complete equation for the pressure drop Equation 6.37 is given as follows:

The approximate entrance and exit loss coefficients are calculated using Equations 6.38, 6.39, as

Fluid	K_c	K_e
Air side	0.278	0.309
Gas side	0.279	0.310

From previous calculations,

$$\left(\frac{A}{A_c}\right)_a = \left(\frac{L}{r_h}\right)_a = 1648$$

$$\left(\frac{A}{A_c}\right)_g = \left(\frac{L}{r_h}\right)_g = 1190$$

Air side, the total pressure drop is calculated as per Equation 6.37.

$$\Delta P_{air} = \frac{26.18^2}{2*9.81}*0.142\left[\left(0.278+1-0.445^2\right)+2\left(\frac{0.208}{0.142}-1\right)+\left(0.0375*1645*\frac{(0.142+0.208)}{2*0.142}\right)\right.$$
$$\left.-\left(1-0.309-0.445^2\right)*\left(\frac{0.208}{0.142}\right)\right] = 383.5\ \text{kgf/m}^2$$

Gas side, the total pressure drop is calculated as per Equation 6.37.

$$\Delta P_{air} = \frac{13.36^2}{2*9.81}*1.98\left[\left(0.279+1-0.443^2\right)+2\left(\frac{1.47}{1.98}-1\right)+\left(0.0155*1190*\frac{(1.98+1.47)}{2*1.98}\right)\right.$$
$$\left.-\left(1-0.31-0.443^2\right)*\left(\frac{1.47}{1.98}\right)\right] = 293.07\ \text{kgf/m}^2$$

These calculations illustrate a procedure for predicting the performance of a given heat exchanger core. No allowance has been made for additional heat transfer resistance due to fouling or to increased pressure drop due to header design. However, the ineffectiveness due to LHC and FN has been considered. The choice of surfaces and core dimensions does not necessarily represent an optimum design. Considering the allowable pressure drops, the fin surfaces can be changed and design can be optimized.

Mechanical design (core design):

Fin minimum thickness (t): Using Equation 6.79, t, is calculated for the following conditions:

P = relevant stream design pressure, Pa
=1.5 times design pressures
n_f = number of fins per metre, 1/m
$n_{f,\text{air}}$ = 239 fins/metre
p_h = pitch of holes in corrugation if fin is perforated, m
= no holes
d_{hole} = diameter of holes if fin is perforated
=0 m
σ_f = allowable tensile stress for fin material at design temperature, Pa
=275 MPa

$$t_{air} = \left(\frac{P}{n_f \sigma_f}\right) \times \left\{\frac{p_h}{(p_h - d_{hole})}\right\},$$
$$= 1.5 \times 1.03 \times 10^5 / 437 \times 275 \times 10^6$$
$$= 0.0013\,\text{mm (it is recommended as 0.075 mm from manufacturing point of view)}$$

$$T_{gas} = \left(\frac{P}{n_f \sigma_f}\right) \times \left\{\frac{p_h}{(p_h - d_{hole})}\right\},$$
$$= 1.5 \times 9.1 \times 10^5 / 239 \times 275 \times 10^6$$
$$= 0.021\,\text{mm (it is recommended as 0.075 mm from manufacturing point of view)}$$

Parting/separating and cap sheets minimum thickness:
The maximum value of t from the following equations should be used:
Simple tensile stress (t_1):

σ_s = allowable tensile stress for sheet material at design temperature (Pa)
=275 MPa

for cap sheets:

$\mathbf{b}'$ = gap between cap sheet and first or last parting sheet (m)
=6.35 mm
P_m = design pressure of streams in first or last layers (Pa)
=1.5 times design pressures

$$t_1 = \mathbf{b}'\frac{P_m}{\sigma_s} \qquad \text{(m)}$$

$$= .00635 \times 1.5 \times 9.1 \times 10^5 / 275 \times 10^6$$

$$= 0.0315\,\text{mm (it is recommended as 0.254mm from structural rigidity point of view)}$$

and for parting sheets:
Since the nth and (n+1)th layers having the same pressure, we get the same thickness for parting sheets, but it is recommended to use a parting sheet thickness of 0.254 mm from the structural rigidity of view.

Flexural stress:

$$t_2 = \left(\frac{P_m}{2\,n_f^2\sigma_s}\right)^{0.5} \quad \text{(m)}$$

$$= \left(1.5 \times 9.1 \times 10^5 / 2 \times 323.17^2 \times 275 \times 10^6\right)^{0.5}$$

$$= 0.154\text{mm(which is closer to the recommended value of 0.254mm)}$$

$$\text{n}_\text{f} = \left(n_{(fn)} * n_{(fn+1)}\right)^{0.5}$$

$$= (239 \times 437)^{0.5} = 323.17$$

Shear stress:

τ_s = allowable shear stress for sheet material at design temperature (Pa)
=32 MPa

$$t_3 = \frac{P_m}{\left(2n_f\tau_s\right)}, \qquad \text{(m)}$$

$$= 1.5 \times 9.1 \times 10^5 / 2 \times 323.17 \times 32 \times 10^6$$

$$= 0.0666\,\text{mm (it is recommended as 0.254mm from structural rigidity point of view)}$$

Side bar minimum thickness:
where

b_b = gap between sheets (m)
=0.00635 m

σ_b = allowable tensile stress for bar material at design temperature
=275 × 10^6 Pa
P = the relevant stream design pressure (Pa)
=9.1 × 10^5 Pa

$$t_b = b_b\left(\frac{1.25P}{\sigma_b}\right)^{0.5} \quad (\text{m})$$

$$= 6.35 \times 10^{-3}\left(1.25 \times 1.5 \times 9.1 \times 10^5 / 275 \times 10^6\right)0.5 = 0.5\,\text{mm}$$

However, it is recommended to have a 4 mm thickness from structural integrity of the unit)

Headers minimum thickness:

Using Equation 6.88, the minimum thickness t is calculated for the following conditions:

P = max. allowable working pressure (kgf/m^2)
= taken as 1.5 times normal working pressure in respective streams
R = radius or arc of header (m)
=2.29/2 = 1.145 m
S = max. allowable working stress (kgf/m^2)
=275 MPa = 2750 bars = 2750 × 1.02 × 10,000 kgf/m^2
E = joint efficiency = 0.85

$$t_{air} = \frac{PR}{2SE - 0.2P} = 92805 \times 1.5 \times (2.29/2)/(5500 \times 1.02 \times 10{,}000 - 0.2 \times 92805.8)$$
$$= 2.84\,\text{mm} \sim 2.85\,\text{mm}$$

$$t_{gas} = \frac{PR}{2SE - 0.2P} = 10475.7 \times 1.5 \times (2.29/2)/(5500 \times 1.02 \times 10{,}000 - 0.2 \times 10{,}475.7)$$
$$= 0.320\,\text{mm} \sim 1.0\,\text{mm}.$$

(However, it is recommended to have 1 mm thickness from structural integrity of the unit.)

Weight calculations:

Density of high temperature steel = 7850 kg/m^3
Volume of the total Heat exchanger = 2.29 × .91 × 1.83 = 3.811 m^3

Air side fins:

Fin thickness: 0.075 mm on air side
Fin density air side: 840 m^2 /m^3
Surface area of air side fin: 840 × 3.811 = 3201.97 m^2
Total weight of air side fins = surface area × thickness × density
= 3201.97 × 0.075/1000 × 7850 = 1885.16 kg ~ 1885 kg

Gas side fins:

Fin Thickness: 0.075 mm on gas side
Fin Density gas side: 1204 m^2 /m3
Surface Area of gas side fin: 1204 × 3.811 = 4588.44 m^2

Total Weight of gas side fins = surface area × thickness × density
= 4588.44 × 0.075/1000 × 7850 = 2701.44 kg ~ 2700 kg

Partition sheet:

No of partition sheet: 352
Surface area of partition sheet: 1.83 × 0.91 m^2
Thickness of partition sheet: 0.254 mm
Weight of partition sheet = 352 × 1.83 × .91 × (0.254/1000) × 7850 = 1168.8 ~ 1170 kg

Top and bottom sheets:

Surface area = 1.83 × 091
Thickness = 2 mm
Weight = 1.83 × 0.91 × 2/1000 × 7850 = 26.14 kg

Side bar weight:
Air side:

Height: 6.35 mm
Thickness: 4 mm
Length: 1.83 m
No of side bars on each side = 176
Weight = 6.35/1000 × 4/1000 × 1.83 × 176 × 2 × 7850 = 128.43 kg

Gas side:

Height: 6.35 mm
Thickness: 4 mm
Length: 0.91 m
No of side bars on each side = 176
Weight = 6.35/1000 × 4/ 1000 × 0.91 × 176 × 2 × 7850 = 63.86 kg

Header:
Air side:

Surface area = π × r × h = π × 0.91 × 2.29
Thickness = 3 mm
Weight = π × 0.91 × 2.29 × 3/1000 × 7850 = 146.49 kg

Gas side:

Surface area = π × r × h = π × 1.83 × 2.29
Thickness = 1 mm
Weight = π × 1.83 × 2.29 × 1/1000 × 7850 = 103.34 kg
Total weight of heat exchanger = 1885 +2700 +1170+ 26.14+128.43+ 63.86+146.49 +103.34 ~ 6223 kg

6.10.2 Example 2: Two-Pass Cross Flow Heat Exchanger

A typical two-pass cross flow compact heat exchanger has to be designed taking the following available data for aerospace applications:

1) HE hot side inlet pressure = P_{hi} = 6.5 bar
2) HE hot side outlet pressure = P_{ho} = 5.5 bar
3) HE cold side inlet pressure = P_{ci} = 1.55 bar

4) HE cold side outlet pressure = P_{co} = 1.3 bar
5) HE hot side inlet temperature = T_{hi} = 733 K =460 °C
6) HE hot side outlet temperature = $T_{ho} = T_{hi} - \varepsilon(T_{hi} - T_{ci}) = 462.16\,K = 189°C$
7) HE cold side inlet temperature = T_{ci} = 363 K = 90 °C
8) HE cold side outlet temperature = T_{co} = 410.43 K = 137.28 °C
9) Mass flow rate across hot side of HE = 0.4166 kg/s
10) Mass flow rate across cold side of HE = 2.5 kg/s
11) Envelope limitations: Core dimensions:
$V = l*b*h = 0.28\,m \times 0.2\,m \times 0.27\,m = 0.01512\,m^3$

i) **Thermal design:**

a) **Assumptions**

1) Heat exchanger effectiveness = ε = 0.732 (which can be iterated)
2) Pressured drop across both hot side of HE = 1000 mbar
3) Pressured drop across both cold side of HE = 200 mbar
4) Thickness between the hot and cold sides of the plate = a = 0.254 mm
5) Thermal *Con*ductivity of the fin material (*Stainless Ste*el 347) = k = 19 W/mK

b) **Fluid properties at the average bulk mean temperature**

	1st pass		2nd pass	
Property	Hot side	Cold side	Hot side	Cold side
Dynamic viscosity μ	30.2×10^{-6} Ns/m^2	22.25×10^{-6} Ns/m^2	24.6×10^{-6} Ns/m^2	23.5×10^{-6} Ns/m^2
Prandtl number Pr	0.693	0.7092	0.700	0.7041
$N_{Pr}^{(2/3)}$	0.7835	0.7952	0.7883	0.7914
Specific heat C_p	1.05 kJ/kg·K	1.0175 kJ/kg·K	1.01 kJ/kg·K	1.013 kJ/kg·K

c) **Surfaces**
Surface characteristics selected as a first approximation are:

Sl no.	Surface characteristics	Core details	
1	Type of fin	Lanced and offset fin/strip fin	Wavy fin
2	Plate spacing b	27 FPI (0.9071 mm)	18 FPI (1.5393 mm)
4	Fin thickness t_f	0.0762 mm	0.0762 mm
5	Fin height, h	6.3 mm	10.5 mm
6	Fin/strip length l	3.175 mm	9.525 mm 2A= 1.968 mm

d) **Results** The detailed design estimations are carried out as per the procedure given in para 6.9 and similar to the worked example 1 listed in para 6.10.1 Results are tabulated below against each design/simulation parameter:

Parameters	Equations	1st pass		2nd pass	
		Hot side	Cold side	Hot side	Cold side
Frontal area, m^2	Cross section	0.14 × 0.27	0.2 × 0.27	0.14 × 0.27	0.2 × 0.27
Fin area / total area	$\frac{A_f}{A_s} = \frac{(2*FH*FL)}{(2*FH*FL)+(2*FS*FH)}$	0.8793	0.8872	0.8793	0.8872
Fin spacing (m)	$s = (b-FT)*10^{-3}$	0.0008645	0.0013348	0.0008645	0.0013348
Hydraulic diameter (m)	$D_{h_h} = \frac{\left[2*\left(\frac{0.0254}{FPI}-FT\right)\right]*(FH-FT)}{\frac{0.0254}{FPI}+FH-2FT}$	0.001518	0.002366	0.001518	0.002366
β (m^2/m^3)	$\beta = \left[2*\left(\frac{0.0254}{FPI}+FH-2FT\right)\right]*\left(\frac{FPI}{(0.0254*FH)}\right)$	2392.01	1587.22	2392.01	1587.22
α (m^2/m^3)	$\alpha = \frac{(b_1*\beta)}{(b_1+b_2+2a)}$	786.84	783.17	786.84	783.17
Total heat transfer area (m^2)	$A_T = \alpha*V$	5.94	5.92	5.94	5.92
σ	$\sigma = \frac{A_{ff}}{A} = \alpha*r_h$	0.2986	0.4633	0.2986	0.4633
Free flow area A_{ff} (m^2)	$A_{ff} = \sigma*A$	0.01128	0.02502	0.01128	0.02502
Specific volume (m^3/kg)	$v = \left(\frac{1}{\rho}\right) = \left(\frac{R*T}{X_{dw}*P}\right)$	0.3269 (in) 0.2436 (out)	0.7016 (in) 0.8095 (out)	0.2436 (in) 0.2733 (out)	0.8095 (in) 1.0817 (out)
Mass velocity (kg/s·m^2)	$G_h = \frac{W_h}{A_{ff_h}}$	36.90	99.90	36.90	99.90

Reynolds no	$N_R = \frac{(4*r_h*G)}{\mu}$	1857.03	10639.16	2276.03	10083.38
Fanning friction factor (lanced and offset)	$f = 2.237\, R_e^{-0.236} \left(\frac{s}{h}\right)^{-0.347} \left(\frac{t}{s}\right)^{0.151} \left(\frac{t}{L}\right)^{0.639}$	0.048218		0.045958	
Colburn factor (lanced and offset)	$j = 0.185\, R_e^{-0.396} \left(\frac{s}{h}\right)^{-0.178} \left(\frac{t}{s}\right)^{-0.403} \left(\frac{t}{L}\right)^{0.29}$	0.012067		0.011133	
Fanning friction factor (wavy fin)	$f = 10.628\, R_e^{-0.359} \left(\frac{h}{s}\right)^{0.264} \left(\frac{2A}{s}\right)^{-0.848} \left(\frac{L}{2A}\right)^{-1.931}$		0.022483		0.022921
Colburn factor (wavy fin)	$j = 0.242\, R_e^{-0.375} \left(\frac{h}{s}\right)^{0.235} \left(\frac{2A}{s}\right)^{-0.288} \left(\frac{L}{2A}\right)^{-0.553}$		0.004538		0.004631
Unit film conductance (W/m^2·K)	$H = \frac{j*G*c_p}{N_{Pr}^{2/3}}$	608.18	586.86	540.42	603.46
m	$m = \sqrt{\frac{2*H}{k*\delta}}$	916.59	900.38	864.02	913.03
l	$l = \frac{b}{2}$	0.00315	0.00525	0.00315	0.00525
Fin effectiveness	$\eta_f = \frac{\tanh ml}{ml}$	0.34420	0.21151	0.36425	0.20858
Surface effectiveness	$\eta_o = 1 - \left(\frac{A_f}{A_s} * \left(1 - \eta_f\right)\right)$	0.42333	0.30044	0.44096	0.29784
Overall heat transfer coefficient W/m^2	$\frac{1}{U} = \left(\frac{1}{(\eta_{o_h} * H_h)}\right) + \left(\frac{1}{\left(\frac{A_c}{A_h}\right) * \eta_{o_c} * H_c}\right)$	83.430		82.344	
NTU_{max}	$NTU_{max} = \frac{U*A_T}{C_{min}}$	1.113		1.133	

(Continued)

Parameters	Equations	1st pass		2nd pass	
		Hot side	Cold side	Hot side	Cold side
Entrance loss co-efficient K_c	$\xi = 0.5(1 - \sigma)$	0.2775	0.2785	0.2775	0.2785
Exit loss co-efficient K_e	$\xi = (1 - \sigma)^2$	0.3080	0.3102	0.3080	0.3102
Pressure drop kg_{f/m^2}	$\frac{\Delta P}{P_{h_i}} = \frac{G^2 v_1}{2 g_c P_1}\left[(K_c + 1 - \sigma^2) + 2\left(\frac{v_2}{v_1} - 1\right) + f\left(\frac{A}{A_c} * \frac{v_m}{v_1}\right) - (1 - \sigma^2 - K_e)\left(\frac{v_2}{v_1}\right)\right]$	50.7627	229.9771	44.7448	310.5397
Overall NTU				2.246	
C_{min}/C_{max}	$(mCp)_{hot}/(mC_p)_{cold}$			0.1733	
Overall Effectiveness (using equation 2.43)	$\varepsilon = \frac{C_{max}}{C_{min}}\left(1 - e^{-\Gamma * \left(\frac{C_{min}}{C_{max}}\right)}\right)$ where $\Gamma = 1 - e^{-NTU}$			0.829	
Exit temp hot side	$T_{ho} = T_{hi} - (T_{hi} - T_{ci})$			426 K	
Exit temp cold side	$T_{co} = T_{ci} + \frac{C_{max}}{C_{min}} * (T_{hi} - T_{ho})$				416.1 K

ii) **Mechanical design (core design):**

a) **Fin minimum thickness (t):** Using Equation 6.79, t, is calculated for the following conditions:

P = relevant stream design pressure, Pa
=1.5 times design pressures
n_f = number of fins per meter, 1/m
$n_{f,air}$ = 1063 fins/metre
p_h = pitch of holes in corrugation if fin is perforated, m
= no holes
d_{hole} = diameter of holes if fin is perforated
=0 m
σ_f = allowable tensile stress for fin material at design temperature, Pa
=275 MPa

$$t_{air\ hot} = \left(\frac{P}{n_f \sigma_f}\right) \times \left\{\frac{p_h}{(p_h - d_{hole})}\right\},$$

$$= 1.5 \times 6.5 \times 10^5 / 1063 \times 275 \times 10^6$$

$$= 0.0033\,\text{mm (assume } 0.0762\,\text{mm)}$$

$$t_{air\ cold} = \left(\frac{P}{n_f \sigma_f}\right) x \left\{\frac{p_h}{(p_h - d_{hole})}\right\},$$

$$= 1.5 \times 1.5 \times 10^5 / 708 \times 275 \times 10^6$$

$$= 0.001155\,\text{mm (assume } 0.0762\,\text{mm)}$$

b) **Parting/separating and cap sheets minimum thickness:**
The maximum value of t from the following equations should be used:
Simple Tensile Stress (t_1):
σ_s = allowable tensile stress for sheet material at design temperature (Pa)
=275 MPa
Cap sheets:

b' = gap between cap sheet and first or last parting sheet (m)
=6.35 mm
P_m = design pressure of streams in first or last layers (Pa)
=1.5 times design pressures

$$t_1 = \mathbf{b}' \frac{P_m}{\sigma_s} \quad \text{(m)}$$

$$= 0.00635 \times 1.5 \times 6.5 \times 10^5 / 275 \times 10^6$$

$$= 0.0225\,\text{mm (assume } 1\,\text{mm)}$$

Parting sheets:
Since nth and $(n+1)$th layers having the same pressure, we get the same thickness for parting sheets, but it is recommended that parting sheet thickness is 0.254 mm from structural rigidity of view.

Flexural stress:

$$t_2 = \left(\frac{P_m}{\left(2\, n_f^2 \sigma_s\right)}\right)^{0.5} \quad (\text{m})$$

$$= \left(1.5 \times 6.5 \times 10^5 / 2 \times 867.52^2 \times 275 \times 10^6\right)^{0.5}$$

$$= 0.0000485\,\text{mm}$$

$$n_f = \left(n_{fn} n_{fn+1}\right)^{0.5}$$

$$= (1063 \times 708)^{0.5} = 867.52$$

Shear stress:

τ_s = allowable shear stress for sheet material at design temperature (Pa)
=32 MPa

$$t_3 = \frac{P_m}{2 n_f \tau_s} \quad (\text{m})$$

$$= 1.5 \times 6.5 \times 10^5 / 2 \times 867.5 \times 32 \times 10^6$$

$$= 0.01757\,\text{mm (assume } 0.254\,\text{mm)}$$

c) **Side bar minimum thickness:**
where
b_b = gap between sheets (m)
=0.00635 m
σ_b = allowable tensile stress for bar material at design temperature
=275 × 10^6 Pa
P = the relevant stream design pressure (Pa)
=6.5 × 10^5 Pa

$$t_b = b_b \left(\frac{1.25P}{\sigma_b}\right)^{0.5} \quad (\text{m})$$

=6.35 × 10^{-3} (1.25 × 1.5 × 6.5 × 10^5 / 275 × $10^{6)0.5}$
=0.422 mm (assume 2 mm thickness from structural integrity of the unit)

d) **Headers minimum thickness:**
Using Equation 6.87, the minimum thickness, t, is calculated for the following conditions:
P = max. allowable working pressure (kgf/m^2)
= taken as 1.5 times normal working pressure in respective streams.
R = radius or arc of header (m)
=0.3 m
S = max. allowable working stress (kgf/m^2)
=275 MPa = 2750 bars = 2750 × 1.02 × 10,000 kgf/m^2

E = joint efficiency = 0.85

$$t_{air} = \frac{PR}{2SE - 0.2P} = 650{,}000 \times 1.5 \times 0.3 /$$
$$(2 \times 275 \times 100000 \times 0.85 - 0.2 \times 1.5.6.5 \times 100000) = 0.626\text{mm}$$

(assume 1.0 mm thickness from structural integrity of the unit)

e) **Weight calculations:**

Density of high temperature steel = 7850 kg/m^3
Volume of the total heat exchanger = 0.2 × 0.28 × 0.27 = 0.01512 m^3

Hot side fins:

Fin thickness: 0.0762 mm on air side
Fin density ratio : 786.84 m^2/m^3
Surface area of air side fin: 786.84 × 0.01512 = 11.897 m^2
Total weight of air side fins = surface area × thickness × density
= 11.897 × 0.0762 / 1000 × 7850 = 7.1164 kg

Cold side fins:

Fin thickness: 0.0762 mm on gas side
Fin density ratio: 783.17 m^2 /m3
Surface area of gas side fin: 783.17 × 0.01512 = 11.84 m^2
Total weight of gas side fins = surface area × thickness × density
= 11.84 × 0.0762 / 1000 × 7850 = 7.0832 kg

Partition sheet:

No of partition sheet: 32
Surface area of partition sheet: 0.2 × 0.28 m^2
Thickness of partition sheet: 0.254 mm
Weight of partition sheet = 32 × 0.056 × (0.254 / 1000) × 7850 = 3.56 kg

Top and bottom sheets:

Surface area = 0.2 × 0.28
Thickness = 1 mm
Weight = 0.2 × 0.28 × 1 / 1000 × 7850 = 0.4396 kg

Side bar weight:

Hot side:

Height: 6.35 mm
Thickness: 2 mm
Length: 0.14 m
No of side bars on each side = 16
Weight = (6.35 × 0.2 + 10.5 × 0.28) / 1000 × 2 / 1000 × 32 × 7850 = 2.115 kg

Header:

Hot side:

Surface area = $\pi \times r \times h = \pi \times 0.3 \times (0.2 + 0.28) \times 2$
Thickness = 1 mm
Weight = π × 0.3 × (0.2 + 0.28) × 2 × 1 / 1000 × 7850 = 7.105 kg
Assume weight of lagging, nozzles and brackets = 2.0 kg
Total weight of heat exchanger = 29.42 kg

6.10.3 Example 3: Compact Evaporator Design

Calculate the convective boiling coefficient for compact plate-fin exchanger for R134a for the following conditions and take measured data in the experiments (see Section 5.7.4):

- Temperature T_{sat} = 272.5 K
- Molecular weight, M =102.03 kg/kmol
- Liquid:
  Density, ρ_l = 1297.1 kg/m^3
  Sp. heat, c_{pl} = 1.34 kJ/kgK
  Viscosity, μ_l = 290.0 μPa·s
  Conductivity, k_l = 0.094 W/m°K
  Surface tension, σ = 0.0119 N/m
  Latent heat, h_{lg} = 199.0 kJ/kg
  Sat. pressure, p_{sat} = 0.6073 bars
  Critical pressure, p_{cri} =40.59 bars
  Prandtl no., Pr_l = 3.5
- Vapour:
  Density, ρ_g = 47.125 kg/m^3
  Sp. heat, c_{pg} = 0.88 kJ/kg·K
  Viscosity, μ_g = 47.125 μ Pa·s
  Gas constant, R = 8314.3 kj/kmol·K
- Surface data:
  Mass flux, G = 63.0 kg/m^2·s
  Quality, x_g = 0.596
  Wall superheat =T_{wall}-T_{sat} = 1.0
  Hyd. diameter, d_h = 0.001189 m
  Fin type = Offset
  Fin length = 0.15 m
  Critical temperature = 101.06 °C

Solution:

First calculate saturation pressure difference from the Clausius–Clapeyron equation – Hesselgreaves [3]

$$\Delta P_{sat} = P_{sat}\left\{ \exp\left[\frac{h_{lg}M}{R}\frac{1}{T_{sat}} - \frac{1}{T_{wall}}\right] - 1\right\} = 2022.87\text{ Pa}$$

Evaluate nucleate boiling coefficient using Equation 5.71 as given by Ranganayakulu and Kabelac [17]

$$H_{nb} = \frac{.00916\Delta T_{sat}^{0.24}\Delta P_{sat}^{0.75}\, C_{pl}^{0.45}\, \rho_l^{0.49}\, k_l^{0.79}}{\sigma^{0.5}\, h_{fg}^{0.24}\, \mu_l^{0.29}\, \rho_g^{0.24}} = 1736.45\text{W/m}^2\text{K}$$

For convective boiling, evaluate the Lockhart–Martinelli parameter, X from the following Equation (Hesselgreaves [3]), taking friction factor Re exponent n= 0.5,

$$X = \left(\frac{1-x_g}{x_g}\right)^{\left(1-\frac{n}{2}\right)} \left(\frac{\rho_g}{\rho_l}\right)^n \left(\frac{\mu_l}{\mu_g}\right)^{\frac{n}{2}} = 1.02$$

and convective enhancement factor, F from Table 5.7 for offset fin,

$$F = \left[1 + \frac{32}{X}\right]^{0.5} = 5.688$$

$$Re = Gd_h/\mu_l = 258.38$$

$$Re_l = Re\left(1 - x_g\right) = 104.38$$

Calculate single-phase heat transfer coefficient using suitable Equation in Table 5.13 as

$$j = 0.1767 \times Re^{-0.45} = 0.0145 \text{ and}$$

$$H_w = jG\,c_p/Pr^{\frac{2}{3}} = 487.999\,\text{W/m}^2\cdot\text{K}$$

Calculate two phase Re for suppression factor, S from Bennett and Chen [42] and Chen [43] as

$$Re_s = Re_l F^{1.25} = 917.12$$

$$S = \frac{1}{1 + 2.53x10^{-6} R\,e_s^{1.17}} = 0.99265$$

Calculate total boiling heat transfer coefficient, H as

$$H = FH_{cb} + SH_{nb} = 4499.83\,\text{W/m}^2\cdot\text{K}$$

Estimate pressure drop as follows:
Two-phase frictional pressure gradient can be expressed as (using Equation 5.82):
Use suitable equation from Table 5.8 as

$$\phi_f{}^2 = 1 + \frac{29}{X} + \frac{3.5}{X^2} = 32.78$$

and

$$\Delta P_F = \frac{2v_l.G^2(1-x)^2 f \phi_f{}^2 L}{D_h} = 4129.76\,\text{Pa}$$

Once the heat transfer coefficient is known, sizing of evaporator can be done similar to Example 1 above.

6.10.4 Example 4: Compact Condenser Design

In this exercise, using additional data from the experiments (Section 5.7.4) from the following table, design a condenser by calculating the heat transfer coefficient and pressure drop as per the procedure from Hesselgreaves [3]:

Parameter	Unit	Value	
		DM water side	Refrigerant side
Type		Serrated (lanced & offset)	
Material		Aluminium	

Fin frequency	FPI		28	
Fin height	mm	5		5
Fin thickness	mm		0.127	
Serration length	mm	3.175		3.175
No. of layers		2		1
Total heat transfer area	m^2	0.4616		0.2286
Hydraulic diameter	mm	1.3450		1.3450

Solution:

1) Refrigerant mass flow rate (m_r): 0.0203889 kg/s
2) Condenser inlet pressure (P_{rin}): 9.61021735 bar (abs)
3) Calculated temperature (T_{calin}):

$$T_{calin} = -0.168 \times P_{rin}^2 + 7.11 \times P_{rin} - 14.81$$

$$T_{calin} = -0.168 \times 9.6102^2 + 7.11 \times 9.6102 - 14.81$$

$$T_{calin} = 38.0027772°C$$

4) Condenser inlet temperature (T_{rin}): 37.96087016 °C
5) Temperature difference $= T_{rin} - T_{calin} = -0.041907$ °C
6) Vapour enthalpy is calculated as

$$h_g = 0.007 \times P_{rin}^3 - 0.324 \times P_{rin}^2 + 5.88 \times P_{rin} + 385.3$$

$$h_g = 0.007 \times 9.6102^3 - 0.324 \times 9.6102^2 + 5.88 \times 9.6102 + 385.3 = 418.0975816\ {}^{kJ}/_{kg}$$

7) Condenser outlet pressure (P_{rout}) = 9.6741 bar (abs)
8) Calculated outlet temperature (T_{calout}) as

$$T_{calout} = -0.168 \times P_{rout}^2 + 7.11 \times P_{rout} - 14.81$$

$$T_{calout} = -0.168 \times 9.674^2 + 7.11 \times 9.674 - 14.81 = 38.2501°C$$

9) Liquid enthalpy is calculated as

$$h_f = -0.208 \times P_{rout}^2 + (9.744 \times P_{rout}) + 178.9$$

$$h_f = -0.208 \times 9.6741^2 + 9.744 \times 9.6741 + 178.9 = 253.6982521\ {}^{kJ}/_{kg}$$

10) $\Delta T_1 = 13.563647$ °C
11) $\Delta T_2 = 9.4835606$ °C
12) Water side heat load measured as 3.4 kW
13) Refrigerant heat transfer area (A_{fr})

$$A_{fr} = A_{rprim} + A_{rsecd} \times \eta_{oer} = 0.230048\,m^2$$

14) Refrigerant mass flux

$$G_r = \frac{m_r}{A_{fr}} = 34.26708\,kg/m^2\,s$$

15) Refrigerant heat flux

$$q_r = \frac{Q_r}{A_{hr}} = 14.5\ {}^{\mathrm{kW}}/_{\mathrm{m}^2}$$

16) Calculate, Re_{eq} as,

$$Re_{eq} = \frac{G\left[(1-x) + x\left(\frac{\rho_l}{\rho_g}\right)^{0.5}\right]}{\mu_l} d_h = 991.42$$

17) Then the convective heat transfer coefficient using Equation 5.101

$$H_r = 1.786 \times 10^{-4} \times \left(\frac{k_L}{D_h}\right)^{0.271} \left(Re_{eq}\right)^{1.5947} \times (Pr_l)^{4.105}$$

$$= 4351.59\,\mathrm{W/m^2K}$$

Refrigerant two phase frictional pressure drop:

1) Total differential pressure (Δp_t) = 0.81 psi (5.589 kPa)
2) Density of gas (ρ_g) = 47.125 kg/m^3
3) Density of liquid (ρ_l) = 1154.2743 kg/m^3
4) Mean volume fraction (ΔX) = 1
5) The pressure drop due to momentum is calculated using Equation 5.92

$$\Delta Pa = G_r^{\,2}\left(v_g - v_l\right)|\Delta X| = 0.0239182\,\mathrm{kPa}$$

6) Mean density using Equation 5.94

$$\rho_m = \left(\frac{x}{\rho_g} + \frac{1-x}{\rho_l}\right)^{-1} = 47.1225\,\mathrm{kg/m^3}$$

7) Flow length (L) = 0.15 m
8) The pressure drop due to gravity is calculated using Equation 5.93

$$\Delta P_g = g\rho_m L = 0.069345\,\mathrm{kPa}$$

9) The pressure drop due to manifold is calculated using Equation 5.95

$$\Delta P_c = \frac{1.5G_r^{\,2}}{2\rho_m} = 0.0187\,\mathrm{kPa}$$

10) The core frictional pressure drop is calculated using Equation 5.91

$$\Delta p_f = \left((68.95 \times 100 \times \Delta p_t) - \Delta p_c + \Delta p_a + \ \Delta p_g\right)/100\ = 5.345\,\mathrm{kPa}$$

11) Kinetic energy per unit volume

$$KE/V = \frac{G_r^{\,2}}{2 \times \rho_m} = 12.468\ \ {}^{\mathrm{J}}/_{\mathrm{m}^3}$$

12) Alternatively frictional pressure drop can be calculated using Equation 5.98 as
$\Delta P_f = 1.2556(KE/V)^{00.4987} = 4.42\,\mathrm{kPa}$ (which is close to the above value)

Once the heat transfer coefficient is known, sizing of the condenser can be carried out similar to Example 1 above.

Nomenclature

A	Exchanger total heat transfer area on one side (m^2)
A_f	Exchanger total fin area on one side (m^2)
A_{fr}	Exchanger frontal area (m^2)
$2A$	wavy fin amplitude (m)
A_s	total surface area (m^2)
A_c	flow area (m^2)
A_m	arithmetic mean dia, m
a	plate thickness, m
b	effective plate gap (m)
c_p	specific heat at constant pressure (J/kgK)
C	heat capacity rate (W/K)
C^*	ratio of heat capacity rates
d_h	hydraulic diameter (m)
d_i	tube inside diameter (m)
d_o	tube outside diameter (m)
E	Joint Efficiency
f	Fanning friction factor
F	LMTD correction factor
F	two- phase heat transfer enhancement factor
FL	fin length (m)
FH	fin height of offset surface (m)
FT	fin height of offset surface (m)
FN	Flow Non-uniformity
FS	fin spacing
g_c	gravitational constant in Eq. 6.89, (m/s^2)
G	mass velocity (kg/m^2s)
H	Convective Heat Transfer Coefficient ($W/m^2{\cdot}k$)
HE	heat exchanger
h	passage height (m)
h_{lg}	latent heat of vaporization (J/kg)
j	Colburn factor
K_c	contraction loss coefficient
K_e	expansion loss coefficient
K_d	momentum flux correction factor, dimensionless
k	thermal conductivity, $W/m{\cdot}K$

LHE	longitudinal heat conduction case
l	length (height) of fin in efficiency formula (m)
l_f	length of strip (fin) (m)
L	flow length, louver pitch (m)
L_t	tube length (m)
m	fin efficiency parameter (m^{-1}), mass flow rate, kg/s
m'	velocity head (kg/m^3 sec)
M	molecular weight (kg/kmol)
$N_{c,h}$	Ntu for cold or hot side
N	number of fins/unit length (m^{-1})
N_t	number of tubes in exchanger
NTU	number of transfer units
Nu	Nusselt number
ΔP	pressure drop, Pa
P	porosity, void volume
P	pressure (Pa), effectiveness parameter
p_c	critical pressure (Pa)
p_f	fin pitch (m)
p_m	design pressure (Pa)
p_{sat}	saturation pressure (Pa)
P_r	Prandtl number
Q	heat flow (W)
q	quality
R	ideal gas constant (J/kg-K), capacity rate ratio
R_c	ideal gas constant
Re	Reynolds number
R_f	fouling resistance (m^2K/W)
r_h	hydraulic radius, m
R_w	wall resistance (m^2K/W)
s	fin spacing (i.e. space between fins) (m)
S	nucleate boiling suppression factor
St	Stanton number
t	thickness, m
TN	Temperature Non-uniformity
t	separation plate thickness (m)
t_f	fin thickness (m)

ΔTm temperature difference, °C
T absolute temperature (K)
U overall heat transfer coefficient, W/m^{2}K
V volume, m^3
v specific volume, m^3/kg
W mass flow rate, kg/sec
ΔX mean volume fraction
x_g quality of vapor
x_t transverse tube pitch ratio in a circular tube bank, dimensionless
x_l longitudinal tube pitch ratio in a circular tube bank
$x_{a,w}$ humidity correction factor
$x_{d,w}$ density correction factor

Greek Symbols

α ratio of total transfer area of one side of exchanger to total exchanger volume (W/m^2·K)
β ratio of total transfer area on one side of heat exchanger to the volume between the plates
Δ denotes difference
ε exchanger effectiveness, dimensionless, void age
ε_p effectiveness of one pass of a multipass heat exchanger, dimensionless
λ longitudinal conduction parameter, dimensionless
η_f fin temperature effectiveness, dimensionless
η_o total overall surface temperature effectiveness
μ dynamic viscosity (Ns/m^2 or kg/ms or Pa·s)
η_f fin efficiency
η_o surface effectiveness
λ thermal conductivity (W/m·K)
ρ density (kg/m^3)
σ porosity, ratio of free flow area to frontal area, A_c/A_{fr}
ϕ_f^2 two-phase frictional multiplier
δ fin thickness, m
χ corrected specific heat for humidity
υ correction factor

Subscripts

1	cold stream
2	hot stream
i	inlet
o	outlet
c	cold stream
cb	convective boiling
cf	cross flow
f	fin
g	gas/vapour
gr	gravity
h	hot stream
l	liquid
lm	log mean
m	mean
min	minimum
max	maximum
nb	nucleate boiling
T	tube
Tl	total flow
w	wall

References

1 Sekulic, D.P., Reconsideration of the Definition of a Heat Exchanger, *Int. J. Heat Mass Transfer*, vol. 33, no. 12, 2748–2750, 1990.
2 Kays, W.M and London, A.L., *Compact Heat Exchangers*, 3rd edn., McGraw Hill, 1984.
3 J.E. Hesselgreaves, *Compact Heat Exchangers, Selection, Design and Operation*, 1st edition, Pergamon, 2001.
4 Kays, W.M and Crawford, M.E., *Convective Heat and Mass Transfer*, 3rd edn., McGraw Hill publications, USA, 1993.
5 Smith, E.M., *Thermal Design of Heat Exchangers, a Numerical Approach*, John Wiley and Sons, New York, 1997.
6 Shah, R.K., *Heat Exchanger Basic Design Methods, in Low Reynolds Number Flow Heat Exchangers*, ed. Kakac, S., Shah, R.K. and Bergles, A.E., Hemisphere, Washington DC, pp 21–71, 1983.

7 Bemtsson, T., Franck, P-A. Hilbert, L., Horgby, K., *Learning from Experiences with Heat Exchangers in Aggressive Environments*, CADDET Analysis Series no. 16, CADDET, Sittard, Netherlands.

8 ESDU Fluid Mechanics Internal Flow Sub-series Data Items, Engineering Sciences Data Unit International Ltd., 251–259, Regent Street, London.

9 Taylor, M.A., *Plate-Fin Heat Exchangers: Guide to their Specification and use.* HTFS, Harwell, UK, 1st Edition, 1987.

10 C. Ranganayakulu, Xing Luo, Stephan Kabelac, The single-blow transient testing technique for offset and wavy fins of compact plate-fin heat exchangers, *Journal of Applied thermal Engineering,* no. 111, pp 1588–1595, 2017.

11 C. Ranganayakulu, Steady state and transient analysis of Compact plate-fin heat exchanger fins for generation of design data using CFD, *International J of Numerical Methods for Heat & Fluid Flow,* vol. 26, Iss:2, pp:440–460, 2016.

12 R. Bala Sundar Rao, G. Ranganath & C. Ranganayakulu, Development of Colburn 'j' factor and fanning friction factor 'f' correlations for compact surfaces of the triangular perforated fins using CFD, *International Journal of Heat Transfer Engineering* Volume 37, Issue 2, pp 150–161, 2016.

13 K.V. Ramana Murthy, C. Ranganayakulu, T.P. Ashok Babu Development of heat transfer coefficient and friction factor correlations for serrated fins in water medium using CFD, *International Journal of Physical Science and Application,* 5(3), 238–248, 2015. doi:10.17265/2159-5348/2015/03.006.

14 Amaranatha Raju. Muppala, Ashok Babu T.P, Ranganayakulu Chennu, Development of single-phase heat transfer correlations for water and R134a in rectangular channel with smooth wavy fin, *International Journal of Physical Science and Application,* 5(3), 199–208, 2015. doi:10.17265/2159-5348/2015/03.006.

15 R. Bala Sundar Rao, G. Ranganath & C. Ranganayakulu, Colburn 'j' factor and fanning friction factor 'f' correlations of Triangular Plain fins surface of a compact heat exchanger using CFD *Applied Mechanics and Materials,* vol. 787, pp 207–211, doi:10-4028/ www. Scie- ntific.net/AMM.787.207, Switzerland, 2015.

16 R. Bala Sundar Rao, G. Ranganath & C. Ranganayakulu, Heat transfer and pressure drop correlations of rectangular perforated fin surface of a compact heat exchanger submitted in *International J of Advanced Manufacturing Technology* Annexure II, 7941, ISSN:0286-3768, Springer Verlag, Germany, 2015.

17 C. Ranganayakulu, S Kabelac, Boiling of R134a in a compact plate-fin heat exchanger having offset strip fins, *Trans. ASME Journal of Heat Transfer,* vol. 137, no. 12, 2015, No. HT-14-1235.

18 R. Bala Sundar Rao, G. Ranganath, C. Ranganayakulu, Development of Colburn '*j*' factor and fanning friction factor '*f*' correlations for compact heat exchangers plains fins by using CFD, *International Journal of Heat and Mass Transfer,* Volume 49(7), pp 991–1000, 2013.

19 Pallavi P., Ranganayakulu C., Development of Heat Transfer coefficient and Friction factor correlations for Offset fins using CFD, *Int. Journal of Numerical Methods for Heat and Fluid Flow,* Volume 21, no. 8, pp 935–951, 2011.

20 L. Sheik Ismail, R. Velraj and C. Ranganayakulu, Studies on pumping power in terms of pressure drop and heat transfer characteristics of compact plate-fin heat exchangers – A review *International J of Renewable and Sustainable Energy Reviews,* 14, 478–485, 2010.

21 L. Sheik Ismail, C. Ranganayakulu and Ramesh K. Shah, Numerical study of flow patterns of compact plate-fin heat exchangers and generation of design data for offset and wavy fins, *International J of Heat and Mass Transfer*, 52, 3972–3983, 2009.

22 Friedel L., Improved friction pressure drop correlation for horizontal and vertical two-phase pipe flow, Paper 2, European two-phase flow group meetings, Ispra, Italy, 1979.

23 Amaranatha Raju Muppala, M.E., Ashok Babu T.P., Chennu Ranganayakulu, Boiling heat transfer and pressure drop analysis of R134a in a brazed heat exchanger with offset strip fins J of Heat and Mass Transfer, Springer, accepted, Ref. HAMT-D-16-00515R1, 2017.

24 Amaranatha Raju Muppala, M.E., Ashok Babu T.P., Chennu Ranganayakulu, *Investigation of flow boiling heat transfer and pressure drop of R134a in a rectangular channel with wavy fin International Journal of Refrigeration*, 2017 (under review).

25 K.V. Ramana Murthy, Mahesh Bondhu, C. Ranganayakulu, T.P. Ashok Babu, Condensation heat transfer and pressure drop of R134a saturated vapour inside a brazed plate-fin heat exchanger with serrated fin, Paper ID: 1318, *J of Heat and Mass Transfer*, vol. 53, issue 1, pp 331–341, 2017.

26 K.V. Ramana Murthy, C. Ranganayakulu, T.P. Ashok Babu, Influence of hydraulic diameter on heat transfer and pressure drop during condensation of R134a refrigerant in a brazed plate-fin heat exchangers, *Int. J Engineering and Innovative Technology*, vol. 5, Issue 7, ID: IJEIT201512_4802, 2016.

27 Ferrell J.K., McGee J.W., Two-phase flow through abrupt expansions and contractions, TID-23394, 3, Rayleigh, N.C. Dept. Chem. Eng., North Carolina State University, 213, 1966.

28 Fitzsimmons D.E., Two-phase pressure drop in piping components, H.W. 80970 Rev 1, General Electric Harford Laboratories, Richland, Washington, 1964.

29 Geiger G.E and Rohrer W.M., Sudden contraction losses in two-phase flow, *J. Heat Transfer*, 88(I), 1–9, 1966.

30 Premoli A., Francesco D., Prima A., An empirical correlation for evaluating two-phase mixture density under adiabatic conditions, European flow group meeting, Milan, June 1970.

31 Robertson J.M., Blundell N., Clarke R.N., Characteristics of pressure gradients in downflow condensing of nitrogen in plain, aluminium plate-fin heat exchanger passages, National Heat Transfer Conference, Pittsburgh, 1987.

32 Ranganayakulu C., Seetharamu K.N., The combined effects of wall longitudinal heat conduction and inlet flow non-uniformity on thermal performance of cross flow plate-fin heat exchanger, International *Journal of Heat and Mass Transfer*, 36, 247–256, 2000.

33 Ranganayakulu C., Seetharamu K.N., The combined effects of wall longitudinal heat conduction, inlet flow non-uniformity and temperature non-uniformity on thermal performance of cross flow plate-fin heat exchanger, *International Communications in Heat and Mass Transfer*, 26, 669–678, 1999.

34 Ranganayakulu C., Seetharamu K.N., Sreevatsan K.V., The combined effects of wall longitudinal heat conduction, inlet flow non-uniformity and temperature non-uniformity on thermal performance of cross flow tube-fin heat exchanger, *International Journal of Heat and Mass Transfer*, 42, 263–273, 1999.

35 Ranganayakulu C., Seetharamu K.N., Sreevatsan K.V., The effects of longitudinal heat conduction in compact plate-fin and tube-fin heat exchangers using a finite element model, *International Journal of Heat and Mass Transfer*, 40, 1261–1277, 1997.

36 Ranganayakulu C., Seetharamu K.N., Sreevatsan K.V., The effects of inlet fluid flow non-uniformity on thermal performance and pressure drops in cross-flow plate-fin compact heat exchangers, *International Journal of Heat and Mass Transfer*, 40, 27–38, 1997.

37 Ranganayakulu C., Seetharamu K.N., The combined effects of inlet flow non-uniformity and temperature non-uniformity on thermal performance of cross-flow plate-fin heat exchanger, *Warme und Stoffubertragung*, 32, 375–383, 1997.

38 Chiou, J.P., Thermal performance deterioration in cross flow heat exchanger due to longitudinal heat conduction in the wall, ASME paper no. 76-WA/HT-8, 1976.

39 Chiou, J.P., The effect of longitudinal heat conduction on cross flow heat exchanger, *Trans. ASME J Heat Transfer*, 100, 436–441, 1978.

40 Chiou, J.P., The advancement of compact heat exchanger theory considering the effects of longitudinal heat conduction and flow non-uniformity, in *Compact Heat Exchangers: History, Technical Advancement and mechanical design problems*, HTD-10:101-1-2, ASME, 1980.

41 Kroeger, P.G., Performance deterioration in high effectiveness heat exchangers due to axial heat conduction effects, in *Advances in Cryogenic Engineering*, Volume. 12, 363–372, 1967.

42 Bennett, D.L., Chen, J.C., Forced convective boiling in vertical tubes for saturated pure components and binary mixtures, *AIChE Journal*, vol. 26, no. 3, pp 454–461, 1980.

43 Chen, J.C., Correlation for boiling heat transfer to saturated fluids in convective flow, *Ind. Eng. Chem. Proc. Des. Dev.*, 5, 322, 1966.

7

Manufacturing and Qualification Testing of Compact Heat Exchangers

7.1 Construction of Brazed Plate-Fin Heat Exchanger

The heat exchanger is assembled from a series of flat sheets/separating sheets and corrugated fins in a sandwich construction. Tube plates (i.e. parting sheets) provide the primary heat transfer surface. Parting sheets are positioned alternately with the layers of fins in the stack to provide containment between individual layers. These elements are built into a complete core and then vacuum brazed to form an integral unit. A section through a typical plate-fin heat exchanger core is shown in Figure 7.1. The heat transfer fins provide the secondary heating surface for heat transfer. Fin types, densities and heights can be varied to ensure that exchangers are tailor-made to meet individual customer requirements in terms of heat transfer performance versus pressure drop. Distributor fins collect and distribute the heat transfer fluid from the header tank to the heat transfer fins at the inlet and reverse the process at the outlet. Distribution fins are taken from the same range as the heat transfer fins, but tend to be less dense.

The heat exchanger core is then encased in a welded structure that incorporates headers, support plates and feed/discharge pipes. Most plate-fin heat exchangers are made of aluminium, with a vacuum-brazed core. Corrosion-resistant and heat-resistant brazing alloys can be used; for example, plate-fin heat exchangers can also be assembled in stainless steel, a variety of nickel-based alloys, and some other special alloys.

7.2 Construction of Diffusion-Bonded Plate-Fin Heat Exchanger

Diffusion bonding is a high temperature, solid state joining process that promotes grain growth across metal boundaries, resulting in a join exhibiting parent metal strength and ductility. This joint is significantly stronger than a brazed joint in an aluminium plate-fin heat exchanger, and hence stainless steel, diffusion-bonded exchangers can be designed for much higher pressures. In addition, they are far less susceptible to thermal shock and thermal fatigue than a brazed aluminium plate-fin heat exchanger. Diffusion bonding has a number of advantages over brazing when assembling a compact heat exchanger.

Compact Heat Exchangers – Analysis, Design and Optimization using FEM and CFD Approach, First Edition. C. Ranganayakulu and K.N. Seetharamu.
© 2018 John Wiley & Sons Ltd. Published 2018 by John Wiley & Sons Ltd.

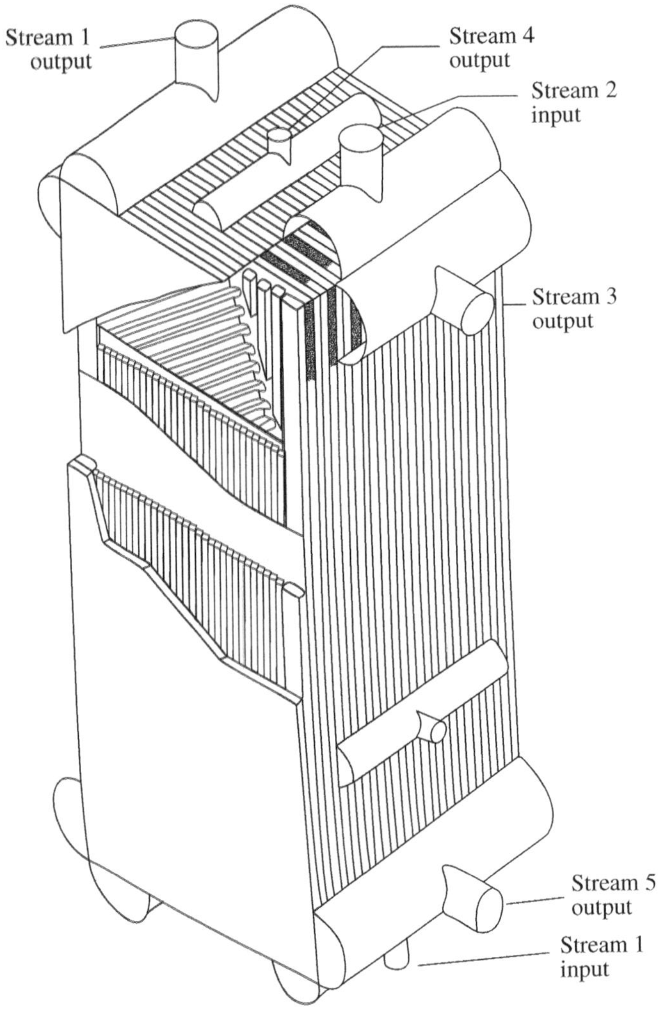

Figure 7.1 Core structure of brazed aluminium plate-fin heat exchanger. (Hesselgreaves [1])

In general, most compact plate-fin heat exchangers use vacuum brazing to assemble the core, with aluminium or stainless steel as the principal core material. Rolls Laval Heat Exchangers Ltd applied a technique used for the cost-effective manufacture of aero-engine components – super plastic forming/diffusion bonding (SPF/DB) – to the construction of plate-fin heat exchangers. This process permits titanium, and potentially stainless steel plate-fin heat exchangers of high integrity to be manufactured, giving superior strength characteristics and enhanced corrosion resistance. The formation of the basic element in the titanium plate-fin heat exchanger, i.e. two parting sheets separated by a secondary surface, involves several stages. Starting with well-prepared parting or separating sheets, a bond inhibitor is deposited on the internal surfaces of the parting sheets such that diffusion bonding only occurs where required between the two sheets (as in roll-bonding) and the third sheet, which forms the secondary surface. The diffusion bonding process is then applied, with the three sheets being held together and subjected to high pressure and temperature. Solid state diffusion bonding takes place between the unmasked surfaces, giving a joint with parent metal properties but without a heat-affected zone or impurities such as flux. The bonded sheets are then placed in a closed die, and controlled internal pressure is applied to plastically super-deform the sandwich. The central sheet stretches to provide the secondary surface, as shown in the super-plastic deformation process, which allows the metal to retain its good mechanical properties. In Figure 7.2, the final stage involves “ironing” the element to ensure flat surfaces that can conform to their neighbours in the heat exchanger matrix. The SPF/DB manufacturing process allows a wide range of internal geometries to be produced, extending beyond conventional finning arrangements such as herringbone and perforated variants. Typical minimum channel heights are about 2 mm, with a maximum of about 5 mm. Unlike brazed plate-fin heat exchangers, the diffusion-bonded unit does not need edge bars. Flow distributors are integrally incorporated during the sandwich deformation process. Modules of up to 41 elements are formed by diffusion bonding the parting sheets of adjacent elements. The modules are then joined at the stream inlets and outlets to form an exchanger block of the required size, to which the headers, nozzles and other external features are welded. Several other commercially significant alloys exhibit super-plasticity, and the technique can be developed for use with both stainless steel and nickel alloys.

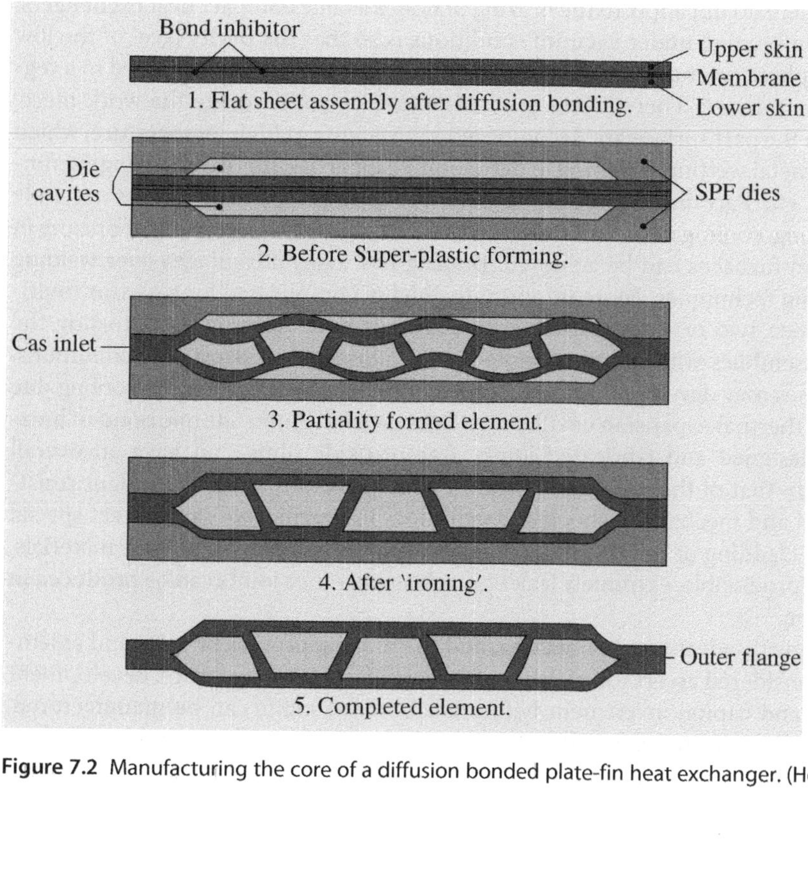

Figure 7.2 Manufacturing the core of a diffusion bonded plate-fin heat exchanger. (Hesselgreaves [1])

7.3 Brazing

The brazing and soldering are accomplished by "wetting" the metals to be joined (parent metals) with a molten metal and subsequent cooling, then the molten metal forms the joint. Generally, if the temperature of the wetting metal is above 450°C, the process is called *brazing*, and the molten metal is referred to as filler metal or braze alloy. If the wetting metal is at temperatures below 450°C, the process is called *soldering* and the molten metal is referred to as solder. However, in welding, both the base metals to be joined are locally melted at the time of joining. Conventional welding can cause distortion of the base parts, and dissimilar metals cannot be joined if they are metallurgically incompatible. For making complicated joints by brazing, generally all the parts are heated to the desired temperature of the filler metal. Pre-cleaning of the joints is essential for good brazing. Cleanliness of base metal, filler metal and flux and the purity of the atmosphere should be as high as practical, to obtain high quality joints. The choice of cleaning process depends on the nature of contamination, specific base metal to be cleaned, degree of cleanliness required for brazing, part configuration and the need to remove or provide a barrier for coating for undesirable elements, such as Al, Ti or Ni. A vacuum or controlled atmosphere is used when the flux-bath is not used to prevent the re-oxidation of the oxide-free surface. Since the base materials do not melt at the brazing temperature, practically any two materials can be joined by brazing. The only requirement is to use a suitable brazing filler metal that is compatible with the base metals, and has a melting point below those of the base metals. Note, however, that the comments here do not apply to the brazing of non-metallic compact heat exchangers.

The reason for brazing under vacuum conditions is so that the purity level of the low pressure atmosphere (vacuum) can be precisely controlled than can be obtained in a regular atmosphere furnace. There is less residual oxygen to contaminate the work-piece. Oxide layers on the part surface are decomposed in a vacuum at high temperature, which improves base metal wetting, resulting in better joint properties: increased strength, minimum porosity, etc. Part distortion is minimized due to heating and cooling at precisely controlled heating/cooling rates. In addition, the repeatability and reliability of brazing in modern vacuum furnaces can be achieved. Brazing has many advantages over welding and other joining techniques. More importantly, highly complicated long and/or multiple joints between two or more parts can be made without burning or distorting the joints. Large assemblies with multiple joints can be fabricated in stress-free conditions. However, stresses may develop in brazed joints of dissimilar materials upon cooling due to the different thermal expansion coefficients. Because of the inter-atomic bond in brazing, properly designed and fabricated joints, free of oxide films, can have an overall strength equal to that of the base metal. Brazed joints generally exhibit excellent resistance to thermal and mechanical shock and vibration. In brazing, we can protect special metal coatings, cladding or important metallurgical characteristics of the base materials. High quality, reproducible, extremely leak tight, close tolerance joints can be produced in mass production.

There are some disadvantages of brazing, and the brazing process of parts and assemblies is often considered an art today. It requires considerable expenditure, development time, research and capital investment before ideal brazed joints can be manufactured, particularly for complicated parts. If any one variable of the brazing process (e.g. flux, filler metals, temperature, brazing cycle time or atmosphere), base materials (e.g. alloying components, surface structure and preparation or joint clearances) or equipment and fixturing is changed, in general, the brazing process needs to be redefined for ideal joints. Chemical and/or mechanical cleaning with or without subsequent fluxing can be used to prepare the surfaces before brazing. In most applications, these fluxes must be removed after brazing. If not properly removed, they can corrode the joined parts, reducing their lifetime significantly. In many brazing processes, the major elements in the filler metal are the same as the base metal except for some additional melting point temperature depressant elements. In the case of aluminium, the brazing is usually done at a temperature which is within 25–70°C of the solidus temperature of the base metal. Hence, close temperature control during brazing is essential. Due to the diffusion of alloying elements, brazing produces an alloyed zone, near the joint, whose mechanical and thermal properties may be different from the base metal. In some systems, brittle inter-metallic compounds are formed at braze–work-piece interfaces, and their growth can have very detrimental effects on mechanical properties.

Different types of vacuum furnaces are available, based on hot-zone construction material, the choice of which depends on the vacuum level requirement, compatibility with base material and braze filler metals, temperature requirements and cooling speed. The most popular insulation and heating element materials are graphite and a refractory metal such as molybdenum. Graphite is a more economical material with lower maintenance requirements. Graphite purity has improved over the past several years, and can now be considered for many vacuum brazing applications. A furnace using graphite requires a larger pumping system to operate at the same vacuum levels as the furnace with a metal hot zone because graphite is hygroscopic. Generally, graphite furnaces heat more slowly compared with one having an all-metal hot zone. Molybdenum does not absorb moisture and heats faster, and it is still recommended in the aerospace industry to process titanium, aluminium and materials having high Ti and Al content. Photographs of a vacuum furnace system along with high and low temperature hot zones are shown in Figures 7.3, 7.4; these are developed by BHEL [4] for the brazing of aerospace compact heat exchangers.

7.3.1 Operations in Brazing

Major operations associated with a brazing process in general, and for compact heat exchangers in particular, are cleaning, assembling and fixturing, fluxing, preheating, brazing, cooling and post brazing treatment. If tubes/plates or fins are clad with braze alloy, as is customary practice with aluminium, a heat exchanger core with flat tubes and fins is first fabricated by stacking fins and tubes and then placing headers on both ends; a plate-fin exchanger with fins and parting sheets is stacked by placing end bars at the layer of the fin, as shown in Figure 7.5. If the braze alloy is going to be applied in the powder/binder, paste, foil or tape form, the parts are first cleaned and the braze alloy is applied during the stacking on individual fins/plates. If braze rings are used in a tube-fin exchanger, fins and braze rings are alternately inserted over tubes.

Cleaning and creating oxide free surfaces are imperative to ensure sound brazed joints of uniform quality. Uniform capillary action is possible only when all grease, oil, dirt and oxides have been removed from both the filler metal and the base metal before brazing. Chemical cleaning methods are more common compared to mechanical cleaning. Chemical cleaning methods include: (a) emulsion cleaning in insoluble

Figure 7.3 Typical vacuum furnace system front end.

Figure 7.4 Typical high temperature/hot zone for SS brazing.

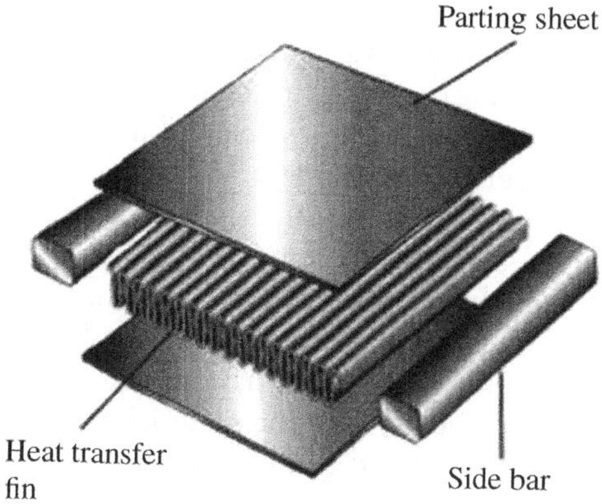

Figure 7.5 Plate-fin exchanger with fins and parting sheets.

hydrocarbons and water is good for removal of oils and cutting fluids, (b) water-base alkaline cleaners are good for removal of oils and cuffing fluids, (c) solvent cleaning in mineral spirits, alcohol, acetone and chlorinated hydrocarbons is good for removal of mineral oils and cutting fluids, but poor for removal of water-soluble oils. Fluorinated solvents are also often used. Solvent cleaning is achieved by either immersing or spraying parts with a petroleum solvent or chlorinated hydrocarbon. Alkaline cleaning is performed by immersing or spraying parts with mixtures of silicates, phosphates, soap, wetting agents, etc. Acid pickling involves cleaning of parts in a weak solution of sulphuric, hydrochloric or nitric acids, or combinations of them. Mechanical cleaning methods, following chemical cleaning, include grit blasting using chilled cast iron, or stainless steel grits or powders and machining or grinding, provided that joint clearances are not disturbed. The presence of contaminants at or near the joins to be brazed would not permit the molten filler metal to wet the surfaces for brazing, or it may restrict or misdirect the filler metal flow. This would result in braze voids, no brazing or reduced mechanical properties of the brazed joints. Mechanical cleaning is generally used for removing heavy scale and more tenacious lubricants such as pigmented drawing compounds.

Assembly and fixturing are essential for ensuring high quality brazed joints. Ideally, assembling of brazing parts is done in such a way that it requires no or minimum fixturing. Assembly may be done by expanding, press fitting or swaging, so that the desired clearance, usually a few thousands of an inch, is obtained for parts to be joined. However, fixturing is used when the above means of assembling are not practical and/or parts are to be properly positioned and maintained during the brazing for the molten filler material to flow in specific directions. The fixture may be two flat metal strips (one at each end of the exchanger core retained by banding), simple baskets or wire stands, clamps, cast supports, etc. Depending upon the thermal expansion of parts to be joined, proper pressure must be applied to ensure ideal joint clearance gaps at braze temperatures. Usually the range of banding pressure is 1–4 atmospheric pressure for plate-fin exchangers. Over-pressure could crush fins. Under-pressure may not produce good brazed joints if the gaps remain large at joints. The most common materials used for fixturing are low carbon steel, stainless steels, wrought and cast heat-resisting alloys, Inconel, etc., depending upon the materials to be joined and the brazing process to be used. Generally, an attempt is made to match the thermal expansion of the fixturing to the thermal expansion of the base metal.

Brazements should be designed so that the detail parts are self-fixturing and self-aligning wherever possible. The use of fixtures in the vacuum furnace increases processing cost and can add to distortion difficulties. Tack welding, poke welding, gravity locating, swaging and stacking can be used when self-jigging is not possible. Maintaining a proper gap clearance is one of the most important factors in the assembly of parts for brazing. Fixtures must be capable of maintaining proper braze clearance, have dimensional stability and have compatible coefficients of thermal expansion with parts being brazed. In designing fixtures, use thin sections with the required rigidity and durability. The total weight of all fixtures should not exceed 50% of the total weight of the assemblies being brazed per furnace run – Shah [2]. Fixtures should be made of materials that are not reactive with the material of the assembly being brazed. Inconel 600, Alloy 230, molybdenum and graphite are recommended, as these materials maintain high strength at elevated temperatures and have good resistance to thermal shock. Dissimilar metals should not be used where differences in thermal expansion could affect assembly dimensions. Bolts and screws should not be used as they may relax upon heating or pressure weld in place. In addition, fixtures should be subjected to the brazing environment prior to use, to ensure stability and relieve stresses, and they must not interfere with the flow of cooling gas.

Fluxing is made to remove, disrupt or dissolve any remaining oxide film from the surfaces to be joined and to protect the freshly cleaned base metal from re-oxidizing. Subsequently, the fluxing permits wetting and flows of molten filler metals to the joints at the brazing temperature. It must be emphasized that the fluxing cannot eliminate proper precleaning. A thorough coating of flux on all surfaces to be brazed is essential to prevent further oxidation of the base metals at elevated temperatures. Also, the flux must melt before the filler metal melts, and remain active throughout the brazing cycle until the filler metal solidifies. If the flux melts at a temperature higher than that for the filler metal, thick oxide films will develop on the molten filler alloy which in turn could inhibit the action of the flux. Fluorides raise the melting point of the flux and are responsible for the etching action. Generally, specific fluxes are developed for each brazing process. The selection of a proper flux for a given brazing process is as important as the selection of other variables of the brazing process. In vacuum brazing, no flux is used.

Preheating of the work-piece is highly desirable for most brazing processes used for brazing of compact heat exchangers. Preheating dries the flux and vaporizes all moisture from the brazement and fixture. Preheating decreases the temperature drop of the braze furnace thus reducing the braze time and premature melting of the brazing filler metal. It can also minimize thermal stresses and warpage when the braze assembly consists of thick and thin parts. The preheating of large work-pieces may be done by radiant heating and/or forced convection. Technical experience suggests that the preheating should be done as uniformly as possible, to within 55 to 80°C of the melting point of the brazing filler metal – Shah [3].

Brazing is achieved when the assembled parts and the filler metal are heated to a sufficiently high temperature that melts the filler metal but not the base metal. In flux brazing, once the filler metal melts it pushes out the molten flux, wets the hot base metals and starts flowing to the joint clearance due to capillary action, and on the way it adapts to sub-microscopic irregularities and dissolves the small high points encountered. Once the joints are filled with the molten filler metal, the brazing is complete. If the brazed parts are now to be removed (such as in dip brazing), care must be taken to ensure no movement of the brazed parts until the filler metal solidifies. The total time required at the brazing temperature may be as little as few seconds to as long as 1–1½ hours. It is dependent upon the filler metal characteristics, how the filler metal is applied (cladding, electroplating, powder, adhesive coated braze tape, etc.) and the size and weight of the brazements. If parts are kept longer than the optimum time at the brazing temperature, (a) the molten filler metal may run away from the joint yielding no brazed joint, (b) dissolution of the base metal may occur weakening the joint due to thinning of the base metal or (c) excessively thick fragile inter-metallic layers may form. Once the brazing temperature is reached, the furnace is generally kept at that temperature for only a short time; the brazement is then taken to the cooling zone or removed from the batch type furnace.

Cooling of brazements may start as soon as the brazing is complete. In a multizone furnace, there is generally a cooling zone that provides a protective atmosphere to eliminate oxidation and discoloration. Here, the cooling rate is generally low, and sometimes heat treatment of the brazements is also carried out at this time. If distortion is not a problem, cooling may be with hot water quenching or rinsing which also removes the flux.

Post-brazing treatment is essential if flux material is present after brazing, which is corrosive to the base metal. The residual flux must be removed completely either by quenching or rinsing with hot or boiling water. Since the dip-brazing flux is highly water-soluble, this is the most efficient method of removing flux. Sometimes, brazements are dipped into a second tank of agitated hot water to remove the remaining flux. Also acid pickling is used for removing the last traces of flux. Post-brazing treatment may follow cleaning and pickling. Post-brazing cleaning is generally not required in controlled atmosphere (inert or reduced gas or vacuum) furnace brazing.

7.3.2 Brazing Filler Metals

A suitable brazing filler metal (BFM) must be able to wet the base metal, must have melting and flow properties to permit distribution by capillary action and must be able to make a strong, sound and metallurgical bond. It must have chemical composition of sufficient homogeneity and stability to minimize separation by liquation in brazing. It should be compatible with the substrate and must be able to produce a braze joint that will meet the specified service requirements and mechanical properties.

7.3.3 Brazing Processes

A compact heat exchanger, particularly the plate-fin, tube-fin or regenerative type, has thousands of joints, many of which may not be easily accessible. The thin fins make brazing difficult due to sagging or distortion. Hence, specialized brazing processes are required which can uniformly heat all interior parts without overheating the external surfaces of the brazements, and the brazements must be protected from oxidation at elevated brazing temperature throughout the brazing cycle. The most commonly used brazing methods for atmosphere are inert or reduced gas furnace brazing.

7.3.4 Vacuum Brazing

It is a common brazing process for medium to high production brazing of compact heat exchangers. In this method, brazing is done in a vacuum with absolute pressures of 1.3 to 1.3×10^{-4} Pa (10^{-2} to 10^{-6} torr) thus preventing serious oxidation of the brazements; special filler metals are used; and usually no flux is used, thus eliminating the need for any post-brazing treatment. It is an environmentally very clean process, and is used where entrapped fluxes cannot be tolerated or reactive metals are brazed. It is used extensively in aerospace, automotive and nuclear fields. Vacuum brazing is not suitable for base metals or filler metals having high vapour pressure or low boiling point constituents. The maintenance of a proper vacuum in the furnace is critical. Operations associated with vacuum brazing are cleaning, assembling and fixturing, loading into the brazing furnace, vacuum brazing using a simultaneous pump-down and the fastest practical continuous radiant heating, back filling, unloading vacuum furnace chamber, cooling and de-fixturing. In a three-zone vacuum furnace, preheating, brazing and cooling zones are separate. Vacuum brazing sheets and/or filler metals must be clean, should not be handled with bare hands (molten filler metal won't wet the base metal at fingerprints) and should be stored in a dry place at a temperature that prevents water condensation from atmosphere.

Three types of vacuum braze furnaces are used: single-pumped retort furnace (a sealed retort of thick material), double-pumped retort furnace (an inner retort contained within a vacuum chamber) and cold-wall vacuum furnaces (a single vacuum chamber with thermal insulation and heating elements inside the chamber). These furnaces can be side-loading with a rectangular work zone or top or bottom loading with a circular work zone. Both batch type and semi-continuous furnaces are used. The semi-continuous furnace could have multizones: preheating, brazing and cooling. While in the batch-type furnace, the parts to be brazed are stationary; they are moving in a semi-continuous furnace. Also, a pressure excursion occurs in the brazing chamber of a multizone furnace. The constituents of gases that make up the vapour pressure in the furnace are more important from the brazing point of view (oxidizing versus non-oxidizing) than the absolute value of the vacuum pressure. In vacuum brazing, the gas from the furnace is analysed before and during brazing to determine its precise composition and to monitor the furnace atmosphere. The furnace is heated electrically and parts are heated by radiation only. Heating elements are made from nichrome or metallic alloys of molybdenum, tungsten or tantalum, depending upon the required temperature range and the furnace size. Other selection criteria are high electrical resistivity, excellent scale stability, high strength at elevated temperatures and low vapour pressure. A variety of metals have been used in compact heat exchangers, depending upon the service temperature and other operating conditions, as well as the need for corrosion, oxidation and high temperature resistance. Brazing of aluminium and its alloys, stainless steels and nickel- and cobalt-based superalloys is briefly discussed below – Shah [2,3].

7.3.4.1 Brazing of Aluminium and its Alloys

Various aluminium and aluminium alloy wrought products and castings are available, depending upon the mechanical, chemical and physical property requirements. Common brazeable wrought aluminium alloys, their composition and melting temperature range are listed in Table 7.1.

Table 7.1 Filler metals for resistance brazing. Shah [2,3].

Sl. No.	Work metal	Brazing filler metal
1	Steels, stainless steels, heat-resisting alloys, copper, copper alloys, nickel alloys.	Silver filler metals (bag type)
2	Aluminium alloys	Al-Si filler metals
3	Copper and copper alloys	Cu-P filler metals

Various aluminium and aluminium alloy wrought products and castings are available, depending upon the mechanical, chemical and physical property requirements. Common brazeable wrought aluminium alloys, their composition and melting temperature range are listed in Table 7.2. The commercially available filler metals for aluminium alloys, as shown in Table 7.3, are also aluminium-based alloys, containing 7–12% silicon, up to 2% other metals (Zn, Cu, Mg) and the rest (92.5–86%) aluminium. For perfect joints and also for small-clearance joints, a short melting range filler metal (e.g. 4047) is most desirable for quick melting, reaching all corners before solidifying, reducing time for the brazement at the brazing temperature and reducing filler metal-base metal diffusion. Filler alloys 4343, 4045, 4004, 4104 and 4047 are commonly used in aluminium brazing and have good corrosion resistance. Aluminium alloys are brazed in the temperature range 549–621°C. The most common brazing processes for aluminium alloys are dip, vacuum and furnace brazing for compact heat exchangers. Except for vacuum brazing, flux is used to promote wetting between the base metal and filler metal. Dip brazing of aluminium has been a common method of brazing of aluminium heat exchangers; vacuum and inert gas furnace brazing are now getting more common.

The brazements are preheated to about 480–575°C before being dipped in the pot. The most commonly used base Al alloy for dip brazing is 3003 and the cladding alloy is 4343. Once the temperature of the work-piece reaches the brazing temperature, the brazing is complete. For small heat exchangers, the immersion time is about 30s to 3 minutes, while for large (cryogenic) heat exchangers weighing more than 450 kg, the immersion time may be as long as 20 minutes (Nippes [5]). Extensive post-brazing treatment is essential to remove the corrosive flux/residues thoroughly and obtain adequate corrosion resistance. Hence, hot water rinsing, pickling and corrosion resistance coating of the pickled surface follow the brazing process and draining of the flux.

In vacuum brazing, either the cladding (e.g. 4004) and/or the base aluminium alloys (e.g. 3005) have magnesium as one of the alloying components in the brazing sheets. Magnesium vaporizes at a high rate above 575°C and reacts with O_2, CO_2 and H_2O at the surface, thus rendering the vacuum environment essentially non-oxidizing. As the aluminium cladding metal approaches the brazing temperature, the brazing surface becomes free of the continuous oxide barrier (layer), the liquid filler metal freely wets the nearby surfaces and flows to the joint areas by capillary action. When the entire furnace load is at a proper temperature, the brazing process is complete and no soaking is required at this brazing temperature. For vacuum brazing of aluminium parts, cold-wall vacuum furnaces (batch type or multizone) are used with electric-resistance radiant heating elements.

Table 7.2 Common brazeable wrought aluminium alloys, their composition and approximate melting range. Shah [2] and Nippes [5].

Alloy	*Nominal composition mainder % remainder Al and impurities*							*Approximate melting range*		Relative brazeability*
	Cu	*Si*	*Mn*	*Mg*	*Zi*	*Ni*	*Cr*	°*C*	°*F*	
1350	99.50% minimum aluminum							646–657	1195–1215	A
1100	99% minimum aluminum							643–657	1190–1215	A
3003	0.12	0.6	1.2	—	—	—	—	643–654	1190–1210	A
3004	—	—	1.2	1.0	—	—	—	629–652	1165–1205	B
3005	—	—	1.2	0.4	—	—	—	638–657	1180–1215	A
5005	—	—	—	0.8	—	—	—	627–652	1160–1205	B
5050	—	—	—	1.2	—	—	—	627–652	1160–1205	B
5052	—	—	—	2.5	—	—	0.25	593–649	1100–1200	C
5154	—	—	—	3.5	—	—	0.25	593–643	1100–1190	D
5456	—	—	0.1	5.2	—	—	0.1	593–652	1100–1205	D
6061	0.25	0.6	—	1.0	—	—	0.25	616–652	1140–1205	B
6063	—	0.4	—	0.7	—	—	—	616–652	1140–1205	A
6151	—	1.0	—	0.6	—	—	0.25	588–649	1090–1200	C
6951	0.25	0.35	—	0.6	—	—	—	616–654	1140–1210	A
7005	—	—	0.45	1.4	4.5	—	0.13	607–646	1125–1195	B
7072	—	—	—	—	1.0	—	—	646–657	1195–1215	A

*A – Alloys readily brazed by all techniques
B – Alloys that can be brazed by all techniques with a little extra care
C – Alloys that require special care and effort
D – Alloys difficult to braze

Table 7.3 Composition and melting range of BFM used in brazing of aluminium alloys. (Nippes [5]).

Alum. assoc. designation	AWS classification	Coposition, weight %							Approximate temperature					
									Solidud		Liquidus		Brazing range	
		Si	*Fe*	*Cu*	*Mn*	*Mg*	*Zi*	*Others*	°F	°C	°F	°C	°F	°C
4343	BAlSi-2	6.8–8.2	0.8	0.25	0.10	—	0.20	—	1070	577	1135	613	1100–1150	593–613
4044	—	7.8–9.2	0.8	0.25	0.10	—	0.20	—	1070	577	1115	602	1100–1135	593–613
4045	BAlSi-5	9.0–11.0	0.8	0.30	0.05	0.05	0.10	0.20Ti	1070	577	1095	591	1090–1120	588–604
4145	BAlSi-3	9.3–10.7	0.8	3.3–4.7	0.15	0.15	0.20	0.15Cr	970	521	1085	585	1060–1120	571–604
4245	—	9.3–10.7	0.8	3.3–4.7	0.07	0.07	9.3–10.7	0.07Cr	960	516	1040	560	1020–1060	549–571
4047	BAlSi-4	11.0–13	0.8	0.30	0.15	0.10	0.20	—	1070	577	1080	582	1080–1120	582–604
4047	BAlSi-7	9.0–10.5	0.8	0.25	0.10	1.0–2.0	0.20	—	1030	554	1105	596	1090–1120	588–604
4004	BAlSi-7	9.0–10.5	0.8	0.25	0.10	1.0–2.0	0.20	—	1030	554	1105	596	1090–1120	588–604
X4104	—	9.0–10.5	0.8	0.25	0.10	1.0–2.0	0.20	0.20–0.20Bi	1030	554	1105	596	1090–1120	588–604
—	BAlSi-B	11.0–13	0.8	0.25	0.10	1.0–2.0	0.20	—	1038	559	1075	579	1080–1120	582–604

Note: a) Composition in per cent maximum unless shown as a range or a minimum and other elements, each 0.055 max.; other elements, total 0.15% max.; remainder, aluminium.

b) Solidus and liquidus temperature are for normal compositions.

7.3.4.2 Brazing of Stainless Steels

All stainless steels have high chromium contents for high heat and corrosion resistance, and it is this high chromium content that makes it difficult to wet stainless steels for brazing. Brazing of stainless steel is best accomplished in a furnace having either a dry hydrogen or dissociated ammonia atmosphere or high vacuum with very low dew point (−50°C) to prevent formation of chromium oxide or refractory oxide, which make wetting difficult.

Four basic groups of stainless steels are: (1) austenitic non-hardenable (2) ferritic non-hardenable (3) martensitic hardenable (4) precipitation hardenable.

- Austenitic steels contain sufficient nickel or nickel and manganese to stabilize the face-centred cubic austenite phase and are designated as AISI 300 and 200 series alloys. The 300 series stainless steels are widely used in plate-fin exchangers with vacuum brazing, because of the high thermal expansion coefficient with vacuum brazing. Also, because of the high thermal expansion coefficient and low thermal conductivity of 300 series stainless steels, care must be exercised to minimize thermal distortion for large exchangers.
- Ferritic non-hardenable stainless steels have low carbon and high chromium content, such as AISI 405, 406, 430, 446. AISI 430 is widely used for plate-fin exchangers as well as rotary regenerators. It should be emphasized that AISI 430 is subject to intergranular corrosion with some silver-based filler metal, which can be avoided by adding a small percentage of nickel to the silver-based filler metal.
- Martensitic hardenable stainless steel have either medium chromium and low carbon (AISI 403, 410 and 416) or high chromium and high carbon (AISI 440A, B and C). This steel is hardenable by heat treatment. A rapid cooling is essential to transform austenites (stable above 960°C or 1760°F) after brazing, and the cooling should be uniform to minimize stresses within the brazements.
- Precipitation hardening stainless steels are chromium-nickel pseudo-austenitic steels having additional alloying elements (such as Al, Ti, Cu, Mo) that result in high strength as well as high heat and corrosion resistance. These steels can be hardened by special heat treatment (due to the addition of alloying elements) and hence the brazing cycle must be compatible with the heat treatment. Nickel plating is done on these steels containing aluminium or titanium to improve wetting in the furnace brazing.

Base metal inclusions (such as oxides, sulphides and nitrides) and surface contaminants (such as oil, grease, graphite, lead, aluminium oxide) are more harmful when brazing stainless steels compared to carbon steels. For fixturing during brazing of stainless steels, materials that can carburize or form nitrides, should not be used since they may reduce the corrosion resistance of the brazements or braze joint quality [5].

The most common filler metals used for brazing stainless steel plate-fin exchangers are generally nickel-based, followed by copper-based when applicable. In order to obtain the maximum joint strength, the brazing temperature, time and joint clearance should be optimized as discussed and outlined by Nippes [5]. Brazing of stainless steel plate-fin exchangers is usually done without flux and in a vacuum furnace or in a controlled-atmosphere furnace. In the latter furnace, either dry hydrogen or dissociated ammonia is used. Dry hydrogen is very effective in reducing chromium oxide, and it protects the base metal surface and furnishes excellent wetting and flow of filler metal without the necessity for flux. However, hydrogen is difficult to dry sufficiently, needs special furnace equipment and is dangerous in storage and handling. Compact plate-fin heat exchangers as large as 1.2 × 1.5 × 3m have been successfully brazed in vacuum and gas furnaces. Vacuum furnaces have either hot walls or cold walls. For externally heated hot wall furnaces, the brazing cycle time is long. The maximum brazing temperature is limited to 1200°C. Since copper and copper alloys have poor oxidation resistance above service temperatures of about 425°C, Nippes [5] evaluated three nickel-based filler metals, and the regenerator is brazed in a vacuum furnace at vacuums lower than 10^{-4} torr. The best filler metal found is BNi–7 at the brazing temperature 1030°C. It resulted in acceptable grain growth in the base metal, acceptable inter-alloying zones and optimal flowability properties.

7.3.4.3 Brazing of Super Alloys

Super alloys have nickel, cobalt or both as base elements. The alloying elements may be chromium, iron, tungsten, molybdenum, etc. These alloys have high strength, corrosion resistance and oxidation resistance at service temperatures 540–1200°C. Hence, compact heat exchangers made from the superalloys are used for high temperature waste heat recovery. Some of the nickel-based superalloys used in plate-fin exchangers are Hastelloy X, Inconel 625 and Inconel 718, and a cobalt-based alloy is Haynes 188. Super alloys are either predominantly precipitation strengthened (such as Inconel 718) or solid-solution strengthened (such as Hastelloy X and Inconel 625). Some superalloys are strengthened by both precipitation and solid solution. The chemistry of those latter alloys must be held tight so that precipitation hardening reactions do not go out of control. Precipitation strengthened alloys owe their high strength to inter-metallic compounds and/or carbides (which are strong and very stable) in precipitated form, which can be observed under a microscope, and these alloys are very strong. Solid-solution alloys are strengthened by alloying elements which remain in total solution form and hence none of them can be seen under a microscope. These alloys have the best corrosion resistance.

The high-temperature strength of some nickel alloys can be further improved by addition of titanium and aluminium (such as in Inconel 718). The oxide of these elements is almost impossible to remove or reduce in a controlled atmosphere (vacuum or hydrogen). These oxides are usually present in precipitation hardenable nickel-based alloys as inclusions, and brazing over these oxides must be avoided. Hence, after the surface is cleaned, a surface pretreatment is applied. It is an electroplate of nickel, referred to as nickel flashing, the required thickness of which increases with increasing content of titanium and aluminium in the alloy. Stress corrosion cracking of precipitation hardenable alloys is highly likely. Hence, careful stress relief during the brazing cycle, or annealing prior to brazing is recommended by Nippes [5]. A very small percentage of boron is usually added to produce grain boundary plugging and to limit the growth of boundary carbides. Nickel and nickel alloys are prone to sulphur and low-melting metal eutectic embrittlement, and hence these elements must be excluded by proper cleaning before brazing. For nickel alloys, human perspiration must be completely avoided, otherwise parts won't braze there. Cleanliness of the base metal and filler metal is essential for nickel-based alloys. Hence, suitable chemical/mechanical cleaning is first performed. If the alloy contains titanium and aluminium (together over about 1%), nickel flashing is necessary. Since the nickel-based filler metals are self-fluxing, they are more forgiving to slight imperfections in cleanliness.

The filler metals used for superalloys are nickel-based for nickel-based base metals and cobalt-based for cobalt-based base metals. Boron (2–3.5%) and/or silicon (3–10%) are used to lower melting point temperature. Phosphorous (0.02–10%) is also used as a melting point temperature depressant for nickel alloys. Chromium, as high as 20%, is added to provide high temperature corrosion and oxidation resistances. Other filler metals for nickel-based superalloys are gold-based for high service temperature applications and copper- or silver-based for low temperature applications. Proprietary filler metals are used for service temperatures above 990°C. Filler metals used for heat exchanger applications are in foil, tape, paste and powder form. If the specific filler metal is available only in powder form, then cements and pastes are used to properly position the filler metal.

Brazing of superalloy plate-fin exchangers is carried out in a vacuum furnace without flux, and may also be carried out in a dry hydrogen atmosphere. Nickel plating should be performed if titanium and aluminium are present. Salt-bath dip brazing is not used with brazing of superalloys due to grain boundary dissolution. The cobalt-based alloy Haynes 188 does not contain titanium and aluminium, and it does not require a very stringent atmosphere as brazing nickel-based alloys do. However, nickel flashing is often performed for better wetting of the filler metal. Filler metal is either a nickel-based or a copper-based alloy. Stress corrosion cracking and liquid-metal embrittlement are again problems during brazing of cobalt-based alloys.

7.3.5 Vacuum Furnace Brazing Cycles

A brazing cycle consists of initial pump-down, initial heating ramp, cement burn-off, stabilizing soak, heating ramp to brazing temperature, brazing soak and cool-down. For easy-to-braze materials, evacuating vacuum to 8×10^{-4} torr is sufficient, but the furnace should be pumped down to below 5×10^{-4} torr, before commencing heating, as Ni-base superalloys containing appreciable amounts of Al and Ti. The initial heating ramp should be 10–16°C per minute. Faster rates are not recommended owing to possible part distortion, spalling of the applied brazing slurry and the likely occurrence of excessive out-gassing with large loads containing an appreciable amount of brazing slurry. For critical materials (e.g. Ni-base superalloys) particularly with heavily applicated parts (e.g. surface braze build-up), a cement burn-off soak is strongly recommended to avoid too high a pressure rise. The soak temperature should be about 28°C below the braze slurry solidus temperature for 15–30 minutes, or until the pressure drops below the desired level, whichever is longer. The soak allows the temperature throughout the load to equalize, so all parts in the load will reach brazing temperature at approximately the same time during the next heating cycle and it ensures that vacuum pressure levels are low enough before proceeding (ramping) to brazing temperature. The final heating rate to brazing temperature is critical. It must be fast enough to avoid excessive liquefaction of the brazing alloy and subsequent alloying with, and erosion of, the base metal. For thin materials (e.g. ≤ 0.25 mm thick), heating rates of 28–40°C per minute are essential.

It is desirable to use the lowest brazing temperature within the recommended brazing range consistent with producing a satisfactory joint. Minimum brazing temperatures are essential in some applications such as when using pure copper brazing filler, filling large gaps with wide-gap nickel alloys, and when brazing very thin materials. For extremely thin metal, fillet type joint (e.g. honeycomb seals), a brazing temperature equal to or slightly below the liquidus temperature is used to avoid excessive flow and erosion.

In general, time at brazing temperature should be long enough to ensure that all sections of a work-piece and all parts within the load reach the desired brazing temperature. Brazing temperature and time are considered critical events. It is strongly recommended that guaranteed soak point be used in the program. For materials that do not require solution heat treatment or hardening, it is recommended that the load be vacuum cooled from brazing temperature to a temperature at least 30°C below the solidus temperature of the brazing slurry before initiating the gas quenching system. If the part requires heat treatment from brazing temperature, then the gas quench must be initiated at the end of the brazing soak period.

7.3.5.1 Vacuum Level during Brazing

The ability to wet and flow freely over a surface is the most important characteristic of any braze alloy. To secure good wetting, parts and vacuum furnace cleanliness and proper vacuum level are essential. Parts require protection when heated, which requires adequate time for thorough purging and evacuation before heating starts. To achieve the high vacuum necessary and ensure that it is maintained through the useful life, it is necessary to outgas/clean the vacuum hot zone periodically. Together with the necessity for low vacuum leak rate, vacuum tightness is very important to achieve quality brazing. A vacuum level of 5 microns or less should be properly controlled. A dew point of −55°C is required to braze stainless steel. Oxygen must also be controlled to suit the process and materials processed. Vacuum conditions in most cases reduce oxides, so braze fillers can wet and flow over the surfaces of the cleaned metal. Turbo molecular or cryo-pumps are used in special applications with more demanding brazing specifications at a 10^{-6} torr vacuum level (Figure 7.6).

A typical brazing cycle, as shown in Figure 7.6, consists an initial pump down to 8×10^{-3} torr is sufficient for relatively easy-to-braze materials, similar to Kowalewski and Szczurek [6]. The furnace should be pumped down to below to 8×10^{-3} torr before commencing heating for difficult-to-braze materials such as Ni-base superalloys containing appreciable amounts of Al and Ti, and a cement burn-off soak during heating cycle is strongly recommended to avoid too high a pressure rise.

7.3.5.2 Cooling Gases

Cooling gases used to shorten the brazing cycle and achieve proper metallurgical properties should be of high purity to prevent the formation of oxides. Inert cooling gases

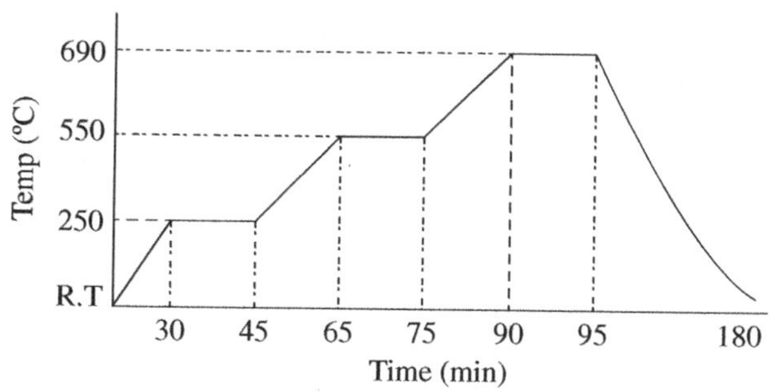

Figure 7.6 Typical brazing cycles.

include argon (99.995% purity with dew point lower than −51°C, inhibits evaporation and forms no compound) and nitrogen (99.998% purity with dew point lower than −51°C should be acceptable, but should be avoided if harmful formation of nitrides occurs).

7.3.5.3 Post Brazing Inspection

A quality control system should be adequate for both general and critical applications. Knowledgeable inspection of finished brazed assemblies includes adequate inspection techniques as applicable to a particular application including visual inspection, leak testing and radiographic examination. Visual inspection is the most widely used non-destructive testing method. Fluorescent penetrant inspection (FPI) is used for machined surfaces only. Parts should be inspected for cracks or voids in the brazed joint or adjacent base material, surface porosity and completeness of brazed joint band around the circumference. Leak testing is most advantageous where gas or liquid tightness of braze joints is required. This technique is limited to low pressure applications. Radiography or ultrasonic inspection should be used for braze joints Class A to detect subsurface and/or internal defects. Some NDT methods have to be used appropriately. For example, FPI is limited to inspecting machined surfaces only, but it is being used to detect defects in braze joint fillets. Similarly, many braze joints should not be radiographically inspected because this technique requires sensitivities that are difficult to achieve (better than 2%).

7.4 Influence of Brazing on Heat Transfer and Pressure Drop

Brazing may decrease or increase the heat transfer rate of an exchanger from the ideal value suggested by Shah and London [7]. Poorly brazed joints result in additional resistance to heat flow and its contribution increases with increasing Reynolds numbers. This may be explained for air (or gas) flow on one side of the exchanger. The added thermal resistance due to poor brazing remains constant regardless of the airflow rate. However, by increasing the airflow rate or its Reynolds number, the air side heat transfer coefficient increase or its thermal resistance decreases. Hence, the relative contribution of the thermal resistance due to poor brazing increases with increasing Reynolds numbers. When experimentally determining heat transfer coefficients, no account is taken of the thermal resistance due to brazing since we do not know exactly what its value is. Effectively, the influence of poor brazing is lumped into the air side heat transfer coefficient. Depending upon the severity of some of the brazing problems discussed earlier, the reduction in the j factor may be as high as 20–30%, or higher at high Reynolds numbers. Usually, it is a standard practice to braze small samples, cut them, take photomicrographs and assess the quality of the brazed joints. It is highly advisable to cut the full size core, after thoroughly testing for heat transfer and pressure drop, at various locations, to determine the quality of the brazing.

When the braze alloy re-solidifies, it can produce a fine-grained rough surface. This is due to a difference in height, i.e. roughening, between those areas of the work-piece surfaces to which the braze alloy spreads and those surfaces to which it does not spread. If tubes or plates are clad, and the fins are not clad, then the fins will have negligible brazing-induced surface roughness. If the fins are clad and tubes are unclad, then the fins will

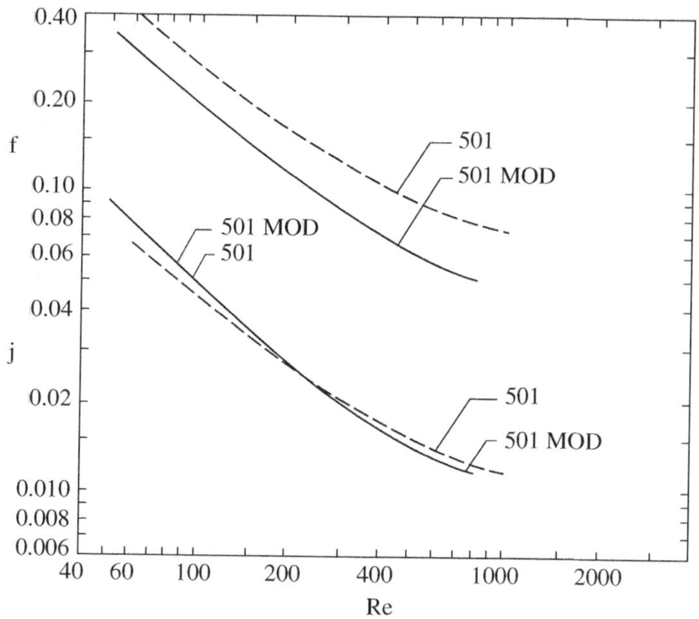

Figure 7.7 Influence of brazing on the *j* and *f* factors for compact offset fin surface. (Shah and London [7])

have a fine-grained surface roughness. If brazing induces surface roughness on the fins, it will have a much more pronounced effect on the pressure drop or friction factor than on the heat transfer coefficient or *j* factor. The relative importance of brazing-induced surface roughness is dependent upon the heat transfer surface area compactness or density.

Shah and London [7] reported *j* and *f* versus *Re* test data for one unbrazed core (referred to as core 501 MOD in that paper) and a dip brazed core (core 501 with plates clad and fins unclad). Both were offset strip in cores with heat transfer surface area density β of about 6000 m^2/m^3 (1850 ft^2/ft^3) and both cores were tested by the transient single-blow test technique. The test results in Figure 7.7 indicate no significant influence of brazing on heat transfer, but there is 35–40% increase in friction factors attributed to the brazing-induced surface roughness primarily on the plates. Generally even though there is no significant influence of brazing on heat transfer, there is 35–40% increase in friction factors attributed to the brazing induced surface roughness primarily on the plates as reported by Shah and London [7]. Some more studies on the influence of brazing on heat transfer and friction factors variations can be referred to in Shah and London [7].

7.5 Testing and Qualification of Compact Heat Exchangers

Heat exchangers are constructed of corrosion resistant materials suitable for the temperatures and pressures encountered. Heat exchangers that are exposed to the direct impact of rain are designed so that they will not be damaged by rainfall of 4.0 inches per hour, where these heat exchangers are designed for aircraft applications. All heat exchangers are designed to satisfactorily pass the pressure and temperature cycling tests wherever required. In applications where hot air is introduced into cooler air upstream from a heat exchanger, the hot air is evenly mixed with the cooler air prior to reaching the inlet of the heat exchanger. The vendor conducts sufficient tests to construct a performance map applicable to the complete range of possible operating conditions. The performance data shows the pressure drop and effectiveness of the heat exchanger for various flow rates and temperatures.

The qualification tests described in this section are intended to demonstrate that a compact heat exchanger is undergoing the tests as per simulated environmental conditions and they are meant to uncover deficiencies in its design and method of manufacturing. The unit undergoing the qualification test is subjected to different stresses under critical environmental conditions, which are expected to be encountered by it throughout the aircraft's service life. The test levels are not introduce any realistic modes of failure. All tests are conducted in accordance with the test procedures laid down in MIL standards or equivalent for any aircraft applications.

All instrumentation used to perform qualification tests are calibrated in accordance with MIL-STD-45662A [8]. All measurements are made using instruments whose accuracies conform to acceptable laboratory standards and that are appropriate for measurement at the environmental conditions involved. The accuracy of instruments and test equipment is verified periodically, in accordance with applicable MIL standard calibration system.

The following acceptance and qualification tests have to be conducted for any aircraft application:

7.5.1 Acceptance Tests

The following acceptance tests, which are required to be conducted on all deliverable, are performed on the heat exchanger in the sequence specified in Table 7.4.

The following are the procedures for conducting different qualification tests on a typical CHE unit.

7.5.1.1 Thermal Performance and Pressure Drop Test

The compact heat exchanger qualification test unit is subjected to a thermal performance and pressure drop test in accordance with MIL-A-83116A [9], Para 4.7.14.1.

There are mainly two types of experimental methods: steady state method and transient technique method for thermal performance and pressure drop measurements for compact heat exchangers for air media. These methods are well established, and procedures are available in various journals and books. However, a brief description of these methods is given here for completeness of testing of compact heat exchangers.

7.5.1.1.1 Steady State Method

The experimental rig comprises the air supply system, the steam supply system, the heat exchanger or the test core and instrumentation/measurement systems, as shown in Figure 7.8. Atmospheric air is drawn by a centrifugal blower through one channel of the heat exchanger and condensation of superheated steam through the other channel of the test unit. The steam can be assumed to be at a uniform temperature when it passes through the channel of the test unit and maintains a uniform temperature without

Table 7.4 Acceptance and qualification tests.

Sl. No.	Name of test	MIL specification
1	Visual examination	
2	Leakage test	
3	Pressure drop test a) fuel circuit b) air circuit	Acceptance tests
4	Proof pressure test	
5	Thermal performance	Para 4.7.14.1 of MIL-A-83116A [9]
6	Vibration	Method 514.3 (Proc-I) of MIL- STD-810D [10].
7	Pressure cycling (endurance)	Para 4.7.14.2 of MIL-A-83116A [9]
8	Thermal shock	Para 4.7.14.3 of MIL-A-83116A [9]
9	Acceleration	Method 513.3 (Proc-II) of MIL- STD-810D [10].
10	Shock	Method 516.3 of MIL- STD-810D [10].
11	Humidity	Method 507.2 of MIL- STD-810D [10].
12	Fungus	Method 508.3 of MIL- STD-810D [10].
13	Salt fog	Method 509.2 of MIL- STD-810D [10].
14	Burst pressure test	Para 4.7.12 of MIL-A-83116A [9].

affecting the air side performance. A part of the steam gets condensed while passing through the test section. The airflow rate, temperature and pressure differential of air across the unit are accurately measured for estimating the rate of heat transfer and pressure drop. Using this data, thermal performance in terms of heat load in case of full heat exchanger, or the j and f factors in case of small test core, are calculated by the usual procedure. The air velocity at the centre of the duct can be measured by using a Pitot static tube or standard orifice flow meter for airflow measurement. The test superheat can be allowed to pass through the known side of the heat exchanger unit. Water is injected into the steam in a chamber called the 'mixer' to de-superheat it. The mixing chamber is followed by a steam separator to remove excess water remaining in the steam. The required degree of superheat is controlled by adjusting the flow rate of water in the mixing chamber. Slightly superheated steam enters the test core through a header at the top of the unit and the blow steam and condensate leave the test unit through a similar header at the bottom. The instrumentation system, composed of velocity, temperature and pressure measurements, is linked with a data acquisition system. In the same test facility, the pressure drops of both streams can be measured using pressure sensors.

7.5.1.1.2 Transient Single-Blow Test Method

The various transient test techniques that are used for the determination of heat transfer characteristics of heat exchanger surfaces are discussed. The transient technique may have the fluid inlet temperature having several options such as a step change, exponential input and periodic or an arbitrary rise/drop of sinusoidal inlet temperature variation.

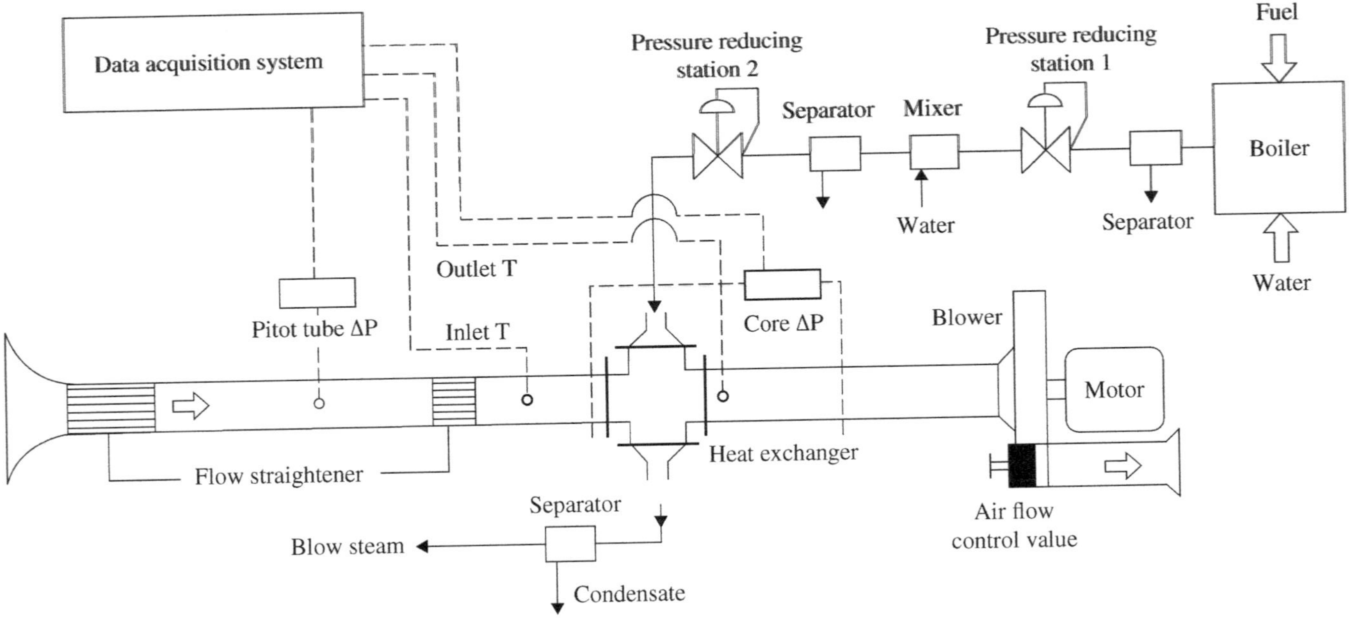

Figure 7.8 Schematic diagram of steady state experimental scheme.

The heat input to the test section will be evaluated by monitoring the fluid temperature at the inlet. The time response of fluid temperature at the outlet depends on the heat exchanger between the fluid and solid materials (fins and plates, in the case of plate-fin heat exchangers) in the test section. A sharp temperature increase can be observed at the inlet immediately after turning on the heater, but the outlet temperature will follow it relatively in mild ways, depending on the fin performance inserted in the test section. Slower response is a sink of larger heat exchange between working fluid and solid parts in the test section.

This test apparatus is an open circuit wind tunnel and it consists of a blower, a heating element, a test segment and devices for temperature, flow and pressure drop measurements. Air is induced to flow through the wind tunnel by a centrifugal air blower. The term "single-blow" came from the fact that the test technique uses only one fluid stream, and heat transfer occurs only between the fluid stream and the heat exchanger surface along which the fluid stream passes. A fluid flows steadily through the heat exchanger surface to be measured. Before the test, the fluid and the solid matrix should have the same uniform initial temperature. Then a variation in the fluid temperature is introduced at the inlet of the test section and fluid temperatures at the inlet and outlet of the test section are measured and recorded continuously. Refer to Figure 7.9 for a schematic diagram of test apparatus.

A sample test section is shown in Figure 7.9. This test sample consists of a few layers of fins separated by partition walls, as per the design of the heat exchanger. The fins and separating plates are stacked and are firmly fixed in the respective test sections. All the sides of the test section are insulated to eliminate the heat losses to the surroundings to the extent possible.

The required quantity of air is induced through the test section by a centrifugal air blower as shown in the schematic diagram of Figure 5.3. A uniform velocity distribution can be obtained in the wind tunnel by installing layers of stainless steel screens in the inlet cross-section. The inlet and outlet temperatures are measured with thermocouple grids uniformly arranged across the inlet and outlet cross-sections of the test section. The uncertainty in temperature measurement is to be maintained less than 0.2 K and the inlet fluid temperature fluctuation is less than 0.2 K. The mass flow rates are measured with an orifice meter, and pressure drops of each test core are recorded

Figure 7.9 Sample test section of compact heat exchanger core.

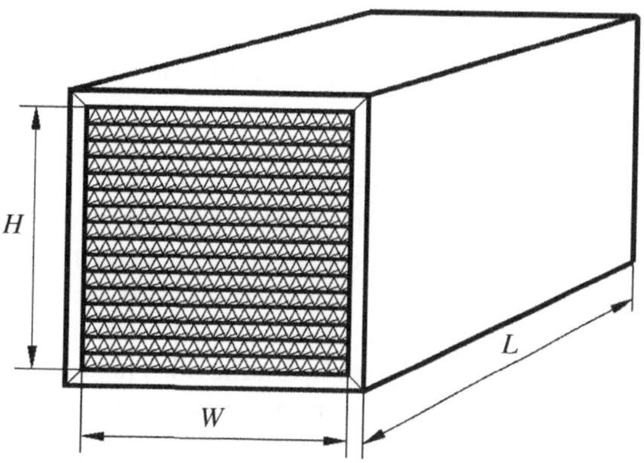

with sensors. An absolute pressure sensor and a differential pressure sensor with an accuracy of 0.2% are to be selected. The inlet temperature pulse (step change) of the airflow is realized with a computer controlled compact electric heater. Initially the heater is turned off. After the inlet and outlet temperatures of the fluid passing through the test core are recorded and initially kept steady, the predetermined power control voltage is applied from the electric power source. The inlet and outlet temperatures and the mass flow rates are measured continuously at a time interval of 0.1 s and saved for evaluation. A sample transient plot of inlet and outlet temperatures were presented in Figure 5.6 for lance and offset fins with 28 FPI and 3.175 mm lance length for reference. From that plot the Colburn *j* factor is estimated using max. slope method and was presented as *Re* vs *j* graphs in Figure 5.8.

7.5.1.2 Pressure Drop Test

The unit is tested individually for a pressure drop test within permissible limits. These tests are conducted at the ambient temperature using the respective fluid.

The inlet and outlet ferrules of the exchanger are connected to the high pressure air line at the pressure drop test rig with the help of suitable interfaces. The air supply to the exchanger is then turned on by opening the respective shut-off valve at the required pressure. The flow, line pressure and the pressure drop across the exchanger are measured and recorded. The pressure drop across any circuit of the heat exchanger at the above flow must pass the requirement. The pressure drop across any circuit of the unit, when tested, must not exceed the design value.

7.5.1.3 Leakage Test

The unit is tested for the absence of any leaks. This test is conducted using air or nitrogen gas at room temperature. The leak test pressure is determined by applying a temperature correction factor to the maximum operating pressure so as to duplicate the corresponding high temperature stress at room (test) temperature.

The unit is filled with working fluid. The inlet port of the unit is connected to the working fluid pump discharge at the test rig through a suitable isolation valve. A pressure gauge of adequate range is connected to the line between the isolation valve and inlet port. After ensuring that no air bubbles remain in the heat exchanger, the outlet port of the exchanger is capped using a suitable blank connector. The working fluid pump in the line is then started and a required pressure is built up in the circuit, by checking the reading on the pressure gauge. The isolation valve is closed as soon as this pressure has built up and the pressure in the circuit is observed for a period of 2 minutes. The unit, when tested as above, must not show any evidence of leakage or sweating at any time during the test.

7.5.1.4 Proof Pressure Test

The unit is tested for withstanding the proof test pressure (1.5 times max operating pressure). The proof test pressure is determined by applying a temperature correction factor to the specified proof test pressure so as to duplicate the corresponding high temperature stress at the ambient (test) temperature.

The unit is filled with hydraulic oil. The inlet port of the unit is connected to a hydraulic power pack inlet hose. A pressure gauge of adequate range is connected to the line between the hydraulic source and the inlet port. After ensuring that no air bubbles remain in either the heat exchanger or the hydraulic line, the outlet port of the exchanger is capped using a suitable blank connector. Hydraulic pressure is then raised gradually to the required level by checking the reading on the pressure gauge. The isolation valve is closed as soon this pressure has built up, and pressure in the circuit is observed for a period of 2 minutes. The unit, when tested as above, must not show any evidence of leakage or sweating, or permanent deformation at any time during the tests.

7.5.2 Qualification Tests

The qualification tests are conducted on one or two representative units which are to be qualified as per the flight envelope of a particular aircraft for aerospace applications. The qualification tests listed in Table 7.4 (Sl. 5–14) are conducted on a compact heat exchanger, which is designed for a typical fighter aircraft, as per the applicable MIL standard. These tests are environmental and endurance tests, and should not be conducted on each deliverable as these tests will consume some of the life of the units.

7.5.2.1 Vibration Test

The unit is subjected to a random vibration test in accordance with MIL-STD-810D, Method 514.3, Procedure-I. The general model vibration control system computer or similar system may be used to control the vibration levels and measure the heat exchanger response procedure for the test.

Prior to the vibration test, the resonant frequencies of the exchanger are identified by conducting a resonance search test, for each of the three axes. Frequencies from 5 Hz to 2000 Hz are scanned at the rate of one octave per minute according to the scheme given in Table 7.5. All resonant frequencies resulting in an amplification factor of 1.5 or above are identified and recorded.

The heat exchanger is properly mounted, utilizing all the mounting points, onto a suitable vibration test fixture, which is then mounted on the vibration test setup. The test is repeated for each of the three principal axes (Figure 7.10) by orienting the exchanger along each axis in turn. One control accelerometer is placed near the mounting location and at least one response accelerometer is placed on the heat exchanger, parallel to the test axis. A broadband random vibration test is conducted according to the random vibration envelope in Figure 7.11, for a duration of 1 hour for each axis. In case of externally mounted equipment, the vibration test is conducted as per Figure 7.12.

At the conclusion of the vibration testing, the heat exchanger is subjected to a leakage test. The unit, when tested as specified above, must not show any evidence of permanent deformation, mechanical failure or damage at any time during or after the vibration tests.

Table 7.5 Resonance search spectrum.

Frequency range	Input level	Input type
5–40 Hz	**0.312** mm	Constant displacement
40–2000 Hz	**2.0** g	Constant acceleration

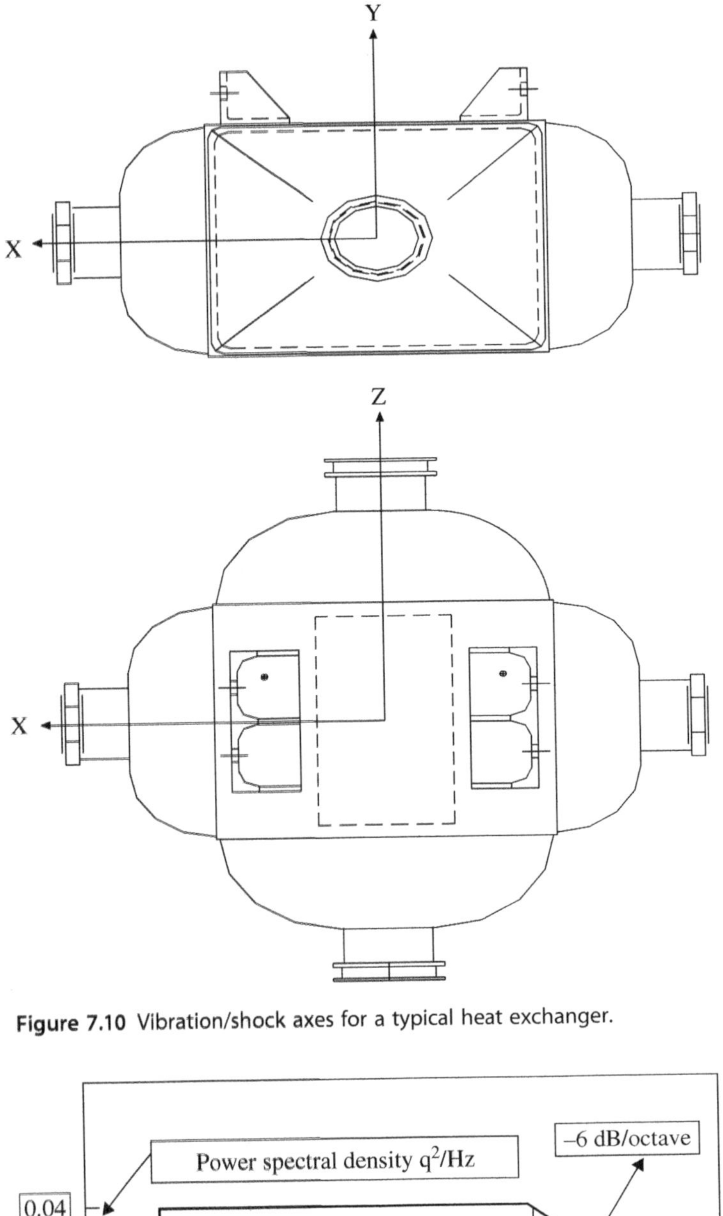

Figure 7.10 Vibration/shock axes for a typical heat exchanger.

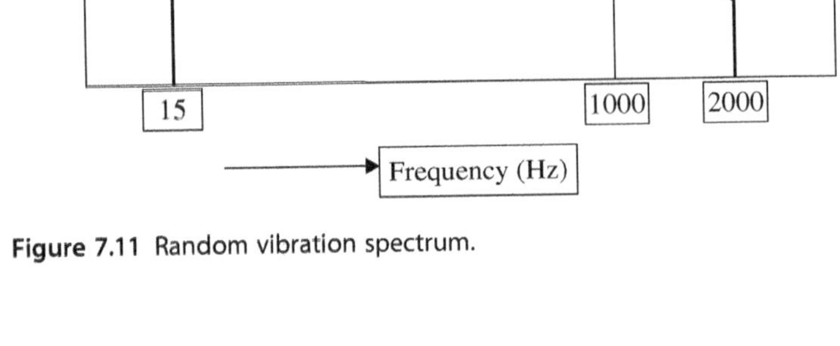

Figure 7.11 Random vibration spectrum.

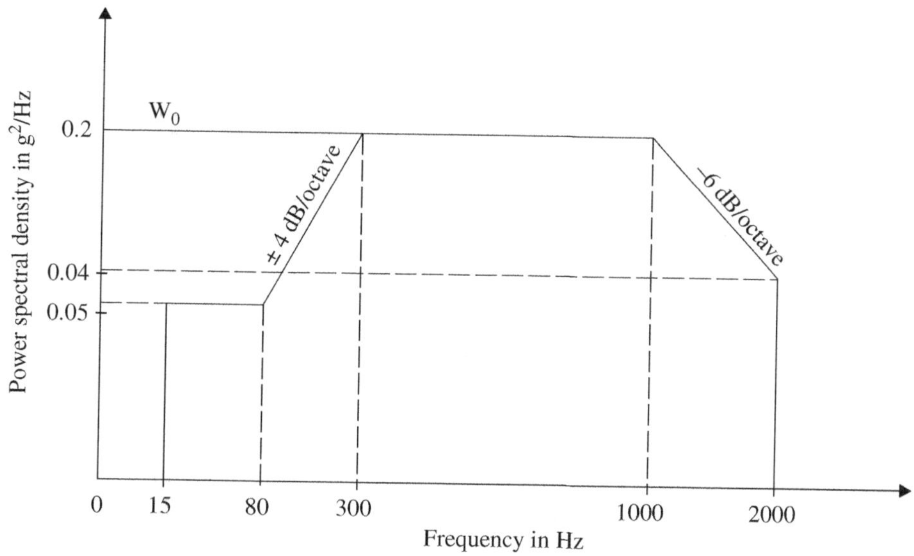

Figure 7.12 Random vibration spectrum (externally mounted equipments).

7.5.2.2 Combined Pressure, Temperature and Flow Cycling

Each heat exchanger of the subsystem is subjected to 5000 pressure-temperature-flow cycles. One cycle consists of: (1) simultaneously introduce hot air and coolant to the heat exchanger at condition I in subparagraph (a) below, (2) increase hot air and coolant flow rate, temperature and pressure at a rate equivalent to operational conditions at the heat exchanger inlet until condition II in subparagraph (b) below is met and outlet temperatures are stabilized, (3) reduce, at a rate equivalent to operational conditions at the heat exchanger inlet, the hot airflow rate, temperature and pressure and the coolant flow rate, temperature and pressure to condition I in subparagraph (a) and maintain conditions until outlet temperatures are stabilized. Condition I and II are follows:

a) Condition I
   1) Hot air:
      Flow rate – engine idle (as defined in the detail specification)
      Temperature – engine idle
      Pressure – engine idle
   2) Coolant:
      Flow rate – engine idle (as defined in the detail specification)
      Temperature – engine idle
      Pressure – engine idle
b) Condition II:
   1) Hot air
      Flow rate – maximum that occurs during maximum hot airflow
      Temperature – value which is associated with maximum flow
      Pressure – value which is associated with maximum flow

2) Coolant:
   Flow rate – quantity that occurs during maximum cold airflow
   Temperature – value that occurs during maximum cold airflow
   Pressure – value that occurs during maximum cold airflow

Following completion of the above test, the high pressure side of the heat exchanger, while exposed to maximum operating temperature, is subjected to 10,000 pressure cycles. One cycle consists of rapidly raising pressure from 0 psig to maximum operating pressure and then rapidly decreasing pressure to 0 psig. Prior to and following completion of the above tests, the heat exchanger is leakage tested.

7.5.2.3 Experimental Evaluation of Endurance Life of Compact Heat Exchanger

To experimentally evaluate the endurance life, the CHE should be subjected to a minimum–maximum–minimum cycle, as given in Figure 7.13, which is derived from a typical aircraft flight envelope as per MIL specification MIL-A-83116A [9]. Figure 7.17 illustrates a typical cycle, wherein the inlet streams are kept in the minimum condition for 23 seconds, raised from minimum to maximum condition in 2 seconds, kept in maximum condition for 23 seconds and brought back to minimum condition in 2 seconds. In the CHE cyclic test facility, the inlet conditions of pressure, temperature and mass flow rate of both the inlet streams (hot and cold) are dynamically varied from minimum to maximum condition and back to minimum conditions using compressors, process equipments such as heaters, coolers, chilling plants, control valves and fast response quick actuating valves. The schematic layout of the facility is given in Figure 7.14.

During the operation of the facility, the compressed air initially passes through the bypass duct until the pressure and the flow are stabilized, so that the required condition to be simulated at the entry to the heat exchanger is achieved. After achieving the required flow, the electric heaters in the appropriate lines are switched on to achieve the required temperature, which would approximately require 1½ hours. Once the temperature and pressure are achieved in all the streams, the cycling is started, using the plant control system. The required number of cycles is logged. Figures 7.15 and 7.16 shows typical input and corresponding output pressures and temperatures recorded during the combined cycling on a typical CHE. The variation of inlet and outlet pressures and temperatures are plotted for both hot and coolant streams. The pressure, temperature and mass flow are kept at minimum condition for 23 seconds and raised from

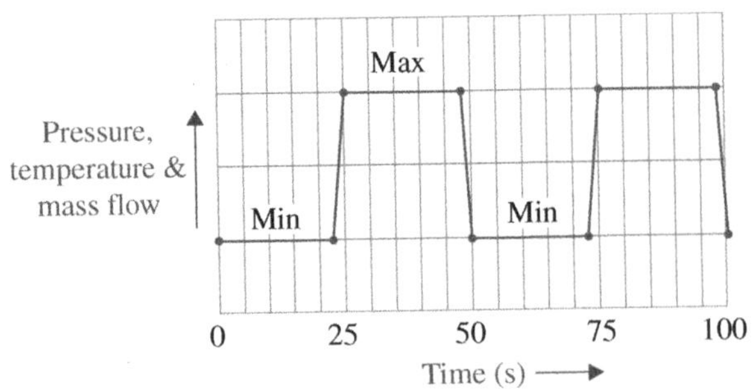

Figure 7.13 Min-max conditions cycles for endurance life.

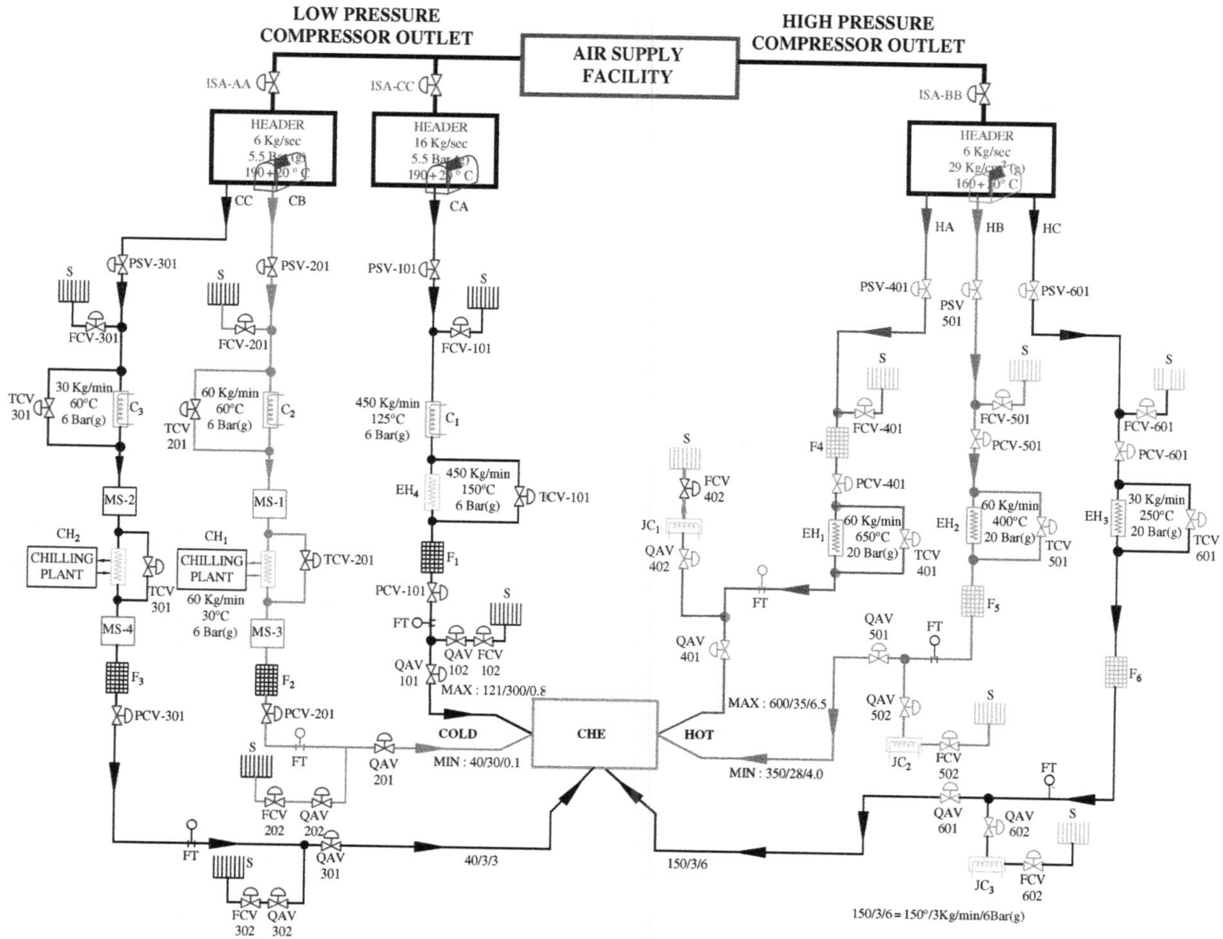

Figure 7.14 Schematic layout of CHE cyclic test facility.

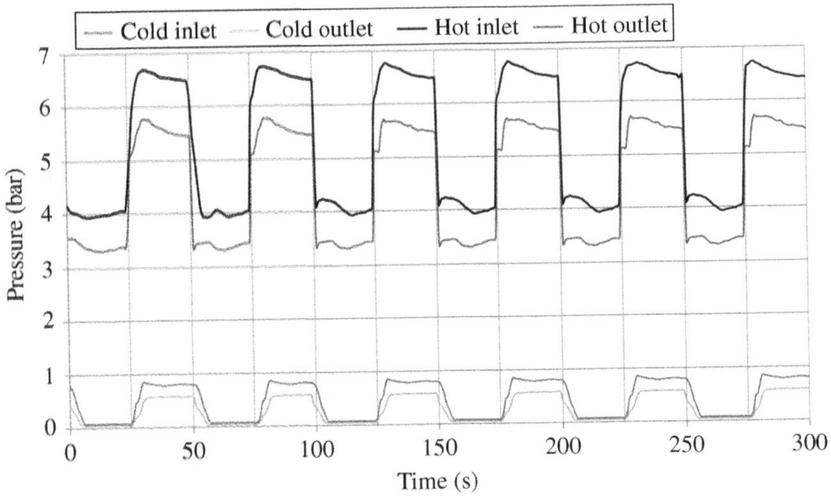

Figure 7.15 Pressure cycles (inlet and outlet streams).

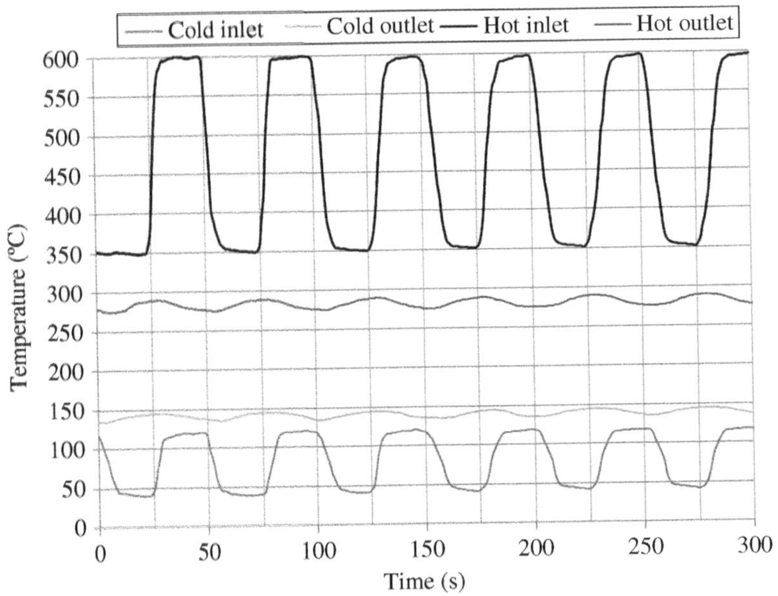

Figure 7.16 Temperature cycles (inlet and outlet streams).

minimum to maximum condition in 2 seconds, kept in maximum condition for 23 seconds, then brought back to minimum condition in 2 seconds. The above-mentioned variation constitutes one cycle for a specific CHE.

7.5.2.4 Pressure Cycling Test

The unit is subjected to a pressure cycling test in accordance with MIL-A-83116A, para 4.7.14.2 [9]. The heat exchanger is subjected to 50,000 pressure cycles by cycling between

Table 7.6 Schedule of one pressure cycle.

Step no.	Operation	Duration
1	Raising pressure from ambient to required pressure	1.0 second
2	Holding pressure at required pressure	1.0 second
3	Reducing pressure from required pressure to ambient	1.0 second
4	Holding pressure at ambient	1.0 second

ambient and specified maximum operating pressure for the circuit. The unit inlet port is connected to the source of pressurized hydraulic oil in the pressure cycling test setup, through a suitable isolation valve. A pressure gauge of adequate range is connected between the isolation valve and exchanger inlet port. Hydraulic oil is circulated through the exchanger and stabilized at required pressure. Pressure in the circuit is then cycled between ambient pressure and required pressure for 50,000 cycles. Each cycle consists of the four steps described in Table 7.6. After completion of every 10,000 cycles, the heat exchanger is subjected to a leakage test in accordance with Section 7.5.1.3.

7.5.2.5 Thermal Shock Test

The unit is subjected to a thermal shock test in accordance with MIL-A-83116A, para 4.7.14.3 [9]. The thermal shock test is performed on the unit by temperature cycling the entire heat exchanger from cold soak at −54°C to the expected maximum inlet fluid temperature. Three cycles are performed on the exchanger. Each cycle consists of (a) soaking at −54°C for one hour, (b) transferring the exchanger from the low temperature chamber to the high temperature chamber maintain maximum operating condition, (c) soaking at high temperature chamber for one hour, (d) transferring exchanger from high temperature back to low temperature chamber. After completion of all cycles, the heat exchanger is subjected to a leakage test as per Section 7.5.1.3.

7.5.2.6 Acceleration Test

The unit is subjected to acceleration test in accordance with MIL-STD-810D, Method 513.3, Procedure-II [10]. The unit is mounted on the acceleration test setup oriented along the test axis on the centrifuge (Figure 7.10) for the required acceleration (g) levels for the orientation being tested. The system is stabilized and the test continued for a period of 1 minute. The heat exchanger is then mounted along the next test axis and the same procedure repeated for each of the other axis orientations. All test data is recorded. After acceleration testing, the heat exchanger is subjected to a leakage test. The unit, when tested as specified above, should not show any evidence of mechanical failure or damage at any time during or after the acceleration test.

7.5.2.7 Shock Test

The unit is subjected to shock test in accordance with MIL-STD-810D, Method 516.3 [10]. The heat exchanger is mounted onto a suitable fixture, which is then mounted on the shock test setup. Three positive and three negative shocks are applied along each of three principal axes (Figure 7.10) by orienting the exchanger along each axis in turn. Shocks are of half-sine wave form with a magnitude of 15 g's (suitably changed depending on the aircraft envelope) over a duration of 11 ± 1 ms. After the shock test, the heat exchanger is subjected to a leakage test. The unit, when tested as specified above, should not show any evidence of mechanical failure or damage following exposure to shock.

7.5.2.8 Humidity Test

The unit is subjected to a humidity test in accordance with MIL-STD-810D, Method 507.2 [10], non-hazardous category. This test is performed to determine resistance of the heat exchanger material to effects of warm and humid environments. The heat exchanger is placed inside the humidity chamber after closing all its inlet and outlet ports. The temperature and humidity inside the chamber are varied according to Figure 7.17 of MIL-STD-810D [10]. The duration of test cycles is as given in Table 7.7 below.

Table 7.7 Humidity Test Cycle durations.

Item	Cycle 1	Conditions
Non-Hazardous	10 cycles	Cycle time : 24 Hours Temp : 30 to 60°C and RH : 85 to 95 %

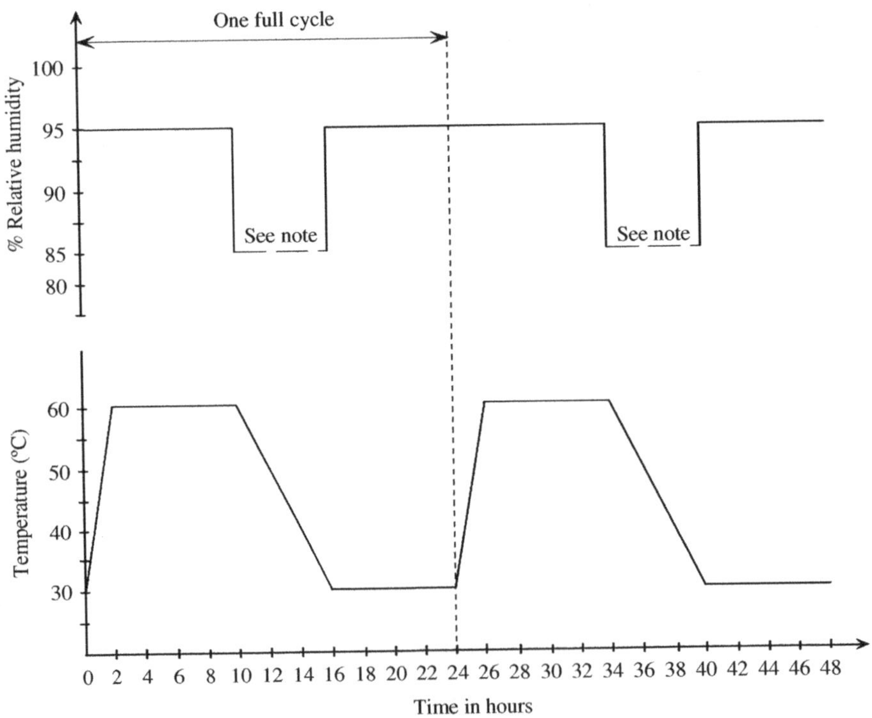

Figure 7.17 Diurnal cycle for temperature–humidity.

The unit, when tested as specified in 7.5.2.8, is not show any evidence of mechanical failure or damage following exposure to humidity.

7.5.2.9 Fungus Test

The unit is subjected to the fungus test in accordance with MIL-STD-810D, Method 508.3. This test is performed to assess the extent to which the exchanger may aid fungal growth or how fungal growth may affect exchanger performance. For performing this test, the procedure specified in MIL-STD-810D, methods 508.3-10 to 508.3-15 [10] are to be followed for preparing and checking the spore suspension and incubation of the exchanger in the test chamber. The minimum test duration of 28 days is used. The unit, when tested as specified above, must not show any evidence of undesirable fungal growth after the test.

7.5.2.10 Salt Fog Test

The negative effects of exposure to salt fog are corrosion due to electrochemical reactions or accelerated stress corrosion, clogging of moving parts, blistering of paints and impairment to protective coatings. The salt fog test is conducted to determine the resistance of the equipment to the effects of a salt atmosphere. In order to accelerate this test and thereby shorten the test time, the specified concentration of moisture and salt is specified greater than what is found in service. The unit is subjected to a salt fog test in accordance with MIL-STD-810D, Method 509.2 [10]. This test is performed to determine resistance of the heat exchanger material to effects such as corrosion and physical degradation caused by saline environments. The heat exchanger is placed inside the salt fog chamber after closing all its inlet and outlet ports. The exchanger is exposed to salt fog for a duration of 48 hours, followed by a drying period of 48 hours. The unit, when tested as above, must not show any evidence of mechanical failure, damage or corroded surfaces following salt fog exposure.

7.5.2.11 Freeze and Thaw

Heat exchangers that are in contact with water are tested to demonstrate their capability to withstand the freeze and thaw conditions expected during service life. The heat exchanger is filled with water and then exposed to an ambient temperature for a period of time which will ensure complete freezing of the water. Thawing is then accomplished by applying heat in a manner and at a rate simulating operational conditions. The number of freeze and thaw cycles is the number expected throughout the life of the air vehicle and is stated in the detail specification. However, in no case is the number be less than 50 cycles. If several possible ways of thawing the water can occur in the particular installation, the number of test cycles is divided in a realistic manner between each of the possible heating methods. Prior to and following completion of the freeze and thaw test, the heat exchanger and water storage tanks must satisfactorily meet the leakage requirements.

7.5.2.12 Rain Resistance

Ram air heat exchangers, which are so located that rain droplets can impact directly on the heat transfer surface, should be subjected to a rain resistance test. This test consists of exposing the heat exchanger for 1 hour to the quantity of rain that would be intercepted by the ram air scoop when the air vehicle is flying at maximum low level speed through rain which is falling at the rate of 10 cm per hour. The average droplet diameter for this test is 3000 microns. The functional capability (heat transfer) of the heat exchanger must not be impaired by this test. Heat exchangers which do not use finned heat transfer surfaces are not required to conduct this test.

7.5.2.13 Sand and Dust

The unit is subjected to the sand and dust test in accordance with MIL-810D Method 510.4 Procedure-I & Procedure-II [10]. Sand and dust in the atmosphere causes abrasion of surfaces, penetration of seals, erosion of surfaces, clogging of openings and filters, fouling of moving parts and degradation of electrical circuits. The sand and dust test is meant for evaluation of equipment to ascertain its ability to resist the ill effects of a dry fine sand laden atmosphere. This test is applicable to all mechanical, electrical, electronic and electromechanical devices for which exposure to the effects of a dry dust (fine sand) laden atmosphere is anticipated.

7.5.2.14 Shock Test (Arrestor Landing)

If the heat exchanger is required to be mounted on a naval variant aircraft, it is necessary to conduct a shock test (arrestor landing) on the any compact heat exchanger (whichever is applicable) to meet navy aircraft requirements as per MIL-STD-810D [10]. The heat exchanger is mounted onto a suitable fixture, which is then mounted on the shock test setup. Fifteen shocks are applied along each of the two principal axes (x and z) by orienting the exchanger along each axis in turn depending on the aircraft operational envelope. Shocks of 21g half sine 40 ms pulse (or) 30g half sine 30 ms pulse (or) 50g half-sine 18 ms pulse (or) 100g half sine 9 ms pulse and duration between two shocks is greater than 10 sec, as per a typical aircraft envelope (suitably changed depending on the aircraft envelope). The unit has to meet the requirements specified in (Table 7.8). The unit, when tested as above, should not show any evidence of mechanical failure or damage during or after the shock (arrestor landing).

7.5.2.15 Gunfire Vibration Test

If heat exchanger is required to be mounted on a military aircraft, where a gun is mounted, it is required to conduct a gun fire vibration test on the compact heat exchanger (as applicable), if the unit is located near to the gun mounting, to meet the gun fire vibration level requirements for a typical military aircraft, which are to be adjusted to suit similar requirements.

Table 7.8 Naval shock test specification.

Sl.No	Test	Specification for navy version
1	Shock (arrestor landing)	i) 21 g half sine 40 ms pulse Or 30 g half sine 30 ms pulse Or 50 g half sine 18 ms pulse Or 100 g half sine 9 ms pulse ii) 15 shocks in each direction of vertical and longitudinal axes iii) Duration between two shocks > 10 sec

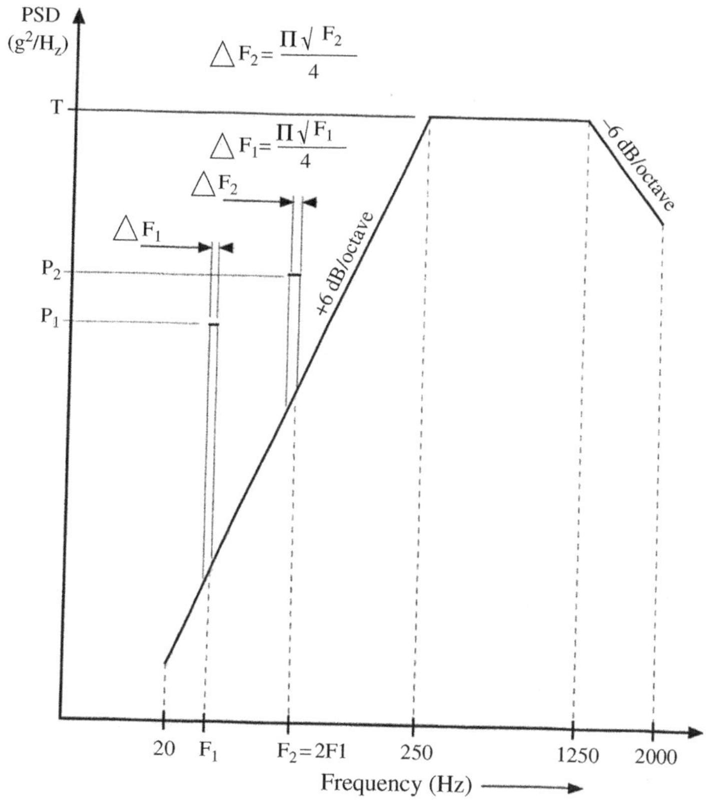

Figure 7.18 Gun fire induced vibration spectrum.

a) Frequency range 20–2000 Hz. Raise up to 250 Hz @6 dB/octave. PSD of 0.064 g^2/Hz within 250–1250 Hz and then falling off @ 6 dB/octave to 2000 Hz.
b) Superimposition: (i) 50 Hz frequency at PSD (P1) 0.009 g^2/Hz, (ii) 100 Hz frequency at PSD (P2) 0.01 g^2/Hz.

Total duration: 7 minutes per axis

The heat exchanger is mounted onto a suitable vibration fixture. As per requirement, all internal and external mounted line replacement units within specified radius from gun muzzle are to be tested for gun fire vibration test as per MIL-STD-810G (Figure 7.18) as follows:

Mount the unit to vibration table using fixture in non-operational mode. Subject the unit to gun fire vibration level in each of three orthogonal axes for specified duration.

7.5.2.16 Burst Pressure Test

The unit is tested at the corresponding burst test pressure. This test is conducted using suitable fluid at ambient temperature. The burst test pressure is determined by first multiplying the maximum operating pressure by 2.5 and then applying a temperature correction factor to this value so as to simulate the corresponding high temperature stress at the ambient (test) temperature.

The unit is filled with suitable fluid (hydraulic fluid). The inlet port of the unit is connected to a source of hydraulic pressure. A pressure gauge of adequate range is connected to the line between the hydraulic pump and the inlet port. After ensuring that no air bubbles remain in either the heat exchanger or the hydraulic line, the outlet port of the exchanger is capped using a suitable blank connector. Hydraulic pressure is then raised gradually to the required level by checking the reading on the pressure gauge. The isolation valve is closed as soon as this pressure is built up and pressure in the circuit is observed for a period of 2 minutes.

The unit, when tested as above, must not rupture or instantaneously burst during the test. However, plastic deformation and core matrix leakage are permitted during this test.

References

1 J.E. Hesselgreaves, *Compact Heat Exchangers, Selection, Design and Operation*, Pergamon, UK, 2001.

2 R.K. Shah, Brazing of Compact Heat Exchangers, in R.K. Shah, A.D. Kraus and D.E Metzger, Editors, *Compact Heat Exchangers – Festschrift* for Prof. A.L. London, Hemisphere Publishing Corp., Washington, DC, pp 491–529, 1990.

3 R.K. Shah, *The art of brazing, in thermo-mechanical aspects of manufacturing and materials processing, An advanced study institute book*, Edited by R.K. Shah, H.Md. Roshan, V.M.K Sastri, K.A. Padmanabhan, Hemisphere Publishing Corporation, New York, pp 193–213, 1992.

4 BHPV Report no. 41011201, dated 22–10-1996.

5 Nippes E.F, *Coordinator, Metals Handbook – Welding, Brazing and Soldering*, Ninth Ed., 6, American Society for Metals, Metals Park, OH, 1983.

6 Janusz Kowalewski and Janusz Szczurek, *Vacuum Brazing, Heat Treating Progress, Volume* **6**, *Issue, May/June 2006 (ASM International).*

7 Shah R.K and London A.L, Influence of brazing on very compact heat exchanger surfaces, ASME Paper no. 71-HT-29, 1971.

8 US MIL standard, MIL-STD-45662A.

9 US MIL standard, MIL-A-83116A.

10 US MIL standard, MIL-STD-810D.

Appendices

A.1 Derivation of Fourier Series Mathematical Equation

The heat exchanger core frontal face is a rectangular domain having edges $2a$, $2b$ in x-y plane as shown in Figure 4.10b and will be a square domain when $a = b$. The sides of the rectangle are along the straight lines $x = a$, $x = -a$, $y = b$ and $y = -b$.

The Navier-Stokes equations are,

$$\rho\left(\frac{\partial u}{\partial t} + u\frac{\partial u}{\partial x} + v\frac{\partial u}{\partial y} + w\frac{\partial u}{\partial z}\right) = -\frac{\partial p}{\partial x} + \mu\left(\frac{\partial^2 u}{\partial x^2} + \frac{\partial^2 u}{\partial y^2} + \frac{\partial^2 u}{\partial z^2}\right) \tag{A-1}$$

$$= -\frac{\partial p}{\partial x} + \mu\left(\nabla^2 u\right) \tag{A-2}$$

$$\rho\left(\frac{\partial v}{\partial t} + u\frac{\partial v}{\partial x} + v\frac{\partial v}{\partial y} + w\frac{\partial v}{\partial z}\right) = -\frac{\partial p}{\partial y} + \mu\left(\frac{\partial^2 v}{\partial x^2} + \frac{\partial^2 v}{\partial y^2} + \frac{\partial^2 v}{\partial z^2}\right) \tag{A-3}$$

$$= -\frac{\partial p}{\partial y} + \mu\left(\nabla^2 v\right) \tag{A-4}$$

$$\rho\left(\frac{\partial w}{\partial t} + u\frac{\partial w}{\partial x} + v\frac{\partial w}{\partial y} + w\frac{\partial w}{\partial z}\right) = -\frac{\partial p}{\partial z} + \mu\left(\frac{\partial^2 w}{\partial x^2} + \frac{\partial^2 w}{\partial y^2} + \frac{\partial^2 w}{\partial z^2}\right) \tag{A-5}$$

$$= -\frac{\partial p}{\partial z} + \mu\left(\nabla^2 w\right) \tag{A-6}$$

Consider that the cold fluid flows in the z-direction, which is parallel to the wall of the rectangular duct. There is no cross flow for the cold fluid. Hence

$$u \equiv 0,\ v \equiv 0,\ w = w(x,y) \tag{A-7}$$

$$\text{and } \frac{\partial w}{\partial z} = 0 \tag{A-8}$$

Inertial terms drop out from the full Navier–Stokes equations. There is a constant pressure gradient in the z-direction and $-\dfrac{\partial p}{\partial z} = \text{constant}$.

Actually the flow of the hot fluid is in the y-direction and is perpendicular to the flow of the cold fluid.

Compact Heat Exchangers – Analysis, Design and Optimization using FEM and CFD Approach, First Edition. C. Ranganayakulu and K.N. Seetharamu.
© 2018 John Wiley & Sons Ltd. Published 2018 by John Wiley & Sons Ltd.

The cold fluid flows under the influence of a constant pressure gradient in the z-direction and its equation is

$$\frac{\partial^2 W_C}{\partial x^2} + \frac{\partial^2 W_C}{\partial y^2} = \frac{1}{\mu}\frac{\partial P_C}{\partial z} = \text{a constant} \tag{A-9}$$

The same equation can be used for other fluid provided the x, y, z coordinates are changed accordingly. Similarly, the hot fluid flows under the influence of a constant pressure gradient in the x-direction and its equation is

$$\frac{\partial^2 W_h}{\partial y^2} + \frac{\partial^2 W_h}{\partial z^2} = \frac{1}{\mu}\frac{\partial P_h}{\partial x} = \text{a constant} \tag{A-10}$$

Only one equation can be shown by removing the suffix c or h in the above Equations A-1,2 to avoid repetitions of the equations. Then the equation for the fluid (cold or hot) at the exchanger inlet duct can be represented as

$$\frac{\partial^2 W}{\partial x^2} + \frac{\partial^2 W}{\partial y^2} = \frac{1}{\mu}\frac{\partial P}{\partial z} = D = \text{a constant} \tag{A-11}$$

On boundary,

$$\text{at } x = \pm a, \quad W(\pm a, y) = 0 \tag{A-12}$$

$$\text{at } y = \pm b, \quad W(x, \pm b) = 0 \tag{A-13}$$

The term $\frac{1}{\mu}\frac{\partial P}{\partial z}$ is a constant that can be regarded as known since the choice of the pressure gradient is in one's control here. As such there is only one unknown, viz., $W = W(x,y)$ controlled by Equations A-1–3 above. The solution based on Fourier series expansion [1,2] is as follows:

Then

$$D = \sum_{m,n=0}^{\infty} D_{m,n} \cos\frac{(2m+1)\pi x}{2a}\cos\frac{(2n+1)\pi y}{2b} \tag{A-14}$$

$$(-a < x < a, \; -b < y < b)$$

$$\int_{-a}^{a} \cos\frac{(2p+1)\pi x}{2a}\cos\frac{(2q+1)\pi x}{2a}\,dx = \begin{cases} 0 & if \;\; p \neq q \\ a & if \;\; p = q \end{cases} \tag{A-15}$$

$$\text{Equation A-14} \Rightarrow \int_{-a}^{a} D \cos\frac{(2p+1)\pi x}{2a}\,dx = \sum_{n=0}^{\infty} (D_{pn})(a)\cos\frac{(2n+1)\pi y}{2b} \tag{A-16}$$

$$LHS = D\frac{2a}{(2p+1)\pi}\left[\sin\frac{(2p+1)\pi x}{2a}\right]_{-a}^{+a} = D\frac{4a}{(2p+1)\pi}\sin\frac{(2p+1)\pi}{2} \tag{A-17}$$

$$\text{Equation A-16} \Rightarrow \quad \frac{4aD}{(2p+1)\pi} \sin\frac{(2p+1)\pi}{2} \int_{-b}^{b} \cos\frac{(2q+1)\pi y}{2b}\, dy = D_{pq}\, ab \tag{A-18}$$

$$LHS = \frac{4a\,D}{(2p+1)\pi} \sin\frac{(2p+1)\pi}{2} \; \frac{2b}{(2q+1)\pi} \; (2) \sin\frac{(2q+1)\pi}{2} \tag{A-19}$$

$$D_{pq} = \frac{16\,D}{(2p+1)\,(2q+1)\pi^2} \sin\frac{(2p+1)\pi}{2} \sin\frac{(2q+1)\pi}{2} \tag{A-20}$$

$$= \frac{16\,D}{(2p+1)\,(2q+1)\pi^2} (-1)^p\,(-1)^q = \frac{16\,D\,(-1)^{p+q}}{(2p+1)\,(2q+1)\prod^2} \tag{A-21}$$

In the Equation A-14, the Fourier coefficients are

$$D_{mn} = \frac{16\,D\,(-1)^{m+n}}{(2m+1)\,(2n+1)\,\pi^2} \tag{A-22}$$

$$(m,n = 0, 1, 2, \cdots\cdots\cdots)$$

$$\frac{\partial^2 W}{\partial x^2} + \frac{\partial^2 W}{\partial y^2} = \sum_{m,n=0}^{\infty} A_{m,n} \cos\frac{(2m+1)\pi x}{2a} \cos\frac{(2n+1)\pi y}{2b} \tag{A-23}$$

$$W = \sum_{p,q} E_{pq} \cos\frac{(2p+1)\pi x}{2a} \cos\frac{(2q+1)\pi y}{2b} \tag{A-24}$$

$$\frac{\partial^2 W}{\partial x^2} = \sum_{p,q} E_{pq} \left\{ -\left[\frac{(2p+1)\pi}{2a}\right]^2 \right\} \cos\frac{(2p+1)\pi x}{2a} \cos\frac{(2q+1)\pi y}{2b} \tag{A-25}$$

$$\frac{\partial^2 W}{\partial y^2} = \sum_{p,q} E_{pq} \left\{ -\left[\frac{(2q+1)\pi}{2b}\right]^2 \right\} \cos\frac{(2p+1)\pi x}{2a} \cos\frac{(2q+1)\pi y}{2b} \tag{A-26}$$

$$\nabla^2\, W = \frac{\partial^2 W}{\partial x^2} + \frac{\partial^2 W}{\partial y^2} \tag{A-27}$$

$$= -\sum_{p,q} E_{pq} \left[\frac{(2p+1)^2\pi^2}{4a^2} + \frac{(2q+1)^2\pi^2}{4b^2}\right] \cos\frac{(2p+1)\pi x}{2a} \cos\frac{(2q+1)\pi y}{2b} \tag{A-28}$$

$$= -E_{pq} \left\{ \frac{(2p+1)^2}{a^2} + \frac{(2q+1)^2}{b^2} \right\} \left(\frac{\pi^2}{4}\right) \equiv D_{pq} \tag{A-29}$$

Comparison of Equation A-11 along with Equation A-14 and Equation A-29 leads to

$$E_{mn} = -\frac{4}{\pi^2} \left[\frac{(2m+1)^2}{a^2} + \frac{(2n+1)^2}{b^2}\right]^{-1} D_{mn} \tag{A-30}$$

$$= \frac{-64D\,(-1)^{m+n}}{(2m+1)\,(2n+1)\pi^4} \; \frac{1}{\left[\frac{(2m+1)^2}{a^2} + \frac{(2n+1)^2}{b^2}\right]} \tag{A-31}$$

the solution to represent the fluid velocity (W) is

$$W(x,y) = \frac{64D}{\pi^4}\sum_{m,n=0}^{\infty}\left\{\frac{(-1)^{m+n+1}}{\left[\frac{(2m+1)^2}{a^2}+\frac{(2n+1)^2}{b^2}\right]}\frac{\cos\frac{(2m+1)\pi x}{2a}\cos\frac{(2n+1)\pi y}{2b}}{(2m+1)(2n+1)}\right\}$$

(A-32)

Verification:

The correctness of the solution A-32 can be verified by partial differentiation with respect to x twice and partial differentiation with respect to y twice and adding the two results.

Over the rectangle $-a<x<a$ and $-b<y<b$, we have the expansion

$$1 = \frac{16}{\pi^2}\sum_{m,n=0}^{\infty}\frac{(-1)^{m+n}}{(2m+1)\ (2n+1)}\cos\frac{(2m+1)\pi x}{2a}\cos\frac{(2n+1)\pi y}{2b} \tag{A-33}$$

The function 1, over the rectangle in Figure 4.10b, satisfies the usual Dirichlet's conditions for the Fourier expansion. This expansion is directly verifiable.

Double Fourier expansion of function 1 in terms of the class of functions $\cos\frac{(2m+1)\pi x}{2a}\ \cos\frac{(2n+1)\pi y}{2b}$ (which is a "complete" set of orthogonal functions) over the rectangular domain ($-a<x<a,\ -b<y<b$.) is certainly possible and the expansion in Equation A-33 is valid over the above rectangular domain.

The functions

$$\cos\frac{\prod x}{2a},\ \cos\frac{3\pi x}{2a},\ \cos\frac{5\pi x}{2a},\ \ldots\ldots\ldots\ldots,\left\{\cos\frac{(2m+1)\pi x}{2a}\ ;\ m=0,\ 1,\ 2\cdots\right\} \tag{A-34}$$

Constitute a complete set of functions over the interval $-a<x<a$.

Similarly, the functions

$$\cos\frac{\pi y}{2b},\ \cos\frac{3\pi y}{2b},\ \cos\frac{5\pi y}{2b},\ \ldots\ldots\ldots\ldots,\left\{\cos\frac{(2n+1)\pi y}{2b}\ ;\ n=0,\ 1,\ 2\cdots\right\} \tag{A-35}$$

forms a complete set over the interval $-b<x<b$

By a direct multiplication of the above two complete sets of functions

$$\cos\frac{(2n+1)\pi x}{2a}\cos\frac{(2m+1)\pi y}{2b}\ ;\ (m,n=0,\ 1,\ 2,\cdots)$$

constitute a complete set of functions.

$$\cos\frac{(2m+1)\pi x}{2a}\cos\frac{(2n+1)\pi y}{2b},\ (m,n=0,1,2,\ldots), \tag{A-36}$$

over the domain $(-a<x<a)\times(-b<y<b)$, i.e. over the rectangular domain referred to earlier.

$$D_{m,n} = \frac{16}{\pi^2}\sum_{m,n=0}^{\infty}\left\{\frac{(-1)^{m+n}}{(2m+1)\ (2n+1)}\right\}D \tag{A-37}$$

The above observations confirm that the function $W(x,y)$ defined in, for example, A-32 does satisfy the equation

$$\frac{\partial^2 W}{\partial x^2} + \frac{\partial^2 W}{\partial y^2} = D$$

Non-slip boundary conditions:
The non-slip boundary condition requires that

$$W = 0 \quad on \quad x = \pm a \text{ for all } y \text{ in the range} -b < y < b \tag{A-38}$$

$$\text{and } W = 0 \quad on \quad y = \pm b \text{ for all } x \text{ in the range} -a < x < a \tag{A-39}$$

and these can be directly checked by noting that

$$\cos(2m+1)\frac{\pi}{2} = 0, \text{ for all integer values of } m \text{ and} \tag{A-40}$$

$$\cos(2n+1)\frac{\pi}{2} = 0, \text{ for all integer values of } n. \tag{A-41}$$

A.2 Molar, Gas and Critical Properties

Properties of various ideal gases at 25 ° C, 100 kPa∗ (SI units)

Gas	Chemical formula	Molecular mass	R kJ/kg · K	ρ kg/m³	C_p* kJ/kg · K	C_p* kJ/kg · K	k
Steam	H_2O	18.015	0.4615	0.0231	1.872	1.41	1.327
Acetylene	C_2H_2	26.038	0.3193	1.05	1.699	1.38	1.231
Air	–	28.97	0.287	1.169	1.004	0.717	1.4
Ammonia	NH_3	17.031	0.4882	0.694	2.13	1.642	1.297
Argon	Ar	39.948	0.2081	1.613	0.52	0.312	1.667
Butane	C_4H_{10}	58.124	0.143	2.407	1.716	1.573	1.091
Carbon monoxide	CO	28.01	0.2968	1.13	1.041	0.744	1.399
Carbon dioxide	CO_2	44.01	0.1889	1.775	0.842	0.653	1.289
Ethane	C_2H_4	30.07	0.2765	1.222	1.766	1.49	1.186
Ethanol	C_2H_5OH	46.069	0.1805	1.883	1.427	1.246	1.145
Ethylene	C_2H_4	28.054	0.2964	1.138	1.548	1.252	1.237
Helium	He	4.003	2.0771	0.1615	5.193	3.116	1.667
Hydrogen	H_2	2.016	4.1243	0.0813	14.209	1.008	1.409

(*Continued*)

Gas	Chemical formula	Molecular mass	R kJ/kg · K	ρ kg/m^3	C_p^* kJ/kg · K	C_p^* kJ/kg · K	k
		Properties of various ideal gases at 25 ° C, 100 kPa∗ (SI units)					
Methane	CH_4	16.043	0.5183	0.648	2.254	1.736	1.299
Methanol	CH_2OH	32.042	0.2595	1.31	1.405	1.146	1.227
Neon	Ne	20.183	0.412	0.814	1.03	0.618	1.667
Nitric oxide	NO	30.006	0.2771	1.21	0.993	0.716	1.387
Nitrogen	N_2	28.013	0.2968	1.13	1.042	0.745	1.4
Nitrous oxide	N_2O	44.013	0.1889	1.775	0.879	0.69	1.274
n-octane	C_8H_{18}	114.23	0.07279	0.092	1.711	1.638	1.044
Oxygen	O_2	31.999	0.2598	1.292	0.922	0.662	1.393
Propane	C_3H_2	44.094	0.1886	1.808	1.679	1.49	1.126
R-12	CCl_2F_2	120.914	0.06876	4.98	0.616	0.547	1.126
R-22	$CHCLF_2$	86.469	0.09616	3.54	0.658	0.562	1.171
R-134a	CF_3CH_2F	102.03	0.08149	4.2	0.852	0.771	1.106
Sulfur dioxide	SO_2	64.059	0.1298	2.618	0.624	0.494	1.263
Sulfur trioxide	SO_3	80.053	0.10386	3.272	0.635	0.531	1.196

A.3 Thermo-Physical Properties of Gases at Atmospheric Pressure

T (K)	ρ (kg/m^3)	c_p (kJ/kg·K)	$\mu.10^7$ (N.s/m^2)	$v.10^6$ (m^2/s)	$k.10^3$ (W/m.K)	$\alpha.10^6$ (m^2/s)	Pr
Air							
100	3.5562	1.032	71.1	2	9.34	2.54	0.786
150	2.3364	1.012	103.4	4.426	13.8	5.84	0.758
200	1.7458	1.007	132.5	7.59	18.1	10.3	0.737
250	1.3947	1.006	159.6	11.44	22.3	15.9	0.72
300	1.1614	1.007	184.6	15.89	26.3	22.5	0.707

T (K)	ρ (kg/m^3)	c_p (kJ/kg·K)	$\mu.10^7$ (N.s/m^2)	$v.10^6$ (m^2/s)	$k.10^3$ (W/m.K)	$\alpha.10^6$ (m^2/s)	Pr
350	0.995	1.009	208.2	20.92	30	29.9	0.7
400	0.8711	1.014	230.1	26.41	33.8	38.3	0.69
450	0.774	1.021	250.7	32.39	37.3	47.2	0.686
500	0.6964	1.03	270.1	38.79	40.7	56.7	0.684
550	0.6329	1.04	288.4	45.57	43.9	66.7	0.683
600	0.5804	1.051	305.8	52.69	46.9	76.9	0.685
650	0.5356	1.063	322.5	60.21	49.7	87.3	0.69
700	0.4975	1.075	338.8	68.1	52.4	98	0.695
750	0.4643	1.087	354.6	76.37	54.9	109	0.702
800	0.4354	1.099	369.8	84.93	57.3	120	0.709
850	0.4097	1.11	384.3	93.8	59.6	131	0.716
900	0.3868	1.121	398.1	102.9	62	143	0.72
950	0.3666	1.131	411.3	112.2	64.3	155	0.723
1000	0.3482	1.141	424.4	121.9	66.7	168	0.726
1100	0.3166	1.159	449	141.8	71.5	195	0.728
1200	0.2902	1.175	473	162.9	76.3	224	0.728
1300	0.2679	1.189	496	185.1	82	238	0.719
1400	0.2488	1.207	530	213	91	303	0.703
1500	0.2322	1.23	557	240	100	350	0.685
1600	0.2177	1.248	584	268	106	390	0.688
1700	0.2049	1.267	611	298	113	435	0.685
1800	0.1935	1.286	637	329	120	482	0.683
1900	0.1833	1.307	663	362	128	534	0.677
2000	0.1741	1.337	689	396	137	589	0.672
2100	0.1658	1.372	715	431	147	646	0.667
2200	0.1582	1.417	740	468	160	714	0.655
2300	0.1513	1.478	766	506	175	783	0.647
2400	0.1448	1.558	792	547	196	869	0.63
2500	0.1389	1.665	818	589	222	960	0.613
3000	0.1135	2.726	955	841	486	1570	0.536

(*Continued*)

T (K)	ρ (kg/m^3)	c_p (kJ/kg·K)	$\mu.10^7$ (N.s/m^2)	$\nu.10^6$ (m^2/s)	$k.10^3$ (W/m.K)	$\alpha.10^6$ (m^2/s)	Pr
Ammonia (NH_3)							
300	0.6894	2.158	101.5	14.7	24.7	16.6	0.887
320	0.6448	2.17	109	16.9	27.2	19.4	0.87
340	0.6059	2.192	116.5	19.2	29.3	22.1	0.872
360	0.5716	2.221	124	21.7	31.6	24.9	0.872
380	0.541	2.254	131	24.2	34	27.9	0.869
400	0.5136	2.287	138	26.9	37	31.5	0.853
420	0.4888	2.322	145	29.7	40.4	35.6	0.833
440	0.4664	2.357	152.5	32.7	43.5	39.6	0.826
460	0.446	2.393	159	35.7	46.3	43.4	0.822
480	0.4273	2.43	166.5	39	49.2	47.4	0.822
500	0.4101	2.467	173	42.2	52.5	51.9	0.813
20	0.3942	2.504	180	45.7	54.5	55.2	0.827
540	0.3795	2.54	186.5	49.1	57.5	59.7	0.824
560	0.3708	2.577	193	52	60.6	63.4	0.827
580	0.3533	2.613	199.5	56.5	63.8	69.1	0.817
Carbon dioxide (CO_2)							
280	1.9022	0.83	140	7.36	15.2	9.63	0.765
300	1.773	0.851	149	8.4	16.55	11	0.766
320	1.6609	0.872	156	9.39	18.05	12.5	0.754
340	1.5618	0.891	165	10.6	19.7	14.2	0.746
360	1.4743	0.908	173	11.7	21.2	15.8	0.741
380	1.3961	0.926	181	13	22.75	17.6	0.737
400	1.3257	0.942	190	14.3	24.3	19.5	0.737
450	1.1782	0.981	210	17.8	28.3	24.5	0.728
500	1.0594	1.02	231	21.8	32.5	30.1	0.725
550	0.9625	1.05	251	26.1	36.6	36.2	0.721
600	0.8826	1.08	270	30.6	40.7	42.7	0.717
650	0.8143	1.1	288	35.4	44.5	49.7	0.712
700	0.7564	1.13	305	40.3	48.1	56.3	0.717
750	0.7057	1.15	321	45.5	51.7	63.7	0.714
800	0.6614	1.17	337	51	55.1	71.2	0.716

T (K)	ρ (kg/m^3)	c_p (kJ/kg·K)	$\mu.10^7$ (N.s/m^2)	$v.10^6$ (m^2/s)	$k.10^3$ (W/m.K)	$\alpha.10^6$ (m^2/s)	Pr
Carbon monoxide (CO)							
200	1.6888	1.045	127	7.52	17	9.63	0.781
220	1.5341	1.044	137	8.93	19	11.9	0.753
240	1.4055	1.043	147	10.5	20.6	14.1	0.744
260	1.2967	1.043	157	12.1	22.1	16.3	0.741
280	1.2038	1.042	166	13.8	23.6	18.8	0.733
300	1.1233	1.043	175	15.6	25	21.3	0.73
320	1.0529	1.043	184	17.5	26.3	23.9	0.73
340	0.9909	1.044	193	19.5	27.8	26.9	0.725
360	0.9357	1.045	202	21.6	29.1	29.8	0.725
380	0.8864	1.047	210	23.7	30.5	32.9	0.729
400	0.8421	1.049	218	25.9	31.8	36	0.719
450	0.7483	1.055	237	31.7	35	44.3	0.714
500	0.67352	1.065	254	37.7	38.1	53.1	0.71
550	0.61226	1.076	271	44.3	41.1	62.4	0.71
600	0.56126	1.088	286	51	44	72.1	0.707
650	0.51806	1.101	301	58.1	47	82.4	0.705
700	0.48102	1.114	315	65.5	50	93.3	0.702
750	0.44899	1.127	329	73.3	52.8	104	0.702
800	0.42095	1.14	343	81.5	55.5	116	0.705
Helium (He)							
100	0.4871	5.193	96.3	19.8	73	28.9	0.686
120	0.406	5.193	107	26.4	81.9	38.8	0.679
140	0.3481	5.193	118	33.9	90.7	50.2	0.676
160	–	5.193	129	–	99.2	–	–
180	0.2708	5.193	139	51.3	107.2	76.2	0.673
200	–	5.193	150	–	115.1	–	–
220	0.2216	5.193	160	72.2	123.1	107	0.675
240	–	5.193	170	–	130	–	–
260	0.1875	5.193	180	96	137	141	0.682
280	–	5.193	190	–	145	–	–

(Continued)

T (K)	ρ (kg/m³)	c_p (kJ/kg·K)	$\mu.10^7$ (N.s/m²)	$v.10^6$ (m²/s)	$k.10^3$ (W/m.K)	$\alpha.10^6$ (m²/s)	Pr
300	0.1625	5.193	199	122	152	180	0.68
350	–	5.193	221	–	170	–	–
400	0.1219	5.193	243	199	187	295	0.675
450	–	5.193	263	–	204	–	–
500	0.09754	5.193	283	290	220	434	0.668
550	–	5.193	–	–	–	–	–
600	–	5.193	320	–	252	–	–
650	–	5.193	332	–	264	–	–
700	0.06969	5.193	350	502	278	768	0.654
750	–	5.193	364	–	291	–	–
800	–	5.193	382	–	304	–	–
900	–	5.193	414	–	330	–	–
1000	0.04879	5.193	446	914	354	1400	0.654
Hydrogen (H_2)							
100	0.24255	11.23	42.1	17.4	67	24.6	0.707
150	0.16156	12.6	56	34.7	101	49.6	0.699
200	0.12115	13.54	68.1	56.2	131	79.9	0.704
250	0.09693	14.06	78.9	81.4	157	115	0.707
300	0.08078	14.31	89.6	111	183	158	0.701
350	0.06924	14.43	98.8	143	204	204	0.7
400	0.06059	14.48	108.2	179	226	258	0.695
450	0.05386	14.5	117.2	218	247	316	0.689
500	0.04848	14.52	126.4	261	266	378	0.691
550	0.04407	14.53	134.3	305	285	445	0.685
600	0.0404	14.55	142.4	352	305	519	0.678
700	0.03463	14.61	157.8	456	342	676	0.675
800	0.0303	14.7	172.4	569	378	849	0.67
900	0.02694	14.83	186.5	692	412	1030	0.671
1000	0.02424	14.99	201.3	830	448	1230	0.673
1100	0.02204	15.17	213	966	488	1460	0.662
1200	0.0202	15.37	226.2	1120	528	1700	0.659
1300	0.01865	15.59	238.5	1279	568	1955	0.655

T (K)	ρ (kg/m^3)	c_p (kJ/kg·K)	$\mu.10^7$ (N.s/m^2)	$v.10^6$ (m^2/s)	$k.10^3$ (W/m.K)	$\alpha.10^6$ (m^2/s)	Pr
1400	0.01732	15.81	250.7	1447	610	2230	0.65
1500	0.01616	16.02	262.7	1626	655	2530	0.643
1600	0.0152	16.28	273.7	1801	697	2815	0.639
1700	0.0143	16.58	284.9	1992	742	3130	0.637
1800	0.0135	16.96	296.1	2193	786	3435	0.639
1900	0.0128	17.49	307.2	2400	835	3730	0.643
2000	0.0121	18.25	318.2	2630	878	3975	0.661
Nitrogen (N_2)							
100	3.4388	1.07	68.8	2	9.58	2.6	0.768
150	2.2594	1.05	100.6	4.45	13.9	5.86	0.759
200	1.6883	1.043	129.2	7.65	18.3	10.4	0.736
250	1.3488	1.042	154.9	11.48	22.2	15.8	0.727
300	1.1233	1.041	178.2	15.86	25.9	22.1	0.716
350	0.9625	1.042	200	20.78	29.3	29.2	0.711
400	0.8425	1.045	220.4	26.16	32.7	37.1	0.704
450	0.7485	1.05	239.6	32.01	35.8	45.6	0.703
500	0.6739	1.056	257.7	38.24	38.9	54.7	0.7
550	0.6124	1.065	274.7	44.86	41.7	63.9	0.702
600	0.5615	1.075	290.8	51.79	44.6	73.9	0.701
700	0.4812	1.098	321	66.71	49.9	94.4	0.706
800	0.4211	1.22	349.1	82.9	54.8	116	0.715
900	0.3743	1.146	375.3	100.3	59.7	139	0.721
1000	0.3368	1.167	399.9	118.7	64.7	165	0.721
1100	0.3062	1.187	423.2	138.2	70	193	0.718
1200	0.2807	1.204	445.3	158.6	75.8	224	0.707
1300	0.2591	1.219	466.2	179.9	81	256	0.701
Oxygen (O_2)							
100	3.945	0.962	76.4	1.94	9.25	2.44	0.796
150	2.585	0.921	114.8	4.44	13.8	5.8	0.766
200	1.93	0.915	147.5	7.64	18.3	10.4	0.737
250	1.542	0.915	178.6	11.58	22.6	16	0.723

(Continued)

T (K)	ρ (kg/m^3)	c_p (kJ/kg·K)	$\mu.10^7$ (N.s/m^2)	$v.10^6$ (m^2/s)	$k.10^3$ (W/m.K)	$\alpha.10^6$ (m^2/s)	Pr
300	1.284	0.92	207.2	16.14	26.8	22.7	0.711
350	1.1	0.929	233.5	21.23	29.6	29	0.733
400	0.962	0.942	258.2	26.84	33	36.4	0.737
450	0.8554	0.956	281.4	32.9	36.3	44.4	0.741
500	0.7698	0.972	303.3	39.4	41.2	55.1	0.716
550	0.6998	0.988	324	46.3	44.1	63.8	0.726
600	0.6414	1.003	343.7	53.59	47.3	73.5	0.729
700	0.5498	1.031	380.8	69.26	52.8	93.1	0.744
800	0.481	1.054	415.2	86.32	58.9	116	0.743
900	0.4275	1.074	447.2	104.6	64.9	141	0.74
1000	0.3848	1.09	477	124	71	169	0.733
1100	0.3498	1.103	505.5	144.5	75.8	196	0.736
1200	0.3206	1.115	532.5	166.1	81.9	229	0.725
1300	0.296	1.125	588.4	188.6	87.1	262	0.721
Water vapour (steam)							
380	0.5863	2.06	127.1	21.68	24.6	20.4	1.06
400	0.5542	2.014	134.4	24.25	26.1	23.4	1.04
450	0.4902	1.98	152.5	31.11	29.9	30.8	1.01
500	0.4405	1.985	170.4	38.68	33.9	38.8	0.998
550	0.4005	1.997	188.4	47.04	37.9	47.4	0.993
600	0.3652	2.026	206.7	56.6	42.2	57	0.993
650	0.338	2.056	224.7	66.48	46.4	66.8	0.996
700	0.314	2.085	242.6	77.26	50.5	77.1	1
750	0.2931	2.119	260.4	88.84	54.9	88.4	1
800	0.2739	2.152	278.6	101.7	59.2	100	1.01
850	0.2579	2.186	296.9	115.1	63.7	113	1.02

A.4 Properties of Solid Materials

Composition	Melting point (K)	Properties at 300 K ρ (kg/m^3)	C_p (J/kg · K)	k (W/m · K)	α . 10^{-6} (m^2/s)	Properties at various temperatures (K) k (W/m.K)/ C_p (J/kg.K) 100	200	400	600	800	1000	1200	1500	2000	2500
Aluminium															
Pure	933	2702	903	237	97.1	302	237	240	231	218					
						482	798	949	1033	1146					
Alloy 2024-T6 (4.5% Cu, 1.5% Mg, 0.6% Mn)	775	2770	875	177	73	65	163	186	186						
						473	787	925	1042						
Alloy 195, Cast (4.5% Cu)		2790	883	168	68.2			174	185						
Beryllium	1550	1850	1825	200	59.2	990	301	161	126	106	90.8	78.7			
						203	1114	2191	2604	2823	3018	3227	3519		
Bismuth	545	9780	122	7.86	6.59	16.5	9.69	7.04							
						112	120	127							
Boron	2573	2500	1107	27	9.76	190	55.5	16.8	10.6	9.6	9.85				
						128	600	1463	1892	2160	2338				
Cadmium	594	8650	231	96.8	48.4	203	99.3	94.7							
						198	222	242							
Chromium	2118	7160	449	93.7	29.1	159	111	90.9	80.7	71.3	65.4	61.9	57.2	49.4	
						192	384	484	542	581	616	682	779	937	

(Continued)

Composition	Melting point (K)	ρ (kg/m^3)	C_p (J/kg · K)	k (W/m · K)	$\alpha \cdot 10^{-6}$ (m^2/s)	Properties at various temperatures (K) k (W/m.K)/ C_p (J/kg.K)									
		Properties at 300 K				100	200	400	600	800	1000	1200	1500	2000	2500
Cobalt	1769	8862	421	99.2	26.6	167	122	85.4	67.4	58.2	52.1	49.3	42.5		
						236	379	450	503	550	628	733	674		
Copper															
Pure	1358	8933	385	401	117	482	413	393	379	366	352	339			
						252	356	397	417	433	451	480			
Commercial bronze	1293	8800	420	52	14		42	52	59						
(90% Cu, 10% Al)							785	460	545						
Phosphor gear bronze	1104	8780	355	54	17		41	65	74						
(89% Cu, 11% Sn)															
Cartridge brass	1188	8530	380	110	33.9	75	95	137	149						
(70% Cu, 30% Zn)							360	395	425						
Constantan	1493	8920	384	23	6.71	17	19								
(55% Cu, 45% Ni)						237	362								
Germanium	1211	5360	322	59.9	34.7	232	96.8	43.2	27.3	19.8	17.4	17.4			
						190	290	337	348	357	375	395			
Gold	1336	19300	129	317	127	327	323	311	298	284	270	255			
						109	124	131	135	140	145	155			
Iridium	2720	22500	130	147	50.3	172	153	144	138	132	126	120	111		
						90	122	133	138	144	153	161	172		

Iron													
Pure	1810	7870	447	80.2	23.1	134	94	69.5	54.7	43.3	32.8	28.3	32.1
						216	384	490	574	680	975	609	654
Armco													
(99.75% pure)		7870	447	72.7	20.7	95.6	80.6	65.7	53.1	42.2	32.3	28.7	31.4
						215	384	490	574	680	975	609	654
Carbon steels													
Plain carbon		7854	434	60.5	17.7			56.7	48	39.2	30		
(Mn ≤ 1%, Si ≤ 0.1%)								487	559	685	1169		
AISI 1010		7832	434	63.9	18.8			58.7	48.8	39.2	31.3		
								487	559	685	1168		
Carbon–silicon		7817	446	51.9	14.9			49.8	44	37.4	29.3		
(Mn ≤ 1%, 0.1% < Si ≤0.6%)								501	582	699	971		
Carbon–manganese–silicon		8131	434	41	11.6			42.2	39.7	35	27.6		
(1% < Mn ≤ 1.65%, 0.1% < Si ≤ 0.6%)								487	559	685	1090		
Chromium (low) steels													
Cr– Mo–Si		7822	444	37.7	10.9			38.2	36.7	33.3	26.9		
(0.18%,0.65% Cr, 0.23% Mo, 0.6% Si)								492	575	688	969		

(*Continued*)

Composition	Melting point (K)	Properties at 300 K				Properties at various temperatures (K) k (W/m.K)/ C_p (J/kg.K)									
		ρ (kg/m^3)	C_p (J/kg · K)	k (W/m · K)	$\alpha \cdot 10^{-6}$ (m^2/s)	100	200	400	600	800	1000	1200	1500	2000	2500
1 Cr– Mo		7858	442	42.3	12.2			42	39.1	34.5	27.4				
(0.16% C, 1% Cr,								492	575	688	969				
0.54% Mo, 0.39% Si)															
1 Cr–V		7836	443	48.9	14.1			46.8	42.1	36.3	28.2				
(0.2% C, 1.02% Cr,								492	575	688	969				
0.15% V)															
Stainless steels															
AISI 302		8055	480	15.1	3.91			17.3	20	22.8	25.4				
								512	559	585	606				
AISI 304	1670	7900	477	14.9	3.95	9.2	12.6	16.6	19.8	22.6	25.4	28	31.7		
						272	402	515	557	582	611	640	682		
AISI 316		8238	468	13.4	3.48			15.2	18.3	21.3	24.2				
								504	550	576	602				
AISI 347		7978	480	14.2	3.71			15.8	18.9	21.9	24.7				
								513	559	585	606				
Lead	601	11340	129	35.3	24.1	39.7	36.7	34	31.4						
						118	125	132	142						
Magnesium	923	1740	1024	156	87.6	169	159	153	149	146					
						649	934	1074	1170	1267					
Molybdenum	2894	10240	251	138	53.7	179	143	134	126	118	112	105	98	90	86
						141	224	261	275	285	295	308	330	380	459

Nickel													
Pure	1728	8900	444	90.7	23	164	107	80.2	65.6	67.6	71.8	76.2	82.6
						232	383	485	592	530	562	594	616
Nichrome	1672	8400	420	12	3.4			14	16	21			
(80% Ni, 20% Cr)								480	525	545			
Inconel X-750	1665	8510	439	11.7	3.1	8.7	10.3	13.5	17	20.5	24	27.6	33
(73% Ni, 15% Cr, 6.7%						–	372	473	510	546	626	–	–
Fe)													
Niobium	2741	8570	265	53.7	23.6	55.2	52.6	55.2	58.2	61.3	64.4	67.5	72.1	79.1
						188	249	274	283	292	301	310	324	347
Palladium	1827	12020	244	71.8	24.5	76.5	71.6	73.6	79.7	86.9	94.2	102	110
						168	227	251	261	271	281	291	307
Platinum													
Pure	2045	21450	133	71.6	25.1	77.5	72.6	71.8	73.2	75.6	78.7	82.6	89.5	99.4
						100	125	136	141	146	152	157	165	179
Alloy 60Pt–40Rh	1800	16630	162	47	17.4			52	59	65	69	73	76	
(60% Pt, 40% Rh)								–	–	–	–	–	–	
Rhenium	3453	21100	136	47.9	16.7	58.9	51	46.1	44.2	44.1	44.6	45.7	47.8	51.9
						97	127	139	145	151	156	162	171	186
Rhodium	2236	12450	243	150	49.6	186	154	146	136	127	121	116	110	112
						147	220	253	274	293	311	327	349	376
Silicon	1685	2330	712	148	89.2	884	264	98.9	61.9	42.2	31.2	25.7	22.7	
						259	556	790	867	913	946	967	992	

(Continued)

		Properties at 300 K				Properties at various temperatures (K) k (W/m.K)/ C_p (J/kg.K)									
Composition	Melting point (K)	ρ (kg/m^3)	C_p (J/kg · K)	k (W/m · K)	α . _10^{-6} (m^2/s)	100	200	400	600	800	1000	1200	1500	2000	2500
Silver	1235	10500	235	429	174	444	430	425	412	396	379	361			
						187	225	239	250	262	277	292			
Tantalum	3269	16600	140	57.5	24.7	59.2	57.5	57.8	58.6	59.4	60.2	61	62.2	64.1	65.6
						110	133	144	146	149	152	155	160	172	189
Thorium	2023	11700	118	54	39.1	59.8	54.6	54.5	55.8	56.9	56.9	58.7			
						99	112	124	134	145	156	167			
Tin	505	7310	227	66.6	40.1	85.2	73.3	62.2							
						188	215	243							
Titanium	1953	4500	522	21.9	9.32	30.5	24.5	20.4	19.4	19.7	20.7	22	24.5		
						300	465	551	591	633	675	620	686		
Tungsten	3660	19300	132	174	68.3	208	186	159	137	125	118	113	107	100	
						87	122	137	142	145	148	152	157	167	
Uranium	1406	19070	116	27.6	12.5	21.7	25.1	29.6	34	38.8	43.9	49			
						94	108	125	146	176	180	161			
Vanadium	2192	6100	489	30.7	10.3	35.8	31.3	31.3	33.3	35.7	38.2	40.8	44.6	50.9	
						258	430	515	540	563	597	645	714	867	
Zinc	693	7140	389	116	41.8	117	118	111	103						
						297	367	402	436						
Zirconium	2125	6570	278	22.7	12.4	33.2	25.2	21.6	20.7	21.6	23.7	26	28.8	33	
						205	264	300	322	342	362	344	344	344	

A.5 Thermo-Physical Properties of Saturated Fluids

Saturated liquids								
T (K)	ρ (kg/m^3)	c_p (kJ/kg · K)	$\mu.10^{-2}$ (N · s/m^2)	$v.10^{-2}$ (m^2/s)	$k.10^{-3}$ (W/m · K)	$\alpha.10^{-7}$ (m^2/s)	Pr	$\beta.10^{-3}$
Engine oil (unused)								
273	899.1	1.796	385	4280	147	0.91	47,000	0.7
280	895.3	1.827	217	2430	144	0.88	27,500	0.7
290	890	1.868	99.9	1120	145	0.872	12,900	0.7
300	884.1	1.909	48.6	550	145	0.859	6400	0.7
310	877.9	1.951	25.3	288	145	0.847	3400	0.7
320	871.8	1.993	14.1	161	143	0.823	1965	0.7
330	865.8	2.035	8.36	96.6	141	0.8	1205	0.7
340	859.9	2.076	5.31	61.7	139	0.779	793	0.7
350	853.9	2.118	3.56	41.7	138	0.763	546	0.7
360	847.8	2.161	2.52	29.7	138	0.753	395	0.7
370	841.8	2.206	1.86	22	137	0.738	300	0.7
380	836	2.25	1.41	16.9	136	0.723	233	0.7
390	830.6	2.294	1.1	13.3	135	0.709	187	0.7
400	825.1	2.337	0.874	10.6	134	0.695	152	0.7
410	818.9	2.381	0.698	8.52	133	0.682	125	0.7
420	812.1	2.427	0.564	6.94	133	0.675	103	0.7
430	806.5	2.471	0.47	5.83	132	0.662	88	0.7
Ethylene glycol [$C_2H_4(OH)_2$]								
273	1130.8	2.294	6.51	57.6	242	0.933	617	0.65
280	1125.8	2.323	4.2	37.3	244	0.933	400	0.65
290	1118.8	2.368	2.47	22.1	248	0.936	236	0.65
300	1114.4	2.415	1.57	14.1	252	0.939	151	0.65
310	1103.7	2.46	1.07	9.65	255	0.939	103	0.65
320	1096.2	2.505	0.757	6.91	258	0.94	73.5	0.65
330	1089.5	2.549	0.561	5.15	260	0.936	55	0.65
340	1083.8	2.592	0.431	3.98	261	0.929	42.8	0.65
350	1079	2.637	0.342	3.17	261	0.917	34.6	0.65
360	1074	2.682	0.278	2.59	261	0.906	28.6	0.65
370	1066.7	2.728	0.228	2.14	262	0.9	23.7	0.65
373	1058.5	2.742	0.215	2.03	263	0.906	22.4	0.65
Glycerin [$C_3H_5(OH)_3$]								
273	1276	2.261	1060	8310	282	0.977	85,000	0.47
280	1271.9	2.298	534	4200	284	0.972	43,200	0.47

(Continued)

Saturated liquids

T (K)	ρ (kg/m^3)	c_p (kJ/kg · K)	$\mu.10^{-2}$ (N · s/m^2)	$v.10^{-2}$ (m^2/s)	$k.10^{-3}$ (W/m · K)	$\alpha.10^{-7}$ (m^2/s)	Pr	$\beta.10^{-3}$
290	1265.8	2.367	185	1460	286	0.955	15,300	0.48
300	1259.9	2.427	79.9	634	286	0.935	6780	0.48
310	1253.9	2.49	35.2	281	286	0.916	3060	0.49
320	1247.2	2.564	21	168	287	0.897	1870	0.5
Refrigerant-134a ($C_2H_2F_4$)								
230	1426.8	1.249	0.04912	0.3443	112.1	0.629	5.5	2.02
240	1397.7	1.267	0.04202	0.3006	107.3	0.606	5	2.11
250	1367.9	1.287	0.03633	0.2656	102.5	0.583	4.6	2.23
260	1337.1	1.308	0.03166	0.2368	97.9	0.56	4.2	2.36
270	1305.1	1.333	0.02775	0.2127	93.4	0.537	4	2.53
280	1271.8	1.361	0.02443	0.1921	89	0.514	3.7	2.73
290	1236.8	1.393	0.02156	0.1744	84.6	0.491	3.5	2.98
300	1199.7	1.432	0.01905	0.1588	80.3	0.468	3.4	3.3
310	1159.9	1.481	0.0168	0.1449	76.1	0.443	3.3	3.73
320	1116.8	1.543	0.01478	0.1323	71.8	0.417	3.2	4.33
330	1069.1	1.627	0.01292	0.1209	67.5	0.388	3.1	5.19
340	1015	1.751	0.01118	0.1102	63.1	0.355	3.1	6.57
350	951.3	1.961	0.00951	0.1	58.6	0.314	3.2	9.1
360	870.1	2.437	0.00781	0.0898	54.1	0.255	3.5	15.39
370	740.3	5.105	0.0058	0.0783	51.8	0.137	5.7	55.24
Refrigerant-22 ($CHClF_2$)								
230	1416	1.087	0.03558	0.2513	114.5	0.744	3.4	2.05
240	1386.6	1.1	0.03145	0.2268	109.8	0.72	3.2	2.16
250	1356.3	1.117	0.02796	0.2062	105.2	0.695	3	2.29
260	1324.9	1.137	0.02497	0.1884	100.7	0.668	2.8	2.45
270	1292.1	1.161	0.02235	0.173	96.2	0.641	2.7	2.63
280	1257.9	1.189	0.02005	0.1594	91.7	0.613	2.6	2.86
290	1221.7	1.223	0.01798	0.1472	87.2	0.583	2.5	3.15
300	1183.4	1.265	0.0161	0.1361	82.6	0.552	2.5	3.51
310	1142.2	1.319	0.01438	0.1259	78.1	0.518	2.4	4
320	1097.4	1.391	0.01278	0.1165	73.4	0.481	2.4	4.69
330	1047.5	1.495	0.01127	0.1075	68.6	0.438	2.5	5.75
340	990.1	1.665	0.0098	0.0989	63.6	0.386	2.6	7.56
350	920.1	1.997	0.00831	0.0904	58.3	0.317	2.8	11.35
360	823.4	3.001	0.00668	0.0811	53.1	0.215	3.8	23.88

Saturated liquids

T (K)	ρ (kg/m^3)	c_p (kJ/kg · K)	$\mu.10^{-2}$ (N · s/m^2)	$v.10^{-2}$ (m^2/s)	$k.10^{-3}$ (W/m · K)	$\alpha.10^{-7}$ (m^2/s)	Pr	$\beta.10^{-3}$
Mercury (Hg)								
273	13,595	0.1404	0.1688	0.124	8180	42.85	0.029	0.181
300	13,529	0.1393	0.1523	0.1125	8540	45.3	0.0248	0.181
350	13,407	0.1377	0.1309	0.0976	9180	49.75	0.0196	0.181
400	13,287	0.1365	0.1171	0.0882	9800	54.05	0.0163	0.181
450	13,167	0.1357	0.1075	0.0816	10,400	58.1	0.014	0.181
500	13,048	0.1353	0.1007	0.0771	10,950	61.9	0.0125	0.182
550	12,929	0.1352	0.0953	0.0737	11,450	65.55	0.0112	0.184
600	12,809	0.1355	0.0911	0.0711	11,950	68.8	0.0103	0.187

Saturated liquid – vapour, 1 atm

Fluid	T_{sat} (K)	h_{fg} (kJ/kg)	ρ_f (kg/m^3)	ρ_g (kg/m^3)	$\sigma \cdot 10^3$ (N/m)
Ethanol	351	846	757	1.44	17.7
Ethylene glycol	470	812	1111c	–	32.7
Glycerin	563	974	1260c	–	63.0c
Mercury	630	301	12,740	3.9	417
Refrigerant R-134a	247	217	1377	5.26	15.4
Refrigerant R-22	232	234	1409	4.7	18.1

A.6 Thermo-Physical Properties of Saturated Water

Temp,	Pressure,	Specific volume (m^3/kg)		Heat of vaporization, h_{fg}	Specific heat ($kJ/kg \cdot K$)		Viscosity ($N.s/m^2$)		Thermal conductivity (W/m.K)		Prandtl number		Surface tension, $\sigma_f \cdot 10^{-3}$	Expansion coefficient, $\beta_f \cdot 10^{-6}$	Temp
T(K)	p (bars)[b]	$v_f \cdot 10^3$	v_g	($kJ/kg \cdot K$)	$c_{p,f}$	$c_{p,g}$	$\mu_f \cdot 10^{-6}$	$\mu_g \cdot 10^{-6}$	$k_f \cdot 10^3$	$k_g \cdot 10^3$	Pr_f	Pr_g	(N/m)	(K^{-1})	T(K)
273.15	0.00611	1	206.3	2502	4.217	1.854	1750	8.02	569	18.2	12.99	0.82	75.5	−68.05	273.15
275	0.00697	1	181.7	2497	4.211	1.855	1652	8.09	574	18.3	12.22	0.82	75.3	−32.74	275
280	0.0099	1	130.4	2485	4.198	1.858	1422	8.29	582	18.6	10.26	0.83	74.8	46.04	280
285	0.01387	1	99.4	2473	4.189	1.861	1225	8.49	590	18.9	8.81	0.83	74.3	114.1	285
290	0.01917	1.001	69.7	2461	4.184	1.864	1080	8.69	598	19.3	7.56	0.84	73.7	174	290
295	0.02617	1.002	51.94	2449	4.181	1.868	959	8.89	606	19.5	6.62	0.85	72.7	227.5	295
300	0.03531	1.003	39.13	2438	4.179	1.872	855	9.09	613	19.6	5.83	0.86	71.7	276.1	300
305	0.04712	1.005	29.74	2426	4.178	1.877	769	9.29	620	20.1	5.2	0.87	70.9	320.6	305
310	0.06221	1.007	22.93	2414	4.178	1.882	695	9.49	628	20.4	4.62	0.87	70	361.9	310
315	0.08132	1.009	17.82	2402	4.179	1.888	631	9.69	634	20.7	4.16	0.88	69.2	400.4	315
320	0.1053	1.011	13.98	2390	4.18	1.895	577	9.89	640	21	3.77	0.89	68.3	436.7	320
325	0.1351	1.013	11.06	2378	4.182	1.903	528	10.09	645	21.3	3.42	0.9	67.5	471.2	325
330	0.1719	1.016	8.82	2366	4.184	1.911	489	10.29	650	21.7	3.15	0.91	66.6	504	330
335	0.2167	1.018	7.09	2354	4.186	1.92	453	10.49	656	22	2.88	0.92	65.8	535.5	335
340	0.2713	1.021	5.74	2342	4.188	1.93	420	10.69	660	22.3	2.66	0.93	64.9	566	340
345	0.3372	1.024	4.683	2329	4.191	1.941	389	10.89	668	22.6	2.45	0.93	64.1	595.4	345
350	0.4163	1.027	3.846	2317	4.195	1.954	365	11.09	668	23	2.29	0.94	63.2	624.2	350
355	0.51	1.03	3.18	2304	4.199	1.968	343	11.29	671	23.3	2.14	0.95	62.3	652.3	355
360	0.6209	1.034	2.645	2291	4.203	1.983	324	11.49	674	23.7	2.02	0.96	61.4	697.9	360

365	0.7514	1.038	2.212	2278	4.209	1.999	306	11.69	677	24.1	1.91	0.97	60.5	707.1	365
370	0.904	1.041	1.861	2265	4.214	2.017	289	11.89	679	24.5	1.8	0.98	59.5	728.7	370
373.15	1.0133	1.044	1.679	2257	4.217	2.029	279	12.02	680	24.8	1.76	0.98	58.9	750.1	373.15
375	1.0815	1.045	1.574	2252	4.22	2.036	274	12.09	681	24.9	1.7	0.99	58.6	761	375
380	1.2869	1.049	1.337	2239	4.226	2.057	260	12.29	683	25.4	1.61	1	57.6	788	380
385	1.5233	1.053	1.142	2225	4.232	2.08	248	12.49	685	25.8	1.53	1	56.6	814	385
390	1.794	1.058	0.98	2212	4.239	2.104	237	12.69	686	26.3	1.47	1.01	55.6	841	390
400	2.455	1.067	0.731	2183	4.256	2.158	217	13.05	688	27.2	1.34	1.03	53.6	896	400
410	3.302	1.077	0.553	2153	4.278	2.221	200	13.42	688	28.2	1.24	1.05	51.5	952	410
420	4.37	1.088	0.425	2123	4.302	2.291	185	13.79	688	29.8	1.16	1.08	49.4	1010	420
430	5.699	1.099	0.331	2091	4.331	2.369	173	14.14	685	30.4	1.09	1.1	47.2		430
440	7.333	1.11	0.261	2059	4.36	2.46	162	14.5	682	31.7	1.04	1.12	45.1		440
450	9.319	1.123	0.208	2024	4.4	2.56	152	14.85	678	33.1	0.99	1.14	42.9		450
460	11.71	1.137	0.167	1989	4.44	2.68	143	15.19	673	34.6	0.95	1.17	40.7		460
470	14.55	1.152	0.136	1951	4.48	2.79	136	15.54	667	36.3	0.92	1.2	38.5		470
480	17.9	1.167	0.111	1912	4.53	2.94	129	15.88	660	38.1	0.89	1.23	36.2		480
490	21.83	1.184	0.0922	1870	4.59	3.1	124	16.23	651	40.1	0.87	1.25	33.9	–	490
500	26.4	1.203	0.0766	1825	4.66	3.27	118	16.59	642	42.3	0.86	1.28	31.6	–	500
510	31.66	1.222	0.0631	1779	4.74	3.47	113	16.95	631	44.7	0.85	1.31	29.3	–	510
520	37.7	1.244	0.0525	1730	4.84	3.7	108	17.33	621	47.5	0.84	1.35	26.9	–	520
530	44.58	1.268	0.0445	1679	4.95	3.96	104	17.72	608	50.6	0.85	1.39	24.5	–	530
540	52.38	1.294	0.0375	1622	5.08	4.27	101	18.1	594	54	0.86	1.43	22.1	–	540
550	61.19	1.323	0.0317	1564	5.24	4.64	97	18.6	580	58.3	0.87	1.47	19.7	–	550
560	71.08	1.355	0.0269	1499	5.43	5.09	94	19.1	563	63.7	0.9	1.52	17.3	–	560

(Continued)

Temp,	Pressure,	Specific volume (m^3/kg)		Heat of vaporization, h_{fg}	Specific heat ($kJ/kg \cdot K$)		Viscosity ($N.s/m^2$)		Thermal conductivity (W/m.K)		Prandtl number		Surface tension, $\sigma_f \cdot 10^{-3}$	Expansion coefficient, $\beta_f \cdot 10^{-6}$	Temp
T(K)	p (bars)[b]	$v_f \cdot 10^3$	v_g	($kJ/kg \cdot K$)	$c_{p,f}$	$c_{p,g}$	$\mu_f \cdot 10^{-6}$	$\mu_g \cdot 10^{-6}$	$k_f \cdot 10^3$	$k_g \cdot 10^3$	Pr_f	Pr_g	(N/m)	(K^{-1})	T(K)
570	82.16	1.392	0.0228	1429	5.68	5.67	91	19.7	548	76.7	0.94	1.59	15	–	570
580	94.51	1.433	0.0193	1353	6	6.4	88	20.4	528	76.7	0.99	1.68	12.8	–	580
590	108.3	1.482	0.0163	1274	6.41	7.35	84	21.5	513	84.1	1.05	1.84	10.5	–	590
600	123.5	1.541	0.0137	1176	7	8.75	81	22.7	497	92.9	1.14	2.15	8.4	–	600
610	137.3	1.612	0.0115	1068	7.85	11.1	77	24.1	467	103	1.3	2.6	6.3	–	610
620	159.1	1.705	0.0094	941	9.35	15.4	72	25.9	444	114	1.52	3.46	4.5	–	620
625	169.1	1.778	0.0085	858	10.6	18.3	70	27	430	121	1.65	4.2	3.5	–	625
630	179.7	1.856	0.0075	781	12.6	22.1	67	28	412	130	2	4.8	2.6	–	630
635	190.9	1.935	0.0066	683	16.4	27.6	64	30	392	141	2.7	6	1.5	–	635
640	202.7	2.075	0.0057	560	26	42	59	32	367	155	4.2	9.6	0.8	–	640
645	215.2	2.351	0.0045	361	90	–	54	37	331	178	12	26	0.1	–	645
647.3c	221.2	3.17	0.0032	0	∞	∞	45	45	238	238	∞	∞	0	–	647.3c

A.7 Solar Radiative Properties of Selected Materials

Description/composition	α_s	ε	α_s/ε	τ_s
Aluminum				
Polished	0.09	0.03	3	
Anodized	0.14	0.84	0.17	
Quartz overcoated	0.11	0.37	0.3	
Foil	0.15	0.05	3	
Brick, red (Purdue)	0.63	0.93	0.68	
Concrete	0.6	0.88	0.68	
Galvanized sheet metal				
Clean, new	0.65	0.13	5	
Oxidized, weathered	0.8	0.28	2.9	
Glass, 3.2 mm thickness				
Float or tempered				0.79
Low iron oxide type				0.88
Metal, plated				
Black sulfide	0.92	0.1	9.2	
Black cobalt oxide	0.93	0.3	3.1	
Black nickel oxide	0.92	0.08	11	
Black chrome	0.87	0.09	9.7	
Mylar, 0.13 mm thickness				0.87
Paints				
Black (Parsons)	0.98	0.98	1	
White, acrylic	0.26	0.9	0.29	
White, zinc oxide	0.16	0.93	0.17	
Plexiglas, 3.2 mm thickness				0.9
Snow				
Fine particles, fresh	0.13	0.82	0.16	
Ice granules	0.33	0.89	0.37	
Tedlar, 0.10-mm thickness				0.92
Teflon, 0.13-mm thickness				0.92

A.8 Thermo-Physical Properties of Fluids

Engine oil

T (°C)	ρ (kg/m^3)	c_p (J/kg·K)	k (W/m·K)	$\mu \times 10^2$ (N.s/m^2)	Pr
0	899	1796	0.147	384.8	47100
20	888	1880	0.145	79.92	10400
40	876	1964	0.144	21.02	2870
60	864	2047	0.14	7.249	1050
80	852	2131	0.138	3.195	490
100	840	2219	0.137	1.705	276
120	828	2307	0.135	1.027	175
140	816	2395	0.133	0.653	116
160	805	2483	0.132	0.451	84

50% Ethylene glycol

T (°C)	ρ (kg/m^3)	c_p (J/kg·K)	k (W/mK)	$\mu \times 10^2$ (N.s/m^2)	Pr
0	1083	3180	0.379	1.029	86.3
20	1072	3310	0.319	0.459	47.6
40	1061	3420	0.404	0.238	20.1
60	1048	3520	0.417	0.139	11.8
80	1034	3590	0.429	0.099	8.3
100	1020	3650	0.442	0.08	6.6
120	1003	3680	0.454	0.066	5.4

Ethylene glycol

T (°C)	ρ (kg/m^3)	c_p (J/kg·K)	k (W/m·K)	$\mu \times 10^2$ (N.s/m^2)	Pr
0	1130	2294	0.242	6.501	615
20	1116	2382	0.249	2.14	204
40	1101	2474	0.256	0.957	93
60	1087	2562	0.26	0.516	51
80	1077	2650	0.261	0.321	32.4
100	1058	2742	0.263	0.215	22.4

Glycerin

T (°C)	ρ (kg/m^3)	c_p (J/kg·K)	k (W/m·K)	$\mu \times 10^2$ (N.s/m^2)	Pr
0	1276	2261	0.282	1060.4	84700
10	1270	2319	0.284	381	31000
20	1264	2386	0.286	149.2	12500
30	1258	2445	0.286	62.9	5380
40	1252	2512	0.286	27.5	2450
50	1244	2583	0.287	18.7	1630

Water

T (°C)	ρ (kg/m^3)	c_p (J/kg·K)	k (W/mK)	$\mu \times 10^2$ (N.s/m^2)	Pr
0	1002	4217	0.552	1792	13.6
20	1000	4181	0.597	1006	7.02
40	994	4178	0.628	654	4.34
60	985	4184	0.651	471	3.02
80	974	4196	0.668	355	2.22
100	960	4216	0.68	282	1.74
120	945	4250	0.685	233	1.45
140	928	4283	0.684	199	1.24
160	909	4342	0.67	173	1.1
180	889	4417	0.675	154	1
200	866	4505	0.665	139	0.94
220	842	4610	0.572	126	0.89
240	815	4756	0.635	117	0.87
260	785	4949	0.611	108	0.87
280	752	5208	0.58	102	0.91
300	714	5728	0.54	96	1.11

Note: Most of the property tables are taken from Internet sites:
www.efunda.com/materials/common_matl/common_matl.cfm
pubs.acs.org/doi/abs/10.1021/ja01611a113
https://en.Wikipedia.org/wiki
mathscinotes.com/wp-content/uploads/2016/01
onlinelibrary.wiley.com/doi/10.1002/9781118131473.app2

References

1 Sokolnikoff I.S, Sokolnikoff E.S., *Higher Mathematics for Engineers and Physicists*, McGraw Hill Book Co., 1949.
2 Spiegel M.R., *Advanced Calculus*, Schaum Publishing Co., 1963.

Index

Compact Heat Exchangers – Analysis, Design and Optimization using FEM and CFD Approach, First Edition. C. Ranganayakulu and K.N. Seetharamu.
© 2018 John Wiley & Sons Ltd. Published 2018 by John Wiley & Sons Ltd.

Printed and bound by CPI Group (UK) Ltd, Croydon, CR0 4YY
17/09/2026
14964271-0001